MATTER & INTERACTIONS I
MODERN MECHANICS

RUTH W. CHABAY

BRUCE A. SHERWOOD

Carnegie Mellon University

JOHN WILEY & SONS, INC.

ACQUISITIONS EDITOR   Stuart Johnson

MARKETING MANAGER   Robert Smith

PRODUCTION EDITOR   Rebecca Rothaug

SENIOR DESIGNER   Maddy Lesure

The cover was printed by Lehigh Press.

To order books or for customer service please, call 1(800)-CALL-WILEY (225-5945).
For instructor supplements visit http://www.wiley.com/college/chabay.

ISBN 0-471-66328-X

Printed in the United States of America

10 9 8 7 6 5 4

Preface to Volume I

Matter & Interactions I: Modern Mechanics focuses on the atomic structure of matter and the interactions that matter undergoes. This two-volume textbook emphasizes the fact that there are only a small number of fundamental principles that underlie the behavior of matter, and that using these powerful principles it is possible to construct models that can explain and predict a wide variety of physical phenomena.

Prerequisites

This book is intended for introductory college physics courses taken by science and engineering students. It is assumed that you have had an introductory physics course in high school. You will need a basic knowledge of derivatives and integrals, which can be obtained by studying calculus concurrently, and a basic knowledge of vectors and vector components. In appendices at the end of this volume are brief reviews of vectors and derivatives.

Modeling

This course places a major emphasis on constructing and using physical "models." A central aspect of science is the modeling of complex real-world phenomena. A physical model is based on what we believe to be fundamental principles; its intent is to predict or explain the most important aspects of an actual situation. Modeling necessarily involves making approximations and simplifying assumptions in order that the model can be analyzed in detail. The principles of classical mechanics and thermal physics that are the focus of Volume I have wide applicability; we can use these principles to model systems as different as molecules and galaxies.

Computer modeling

Computer modeling is important in this course, because it makes it possible for you to analyze complex systems which would otherwise require very sophisticated mathematics or which could not be analyzed at all without a computer. Numerical calculations based on the momentum principle give us the opportunity to watch the dynamical evolution of the behavior of a system. Simple models frequently need to be refined and extended; this can be done straightforwardly with a computer model but is often impossible with a purely analytical model.

Computer modeling has become as important as theory and experiment in contemporary science. We introduce you to serious computer modeling right away to help you build a strong foundation in the use of this important tool.

Desktop experiments

In order to use this book you will need some simple equipment such as weights, string, coffee filters, a stopwatch, a weak spring, and a styrofoam cup. A toy gyroscope is needed for a homework problem in the chapter dealing with gyroscopes. Integrated with discussions of theory are critical "desktop" experiments, experiments to be carried out using simple equipment in a regular classroom, a lab, or even at home. It is important that you apply the same care and attention to carrying out the desktop experiments that you apply to learning the theory, since the theory and the experiments reinforce each other. You will find that by using very simple equipment you can gain insight into rather deep scientific issues.

? Stop and Think

As you read the text, you will frequently come to a paragraph that asks you to stop and think by making a prediction, carrying out a step in a derivation or analysis, or applying a principle. Usually these questions are answered in the following paragraphs, but it is important that you make a serious effort to answer the questions on your own before reading further. Be honest in comparing your answers to those in the text; paying attention to surprising or counterintuitive results can be a useful learning strategy.

Exercises

Small exercises that require you to apply new concepts are found at the end of many sections of the text. These may involve qualitative reasoning or simple calculations. You should work these exercises when you come to them, to consolidate your understanding of the material you have just read. Answers to the exercises are at the end of each chapter.

Worked-out example problems

Following the summary page of each chapter are one or more worked-out example problems. The purpose is to show you how to apply the fundamental principles in complex situations.

Conventions used in diagrams and equations

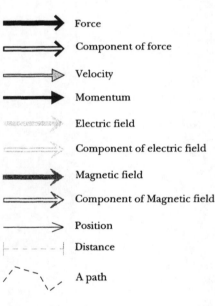

Force

Component of force

Velocity

Momentum

Electric field

Component of electric field

Magnetic field

Component of Magnetic field

Position

Distance

A path

The conventions most commonly used to represent vectors and scalars in diagrams in this text are shown in the adjacent figure. In equations and text, a vector will be written with an arrow above it:

$$\vec{p}$$

The magnitude of $\vec{p}$ is written as $|\vec{p}|$ or simply as p.
The x component of $\vec{p}$ is written as p_x.
A vector may be written in terms of components or in terms of unit vectors along the x, y, and z axes:

$$\vec{p} = <p_x, p_y, p_z> = p_x\hat{\imath} + p_y\hat{\jmath} + p_z\hat{k}$$

There is a brief review of vectors in an appendix at the end of this volume.

To the instructor

The approach to mechanics and thermal physics taken in this volume differs significantly from that in most introductory physics texts. Key emphases of our approach are these:

- Starting from fundamental principles rather than secondary formulas
- Atomic-level description and analysis
- Modeling the real world through idealizations and approximations
- Computer modeling of physical systems
- Qualitative reasoning that complements the quantitative reasoning
- Unification of mechanics and thermal physics
- Tight integration of theory and experiment

The following one-semester introductory course on electricity and magnetism is addressed by *Matter & Interactions II: Electric & Magnetic Interactions*, Ruth Chabay & Bruce Sherwood, Wiley 2002.

Textbook web site

To obtain useful supplements to our textbook, go to the Wiley web site:

http://www.wiley.com/college/chabay

There you will find a daily log of what we do in own classrooms, sample assignment sheets, problem solutions, sample exams, etc. At the start of the daily log is procurement information for the desktop experiments used with this volume. You will also find a link to the authors' web site, which includes additional information of a general character, including useful educational software.

Computer homework problems

Some critically important homework problems are designed for the student to write a computer program. We strongly recommend VPython as the best tool for student use. An adequate subset of the underlying Python programming language can be taught in an hour or two, even to students who have never written a program before. Real-time 3D animations are generated as a side effect of student computations, and these animations provide powerfully motivating and instructive visualizations of fields and motions. VPython supports true vector computations, which encourages students to begin thinking about vectors as powerful tools rather than barriers to understanding. VPython can be obtained at no cost for Windows, Macintosh, and Linux through the authors' web site (accessible from the Wiley web site).

Homework problems

Given the emphasis on physical modeling, it is appropriate for homework to consist of a small number of rather large problems, rather than a large number of small, routine problems. In addition to the large homework problems provided at the end of a chapter, there are many small-scale problems in the form of exercises distributed throughout the chapter, and small review questions at the end of the chapter. You may wish to ask students to write out some exercises or review questions as part of the homework, or to have them work out exercises at intervals during a lecture.

In-line problems

Some of the homework problems are central to the course and cannot be omitted without compromising the enterprise; an example is the in-line computer homework problem in Chapter 2 on planetary orbits. In-line homework problems should normally be assigned unless the associated topic is omitted.

Choosing topics

In our own teaching, with daily 50-minute periods (3 lectures and 2 small-group workshops every week), we are able to complete most of this modern mechanics volume in a 15-week semester. You can read the organizational details in the daily log available at the Wiley web site.

What can be omitted if there is not enough time to do everything? Let us say that the one thing we feel should *not* be omitted is the introduction to entropy in terms of quantum statistical mechanics (Chapter 10). This is a climax of the theme of integrating mechanics and thermal physics. One way to decide what can be omitted is to be guided by what needs to be done to ensure dealing adequately with entropy.

Any starred section (*) can safely be omitted. Material in these sections is not referenced in later work. Next we offer chapter by chapter discussions of what other topics might be omitted, and what the ramifications are of such omissions.

Chapter 1 (Matter & Interactions): Depending on the prior experiences of your students, you might go through the descriptive portions of this chapter rather quickly, but the only section which can definitely be omitted at no cost is the starred section on units. Note the homework problems that engage students right away with concepts of modeling: making approximations, simplifying assumptions, and estimates. It is important to talk about these concepts, because they usually are unfamiliar even to students with strong physics backgrounds.

Chapter 2 (Predicting the Future): You might choose to omit the sections on determinism and why we call them "Newton's" laws. Everything else is central to the enterprise. In our own teaching we don't assign the three-body problem but we demonstrate a three-body program in class. Some students do the three-body problem as a personal project. It is a natural extension of the planetary orbit program that is central to the chapter.

Chapter 3 (The Atomic Nature of Matter): The sections on buoyancy and pressure could be omitted (you could return to these topics during Chapter 11 on gases). You should decide how much time you wish to spend on the class of problems discussed in the section on unknown force laws (roller coasters, etc.). Such homework problems have often been considered to be the heart of the mechanics course, but we have given them lower priority in order to focus clearly on the fundamental issues of the Newtonian synthesis: predicting the future by repetitive updating of momentum and position with known force laws. This time-evolution character of the momentum principle does not come through clearly when students calculate a constant acceleration for a known motion and invoke kinematics equations.

Chapter 4 (Conservation of Energy: Limits on the Possible): This is the foundation for the energy concept, and nothing should be omitted (other than starred sections, of course). The starred section on "a puzzle" is an amusing teaser that hints at field energy. The starred derivations at the end of the chapter are probably of interest to few students but are provided for completeness.

Chapter 5 (Energy in Macroscopic Systems): If you are pressed for time, you might choose to omit the second half of the chapter on energy dissipation, beginning with Section 5.6. We ourselves do not typically have time to discuss the starred sections on resonance.

Chapter 6 (Energy Quantization): None of the unstarred sections of this short chapter should be omitted.

Chapter 7 (Multiparticle Systems): The starred sections on modeling friction are not mathematically difficult but are conceptually challenging. Nothing else in this chapter is a candidate for omission.

Chapter 8 (Collisions: Exploring the Nucleus): A good candidate for omission is the analysis of collisions in the center-of-mass frame.

Chapter 9 (Angular Momentum): In our experience students can much more easily handle situations where the angular momentum doesn't change than situations where there is a net torque. If you are pressed for time you might choose to omit applications involving nonzero torque. The starred sections on gyroscopes are interesting but students find them difficult, presumably because the analysis includes both a nonzero torque and three dimensions.

Chapter 10 (Entropy: More Limits on the Possible): We strongly recommend doing at least the first half of this chapter, on the Einstein solid. To make this work well, exposition and student computing should be interwoven. It might seem odd to ask the students simply to reproduce the results shown in the textbook, but in our experience and that of our students it is enormously educational to be required to make the ideas concrete in a computer program. You find yourself asking "What exactly is q1?" and answering very concretely by actually calculating with the initially unfamiliar quantities. It is the difference between writing an essay about entropy and calculating the actual value of the entropy.

You might possibly omit the lengthy application of the Boltzmann distribution to a gas, but if you do Chapter 11 you will need a few key results from this analysis.

Chapter 11 (The Kinetic Theory of Gases): One should not omit the initial sections, including the relationship of pressure to the results from Chapter 10. We should point out that the important formula for particle flow ($nA\bar{v}$) and the concept of mean free path are both used in Volume II (though they are rederived there). You might decide to omit the sections on macroscopic energy transfers (isothermal and adiabatic processes).

Chapter 12 (The Efficiency of Engines): This entire chapter is another application of the second law of thermodynamics and may be omitted. Another possibility is to omit just the sections on nonzero-power engines.

Acknowledgments for Volume I

We owe much to the unusual working environment provided by the Center for Innovation in Learning and the Department of Physics at Carnegie Mellon, which made it possible during the 1990's to carry out the research and development leading to our two-volume textbook. We are grateful for the open-minded attitude of our colleagues in the Carnegie Mellon physics department toward curriculum innovations.

For the writing of Chapter 10 on quantum statistical mechanics we gratefully acknowledge the inspiration provided in an article by Thomas A. Moore and Daniel V. Schroeder, "A different approach to introducing statistical mechanics," *American Journal of Physics* vol. 65, pp. 26-36 (January 1997). We thank Robert Bauman, Gregg Franklin, and Curtis Meyer for helping us think deeply about energy. Vidhya Ramachandran contributed a useful take-home experiment on specific heat.

We thank David Andersen and David Scherer for the development of tools that enabled us and our students to write associated software. We also thank Stacey Benson, Krishna Chowdary, Thomas Foster, Barry Luokkala, Matthew Kohlmyer, Robert Swendsen, and Hugh Young.

This project was supported, in part, by the National Science Foundation (grants MDR-8953367, USE-9156105, DUE-9554843, and DUE-9972420). Opinions expressed are those of the authors, and not necessarily those of the Foundation.

Ruth W. Chabay
Bruce A. Sherwood
Center for Innovation in Learning and Department of Physics
Carnegie Mellon University
May 2001

Image Credits

Chapter 1: Randall Feenstra (Figure 1.3). Chapter 3: Randall Feenstra (Figure 3.2). Chapter 5: Stacey Benson (Figure 5.23 and Figure 5.24). Judith Harrison (Figure 5.35).

All other figures were created by the authors, using Adobe Illustrator, Adobe Photoshop, VPython, POV-Ray, and Specular Infini-D.

The text was produced by the authors as camera-ready copy, using Adobe FrameMaker. The font is 10-point New Baskerville.

Volume I: Modern Mechanics

Chapter 1

Matter & Interactions

Chapter 1

Matter & Interactions

1.1   Understanding matter and its interactions

This course deals with the nature of matter and its interactions. The variety of phenomena that we will be able to explain and understand is very wide, from the orbit of the Moon to the cooling of a gas.

> The main goal of this course is to have you engage in a process central to science: the attempt to explain in detail a broad range of phenomena using a small set of powerful fundamental principles.

> The specific focus is on learning how to model the nature of matter and its interactions in terms of a small set of physical laws that govern all mechanical interactions, and in terms of the atomic structure of matter.

Physicists try to understand the world by constructing models that incorporate ideas about the nature of matter and the fundamental physical principles governing its interactions. They begin by building very simple models, and they test the models by using them to predict or explain physical phenomena. If the predictions agree reasonably well with experimental observations, then the model may be progressively refined in order to produce better quantitative agreement with experiment. Over time, a model is applied to a wider and wider range of phenomena. If it continues to give good predictions and explanations, it becomes widely accepted in the scientific community. This process of constructing, testing, and refining models, is central to all of science. In this course you will engage in this process.

This first chapter offers a survey of the kinds of matter and the kinds of interactions that we will study in this course. The major topics are:
- The kinds of matter that exist in the Universe.
- How to detect interactions that matter has with other matter.
- The four fundamental interactions: gravitational, electromagnetic, strong, and weak.
- Fundamental principles that apply to all four fundamental interactions.

Some of the material may be familiar to you, but other parts may be entirely new. These novel elements will be discussed in more detail in later chapters, as needed.

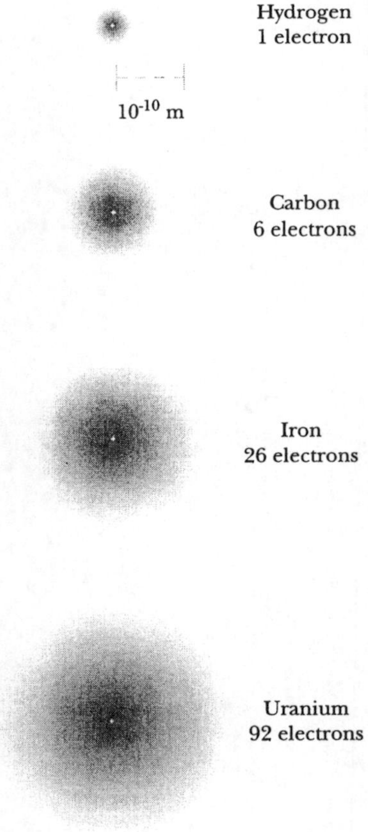

1.2   Kinds of matter

We begin by briefly describing the kinds of matter that we will analyze in this course.

1.2.1   Atoms and nuclei

Ordinary matter is made up of tiny atoms. An atom isn't the smallest type of matter, for it is composed of even smaller objects (electrons, protons, and neutrons), but many of the ordinary everyday properties of ordinary matter can be understood in terms of atomic properties and interactions. As you probably know from studying chemistry, atoms have a very small, very dense core, called the nucleus, around which is found a large cloud of electrons. The nucleus contains protons and neutrons, collectively called nucleons.

Hydrogen
1 electron

10^{-10} m

Carbon
6 electrons

Iron
26 electrons

Uranium
92 electrons

Figure 1.1 Atoms of hydrogen, carbon, iron, and uranium. The white dot shows the location of the nucleus. On this scale, however, the nucleus would be much too small to see.

Electrons are kept close to the nucleus by electric attraction to the protons (the neutrons don't interact with the electrons).

? Recall your previous studies of chemistry. How many protons and electrons are there in a hydrogen atom? In helium or carbon atoms?

Throughout this text you will encounter questions like the preceding one, which ask you to stop and think before reading further. An important part of reading and understanding a scientific text is to ask yourself questions and to try to answer them. You will learn more from reading this text if you try to answer these questions before looking at the discussion in the subsequent paragraph.

See the periodic table on the inside front cover of this textbook. Hydrogen is the simplest atom, with just one proton and one electron. A helium atom has two protons and two electrons. A carbon atom has six protons and six electrons. Near the other end of the chemical periodic table, a uranium atom has 92 protons and 92 electrons Figure 1.1 shows the approximate cloud of electrons for several elements but cannot show the nucleus to the same scale; the tiny dot marking the nucleus in the figure is much larger than the actual nucleus.

The radius of the electron cloud for a typical atom is about 10^{-10} meter. The reason for this size can be understood using the principles of quantum mechanics, a major development in physics in the early 20th century. The radius of a proton is about 10^{-15} meter, very much smaller than the radius of the electron cloud.

Nuclei contain neutrons as well as protons (Figure 1.2). The most common form or "isotope" of hydrogen has no neutrons in the nucleus. However, there exist isotopes of hydrogen with one or two neutrons in the nucleus (in addition to the proton). Hydrogen atoms containing one or two neutrons are called deuterium or tritium. The most common isotope of helium has two neutrons (and two protons) in its nucleus, but a rare isotope has only one neutron; this is called helium-3.

The most common isotope of carbon has six neutrons together with the six protons in the nucleus (carbon-12), while carbon-14 with eight neutrons is an isotope that plays an important role in dating archeological objects. Uranium-235, which can undergo a fission chain reaction, has 92 protons and 143 neutrons, while uranium-238, which does not undergo a fission chain reaction, has 92 protons and 146 neutrons.

1.2.2 Molecules and solids

When atoms come in contact with each other, they may stick to each other ("bond" to each other). Several atoms bonded together can form a molecule—a substance whose physical and chemical properties differ from those of the constituent atoms.

An ordinary-sized rigid object made of bound-together atoms and big enough to see and handle is called a solid, such as a bar of aluminum. A new kind of microscope, the scanning tunneling microscope (STM), is able to map the locations of atoms on the surface of a solid, which has provided new techniques for investigating matter at the atomic level. Two such images appear in Figure 1.3. You can see that atoms in a crystalline solid are arranged in a regular three-dimensional array. The arrangement of atoms on the surface depends on the direction along which the crystal is cut. The irregularities in the bottom image reflect "defects," such as missing atoms, in the crystal structure.

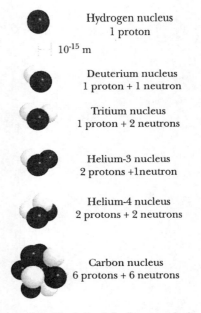

Hydrogen nucleus
1 proton

10^{-15} m

Deuterium nucleus
1 proton + 1 neutron

Tritium nucleus
1 proton + 2 neutrons

Helium-3 nucleus
2 protons + 1 neutron

Helium-4 nucleus
2 protons + 2 neutrons

Carbon nucleus
6 protons + 6 neutrons

Figure 1.2 Nuclei of hydrogen, helium, and carbon.

Figure 1.3 Two different surfaces of a crystal of pure silicon. The images were made with a scanning tunneling microscope.

1.2.3   Liquids and gases

When a solid is heated to a higher temperature, the atoms in the solid vibrate more vigorously about their normal positions. If the temperature is raised high enough, this thermal agitation may destroy the rigid structure of the solid. The atoms may become able to slide over each other, in which case the substance is a liquid. The temperature at which a solid turns into a liquid (or vice versa) is called the melting point. For example, the melting point of water is 0° C (273 Kelvin or 273 K). The melting point of aluminum is 660° C (933 K).

At even higher temperatures the thermal motion of the atoms or molecules may be so large as to break the interatomic or intermolecular bonds completely, and the liquid turns into a gas. In a gas the atoms or molecules are quite free to move around, only occasionally colliding with each other or the walls of their container. The temperature at which a liquid turns into a gas is called the boiling point. The boiling point of water is 100° C (373 K). The boiling point of nitrogen is –196° C (77 K); the boiling point of aluminum is 2057° C (2330 K).

In this course we will learn how to analyze many aspects of the behavior of solids and gases. We won't have very much to say about liquids, because their properties are much harder to analyze. Solids are simpler to analyze than liquids because the atoms stay in one place (though with thermal vibration about their usual positions). Gases are simpler to analyze than liquids because between collisions the gas molecules are approximately unaffected by the other molecules. Liquids are the awkward intermediate state, where the atoms move around rather freely, but always in contact with other atoms. This makes the analysis of liquids very complex.

1.2.4   Hadrons and leptons

Protons and neutrons are not themselves fundamental objects. A great deal of experimental and theoretical work in the latter part of the twentieth century has led to a model in which protons and neutrons are themselves composed of other particles called "quarks." Both protons and neutrons can be thought of as consisting of three quarks, and there also exist "mesons" consisting of quark-antiquark pairs. Many different kinds of mesons and of three-quark particles have been created in particle collisions produced in high-energy accelerators. Particles that are made of quarks and/or antiquarks are called "hadrons."

The electron is not a hadron but is an example of a class of nonquark particles called "leptons" (meaning lightweight particles, although some now known are in fact heavy). Other leptons are the muon and tauon, which behave essentially like heavy electrons, and the neutrino, which has very little mass, probably zero. (There are actually three kinds of neutrinos, associated with the electron, muon, and tauon.)

For both hadrons and leptons, there exist "antiparticles" with the same mass but with other properties that are opposite to the particle properties. For example, the proton carries a positive electric charge and the electron a negative charge, while the antiproton has a negative charge and the antielectron (the positron) has a positive charge. When a particle and its own antiparticle collide, they can annihilate each other, forming a spray of other particles or a burst of high-energy electromagnetic radiation.

Here is a table of properties for all of the leptons and for a few of the many known hadrons (proton, neutron, and pion). Masses are given in terms of kilograms and also in terms of MeV/c^2, a special mass unit used routinely in particle physics (million electron volts per speed of light squared)

Symbol	Name	Mass, kg	Mass, MeV/c^2
Leptons (not made of quarks)			
e^-, e^+	electron, positron	9.11×10^{-31}	0.511
μ^-, μ^+	muon (mu-minus, mu-plus)	1.88×10^{-28}	106
τ^-, τ^+	tauon (tau-minus, tau-plus)	3.18×10^{-27}	1784
$\nu, \bar{\nu}$	neutrino, antineutrino (actually three kinds: ν_e, ν_μ, ν_τ)	0?	0? (may have small mass)
Hadrons (made of quarks)			
$p, \bar{p}$	proton (+), antiproton (−)	1.67×10^{-27}	938.3
$n, \bar{n}$	neutron, antineutron	1.68×10^{-27}	939.6
π^+, π^-	pion (pi-plus, pi-minus)	2.48×10^{-28}	140
π^0	(pi-zero)	2.40×10^{-28}	135

1.2.5 Planets, stars, solar systems, and galaxies

In our brief survey of the kinds of matter that we will study, we make a giant leap in scale from subatomic particles all the way up to planets and stars, such as our Earth and Sun. In this course we will see that many of the same principles that apply to atoms apply to planets and stars. By making this leap we bypass an important physical science, geology, whose domain of interest includes the formation of mountains and continents. We will study objects that are much bigger than mountains, and we will study objects that are much smaller than mountains, but we don't have time in one course to apply the principles of physics to every important kind of matter!

Our Sun and its accompanying planets constitute our Solar System (Figure 1.4). It is located in the Milky Way galaxy, a giant rotating disk-shaped system of stars. On a clear dark night you can see a band of light (the Milky Way) coming from the huge number of stars lying in this disk, which you are looking at from a position about two-thirds of the way out from the center of the disk. Our galaxy is a member of a cluster of galaxies that move around each other much as the planets of our Solar System move around the Sun. The Universe contains many such clusters of galaxies.

1.3 An example of an interaction: electric interactions

Various kinds of matter can participate in different interactions. First we'll consider electric interactions.

1.3.1 Example: Electric interactions

A specific example of an interaction is the electric interaction between "charged" particles such as the protons and electrons found in atoms. As you probably already know, it is observed that two protons repel each other, as do two electrons, while a proton and an electron attract each other (Figure 1.5).

Cluster of galaxies

Galaxy

Solar System

Figure 1.4 Our Solar System exists inside a galaxy, which itself is a member of a cluster of galaxies.

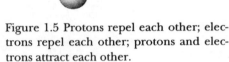

Figure 1.5 Protons repel each other; electrons repel each other; protons and electrons attract each other.

These and other experimental observations are adequately summarized as follows:

- By definition, we assign a "positive electric charge" to protons, and we assign a "negative electric charge" to electrons.
- We can then state that two particles with like charges repel, and two particles with unlike charges attract. The interaction has a direction; it occurs along the line between the particles.
- The strength of the interaction is inversely proportional to the square of the distance between the particles.

This summary of experimental observations is called "Coulomb's law," and the observed repulsions and attractions are called "electric interactions." In addition to protons and electrons, there are many other particles that carry electric charge and obey Coulomb's law. There are also particles that do not carry electric charge and are not attracted or repelled by electrons. One example of such a "neutral" particle is the neutron, which is found inside the nuclei of atoms.

We will use electric interactions (as described and summarized by Coulomb's law) to illustrate some important general properties of interactions. Later we will discuss other kinds of interactions, all of which share some of the basic properties of electric interactions.

1.4 Detecting interactions

How can we detect interactions? In this section we consider various kinds of observations that indicate the presence of interactions.

? Before you read further, take a moment to think about your own ideas of interactions. How can you tell that two objects are interacting with each other?

1.4.1 Change of direction

Suppose you observe a proton moving through a region of outer space, far from almost all other objects. You see the proton moving along a path like the one shown in Figure 1.6. The path of the proton is shown as though a camera had taken multiple exposures at equal time intervals.

? Do you see evidence that the proton is interacting with another object?

Evidently a change in direction is a vivid indicator of interactions. If you observe a change in direction of the motion of a proton, you will find another object somewhere that has interacted with this proton.

? Suppose that the only other object nearby was another proton. What was the approximate initial location of this second proton?

Since two protons repel each other electrically, the second proton must have been located to the right of the bend in the first proton's path.

1.4.2 Change of speed

Suppose that you observe a positron (e^+, an antielectron, which has the same mass as an electron but a positive charge) traveling in a straight line through outer space far from almost all other objects (Figure 1.7). The path of the positron is shown as though a camera had taken multiple exposures at equal time intervals.

? Where is the positron's speed largest? Where is the positron's speed smallest? You may have to look rather closely, or even measure some distances with a ruler, to see the change in speed. It is easy to detect

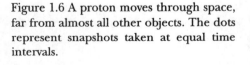

Figure 1.6 A proton moves through space, far from almost all other objects. The dots represent snapshots taken at equal time intervals.

even a small change of direction, but it is often hard to detect even a sizable change of speed by eye.

The speed is largest at the top, where the dots are farther apart. It is smallest at the bottom, where the dots are closer together.

? Suppose that the only other object nearby was a proton. What was the approximate initial location of this proton?

The proton must have been located directly below the starting point of the path.

Evidently a change in speed is an indicator of interactions. If you observe a change in speed of a positron, you will find another object somewhere that has interacted with the positron.

1.4.3 Change of velocity: change of speed or direction

It is useful to combine speed and direction into one quantity called "velocity." Consider an airplane that is flying with a speed of 1000 kilometers/hour in a direction that is due east. We say the velocity is 1000 km/hr, east, where we specify both speed and direction. We can combine the two indicators of interaction, change of speed and change of direction, into one compact statement:

> A change of velocity (speed or direction) indicates the existence of an interaction.

Figure 1.7 A positron moves through space, far from almost all other objects. The dots represent snapshots taken at equal time intervals.

Mathematical quantities called "vectors," which have their own special rules of algebra, can be used to describe velocity. A vector is a quantity that has both a magnitude and a direction. In diagrams, velocity vectors are represented as lines with arrowheads that point in the direction of the motion of interest, and with a length proportional to the speed. Figure 1.8 shows two successive positions of a particle, with velocity vectors indicating a change in speed of the particle. Figure 1.9 shows three successive positions of a particle, with velocity vectors indicating a change in direction but no change in speed.

Figure 1.8 Two successive positions of a particle, with velocity vectors indicating a change in speed of the particle.

Velocity is just one of many physical quantities that have both a direction and a "magnitude" (speed in the case of velocity vectors). Other examples of vector quantities include displacement (change of position, both direction and magnitude), force (direction and magnitude), and the magnetic field of the Earth (which varies in direction and magnitude over the Earth's surface). We will use vectors extensively in our modeling of interacting systems.

Note: Reviewing vectors

You presumably have used vectors in previous math or physics courses. An appendix at the end of this book provides a review of vectors and the vector operations and notation we will use in this course. You should read the vector appendix, paying close attention to any concepts that are not immediately familiar. It is extremely important that you be comfortable working with vectors, because we will use them throughout this course.

Figure 1.9 Three successive positions of a particle, with velocity vectors indicating a change in direction but no change in speed.

1.4.4 Uniform motion—no interaction

Suppose you observe a proton moving along in outer space far from all other objects. We don't know what made it start moving in the first place; presumably a long time ago an electric interaction gave it some velocity and it has been coasting through the vacuum of space ever since.

It is an observational fact that such an isolated particle moves at constant, unchanging speed, in a straight line. Its velocity does not change (neither

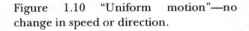

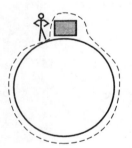

Figure 1.10 "Uniform motion"—no change in speed or direction.

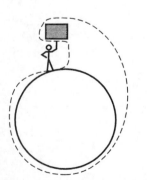

Figure 1.11 When you pick up the block, the (Earth + block) system undergoes a change of shape.

its direction nor its speed changes). We call such motion with unchanging velocity "uniform motion" (Figure 1.10).

Because of this behavior of isolated particles, we conclude that whenever we observe uniform motion, this is an indicator of little or no interaction, or of interactions cancelling each others' effects. An important special case of uniform motion is a particle at rest (zero speed). Constant speed (including zero) and constant direction implies the absence of interaction.

1.4.5   Change of identity

A different indicator of interactions is change of identity. An example of a change of identity is a chemical reaction, such as the formation of water (H_2O) from the burning of hydrogen in oxygen. A water molecule behaves very differently from the hydrogen and oxygen atoms of which it is made.

Another example of change of identity is the radioactive "decay" of a particle. Neutrons are an important and stable constituent of nuclei, but a neutron all by itself outside a nucleus, a "free" neutron, is unstable, with an average lifetime of about 15 minutes. When it decays, it decays into three particles, a (positive) proton, a (negative) electron, and a (neutral) antineutrino:

$$n \rightarrow p^+ + e^- + \bar{\nu}$$

This decay is possible because the neutron has a slightly larger mass than the mass of the proton plus the mass of the electron. (Neutrinos and antineutrinos have very little mass—probably zero.) Neutron decay also happens inside some "radioactive" nuclei, in which case the electron and the antineutrino are emitted from the nucleus and can be detected, but the proton remains bound in the nucleus, which now represents a different element because it has one more proton and one less neutron than before ("nuclear transmutation").

In neutron decay there is a striking change of identity. The neutron ceases to exist, and three new particles are created. Despite the fact that the neutron did not seem to interact with other particles, the notion of "interaction" has been generalized to include particle decays. Neutron decay is an example of what is called the "weak" interaction (more on this in a moment).

1.4.6   Change of configuration

We have mainly been discussing interactions among "point" particles—objects that are very small compared to other distances of interest. If a pointlike particle changes velocity (direction or speed), or changes identity (including disappearing altogether), we know that it is involved in some kind of interaction. In more complicated systems made of many atoms, there can be interactions without visible change in direction or speed. We briefly discuss such effects.

Slowly bend a pen or pencil, then hold it in the bent position. The speed hasn't changed, nor is there a change in the direction of motion (it's not moving!). The pencil has not changed identity. Evidently a change of configuration (or change of shape) can be evidence for interactions. Compressing or stretching a spring is another example of this kind of change.

Depending on how you define a "system," moving one object relative to another can be a change of shape. For example, if you move a block closer to or farther away from the Earth, the system consisting of Earth plus block undergoes a change of shape (Figure 1.11). Your interaction with the (Earth+block) system caused the change of shape.

Other changes in configuration include "phase changes" such as the freezing or boiling of a liquid, brought about by interactions with the surroundings. In different phases (solid, liquid, gas), atoms or molecules are arranged differently. Changes in configuration at the atomic level are another indication of interactions.

1.4.7 Change of temperature

Another indication of interaction in multiparticle systems is change of temperature. Place a pot of water on a hot stove. If you stick your finger into the water you detect a change—the water starts to feel hot. At the macroscopic level the water doesn't look any different. It doesn't seem to be moving faster, or changing direction, or undergoing a change of shape. However, at the atomic level, water molecules are jostling around more vigorously. (Actually, there is a tiny change of shape, because the density of water is slightly different at different temperatures.)

This (nearly) invisible change is conveniently monitored with a thermometer. A change in temperature is an indication of an interaction (in this case, between the hot stove and the initially cool water).

1.4.8 Change of position?

Is a change of position an indicator of an interaction? That depends. If the change of position occurs simply because a particle is moving at constant speed and direction, then a mere change of position is not an indicator of an interaction, since uniform motion is an indicator of *no* interaction.

? If however you observe a particle at rest in one location, and later you observe it again at rest but in a different location, did an interaction take place?

Yes. You can infer that there must have been an interaction to give the particle some velocity to move the particle toward the new position, and another interaction to slow the particle to a stop in its new position.

1.4.9 Indirect evidence for an interaction

Sometimes there is indirect evidence for an interaction. When something doesn't change although you expect a change, this indicates that another interaction is present. Consider a balloon that hovers motionless in the air despite the downward gravitational pull of the Earth. Evidently there is some other kind of interaction that opposes the gravitational interaction. In this case, interactions with air molecules have the net effect of pushing up on the balloon ("buoyancy").

The stability of the nucleus of an atom suggests the existence of another kind of interaction. The nucleus contains positively charged protons that repel each other, yet the nucleus remains intact. We conclude that there must be some other kind of force present, a nonelectric attractive force that overcomes the electric repulsion. This is evidence for the "strong interaction" that acts between nucleons (both protons and neutrons).

1.4.10 Summary: changes as indicators of interactions

Here then are the most common indicators of interactions:
- change of velocity (change of direction and/or change of speed)
- change of identity
- change of shape of multiparticle system
- change of temperature of multiparticle system
- lack of change when change is expected

1.5   Four fundamental interactions

Considering the great complexity of the world around us, it is surprising that all observed interactions can be classified into just four types of fundamental interactions on the basis of their intrinsic strengths:

- gravitational interactions (all objects attract each other gravitationally)
- electromagnetic interactions (electric and magnetic interactions, closely related to each other)
- "strong" interactions (inside the nucleus of an atom)
- "weak" interactions (including in particular neutrino interactions)

Any of these interactions can be indicated by changes in velocity, identity, shape, temperature, or by a lack of change when change is expected.

1.5.1   The gravitational interaction

All objects in the universe gravitationally attract each other. Evidence for the gravitational interaction includes the observations that the speed of a dropped object increases as it falls toward the Earth (change of speed), and planets travel in nearly circular orbits around the Sun (change of direction).

The gravitational interaction is the most familiar interaction in everyday life and seems quite strong, yet in an important sense it is quite weak. It is of course easy to observe the interaction between the Earth and a falling object, but note that it takes the entire Earth to produce the effect! Gravitational interactions between less massive objects are difficult to detect. With very delicately balanced equipment it is possible to observe two metal balls attracting each other gravitationally, but this experiment, first performed by Cavendish (1797-1798), is not easy to do because the interaction is so weak.

1.5.2   The electromagnetic interaction

As we stated earlier, electric interactions affect charged particles. Magnetic interactions affect moving charged particles; in fact, magnetic interactions can be considered to be a relativistic aspect of electric interactions. The second volume of this text, subtitled *Electric & Magnetic Interactions*, focuses on models of the electromagnetic interaction. The close connection between electric and magnetic interactions was established in the nineteenth century, especially as a result of the work of Faraday and Maxwell.

Evidence for electric and magnetic interactions includes the observations that two protons repel each other (this cannot be a gravitational attraction because gravitational interactions are always attractive), and that a moving proton can be deflected by a magnet (change of direction).

The electromagnetic interaction is intrinsically much stronger than gravitation. For example, the electric interaction is responsible for the phenomenon of "static cling," in which two pieces of clothing may interact electrically more strongly with each other than they do gravitationally with the whole Earth. The fact that two protons repel each other implies that their electric repulsion is much stronger than their gravitational attraction. The electric interatomic attraction is evidently much stronger than gravity—otherwise your bones would fall apart.

We can distinguish between gravitational and electromagnetic interactions on the basis of their respective strengths, by the fact that some particles behave as though they have no electric charge and no electric interactions (the neutron and the neutrino are examples of electrically neutral particles), and by the fact that some electric interactions are repulsive.

There are important short-range electric interactions between neutral atoms (which are multiparticle entities). Even though an atom normally contains equal numbers of positively charged protons and negatively charged electrons, two atoms attract each other when they are close but not too close to each other (in certain situations this is called the "van der Waals force"). They repel each other if you try to push them very close together. This behavior springs from the basic electric interaction, but quantum mechanics is required in order to understand the details.

So far we can rank the strengths of the interactions thus:

electromagnetic > gravitational

1.5.3 The strong interaction

The nucleus of an atom is composed of positively charged protons and uncharged neutrons (Figure 1.12). Protons and neutrons are themselves composed of fundamental particles called quarks, which are held together by an interaction called "the strong interaction." Evidence for the strong interaction includes the observations that nuclei don't normally fly apart despite the mutual electric repulsions of the protons (lack of change when change is expected), and that a moving neutron (with no electric charge) can be deflected by a nucleus (change of velocity).

This interaction is called the "strong" interaction because it is intrinsically much stronger than the electromagnetic interaction (since nuclei don't normally fly apart). We have already seen that the electromagnetic interaction is much stronger than the gravitational interaction, so now our ranking of the three interactions looks like this:

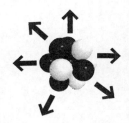

Figure 1.12 The protons in the nucleus of an atom exert repulsive electric forces on each other.

strong > electromagnetic > gravitational

The strong interaction has the property that nucleons (protons or neutrons) attract each other extremely strongly when they are in contact with each other, but the interaction has a very short range. If the nucleons are not quite touching, the interaction is very small, and then the electric interaction between two protons can actually be much larger than the strong interaction.

This allows us to understand the explosive fission of uranium. It is a special property of one form of uranium (U-235) that if the uranium nucleus absorbs an extra neutron, the nucleus may elongate into a dumbbell shape (Figure 1.13). This elongation decreases the contact among protons and neutrons, decreasing the magnitude of the strong interaction between them. The mutual electric repulsion of the two positively charged fragments of the nucleus then blows the fragments apart, sending them hurtling at high speed through the block of uranium, and heating the metal to a high temperature.

In the fission process there is a change of identity: a uranium nucleus is changed into two other nuclei. The strong interaction can also produce a change of identity of individual particles. Particle accelerators produce high-energy beams of particles that interact through the strong interaction. Among these particles are positive, negative, or neutral particles called "pi mesons" or "pions," which can produce a change of identity when they hit a nucleon. Here is a typical reaction, in which a negative pion π^- (pronounced "pi-minus") collides with a proton p^+ and produces a neutron n^0 and a neutral pion π^0 (pronounced "pi-zero"):

Figure 1.13 A U-235 nucleus undergoing fission.

$$\pi^- + p^+ \rightarrow n^0 + \pi^0$$

This is a high-probability process, which is another of the many reasons why the strong interaction is called "strong."

As we mentioned earlier, experimental and theoretical work have led us to believe that the strongly interacting particles (including protons, neutrons, and pions, but not electrons) are themselves composed of other particles called "quarks." The proton is believed to consist of two "up" or u quarks (each with charge $+\frac{2}{3}e$, where $+e$ is the charge on a proton), and one "down" or d quark (with charge $-\frac{1}{3}e$). The proton is described as uud for short, with a net charge of $(+\frac{2}{3}e) + (+\frac{2}{3}e) + (-\frac{1}{3}e) = +e$.

Particles called pi mesons are believed to consist of quark-antiquark pairs involving an up or down quark and an up or down antiquark. The up antiquark, $\bar{u}$, carries an electric charge of $-\frac{2}{3}e$, the negative of the charge carried by an up quark. Similarly, the down antiquark $\bar{d}$ has an electric charge of $+\frac{1}{3}e$, the negative of the charge of the down quark.

Exercises

Often at the end of a section you will find exercises such as those that follow. It is extremely important that you work through exercises as you come to them, to make sure that you can apply what you have just read. Passive acquaintanceship with the ideas isn't enough—you must be able to use the ideas productively.

Answers to exercises may be found at the end of the chapter, but it is very important that you make a serious attempt to do an exercise before checking your answer. If you don't make a serious effort on your own, you will have substituted inefficient and wasteful passive reading for active engagement with the ideas, and you won't learn very much.

Ex. 1.1 Like the proton, the neutral neutron is thought to be made of three quarks of the u and d varieties. How many u quarks and how many d quarks must there be in a neutron?

Ex. 1.2 What is the quark/antiquark composition of the positive pion (π^+)? Of the negative pion (π^-)?

1.5.4 The weak interaction

The "weak" interaction is seen in rather exotic phenomena, often involving a change of identity. An example is the collision of an antineutrino with a proton to produce a neutron and a positron (change of identity):

$$\bar{\nu} + p^+ \rightarrow n + e^+ \quad \text{(weak interaction; low rate of reaction)}$$

This reaction is observed when a beam of antineutrinos is shot at a target of liquid hydrogen (whose nuclei are protons), and positrons are observed to emerge at high speed from the hydrogen (Figure 1.14). The rate at which this reaction occurs is extremely low—almost all of the antineutrinos pass right through the liquid hydrogen without interacting. The neutral antineutrino doesn't interact electrically with the proton, and the gravitational interaction is negligible. The interaction of neutrinos with matter is so weak that most of the neutrinos coming from the Sun pass right through the entire Earth without interacting!

Clearly there is some kind of interaction with the proton, but it happens so rarely that it is not an example of the "strong" interaction. We call it a "weak" interaction. (In contrast, most pi mesons entering the liquid hydrogen would interact with the protons, through the "strong" interaction.)

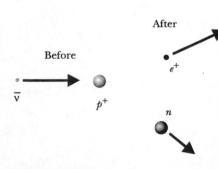

Figure 1.14 An antineutrino reacts with a proton, producing a high speed positron and a neutron.

This antineutrino reaction is closely related to the decay of a free neutron into a proton, an electron, and an antineutrino:

$$n \rightarrow p^+ + e^- + \bar{\nu} \quad \text{(weak interaction; decays in about 15 minutes)}$$

This decay, with an average lifetime of about 15 minutes, is extremely slow by the standards of the strong or the electromagnetic interactions. In contrast, there is an excited state of the proton called the Δ^{++} (delta), which can be produced in a strong interaction when a positive pion collides with a proton. The Δ^{++} decays back into a proton and a positive pion in about 10^{-23} seconds!

$$\pi^+ + p^+ \rightarrow \Delta^{++} \quad \text{(strong interaction creates a } \Delta^{++})$$

$$\Delta^{++} \rightarrow p^+ + \pi^+ \quad \text{(strong interaction; decays in about } 10^{-23} \text{ seconds)}$$

In addition to neutron decay, many other examples of extremely low-rate processes were discovered and were found to share other properties in addition to their low rates. In particular, many of these phenomena (though not all) involve neutrinos or antineutrinos, electrically uncharged particles that have a very small mass (probably zero; if so, they travel at the speed of light).

After these low-rate interactions were discovered and put into a class of their own, other properties of the weak interaction were discovered that were quite special. For example, we say that the weak interaction violates "parity conservation," whereas the other interactions do not. If you observe a strong, electromagnetic, or gravitational interaction reflected in a mirror, the interaction looks completely normal and entirely possible. In contrast, a mirror image of a weak interaction can turn out to correspond to something that never happens! Antineutrinos have "spin" that is always oriented in the direction of motion, and to describe this behavior we say that antineutrinos are always right-handed, in the sense of an ordinary right-hand screw. Handedness flips in a mirror (Figure 1.15), and the mirror image of neutron decay would involve a left-handed antineutrino, whose spin would be opposite to its direction of motion. This has never been observed.

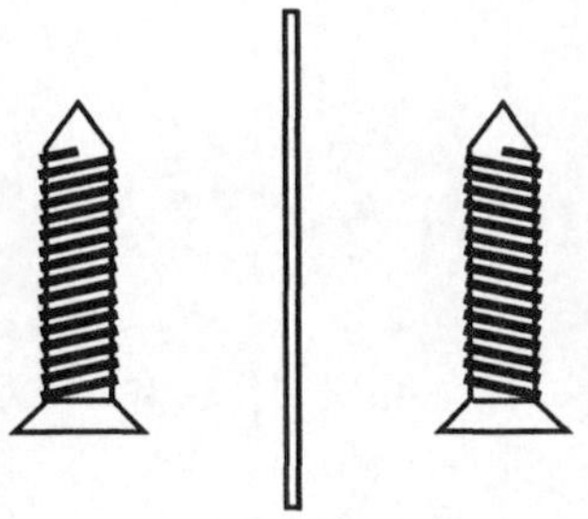

Another oddity of the weak interaction is that if you not only reflect in a mirror but also change particles into antiparticles (and vice versa), you do get a possible event. For example, an antineutron decays into an antiproton, a positron, and a left-handed neutrino (instead of the right-handed antineutrino emitted in ordinary neutron decay):

$$\bar{n} \rightarrow p^- + e^+ + \nu$$

Figure 1.15 A right-handed screw when reflected in a mirror looks like a left-handed screw.

Despite its weakness, the weak interaction is intrinsically much stronger than the gravitational interaction. Our final order of strengths is this:

strong > electromagnetic > weak > gravitational

Gravitational interactions between individual particles can almost always be neglected compared to their other interactions. It is only when one of the interacting bodies is huge, such as the Earth, that gravitational interactions become significant.

While the four different kinds of fundamental interactions were initially identified on the basis of their respective strengths, each interaction is also characterized by the kinds of particles that participate in the interaction and by the detailed nature of the interactions. For example, electrons and neutrinos are unaffected by the strong interaction, while neutrons have no electric interaction. The weak interaction fails to conserve parity, but the other three interactions do conserve parity. The gravitational interaction is solely attractive, but the electric interaction can be repulsive.

Unifying the weak and the electromagnetic interactions

In 1979 the Nobel prize for physics was awarded to Sheldon Glashow, Abdus Salam, and Steven Weinberg for unifying the weak interaction and the electromagnetic interaction. They showed that these interactions could be considered different manifestations of the same underlying type of interaction, and that in the extremely high-temperature conditions of the very early Universe, just after the "Big Bang," the strength of the "weak" interaction would have been the same as the strength of the electromagnetic interaction. It is only in the current low-temperature state of the Universe that the weak interaction appears "weak."

Because of this unification, we may choose to say that there are just three fundamental types of interaction: gravitational, strong, and "electroweak." Nevertheless, in typical low-energy phenomena the properties and strengths of the weak interaction and the electromagnetic interaction are quite different, and it is common practice to say that there are four different fundamental interactions.

Further unifications?

Some physicists are currently working on "grand unified theories" which they hope will make it possible to treat the strong and electroweak interactions as closely related aspects of a single unified interaction, just as the electromagnetic and weak interactions can be viewed as different aspects of the electroweak interaction. There is even hope of a "theory of everything" that would unify gravitational interactions with the others, though at the time of writing, this dream seems more remote than the hope of bringing the strong and electroweak interactions together.

Ex. 1.3 In everyday life you frequently interact with objects by touching them. You observe other objects interacting by contact; for example, a bowling ball collides with a bowling pin and knocks it over. Is "contact interaction" a fifth fundamental kind of interaction?

Ex. 1.4 Suppose a scientist claims to have discovered a fifth type of fundamental interaction. Think about the considerations that led to classifying interactions into four different kinds. What would lead a scientist to propose that a fifth type of interaction exists?

1.6   Principles applying to all interactions

Despite the fact that there are four fundamentally different kinds of interactions (or three, depending on how you count), there are general principles which apply to all interactions. In this section we will consider some important general properties that hold true for all kinds of interactions (strong, electromagnetic, weak, and gravitational).

1.6.1   Newton's first law of motion

The basic relationship between change of velocity and interaction is summarized qualitatively by Newton's "first law of motion":

NEWTON'S FIRST LAW OF MOTION

An object moves in a straight line and at constant speed except to the extent that it interacts with other objects.

The words "to the extent" imply that the stronger the interaction, the more change there will be in direction and/or speed. The weaker the interaction, the less change. If there is no interaction at all, the direction doesn't change and the speed doesn't change; this important case is called "uniform motion." This case can also be called "uniform velocity" or "constant velocity," since velocity refers to both speed and direction. Included in the meaning of constant velocity is an object that is not moving, having a constant velocity of zero.

Newton's first law of motion is only qualitative, because it doesn't give us a way to calculate quantitatively how much change in speed or direction will be produced by a certain amount of interaction, a subject we will take up in Chapter 2. Nevertheless, Newton's first law of motion is important in providing a conceptual framework for thinking about the relationship between interaction and motion.

Can Newton's first law of motion be violated?

Newton's first law of motion doesn't seem to apply to many everyday situations. To push a chair across the floor at constant speed, you have to keep pushing all the time. If you stop pushing, the chair stops. But doesn't Newton's first law of motion say that the chair should keep moving in a straight line at constant speed?

? Is this a violation of Newton's first law of motion? Try to answer this question before reading farther.

The complicating factor here is that your hands aren't the only objects that are interacting with the chair. The floor also interacts with the chair, in a way that we call friction. If you push hard enough to compensate exactly for the floor friction, the sum of all the interactions is zero, and the chair moves at constant speed as predicted by Newton's first law. The interaction with the floor is identified by "lack of change when change is expected." The chair moves at constant velocity despite your pushing on it, which means that something else must also be interacting with it (the floor).

It is difficult to observe motion without friction in everyday life, because objects almost always interact with many other objects, including air, flat surfaces, etc. One example of a nearly friction-free situation is a hockey puck sliding on ice. The puck slides a long way at nearly constant speed in a straight line (constant velocity) because there is little friction with the ice. An even better example is the uniform motion of an object in outer space, far from all other objects.

Ex. 1.5 Experiment: There is a very simple experiment that you can do to illustrate Newton's first law of motion. Place a ball on a book and move the book or walk with the book in uniform motion. Note that you don't really have to do anything to the ball to keep the ball moving with constant velocity (relative to the ground) or to keep the ball at rest (relative to you). Then stop suddenly, or abruptly change your direction or speed. If there is little interaction between the book and the ball, the ball tends to continue with its original uniform velocity in accordance with Newton's first law of motion.

Ex. 1.6 Some science museums have an exhibit called a Bernoulli blower, in which a basketball hangs suspended in a column of air blown upward by a strong fan. If you saw a ball suspended in the air but didn't know the blower was there, why would Newton's first law of motion make you suspect that something must be holding the ball up?

Ex. 1.7 A spaceship far from all other objects uses its impulse power system to attain a speed of 10^4 m/s. The crew then shuts off the power. According to Newton's first law, what will happen to the motion of the spaceship from then on?

Ex. 1.8 Why do we use a spaceship to illustrate Newton's first law? Why not a car or a train?

1.6.2 The principle of relativity

Newton's first law of motion has a larger context—there are principles even more global and more fundamental. For example, a great variety of experimental observations has led to the establishment of the following principle:

THE PRINCIPLE OF RELATIVITY

Physical laws work in the same way for observers in uniform motion as for observers at rest.

This principle is called "the principle of relativity." (Einstein's extensions of this principle are known as "special relativity" and "general relativity.") Phenomena observed in a room in uniform motion (for example, on a train moving with constant speed on a smooth straight track) obey the same physical laws in the same way as experiments done in a room that is not moving. According to this principle, Newton's first law of motion should be true both for an observer moving at constant velocity and for an observer at rest.

The cosmic microwave background

Technically, the principle of relativity applies only to observers who have a constant speed and direction (or zero speed) relative to the cosmic microwave background, which provides the only backdrop and frame of reference with an absolute, universal character. It used to be that the basic reference frame was loosely called "the fixed stars," but stars and galaxies have their own individual motions within the Universe and do not constitute an adequate reference frame with respect to which to measure motion.

The cosmic microwave background is low-intensity electromagnetic radiation with wavelengths in the microwave region, which pervades the Universe, radiating in all directions. Measurements show that our galaxy is moving through this microwave radiation with a large, essentially constant velocity, toward a cluster of a large number of other galaxies. The way we detect our motion relative to the microwave background is through the "Doppler shift" of the frequencies of the microwave radiation, toward higher frequencies in front of us and lower frequencies behind. This is essentially the same phenomenon as that responsible for a fire engine siren sounding at a higher frequency when it is approaching us and a lower frequency when it is moving away from us.

The discovery of the cosmic microwave background provided major support for the "Big Bang" theory of the formation of the Universe. According to the Big Bang theory, the early Universe must have been an extremely hot mixture of charged particles and high-energy, short-wavelength electromagnetic radiation (visible light, x-rays, gamma rays, etc.). Electromagnetic radiation interacts strongly with charged particles, so light could not travel very far without interacting, making the Universe essentially opaque. Also, the Universe was so hot that electrically neutral atoms could not form without the electrons immediately being stripped away again by collisions with other fast-moving particles.

As the Universe expanded, the temperature dropped. Eventually the temperature was low enough for neutral atoms to form. The interaction of electromagnetic radiation with neutral atoms is much weaker than with individual charged particles, so the radiation was now essentially free, dissociated from the matter, and the Universe became transparent. As the Universe continued to expand (the actual space between clumps of matter got bigger!), the wavelengths of the electromagnetic radiation got longer, until today this fossil radiation has wavelengths in the relatively low-energy, long-wavelength microwave portion of the electromagnetic spectrum.

Inertial frames of reference

It is an observational fact that in reference frames that are in uniform motion with respect to the cosmic microwave background, far from other objects (so that interactions are negligible), an object maintains uniform motion. Such frames are called "inertial frames" and are reference frames in which Newton's first law of motion is valid.

? Is the surface of the Earth an inertial frame?

No! The Earth is rotating on its axis, so the velocity of an object sitting on the surface of the Earth is constantly changing direction, as is a coordinate frame tied to the Earth (Figure 1.16). Moreover, the Earth is orbiting the Sun, and the Solar System itself is orbiting the center of our Milky Way galaxy, and our galaxy is moving toward other galaxies. So the motion of an object sitting on the Earth is actually quite complicated and definitely not uniform with respect to the cosmic microwave background.

However, for many purposes the surface of the Earth can be considered to be (approximately) an inertial frame. For example, it takes 6 hours for the rotation of the Earth on its axis to make a 90° change in the direction of the velocity of a "fixed" point. If a process of interest takes only a few minutes, during these few minutes a "fixed" point moves in nearly a straight line at constant speed due to the Earth's rotation, and velocity changes in the process of interest are typically much larger than the very small velocity change of the approximate inertial frame of the Earth's surface.

Similarly, although the Earth is in orbit around the Sun, it takes 365 days to go around once, so for a period of a few days or even weeks the Earth's orbital motion is nearly in a straight line at constant speed. Hence for many purposes the Earth represents an approximately inertial frame despite its motion around the Sun.

Some consequences

Suppose you're riding on a train, looking at a glass on your table (Figure 1.17). To you, the glass appears to be at rest (zero speed). You certainly don't have to push on it to keep it going—it just sits there. You say that the glass obeys Newton's first law of motion.

Think about how this looks to someone on the ground, looking through the train window at the glass on the table (Figure 1.18). For that observer, the glass is traveling at a constant high speed v, and it is clear to the observer that you aren't doing anything active to keep the glass moving.

Both of you correctly conclude that uniform motion, whether at zero speed or nonzero speed, does not require any effort or interaction. Both of you agree that Newton's first law of motion correctly describes the behavior of the glass. Your observations and conclusions are in accord with the principle of relativity; physical laws work in the same way both for the observer in uniform motion and for the observer at rest.

? Consider a rock at rest in outer space, far from all other objects. You are in a spaceship that is also at rest, far from the rock, and you continually use your radar to measure the position of the rock with respect to your own frame of reference (Figure 1.19). Do your

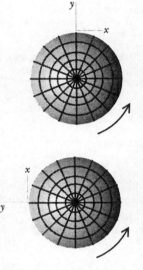

Figure 1.16 Axes tied to the Earth rotate through 90° in a quarter of a day (6 hours).

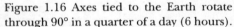

Figure 1.17 You're riding on a train, looking at a glass at rest on your table.

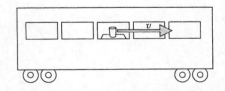

Figure 1.18 An observer on the ground beside the train sees the glass go by at (constant) high speed.

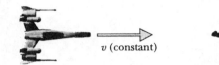

Figure 1.19 Both your spaceship and a rock are at rest (relative to the cosmic microwave background).

v (constant)

Figure 1.20 Your spaceship coasts toward the rock with constant velocity.

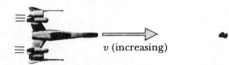

v (increasing)

Figure 1.21 With rockets blasting, your spaceship moves toward the rock with increasing speed.

measurements of the rock's position vs. time agree with Newton's first law of motion?

According to your measurements, the rock remains always at rest, which is in agreement with Newton's first law.

? Now suppose your ship is coasting at a constant velocity toward the rock. Again, you continually use your radar to measure the position of the rock with respect to your own frame of reference (Figure 1.20). Do your measurements of the rock's position vs. time agree with Newton's first law of motion?

Your measurements give a picture of a rock that has constant speed and constant direction of motion, which is consistent with Newton's first law. Nothing is interacting with the rock, so it maintains uniform motion. The principle of relativity says that the two situations, the rock at rest with respect to you, or the rock moving with constant velocity with respect to you, are in a deep sense really the same kinds of phenomena.

? However, now consider what you would see if you turn on your rockets so that your spaceship accelerates toward the rock—that is, it moves faster and faster (Figure 1.21). Would the position of the rock vs. time, as measured by your radar, agree with Newton's first law? Does the principle of relativity apply for an observer who is accelerating?

Your measurements would now indicate that the rock, which is really at rest (with respect to the cosmic microwave background), is moving faster and faster (relative to you), without any interaction to cause this change of speed. Evidently Newton's first law of motion is not valid in an accelerating frame of reference. We say that the accelerating spaceship is a "non-inertial frame" of reference.

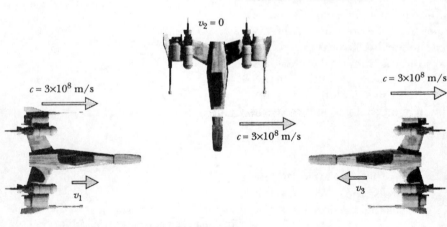

$v_2 = 0$

$c = 3 \times 10^8$ m/s

v_1

$c = 3 \times 10^8$ m/s

$c = 3 \times 10^8$ m/s

v_3

Figure 1.22 Light emitted by the spaceship on the left is measured to have the same speed by observers in all three ships.

Similarly, if you drive your spaceship along a curving trajectory (non-uniform motion), your radar measurements would display a curving path for the rock in the absence of interactions affecting the rock. But your spaceship is not an inertial frame of reference, because it does not have a uniform motion relative to the cosmic microwave background.

The special theory of relativity

Einstein's special theory of relativity (published in 1905) built on the basic principle of relativity but added the conjecture that the speed of a beam of light must be the same as measured by observers in different frames of reference in uniform motion with respect to each other. In Figure 1.22, observers on each spaceship measure the speed of the light c emitted by the ship on the left to be the same ($c = 3 \times 10^8$ m/s), despite the fact that they are moving at different velocities.

This additional condition seems peculiar and has far-reaching consequences. After all, the glass on the train or the rock in outer space have different speeds relative to an observer, depending on the motion of the observer. Yet a wide range of experiments has confirmed Einstein's conjecture; all observers measure the same speed for the same beam of light, $c = 3 \times 10^8$ m/s. (The color of the light is different for the different observers, but the speed is the same.)

On the other hand, if someone on the ship on the left throws a ball or a proton or some other piece of matter, the speed of the object will be different for observers on the three ships; it is only light (and probably neutrinos) whose speed is independent of the observer.

Einstein's theory has interesting consequences. For example, it predicts that time will run at different rates in different frames of reference. These predictions have been confirmed by many experiments. These unusual effects are large only at very high speeds (a sizable fraction of the speed of light), which is why we don't normally observe these effects in everyday life, and why we can use nonrelativistic calculations for low-speed phenomena.

1.7 Quantifying the effects of interactions

Changes are symptoms of interactions—changes in speed, direction, shape, temperature, or identity. Newton's first law of motion gives us a qualitative connection between interactions and their effects on motion (speed and direction; note that this law does not deal with changes in shape, temperature, or identity). In order to make possible quantitative predictions or explanations of physical phenomena, we need a quantitative measure of interactions and a quantitative measure of effects of those interactions.

Newton's first law of motion was phrased in this way:

> **An object moves in a straight line and at constant speed except to the extent that it interacts with other objects.**

In other words, "the stronger the interaction, the bigger the change in the motion." Implicit in this statement is the important idea that if there is no interaction, an object's motion will not change. If an object is moving it will continue to move in a straight line, with no change of direction or speed. If it is not moving, it will remain at rest. A quantitative version of this law would provide a means of explaining or predicting the behavior of an object, if we could enumerate all of its interactions with other objects.

1.7.1 Momentum

Consider what makes change in motion difficult. If two objects have the same velocity but one is much heavier than the other, it is more difficult to change the heavy object's speed or direction. It is easier to stop a baseball traveling at a hundred miles per hour than to stop a car traveling at a hundred miles per hour! It is easier to change the direction of a moving canoe than to change the direction of a large, massive ship; the doomed ship Titanic couldn't change course quickly enough after the iceberg was spotted.

To take into account both motion and mass, we might define a vector quantity $\vec{p}$ called "momentum" that is the product of the mass m and the vector velocity $\vec{v}$ (the triple equal sign means "is defined as"): $\vec{p} \equiv m\vec{v}$. Instead of saying "the stronger the interaction, the bigger the change in the motion," we will say "the stronger the interaction, the bigger the change in the momentum."

However, experiments on particles moving at very high speeds, close to the speed of light c (3×10^8 m/s), show that changes in $m\vec{v}$ are not proportional to the strength of the interactions. As you keep applying a force to a particle near the speed of light, the speed of the particle barely increases. What does increase? It is observed that changes in the following quantity are proportional to the amount of interaction:

DEFINITION OF MOMENTUM

$$\vec{p} \equiv \frac{m\vec{v}}{\sqrt{1 - v^2/c^2}}$$

This is the "relativistic" definition of momentum. Einstein in his special theory of relativity predicted that this would be the appropriate definition for momentum at high speeds, a prediction that has been abundantly verified in a wide range of experiments. The m in the definition is called the "rest mass" of the particle, because it is the mass of the particle when it is at rest.

If the speed v is much smaller than the speed of light c ($v \ll c$), the denominator is nearly equal to 1, so for low speeds the momentum is approximately $m\vec{v}$, which is the approximate "nonrelativistic" momentum. But if the speed is nearly as large as the speed of light, the denominator is very small, and the magnitude of the momentum is much larger than mv.

We will repeatedly emphasize the role of momentum throughout this course because of its fundamental importance not only in classical (prequantum) mechanics but also in quantum mechanics. Also, the use of momentum clarifies the physics analysis of certain complex processes such as collisions, including collisions at speeds approaching the speed of light.

Another interpretation

The quantity $m/(\sqrt{1 - v^2/c^2})$ can be thought of as the "relativistic mass" of a moving particle, and we could choose to have the symbol m stand for the relativistic mass, because it is valid to think of the mass as increasing at high speed. However, scientists who routinely work with relativistic interactions (particle physics), find it clearer and more convenient to use the symbol m to refer to the rest mass (a constant independent of speed), and we will follow this practice.

Vector notation

Momentum is expressed as a vector, $\vec{p}$, having magnitude and direction (see the appendix on vectors at the end of this volume). Here are conventions we will use for describing a vector:

$|\vec{p}|$ is the magnitude of the vector, also written simply as p

$\vec{p} = \langle p_x, p_y, p_z \rangle$ is the vector written as a list of its components

$\vec{p} = p_x\hat{i} + p_y\hat{j} + p_z\hat{k}$ is the vector written in terms of "unit vectors"

Example: Show that when the speed is one-tenth the speed of light ($v = 0.1c$), the ratio of the correct relativistic momentum to the approximate nonrelativistic momentum mv is very nearly equal to 1. Yet this is a very high speed, 3×10^4 kilometers per second or about 10^8 km/hour or about 6×10^7 miles/hour!

Solution: $\dfrac{p}{mv} = \left(\dfrac{1}{mv}\right)\dfrac{mv}{\sqrt{1 - v^2/c^2}} = \dfrac{1}{\sqrt{1 - (0.1)^2}} = 1.00005 \approx 1$

Ex. 1.9 Show that when the speed is within one percent of the speed of light ($v = 0.99c$), the ratio of the correct relativistic momentum to the approximate nonrelativistic momentum mv is quite large. Such speeds are attained in particle accelerators.

Ex. 1.10 If $\vec{p} = \langle 4, -5, 2 \rangle$ kg·m/s, what is the magnitude $|\vec{p}|$ of the momentum?

1.7.2 The momentum principle

Newton's first law, "the stronger the interaction, the bigger the change in the momentum," states a qualitative relationship between momentum and interaction. The momentum principle states this relation in a powerful, quantitative way, which can be used to predict the behavior of objects.

THE MOMENTUM PRINCIPLE

$$\Delta\vec{p} = \vec{F}_{net}\Delta t \text{ (for a short time interval } \Delta t)$$

The capital Greek letter delta (Δ) means "change of" (something). In following sections we will explain in detail what we mean by the net force $\vec{F}_{net}$. In words, we read the momentum principle like this:

Change of Momentum is equal to
Net Force times **Duration of Interaction**
or
Change of Momentum is proportional to **Extent of Interaction**.

The momentum principle written in terms of vectors really represents three ordinary equations, for components of the motion along the x, y, and z axes:

$$\Delta p_x = F_{net,\,x}\Delta t$$

$$\Delta p_y = F_{net,\,y}\Delta t$$

$$\Delta p_z = F_{net,\,z}\Delta t$$

The momentum principle has been experimentally verified in a very wide range of phenomena. However, it does not apply to some aspects of the very small: understanding the internal structure of atoms and nuclei requires quantum mechanics. Nor does it apply to some aspects of the very large: understanding the large-scale structure of the Universe (cosmology), and the exotic nature of black holes, requires general relativity.

Historically, the momentum principle is often known as "Newton's second law of motion." We will refer to it as the momentum principle to emphasize the key role played by momentum in physical processes.

Derivative form of the momentum principle

The form of the momentum principle given above is the form that is most useful in computer calculations. Another important form is obtained by dividing by the time interval Δt:

$$\frac{\Delta\vec{p}}{\Delta t} = \vec{F}_{net}$$

When the net force acting on an object is varying during a time interval, for accuracy we need to use small values of the time interval Δt in calculations. In the mathematical limit, we should let the time interval Δt approach zero (an infinitesimal time interval), and the ratio of the infinitesimal momentum change (written as $d\vec{p}$) to the infinitesimal time interval (written as dt) is the time derivative of the momentum. In this way we can obtain a derivative form of the momentum principle.

As $\Delta t \to 0$, $\dfrac{\Delta\vec{p}}{\Delta t} \to \dfrac{d\vec{p}}{dt}$, and we obtain the following form:

DERIVATIVE FORM OF THE MOMENTUM PRINCIPLE

$$\frac{d\vec{p}}{dt} = \vec{F}_{net}$$

In words, "the time rate of change of the momentum of an object is equal to the net force acting on the object." Or in calculus terms, "the derivative of the momentum with respect to time is equal to the net force acting on the object."

A nonrelativistic form of the momentum principle

In simple cases the derivative form of the momentum principle reduces to a form that is probably more familiar. If the mass is constant (the usual situation), and the speed is low enough that the momentum is well approximated by $\vec{p} \approx m\vec{v}$, we have this:

$$\frac{d(m\vec{v})}{dt} = m\frac{d\vec{v}}{dt} = \vec{F}_{net} \quad \text{(nonrelativistic form; constant mass)}$$

This form of the momentum principle (Newton's second law) is probably familiar to you: mass times acceleration (time rate of change of velocity) is equal to the net force.

There do exist situations where the mass isn't constant. One example is a rocket that throws exhaust out the back and as a result has decreasing mass. In such cases the momentum-based formula $d\vec{p}/dt = \vec{F}_{net}$ gives the correct results, whereas the more familiar constant-mass formula $md\vec{v}/dt = \vec{F}_{net}$ cannot be used. When the principle is written in terms of momentum, it is valid even for objects whose mass changes or which are moving at speeds close to the speed of light, as long as we use the relativistically correct definition of momentum. So $d\vec{p}/dt = \vec{F}_{net}$ is the more general form.

In the next sections we will examine in detail the meaning of the momentum principle in the form $\Delta\vec{p} = \vec{F}_{net}\Delta t$, and in the next chapter we will see how we can use it in complex situations to predict the future of moving objects.

1.7.3 Change of momentum

Earlier we discussed two different kinds of changes in motion: change of speed and change of direction. A mathematical formulation of "change of momentum" must include either a change in the magnitude of the momentum, or a change in the direction of the momentum, or both.

Change in magnitude

It's easy to come up with a reasonable measure for quantifying change of speed, which is what we call the magnitude of the velocity. In standard "SI" units ("Systeme International"), speed is measured in meters per second, abbreviated "m/s".

? Suppose you are driving a car, and you speed up from 20 m/s to 25 m/s to pass a truck. What is the change of speed? (20 m/s is about 45 miles per hour or 72 km per hour.)

Clearly the change in the speed is +5 m/s. More formally, we write the following:

$$\Delta v = v_2 - v_1$$

Here, the equation means "the change of the speed v was $v_2 - v_1$," where the initial speed v_1 was 20 m/s and the final speed v_2 was 25 m/s. Again, we point out that in physics the capital Greek letter delta (Δ) is often used to mean "change of" (something).

These speeds are very small compared to the speed of light, so we can use the approximate nonrelativistic formula for momentum. If the car has a mass of 1000 kg, the change in the magnitude p of the momentum is

$$\Delta p = p_2 - p_1 = (1000 \text{ kg})(25 \text{ m/s}) - (1000 \text{ kg})(20 \text{ m/s})$$

$$\Delta p = 5000 \text{ kg·m/s}$$

Change in magnitude at high speed

There is a striking consequence of the relativistic formula for momentum. When a force increases the momentum so much that v approaches the speed of light c, further effort makes only tiny increases in the speed. Although the momentum increases, the increase in speed is very small; the increase in momentum is explained by the decrease in the denominator $\sqrt{1 - v^2/c^2}$ as v approaches c. No matter how long you keep applying the force, and no matter how large the momentum becomes, the speed never gets quite as big as c, which is a cosmic speed limit that can never be exceeded.

Change of direction

There are various ways to specify a change in the direction of motion. For example, if you use compass directions, you could say that an airplane changed its direction from 30° east of North to 45° east of North: a 15° clockwise change. One can imagine various other schemes, involving other kinds of coordinate systems. The standard way to deal with this is to use vectors.

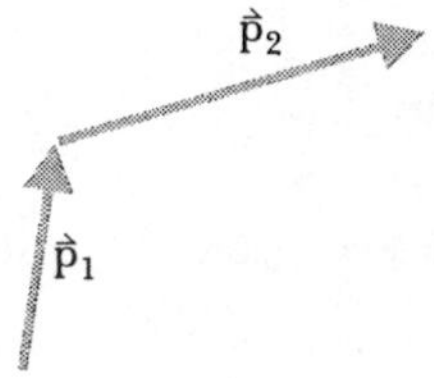

Change of a vector

Momentum (or velocity or position) can be described by a vector, a mathematical quantity that has both a magnitude and a direction. Vectors have their own special rules of arithmetic, and they must be added and subtracted as vectors, not simply as ordinary numbers. In this course it will be important for you to be able to do vector arithmetic both graphically, for qualitative analysis of a situation, and by resolving vectors into components along chosen coordinate axes, for quantitative modeling. Vectors are reviewed in an appendix at the end of this volume, which you should study on your own. We will use vectors a lot.

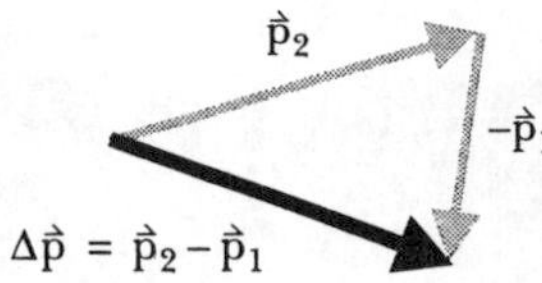

The change in the momentum $\Delta\vec{p}$ during a time interval is also a vector: $\Delta\vec{p} = \vec{p}_2 - \vec{p}_1$. This vector expression captures both changes in magnitude and changes in direction. Figure 1.23 is a graphical illustration of a change from an initial momentum $\vec{p}_1$ to a final momentum $\vec{p}_2$. To calculate the momentum change $\Delta\vec{p} = \vec{p}_2 - \vec{p}_1$, we form the vector $-\vec{p}_1$ (the opposite of $\vec{p}_1$) and add it to $\vec{p}_2$, yielding $\vec{p}_2 + (-\vec{p}_1) = \vec{p}_2 - \vec{p}_1$ (Figure 1.23).

Alternatively, you can think of $\Delta\vec{p}$ as that vector which you add to the initial momentum $\vec{p}_1$ in order to obtain the final momentum $\vec{p}_2 = \vec{p}_1 + (\vec{p}_2 - \vec{p}_1)$, as shown in Figure 1.24.

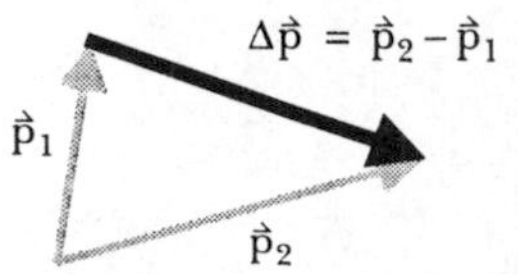

1.7.4 Extent of interaction

The term "extent of interaction" includes both a measure of the strength of the interaction and of its duration.

Strength of interaction

Physicists employ the concept of "force" to quantify interactions between two objects. Like momentum or velocity, force can be described by a vector, since a force also has a magnitude and is exerted in a particular direction. Measuring the magnitude of the velocity of an object (in other words, measuring its speed) is a familiar task, but how do we measure the magnitude of a force? One approach involves measuring the effect of a particular interaction between two known objects, and comparing the effects of other interactions to this effect.

For example, suppose we use a spring to measure the magnitude of the force exerted by the Earth on a metal block. We place the block on top of the spring, and note that the spring is compressed a distance s (Figure 1.25). When we place two such blocks on top of the spring, we see

Figure 1.23 Calculation of $\Delta\vec{p}$.

Figure 1.24 Alternative computation of $\Delta\vec{p}$.

$$\Delta\vec{p} = \vec{F}_{net}\Delta t$$

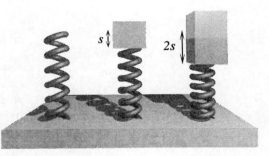
Figure 1.25 Compression of a spring is a measure of force.

that the spring is compressed twice as much. By experimentation, we find that any spring made of the same material and produced to the same specifications behaves in the same way.

Similarly, we can observe how much the spring stretches when the same blocks are suspended from it. We find that one block stretches the spring by the same distance *s*, and two blocks stretch it by 2*s* (Figure 1.26). We can use such a spring to make a scale for measuring forces, calibrating it in terms of what force is represented by what stretch.

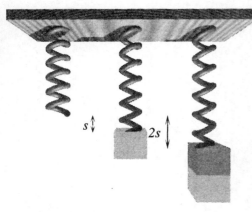

Figure 1.26 Extension of a spring is also a measure of force.

Net force

Although the spring is exerting a force on the block hanging from it, the momentum of the block is not changing. How can this be? Does this imply that the block is not interacting with anything? That would not make sense. Apparently forces exerted by different objects can add up to produce no net effect at all. The spring exerts an upward force on the block, but the Earth exerts a downward gravitational force on the block, and the net result is that the block's momentum does not change. Since the effect we observe appears to be due to all the objects exerting forces on the spring, it must be the total or *net* force acting on the spring that determines its $\Delta \vec{p}$. We refer to this total force as $\vec{F}_{net}$.

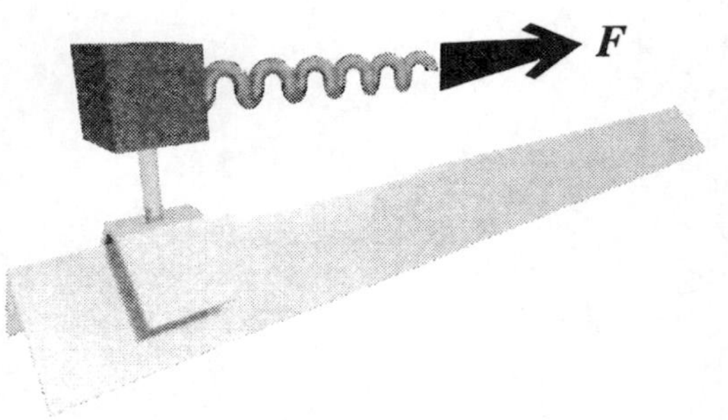

We find experimentally that the magnitude of the net force acting on an object affects the magnitude of the change in its momentum. Suppose we place a block on a long air track, where it rests on a cushion of air that provides a nearly frictionless surface, and attach our calibrated spring to it (Figure 1.27). If we pull with a force *F* (a force that stretches the spring by a distance *s*) for a short time, the momentum of the block increases from zero to an amount *mv* (if *v* << *c*). If we pull with a force 2*F* (stretching the spring a distance 2*s*), the block's momentum increases to 2*mv* in the same amount of time. Apparently the magnitude of the change in momentum is proportional to the magnitude of the net force applied to the object:

$$|\Delta \vec{p}| \propto |\vec{F}_{net}|$$

Figure 1.27 Apply a constant force to a block on a low-friction air track.

Duration of interaction

A simple experiment will show us that not only the magnitude of the force, but the length of time during which it acts on an object, affects the change of momentum of the object. Suppose we again place a block on an air track and attach our spring to it. We pull on the block for a time Δt with a force *F*, and we observe that the momentum of the block increases from zero to an amount *mv*. However, if we repeat the experiment but pull for a time twice as long with the same force *F*, the block's momentum increases to 2*mv*; the change in its momentum is twice as great. We observe that the magnitude of the change in momentum is directly proportional to the length of time $\Delta t = t_2 - t_1$ during which the force acts on the object:

$$\Delta \vec{p} = \vec{F}_{net} \Delta t$$

$$|\Delta \vec{p}| \propto \Delta t$$

Effect of mass

One would expect that a force of a given magnitude would have more of an effect on the velocity of an object whose mass was small than on a more massive object. This inverse relation between an object's mass and the change in its velocity can easily be verified by experiment. For example, if we apply a force of magnitude *F* for a time Δt to a block of mass *m* that initially is moving at a speed *v*, we observe that the block's speed increases by an amount Δv, whereas if the same force is applied for the same time to a block of mass

$2m$, the speed increase is only half as much. $|\Delta \vec{v}|$ appears to be inversely proportional to the mass of the object.

$$|\Delta \vec{v}| \propto \frac{1}{m} \text{ (for speeds small compared to the speed of light } c\text{)}$$

The results of such experiments are the reason why we focus on changes in momentum, a quantity that incorporates both mass and velocity, the two properties of an object that determine its response to a net force.

1.7.5   The momentum principle is valid only in an inertial frame

Earlier we discussed the fact that Newton's first law is valid only in an "inertial frame" of reference, one in uniform motion (or at rest) with respect to the pervasive cosmic microwave background. Since the momentum principle is a quantitative version of Newton's first law, we expect the momentum principle to be valid only in an inertial frame. Let's check that this is true.

If you view some objects from a space ship that is moving uniformly with velocity $\vec{v}_s$ with respect to the cosmic microwave background, all of the velocities of those objects have the constant $\vec{v}_s$ subtracted from them, as far as you are concerned. With a constant spaceship velocity, we have $\Delta \vec{v}_s = 0$, and the momentum principle reduces to the following (for speeds small compared to c):

$$\Delta [m(\vec{v} - \vec{v}_s)] = \Delta(m\vec{v}) = \vec{F}_{net} \Delta t$$

Therefore, if the velocity of the space ship doesn't change (it represents an inertial frame of reference), the form (and validity) of the momentum principle is unaffected by the motion of the space ship.

However, if your space ship increases its speed toward an isolated object, $\Delta \vec{v}_s \neq 0$, and the object's motion relative to you changes without any force acting on it. In that case the momentum principle is not valid for the object, because you are not in an inertial frame. Although the Earth is not an inertial frame because it rotates, and goes around the Sun, it is close enough to being an inertial frame for many everyday purposes.

1.7.6   Units

The SI unit of mass is the kilogram (kg), the unit of distance is the meter (m), the unit of time is the second (s), and the unit of force is the newton (N). A one newton force applied for one second to a block will increase its momentum by an amount of one kilogram·meter per second, which are the units of momentum. If the block has a mass one kilogram, the speed of the block will increase by 1 meter per second. A note at the end of this chapter provides details on the rationale for this definition, and on the way that mass is measured.

One newton is a rather small force. One newton is approximately the gravitational force of the Earth on a small apple, or about a quarter of a pound.

1.7.7   Impulse

The product of a force and a time interval is called "impulse":

DEFINITION OF IMPULSE

Impulse is defined as $\vec{F} \Delta t$ (for small Δt)

With this definition of impulse we can state the momentum principle like this:

**The change of momentum of an object
is equal to the net impulse applied to it.**

$$\Delta \vec{p} = \vec{F}_{net} \Delta t \text{ (net impulse due to the net force)}$$

1.7.8   Applying these concepts

You should now do the following simple exercises to make sure that you can apply what you have just studied. Do these exercises before checking the answers given at the end of this chapter. If you find some mistakes, make sure that you resolve the issues right away, because we will be building on these basic concepts.

Example: You were driving a car with velocity <25, 15, 0> m/s; that is, your *x* component of velocity was +25 m/s, your *y* component of velocity was +15 m/s, and your *z* component of velocity was 0. You quickly turned and braked, and your velocity became <10, 18, 0> m/s. The mass of the car was 1000 kg. What was the (vector) change in momentum $\Delta \vec{p}$ during this maneuver? Pay attention to signs. What was the (vector) impulse applied to the car by the ground?

  Solution:

  $\vec{p}_i = (1000 \text{ kg})<25, 15, 0> \text{ m/s} = <25000, 15000, 0> \text{ kg·m/s}$

  $\vec{p}_f = (1000 \text{ kg})<10, 18, 0> \text{ m/s} = <10000, 18000, 0> \text{ kg·m/s}$

  $\Delta \vec{p} = \vec{p}_f - \vec{p}_i = <-15000, 3000, 0> \text{ kg·m/s} = \text{impulse by ground}$

Ex. 1.11 In the previous exercise, if the maneuver took 3 seconds, what was the average net (vector) force $\vec{F}_{net}$ that the ground exerted on the car?

Ex. 1.12 A hockey puck is sliding along the ice with nearly constant momentum <10, 5, 0> kg·m/s when it is suddenly struck by a hockey stick with a force $\vec{F} = <0, 2000, 0>$ N that lasts for only 3 milliseconds (3×10^{-3} s). What is the new momentum of the puck?

Ex. 1.13 A 50 kg child is riding on a carousel (merry-go-round) at a constant speed of 5 m/s. What is the magnitude of the change in the child's momentum $|\Delta \vec{p}|$ in going all the way around (360°)? In going halfway around (180°)? (It helps to draw a diagram.)

1.8   Updating the momentum and the position

It is often the case that we know the force on an object. For example, when two stars orbit each other they exert known gravitational forces on each other, as we will see in detail in the next chapter. If we know the momentum of an object at a particular instant, and we know the net force acting on it, we can calculate what the momentum will be a short time Δt later, as you saw in Ex. 1.12:

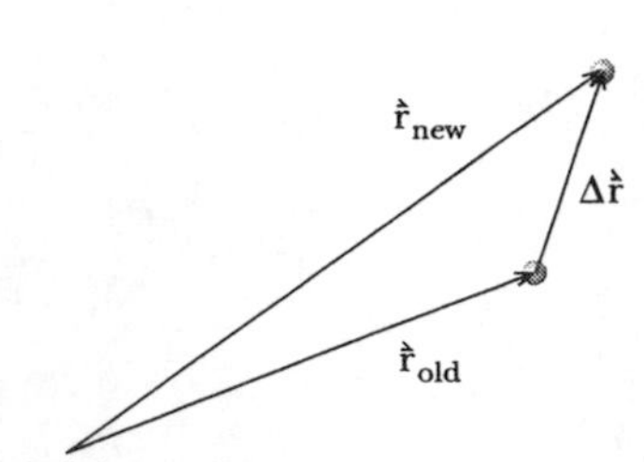

UPDATING THE MOMENTUM

$$\vec{p}_{new} = \vec{p}_{old} + \vec{F}_{net} \Delta t \text{ (for small } \Delta t\text{)}$$

Figure 1.28 Position vectors describing the old and new positions of the star.

This is part of what is needed to predict the motion of a star. Another part is the calculation of where the star has moved to in the short time Δt. With respect to some origin, the old position of the star is represented by a "position vector" $\vec{r}_{old}$, and the new position by a position vector $\vec{r}_{new}$ (Figure 1.28). The change in the position of the star is

$$\Delta\vec{r} = \vec{r}_{new} - \vec{r}_{old} = \vec{v}\Delta t \text{ (for small } \Delta t)$$

This follows from the definition of velocity, in the direction of the motion:

$$\vec{v} = \frac{\Delta\vec{r}}{\Delta t} \rightarrow \frac{d\vec{r}}{dt} \text{ in the limit of very small } \Delta t$$

Therefore, the way to update the position vector is this:

$$\vec{r}_{new} = \vec{r}_{old} + \vec{v}\Delta t \text{ (for small } \Delta t)$$

If the object is moving with a speed $v \ll c$ (that is, a speed much smaller than the speed of light), the momentum is approximately $m\vec{v}$, so we have this:

UPDATING THE POSITION NONRELATIVISTICALLY

$$\vec{r}_{new} = \vec{r}_{old} + \left(\frac{\vec{p}}{m}\right)\Delta t \text{ (for } v \ll c \text{ and small } \Delta t)$$

It doesn't much matter whether we use $\vec{p}_{new}$ or $\vec{p}_{old}$ to update the position as long as Δt is quite small, because the momentum doesn't change much during a short time interval.

? What if v is comparable to c? Try to do the algebra to find $\vec{v}$ from $\vec{p}$.

Here is a way to solve algebraically for $\vec{v}$ in terms of $\vec{p}$:

$$p = \frac{mv}{\sqrt{1 - v^2/c^2}}$$

$$\frac{p^2}{m^2} = \frac{v^2}{1 - v^2/c^2}$$

$$\frac{p^2}{m^2} - \left(\frac{p^2}{m^2 c^2}\right)v^2 = v^2$$

$$\left(1 + \frac{p^2}{m^2 c^2}\right)v^2 = \frac{p^2}{m^2}$$

$$v = \frac{p/m}{\sqrt{1 + \left(\frac{p}{mc}\right)^2}}$$

But since $\vec{p}$ and $\vec{v}$ are in the same direction, we can write this:

$$\vec{v} = \frac{\vec{p}/m}{\sqrt{1 + \left(\frac{p}{mc}\right)^2}}$$

UPDATING THE POSITION RELATIVISTICALLY

$$\vec{r}_{new} = \vec{r}_{old} + \frac{1}{\sqrt{1 + \left(\frac{p}{mc}\right)^2}}\left(\frac{\vec{p}}{m}\right)\Delta t \text{ (for small } \Delta t)$$

The velocity $\vec{v}$ reduces to $\vec{v} \approx \vec{p}/m$ if $v \ll c$:

$$\text{If } \frac{p}{mc} \approx \frac{mv}{mc} = \frac{v}{c} \ll 1, \text{ then } \sqrt{1 + \left(\frac{p}{mc}\right)^2} \approx 1.$$

Example: At time 5.23 s an object is located at <5, 3, –2> m, and its velocity at this instant is <–10, 20, 4> m/s. Approximately, where is the object at time 5.25 s?

  Solution: We'll assume that the time interval is short enough that the velocity doesn't change much during this time interval: $\Delta t = (5.25 - 5.23) \text{ s} = 0.02 \text{ s}$

$\vec{r}_{new} \approx$ <5, 3, –2> m + (<–10, 20, 4> m/s)(0.02 s)

$\vec{r}_{new} \approx$ <5, 3, –2> m + <–0.2, 0.4, 0.08> m

$\vec{r}_{new} \approx$ <4.8, 3.4, –1.92> m

Ex. 1.14 At time $t_1 = 14.0$ s, a car with mass 1200 kg is located at <100, 0, 25> m and has momentum <6000, 0, –3600> kg·m/s. At time $t_2 = 14.5$ s, what is the approximate position of the car?

Ex. 1.15 If p/m is $0.9c$, what is v in terms of c?

1.9   *Definitions, measurements, and units

Using the momentum principle requires a consistent way to measure length, time, mass, and force, and a consistent set of units. We state the standard definitions of the standard units, and we briefly discuss some subtle issues underlying this choice of units.

Units: meters, seconds, kilograms, and newtons

Originally the meter was defined as the distance between two scratches on a platinum bar in a vault in Paris, and a second was 1/86,400th of a "mean solar day." Now however the second is defined in terms of the frequency of light emitted by a cesium atom, and the meter is defined as the distance light travels in 1/299,792,458th of a second, or about 3.3×10^{-9} seconds (3.3 nanoseconds). The speed of light is defined to be exactly 299,792,458 m/s. As a result of these modern redefinitions, it is really speed (of light) and time that are the internationally agreed-upon basic units, not length and time.

   By international agreement, one kilogram is the mass of a platinum block kept in that same vault in Paris. As a practical matter, other masses are compared to this standard kilogram by using a balance-beam or spring weighing scale (more about this in a moment). The newton, the unit of force, is defined as that force which imparts to 1 kilogram a velocity change of 1 m/s in 1 second. We can make a scale for force by calibrating the amount of stretch of a spring in terms of newtons.

Some subtle issues

What we have just said about units is sufficient for practical purposes to predict the motion of objects, but here are some questions that might bother you. Is it legitimate to measure the mass that appears in the momentum principle by seeing how that mass is affected by gravity on a balance-beam scale? Is it legitimate to use the momentum principle to define the units of force, when the concept of force is itself associated with the same law? Is this all circular reasoning, and the momentum principle merely a definition with little content? We present a chain of reasoning that addresses these issues.

Measuring inertial mass

When we use balance-beam or spring weighing scales to measure mass, what we're really measuring is the "gravitational mass," that is, the mass that appears in the law of gravitation and is a measure of how much this object is affected by the gravity of the Earth. In principle, it could be that this "gravitational mass" would be different from the so-called "inertial mass"—the mass that appears in the momentum principle. It is possible to compare the inertial masses of two objects, and we find experimentally that inertial and gravitational mass seem to be entirely equivalent.

Here is a way to compare two inertial masses directly, without involving gravity. Starting from rest, pull on the first object with a spring stretched by some amount s for an amount of time Δt, and measure the increase of speed Δv_1. Then, starting from rest, pull on the second object with the same spring stretched by the same amount s for the same amount of time Δt, and measure the increase of speed Δv_2. We *define* the ratio of the inertial masses as $m_1/m_2 = \Delta v_2/\Delta v_1$. Since one of these masses could be the standard kilogram kept in Paris, we now have a way of measuring inertial mass in kilograms. Having defined inertial mass this way, we find experimentally that the momentum principle is obeyed by both of these objects in all situations, not just in the one special experiment we used to compare the two masses.

Moreover, we find to extremely high precision that the inertial mass in kilograms measured by this comparison experiment is exactly the same as the gravitational mass in kilograms obtained by comparing with a standard kilogram on a balance-beam scale (or using a calibrated spring scale). This justifies the convenient use of ordinary weighing scales to determine inertial mass.

Is this all circular reasoning?

The definitions of force and mass may sound like circular reasoning, and the momentum principle may sound like just a kind of definition, with no real content, but there is real power in the momentum principle. Forget for a moment the definition of force in newtons and mass in kilograms. The experimental fact remains that any object if subjected to a single force by a spring with constant stretch experiences a change of momentum (and velocity) proportional to the duration of the interaction. Note that it is not a change of *position* proportional to the time (that would be a constant speed), but a change of *velocity*. That's real content. Moreover, we find that the change of velocity is proportional to the amount of stretch of the spring. That too is real content.

Then we find that a different object undergoes a different rate of change of velocity with the same spring stretch, but after we've made one single comparison experiment to determine the mass relative to the standard kilogram, the momentum principle works in all situations. That's real content.

Finally come the details of setting standards for measuring force in newtons and mass in kilograms, and we use the momentum principle in helping set these standards. But logically this comes after having established the law itself.

1.10 Summary

Fundamental Physical Principles

All interactions are subject to the following law and principles:
- Newton's first law of motion: An object moves in a straight line and at constant speed except to the extent that it interacts with other objects. (Valid for "inertial frames" of reference in uniform motion relative to the cosmic microwave background.)
- Principle of relativity: Physical laws work in the same way for observers in uniform motion as for observers at rest.

THE MOMENTUM PRINCIPLE

$$\Delta \vec{p} = \vec{F}_{net}\Delta t \text{ for small } \Delta t, \text{ or } \frac{d\vec{p}}{dt} = \vec{F}_{net}$$

New concepts

Kinds of matter to be studied in this course:
- nuclei, atoms, and molecules
- solids and gases (not much about liquids)
- planets, stars, and galaxies

Interactions are indicated by
- change of velocity (change of direction and/or change of speed)
- change of identity
- change of shape of multiparticle system
- change of temperature of multiparticle system
- lack of change when change is expected

Four fundamental types of interaction have been identified:
- "strong" interactions (inside the nucleus of an atom)
- electromagnetic interactions (electric and magnetic interactions, closely related to each other)
- "weak" interactions (including in particular neutrino interactions)
- gravitational interactions (all objects attract each other gravitationally)

Momentum is defined as

$$\vec{p} \equiv \frac{m\vec{v}}{\sqrt{1 - v^2/c^2}} \quad (m \text{ is the rest mass})$$

which reduces to $\vec{p} \approx m\vec{v}$ at speeds such that $v \ll c$.

Change of magnitude Δp; change of magnitude and direction $\Delta \vec{p}$

Force is a quantitative measure of interactions; units are newtons.

Impulse is the product of force times time $\vec{F}\Delta t$; momentum change equals net impulse (the impulse due to the net force).

Results

UPDATING THE MOMENTUM

$$\vec{p}_{new} = \vec{p}_{old} + \vec{F}_{net}\Delta t \text{ (for small } \Delta t)$$

UPDATING THE POSITION NONRELATIVISTICALLY

$$\vec{r}_{new} = \vec{r}_{old} + \left(\frac{\vec{p}}{m}\right)\Delta t \ \ (\text{for } v \ll c \text{ and small } \Delta t)$$

UPDATING THE POSITION RELATIVISTICALLY

$$\vec{r}_{new} = \vec{r}_{old} + \frac{1}{\sqrt{1 + \left(\frac{p}{mc}\right)^2}}\left(\frac{\vec{p}}{m}\right)\Delta t \ \ (\text{for small } \Delta t)$$

1.11 Example problem: Position and momentum of a ball

In each chapter you will find a worked-out example problem which shows in detail how to apply fundamental physics principles to the analysis of significant phenomena.

At a certain instant a small steel ball of mass 0.25 kg is located at position $<15, -35, 12>$ m and has momentum $<2, 8, 5>$ kg·m/s. At this instant the ball is acted on by a net force $<20, 50, -85>$ N. What is its approximate momentum and position 3 milliseconds later $(3 \times 10^{-3}$ s$)$? What approximations or simplifying assumptions did you make in your analysis?

Solution

Always start an analysis from fundamental principles. In this case we can update the momentum using the momentum principle:

$$\vec{p}_{new} = \vec{p}_{old} + \vec{F}_{net}\Delta t$$

$$\vec{p}_{new} = (<2, 8, 5> \text{ kg·m/s}) + (<20, 50, -85> \text{ N})(3 \times 10^{-3} \text{ s})$$

$$\vec{p}_{new} = (<2, 8, 5> \text{ kg·m/s}) + (<0.06, 0.15, -0.255> \text{ N·s})$$

$$\vec{p}_{new} = <2.06, 8.15, 4.745> \text{ kg·m/s}$$

It is clear that $v \ll c$, so we can update the position nonrelativistically:

$$\vec{r}_{new} = \vec{r}_{old} + \left(\frac{\vec{p}}{m}\right)\Delta t$$

The momentum didn't change much, so let's simply use the old momentum to update the position:

$$\vec{r}_{new} = (<15, -35, 12> \text{ m}) + \frac{(<2, 8, 5> \text{ kg·m/s})}{(0.25 \text{ kg})}(3 \times 10^{-3} \text{ s})$$

$$\vec{r}_{new} = (<15, -35, 12> \text{ m}) + (<0.024, 0.096, 0.06> \text{ m})$$

$$\vec{r}_{new} = <15.024, -34.904, 12.06> \text{ m}$$

We made the simplifying assumption (in the absence of additional information) that the force didn't change much during 3 ms. Or to put it another way, we assumed that a time interval of 3 ms was sufficiently short that we could update the momentum fairly well assuming a constant force during that short time interval.

In updating the momentum we made the approximation that the momentum didn't change much during the 3 ms time interval, and we used the momentum at the beginning of the interval. We could instead have chosen to use the momentum at the end of the time interval as representative of the average momentum during the time interval.

A better approximation to use in updating the position might be to average the old and new momenta:

$$\vec{p}_{avg} \approx \frac{\vec{p}_{new} + \vec{p}_{old}}{2} = <2.03, 8.075, 4.8725> \text{ kg·m/s}$$

1.12 Review questions

The purpose of review questions is to help you reflect on the most important concepts in the chapter. Try to answer the questions without flipping through the chapter looking for an answer, but think about what you know already from having done the exercises throughout the chapter. If you are stumped, look at the "Summary" on the preceding page.

If you are still stumped after looking at the chapter summary, make a note in your notebook, as a reminder that you may need to spend some extra time studying this particular concept.

Detecting interactions

RQ 1.1 Give two examples (other than those discussed in the text) of interactions that may be detected by observing:
  (a) change in velocity
  (b) change in temperature
  (c) change in shape
  (d) change in identity
  (e) lack of change when change is expected

RQ 1.2 In which of the following situations are two objects interacting? How can you tell?
  (a) a book rests on a table
  (b) a baseball that was hit by a batter flies toward the outfield
  (c) water freezes in an ice cube tray in the freezer
  (d) a communications satellite orbits the earth
  (e) a space probe leaves the solar system traveling at constant speed toward a distant star
  (f) a neutrino passes through the earth
  (g) a charged particle leaves a curving track in a particle detector

RQ 1.3 Which one of the following is *not* conclusive evidence of an interaction?
  (a) Change of velocity, either change of direction or change of speed.
  (b) Change of shape or configuration without change of velocity.
  (c) Change of position without change of velocity.
  (d) Change of identity without change of velocity.
  (e) Change of temperature without change of velocity.
  Explain your choice.

Fundamental interactions

RQ 1.4 Which interaction between two protons is stronger, the gravitational attraction between them or the electric repulsion between them? If two protons are at rest, will they begin to move toward each other or away from each other?

Relativity

RQ 1.5 Which of the following observers might observe something that appears to violate Newton's first law of motion? Explain why.

  (a) a person standing still on a street corner

  (b) a person riding on a roller coaster

(continued on next page)

(c) a passenger on a starship travelling at $0.75c$ toward the nearby star Alpha Centauri

(d) an airplane pilot doing aerobatic loops

(e) a hockey player coasting across the ice

Newton's first law of motion

RQ 1.6 Moving objects left the traces labelled *A - F.* The dots were deposited at equal time intervals (for example, one dot each second). Which trajectories show evidence that the moving object was interacting with another object somewhere? In each case the object starts from the square.

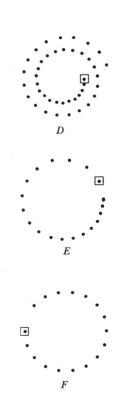

D

E

F

A

B

C

The momentum principle

RQ 1.7 At a certain instant, a particle is moving in the $+x$ direction with momentum $+10$ kg·m/s. During the next 0.1 s, a constant force acts on the particle: $F_x = -6$ N, and $F_y = +3$ N. What is the magnitude of the momentum of the particle at the end of this 0.1 s interval?

RQ 1.8 At $t = 12.0$ seconds an object with mass 2 kg was observed to have a velocity of <10, 35, –8> m/s. At $t = 12.3$ seconds its velocity was <20, 30, 4> m/s. What was the average (vector) net force acting on the object?

RQ 1.9 A proton has mass 1.7×10^{-27} kg . What is the magnitude of the impulse required to increase its speed from $0.990c$ to $0.994c$?

1.13 Problems

Problem 1.1 A hockey puck

An 0.4 kg hockey puck is sliding along the ice with velocity <20, 0, 0> m/s. As the puck slides past location <5, 3, 0> m on the rink, a player strikes the puck with a sudden force in the $+y$ direction, and the hockey stick breaks. Some time later, the puck's position on the rink is <13, 21, 0> m. When we pile weights on the side of a hockey stick we find that the stick breaks under a force of about 1000 N.

(a) For approximately how much time $t_{contact}$ was the hockey stick in contact with the puck? Be sure to show clearly the steps in your analysis.

(b) What approximations and/or simplifying assumptions did you make in your analysis?

Problem 1.2 A student collision

Two students are running to classes in opposite directions. They run into each other head-on and stop abruptly. Using physics principles, estimate the force that one student exerts on the other during the collision. You will need to estimate some quantities; give reasons for your choices and provide checks showing that your estimates are physically reasonable.

The remaining problems are intended to introduce you to using a computer to model matter, interactions, and motion. You will build on these small calculations to build models of physical systems in later chapters.

Some parts of these problems can be done with almost any tool (spreadsheet, math package, etc.). Other parts are most easily done with a programming language. We recommend the free 3D programming language VPython. Your instructor will introduce you to an available computational tool and assign problems, or parts of problems, that can be addressed using the chosen tool.

Problem 1.3 Move an object across a computer screen

(a) Write a program that makes an object move from left to right across the screen at speed v. Make v a variable, so you can change it later. Let the time interval for each step of the computation be a variable dt, so that the position x increases by an amount $v*dt$ each time. If it can be done with the tool you are using, erase the previous position of the object as you go, to simulate the actual motion of an object.

(b) Modify a copy of your program to make the object run into a wall and reverse its direction.

(c) Make a modification so that the object's speed v is no longer a constant but changes smoothly with time. Is the speed change clearly visible to an observer? Try to make one version in which the speed change is clearly noticeable, and another in which it is not noticeable.

(d) Corresponding to part (c), make a computer graph of x vs. t, where t is the time.

(e) Corresponding to part (c), make a computer graph of v vs. t, where t is the time.

Turn in your programs for parts (c), (d), and (e).

Problem 1.4 Move an object at an angle

(a) Write a program that makes an object move at an angle.

(b) Change the component of velocity of the object in the x direction but not in the y direction, or vice versa. What do you observe?

(c) Start the object moving at an angle and make it bounce off at an appropriate angle when it hits a wall.

Turn in your answer to part (b), and the final version of your computation, part (c).

Problem 1.5 Move an object, leave a trail

Write a program that makes an object move smoothly from left to right across the screen at speed v, leaving a trail of dots on the screen at equal time intervals. If the dots are too close together, leave a dot every N steps, and adjust N to give a nice display.

1.14 Answers to exercises

1.1 (page 12) udd

1.2 (page 12) $u\bar{d}$; $\bar{u}d$

1.3 (page 14)   No, this is just the electric force between neutral atoms (due to quantum effects).

1.4 (page 14)   It would have to have a different strength and/or not interact with some particles.

1.6 (page 15)   One expects the ball to fall due to the gravitational pull of the Earth. Since it doesn't, there must be some other interaction that counteracts the gravitational interaction ("lack of change when change expected").

1.7 (page 16)   Drift at constant speed of 10^4 m/s.

1.8 (page 16)   In space, far from other objects, presumably interactions are negligible, whereas car and train interact with the ground.

1.9 (page 20)   7.1

1.10 (page 20) $\sqrt{(4)^2 + (-5)^2 + (2)^2} = 6.7$ kg·m/s

1.11 (page 26)   Since impulse $= \vec{F}\Delta t$, we find $\vec{F} = <5000, -1000, 0>$ N

1.12 (page 26)   Since $\Delta\vec{p} = \vec{p}_f - \vec{p}_i = \vec{F}_{net}\Delta t$, $\vec{p}_f = \vec{p}_i + \vec{F}_{net}\Delta t$, so

$$\vec{p}_f = (<10, 5, 0> \text{ kg·m/s}) + (<0, 2000, 0> \text{ N})(3 \times 10^{-3} \text{ s})$$

$$\vec{p}_f = <10, 11, 0> \text{ kg·m/s}$$

1.13 (page 26)   0 all the way around; 500 kg·m/s halfway around

1.14 (page 28) $<102.5, 0, 23.5>$ m

1.15 (page 28) $0.67c$

Chapter 2

Predicting the Future

Chapter 2

Predicting the Future

Using the momentum principle (Newton's second law), we can carry out quite ambitious quantitative predictions. As an example, we will focus on the gravitational interaction. We will predict the motion of planets around a star and the motion of binary stars, and explore what kinds of paths are possible.

2.1 The gravitational force law

The momentum principle gives us a quantitative expression for the effect of interactions on the momentum of an object if we know what net force acts on the object. In this chapter we are going to apply the momentum principle to the motion of planets and stars. In order to do this, we need to be able to describe mathematically the net force acting on an object.

In the 1600's Isaac Newton deduced that there must be an attractive force associated with a binary gravitational interaction (a gravitational interaction between two objects). The force F_g acts along a line connecting the two objects (Figure 2.1), and its magnitude is proportional to the mass m_1 of the first object and to the mass m_2 of the second object, and inversely proportional to the square of the distance r between the two objects (G is a proportionality constant):

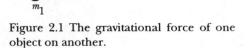

Figure 2.1 The gravitational force of one object on another.

THE GRAVITATIONAL FORCE LAW

$$F_g = G\frac{m_1 m_2}{r^2}$$

Note that the gravitational force law does not depend on the momentum or velocity of the objects. In principle, the nature of the gravitational force could have been much more complicated!

The gravitational force law applies to objects that are "point-like" (very small compared to the distance between the objects). In Volume II of this textbook we will be able to show that any hollow spherical shell with uniform density acts gravitationally on external objects as though all the mass of the shell were concentrated at its center. The density of the Earth is not uniform, because the central iron core has higher density than the outer layers. But by considering the Earth as layers of hollow spherical shells, each of nearly uniform density, we get the result that the Earth can be modeled as a point mass located at the center of the Earth. Similar statements can be made about other planets and stars. In Figure 2.2, the gravitational force is correctly calculated using the center-to-center distance r.

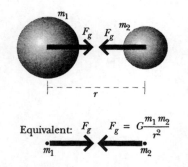

Figure 2.2 A sphere whose density depends only on distance from the center acts like a point particle with all the mass at the center.

This is not an obvious result. After all, in Figure 2.2 some of the atoms are closer and some farther apart than the center-to-center distance r, but the net effect after adding up all the interactions of the individual atoms is as though the two objects had collapsed down to points at their centers. This is a very special property of $1/r^2$ forces, both gravitational and electric, and is not true for forces that have a different dependence on distance.

Using the gravitational force law, Newton was able to explain very accurately the motion of planets and moons in the Solar System. Later scientists have found that it is possible to explain even galactic motions with the same formula, so it appears that this is truly a universal law of gravitation, applying to all pairs of masses everywhere in the universe.

Measuring the proportionality constant G

In order to make quantitative predictions and analyses of physical phenomena involving gravitational interactions, it is necessary to know the proportionality constant "G." In 1797-1798 Henry Cavendish performed the first experiment to determine a precise value for G (Figure 2.3). In this kind of experiment, a bar with metal spheres at each end is suspended from a thin quartz fiber which constitutes a "torsional" spring. From other measurements, it is known how large a tangential force measured in newtons is required to twist the fiber through a given angle. Large balls are brought near the suspended spheres, and one measures how much the fiber twists due to the gravitational interactions between the large balls and the small spheres.

If the masses are measured in kilograms, the distance in meters, and the force in newtons, the gravitational constant G has been measured in such experiments to be

$$G = 6.7 \times 10^{-11} \frac{\text{N} \cdot \text{m}^2}{\text{kg}^2}$$

This extremely small number reflects the fact that gravitational interactions are inherently very weak compared with electromagnetic interactions or the strong interaction. The only reason that gravitational interactions are significant in our daily lives is that objects interact with the entire Earth, which has a huge mass. It takes sensitive measurements such as the Cavendish experiment to observe gravitational interactions between two ordinary-sized objects.

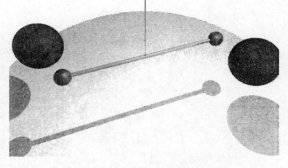

Figure 2.3 The Cavendish experiment.

Ex. 2.1 A 3 kg ball and a 5 kg ball are 2 m apart, center to center. What is the magnitude of the gravitational force that the 3 kg ball exerts on the 5 kg ball? What is the magnitude of the gravitational force that the 5 kg ball exerts on the 3 kg ball?

Ex. 2.2 Masses M and m attract each other with a gravitational force F. Mass m is replaced with a mass $3m$, and it is moved four times farther away. Now what is the force?

Ex. 2.3 Measurements show that the Earth's gravitational force on a mass of 1 kg near the Earth's surface is 9.8 N. The radius of the Earth is 6400 km (6.4×10^6 m). From these data determine the mass of the Earth.

2.1.1 Reciprocity (Newton's third law of motion)

An important aspect of the gravitational force law is that the force that object 1 exerts on object 2 is equal and opposite to the force that object 2 exerts on object 1 (Figure 2.4). This is clear from the algebraic form of the law, because $m_1 m_2 = m_2 m_1$. Electric forces (including the forces of atoms in contact with each other) also have this property, which is called "reciprocity" or "Newton's third law of motion":

RECIPROCITY

$$\vec{F}_{\text{on 1 by 2}} = -\vec{F}_{\text{on 2 by 1}}$$

The force that the Earth exerts on the massive Sun is just as big as the force that the Sun exerts on the Earth, so in the same time interval the momentum changes are equal in magnitude and opposite in direction:

$$\Delta \vec{p}_2 = \vec{F}_{\text{on 2 by 1}} \Delta t = -\Delta \vec{p}_1$$

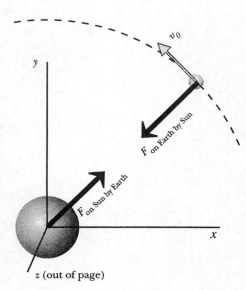

Figure 2.4 The Sun and the Earth exert equal and opposite forces on each other.

However, the velocity change $\Delta\vec{v} = \Delta\vec{p}/m$ of the Sun will be extremely small compared to the velocity change of the Earth, because the mass of the Sun is enormous in comparison with the mass of the Earth. The mass of our Sun, which is a rather ordinary star, is 2×10^{30} kg. This is enormous compared to the mass of the Earth (6×10^{24} kg; see Ex. 2.3) or even to the mass of the largest planet in our Solar System, Jupiter (2×10^{27} kg). Nevertheless, very accurate measurements of small velocity changes of distant stars have been used to infer the presence of planets orbiting those stars.

2.1.2 Gravitational force near the Earth's surface

Next we will develop a convenient way of expressing approximately the gravitational force that the Earth exerts on an object of mass m near the Earth's surface (Figure 2.5). The magnitude of this force is

$$F_g = G\frac{M_E m}{(R_E + y)^2}$$

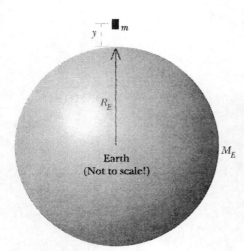

Figure 2.5 Determining the gravitational force of the Earth on an object a height y above the surface.

where y is the distance of the object above the surface of the Earth, R_E is the radius of the Earth, and M_E is the mass of the Earth.

As we mentioned earlier, a spherical object of uniform density can be treated as if all its mass were concentrated at its center. As a result, we can treat the effect of the Earth on the object as though the Earth were a tiny, very dense ball a distance $R_E + y$ away.

The gravitational force exerted by the Earth on an atom at the top of the object is slightly different from the force on an atom at the bottom of the object, because each atom is a slightly different distance from the center of the Earth. How much can this difference really matter in an analysis? Suppose the height of the object is a meter, and the bottom of the object is one meter above the surface. The radius of the Earth is about 6.4×10^6 m. Then

$$F_{\text{top}} = G\frac{M_E m}{(6400002 \text{ m})^2} \quad \text{whereas} \quad F_{\text{bottom}} = G\frac{M_E m}{(6400001 \text{ m})^2}$$

For most purposes this difference is not significant. In fact, for all interactions of objects near the surface of the Earth, it makes sense to use the same approximate value, R_E, for the distance from the object to the center of the Earth. This simplifies calculation of gravitational forces by allowing us to combine all the constants into a single lumped constant, g:

$$F_g \approx \left(G\frac{M_E}{R_E^2} \right)m = gm, \text{ so that } g = G\frac{M_E}{R_E^2}$$

The constant g, called "the magnitude of the gravitational field," has the value 9.8 newtons/kilogram near the Earth's surface. The "gravitational field" $\vec{g}$ at a location in space is defined to be the (vector) gravitational force that would be exerted on a 1 kg mass placed at that location. A 2 kg mass would experience a force twice as large. In general, a mass m will experience a force $m\vec{g}$ of magnitude mg.

Note that g is a *positive* number, the magnitude of the gravitational field. We can use this approximate formula for gravitational force in our analysis of any interactions occurring near the surface of the Earth.

Ex. 2.4 Show that GM_E/R_E^2 equals 9.8 N/kg.

Ex. 2.5 At what height above the surface of the Earth is there a 2% difference between the approximate gravitational force mg and the actual gravitational force on an object?

2.2 The superposition principle

What if there are several stars interacting gravitationally with each other? The superposition principle tells us how to do deal with such situations:

THE SUPERPOSITION PRINCIPLE

The net force on an object is the vector sum of the individual forces exerted on it by all other objects.

Each individual interaction is unaffected by the presence of other interacting objects.

The superposition principle is completely general. It applies to all kinds of interactions: gravitational, electromagnetic, weak, and strong interactions.

The implications of this principle are not always intuitively obvious. For example, this principle implies that the force exerted by the Sun on the Earth at a particular distance will always be the same, regardless of how many other planets there are that also interact with the Sun and the Earth. The presence of other objects and interactions does not affect the interactions between each pair of objects! An interaction doesn't get "used up," and the interaction between one object and another is unaffected by the presence of a third object. To take a silly example, hold an object over a table and then let go (Figure 2.6). The object falls, so evidently the table doesn't block the distant (non-contact) interaction with the Earth!

Figure 2.6 The presence of the table does not change the interaction of the glass with the Earth.

2.3 Graphical prediction of motion

It is very instructive to apply the momentum principle qualitatively and graphically to a slightly complicated gravitational situation, to see how the momentum principle determines the motion. Consider a case of an encounter between two asteroids. Some time ago the asteroids were far apart and hardly interacted. We'll begin our study when they have gotten close enough to each other that their mutual gravitational attractions are starting to have a noticeable effect on their motions. In Figure 2.7 we draw vectors to show the momenta and forces at this instant; the forces are calculated from the gravitational force law. Asteroid 2 has twice the mass of asteroid 1.

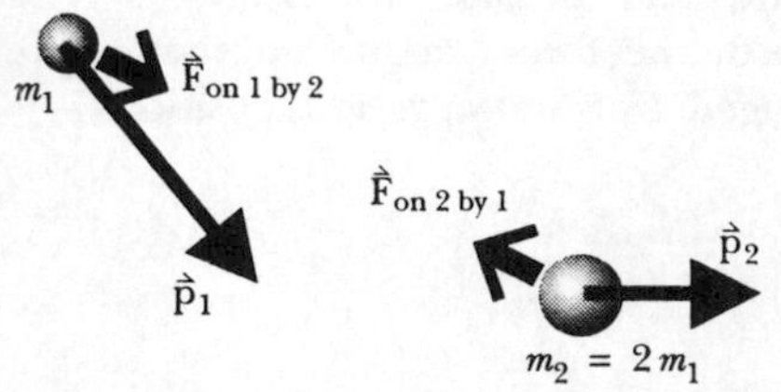

? What does the gravitational force law say about the relative magnitudes of the two forces? Which force will cause a larger change in momentum? Which force will cause a larger change in velocity?

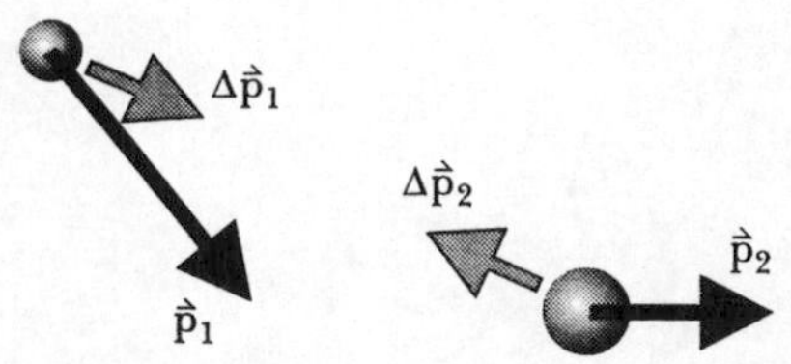

Because the gravitational force is proportional to $m_1 m_2$, the two forces have exactly the same magnitude (with opposite directions), which is an example of reciprocity (Newton's third law of motion). Therefore the magnitude of the change in momentum will be the same for each asteroid. However, the force acting on the lighter asteroid (m_1) will have twice as big an effect on the velocity.

The momentum principle predicts that in a short time interval Δt, the momentum of an asteroid will change by an amount $\Delta \vec{p} = \vec{F}_{net}\Delta t$ which we can calculate, because we know the force acting on each asteroid. Figure 2.8 shows the momentum changes that will occur during the time interval Δt. Note that the momentum changes of the two asteroids are equal in magnitude but opposite in direction, because the gravitational forces are equal and opposite.

Now that we have determined the momentum change $\Delta \vec{p}$ for each asteroid, we can add the momentum change to the present momentum to determine the new momentum $\vec{p}_{new} = \vec{p}_{old} + \Delta \vec{p}$, as in Figure 2.9.

Figure 2.7 Initial momenta of the two asteroids, and forces exerted on each.

Figure 2.8 The change in momentum of each asteroid due to the force acting on it.

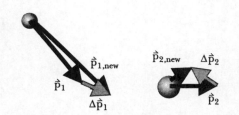

Figure 2.9 The new momentum is the vector sum of the old momentum and the change in momentum during the time interval Δt.

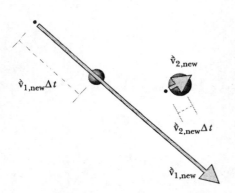

Figure 2.10 Each asteroid moves a distance $v\Delta t$ in the direction of its velocity. Dots show previous positions.

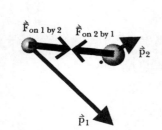

Figure 2.11 Because the asteroids have moved, the forces acting on them have changed. Dots show previous positions.

How do we find the new position of the asteroid? During the short time interval Δt, an asteroid will move an amount $\vec{v}\Delta t$, in the direction of its new velocity $\vec{v} = \vec{p}/m$ (Figure 2.10; we assume that $v \ll c$). The velocity is changing during this time interval, but if Δt is quite small, the change in the velocity is quite small, and it doesn't matter very much whether we use the velocity at the start or end of the time interval, or some kind of average. (If in some process the speed v is so high that $m\vec{v}$ is a poor approximation to the momentum, we can't use $\vec{v} = \vec{p}/m$ but must calculate $\vec{v}$ from the full relativistic formula for $\vec{p}$.)

Figure 2.11 shows the new positions and momenta and forces after one time interval Δt. The two asteroids are closer together now, so the gravitational forces that we calculate are larger.

We are now in a situation just like the one we started with—we know the positions of the asteroids, which lets us calculate the gravitational forces, and we know the momenta of the asteroids, to which we can add the momentum change $\Delta\vec{p} = \vec{F}_{net}\Delta t$ for each asteroid in the next time interval. We can repeat the procedure, one time interval after another, and predict the motion of both asteroids.

Try it! It is useful to work out graphically two more steps in the approximate trajectories of the two asteroids. In this chapter you will learn how to automate this process by having a computer carry out the detailed calculations.

Why not just use calculus?

You might wonder why we don't simply use calculus to predict the motion of the two asteroids. One answer is that we do use calculus. Step by step, we add up a large number of tiny increments of the momentum of an asteroid to calculate a large change in its momentum, and this corresponds to an approximate evaluation of an integral (which is an infinite sum of infinitesimal amounts).

A more interesting answer is that the motion of most physical systems *cannot* be predicted using calculus in any way other than by this step-by-step approach. In a few cases calculus does give a general result without going through this procedure. For example, an object subjected to a constant force has a constant acceleration, and calculus can be used to obtain a prediction for the position (involving $\frac{1}{2}at^2$). The elliptical orbits of two stars around each other can be predicted mathematically, although the math is quite challenging. But the general motion of three stars around each other has never been successfully analyzed in this way. The basic problem is that it is usually relatively easy to take the derivative of a known function, but it is often impossible to determine in algebraic form the integral of a known function, which is what would be involved in long-term prediction.

It isn't a question of taking more math courses in order to be able to solve the "three-body" problem: there is no general mathematical solution. However, a step-by-step procedure of the kind we carried our for the two asteroids can easily be extended to three bodies, or even more, as we will see later in this chapter. For this reason we will study the step-by-step prediction method in detail, because it is a powerful technique of increasing importance in modern science, thanks to the availability of powerful computers to do the repetitive work for us.

2.4 Conservation of momentum

For each step in the process you carried out, the change in the momentum for the two asteroids was equal and opposite, because the forces were equal and opposite. Consider the two asteroids together as a combined system.

? For each step, what is the change in the total momentum of the combined system, $\vec{p}_1 + \vec{p}_2$?

The changes in the momenta are equal and opposite, $\Delta\vec{p}_1 + \Delta\vec{p}_2 = 0$, which we can write as $\Delta(\vec{p}_1 + \vec{p}_2) = 0$. This shows that the total momentum $\vec{p}_1 + \vec{p}_2$ does not change.

Here's another way to see this. Let $\vec{F}$ be the force exerted on asteroid 1 by asteroid 2, so $-\vec{F}$ is the force exerted on asteroid 2 by asteroid 1. After a short time interval Δt the new total momentum is this:

$$(\vec{p}_1 + \vec{F}\Delta t) + (\vec{p}_2 - \vec{F}\Delta t) = \vec{p}_1 + \vec{p}_2$$

Whatever momentum is lost by one asteroid is gained by the other asteroid. This is a simple example of an important principle, the "conservation of momentum." In this case the net force acting on the two-asteroid system is zero, because the only forces are the gravitational forces that the asteroids exert on each other, and these are equal and opposite, so they add up to zero. Zero net impulse ($\vec{F}_{net}\Delta t$), zero net momentum change. One of the asteroids can gain momentum (due to a force acting on it), but only if the other asteroid loses the same amount. In Chapter 8 we will apply conservation of momentum to collisions.

Relativistic momentum conservation

Particle accelerators produce beams of particles such as electrons, protons, and pions at speeds very close to the speed of light. When these high-speed particles interact with other particles, experiments show that the total momentum is conserved, but only if the momentum of each particle is defined in the way Einstein proposed, $\vec{p} = m\vec{v}/\sqrt{1-(v^2/c^2)}$. It is found that using the low-speed approximation $\vec{p} \approx m\vec{v}$ there is no conservation of momentum when the speed v approaches the speed of light c.

Ex. 2.6 In the interaction of the two asteroids, was the momentum of asteroid 1 constant? Why or why not? Was the momentum of asteroid 2 constant? Why or why not? Was the total momentum of the two asteroids constant? Why or why not?

2.5 The momentum principle for multiparticle systems

We treated the asteroids like simple "point" particles. Is this valid? Don't the atoms that make up an asteroid interact with each other? Starting from the momentum principle for a point particle, and using the principle of reciprocity, we can derive an extremely important equation that predicts the overall motion of a complex, multiparticle system such as an asteroid. This equation looks surprisingly like the momentum principle for a single point particle. The change of the total momentum $\vec{P}_{tot}$ of a system is equal to the net "external" impulse acting on the system:

THE MOMENTUM PRINCIPLE FOR MULTIPARTICLE SYSTEMS

$$\Delta\vec{P}_{tot} = \vec{F}_{net,ext}\Delta t, \text{ or in derivative form } \frac{d\vec{P}_{tot}}{dt} = \vec{F}_{net,ext}$$

Definition of total momentum: $\vec{P}_{tot} = \vec{p}_1 + \vec{p}_2 + \vec{p}_3 + ...$

In Chapter 7 we will show that in cases where $v \ll c$ and mass is constant the new equation is equivalent to

$$M\frac{d\vec{v}_{cm}}{dt} = \vec{F}_{net,ext} \text{ (if } v << c \text{, and constant mass)}$$

where M is the total mass of the system and $\vec{v}_{cm}$ is the velocity of the "center of mass" of the system. The validity of the multiparticle form of the momentum principle justifies our frequent treatment of an asteroid or a spacecraft or a planet as though it were a single particle.

We will now derive the multiparticle version of the momentum principle. To be as concrete as possible, we'll consider a system consisting of just three particles. It is easy to see how this generalizes to larger systems, including a star consisting of an astronomically huge number of atoms.

2.5.1 Derivation of the multiparticle version of the momentum principle

In a three-particle system (Figure 2.12) we show all of the forces acting on each particle, where the lower case $\vec{f}$'s are forces the particles exert on each other (so-called "internal" forces), and the upper case $\vec{F}$'s are forces exerted by other objects that are not shown and are not part of our chosen system (these are so-called "external" forces, such as the gravitational attraction of the Earth, or a force that you exert by pulling on one of the particles).

We will use the following shorthand notation: $\vec{f}_{1,3}$ will denote the force exerted on particle 1 by particle 3; $\vec{F}_{2,ext}$ will denote the force on particle 2 exerted by external objects, and so on. Here is the momentum principle for each of the three particles:

$$\frac{d\vec{p}_1}{dt} = \vec{F}_{1,ext} + \vec{f}_{1,2} + \vec{f}_{1,3}$$

$$\frac{d\vec{p}_2}{dt} = \vec{F}_{2,ext} + \vec{f}_{2,1} + \vec{f}_{2,3}$$

$$\frac{d\vec{p}_3}{dt} = \vec{F}_{3,ext} + \vec{f}_{3,1} + \vec{f}_{3,2}$$

Nothing new so far. But now we add up these three equations. That is, we create a new equation by adding up all the terms on the left sides of the three equations, and adding up all the terms on the right sides, and setting them equal to each other:

$$\frac{d\vec{p}_1}{dt} + \frac{d\vec{p}_2}{dt} + \frac{d\vec{p}_3}{dt} = \vec{F}_{1,ext} + \vec{f}_{1,2} + \vec{f}_{1,3} + \vec{F}_{2,ext} + \vec{f}_{2,1} + \vec{f}_{2,3} + \vec{F}_{3,ext} + \vec{f}_{3,1} + \vec{f}_{3,2}$$

Many of these terms cancel. By the principle of reciprocity (Newton's third law of motion), which is obeyed by gravitational and electric interactions, we have the following:

$$\vec{f}_{1,2} = -\vec{f}_{2,1}$$

$$\vec{f}_{1,3} = -\vec{f}_{3,1}$$

$$\vec{f}_{2,3} = -\vec{f}_{3,2}$$

Thanks to reciprocity, all that remains after the cancellations is this:

$$\frac{d\vec{p}_1}{dt} + \frac{d\vec{p}_2}{dt} + \frac{d\vec{p}_3}{dt} = \vec{F}_{1,ext} + \vec{F}_{2,ext} + \vec{F}_{3,ext}$$

The total momentum of the system is $\vec{P}_{tot} = \vec{p}_1 + \vec{p}_2 + \vec{p}_3$, so we have:

$$\frac{d\vec{P}_{tot}}{dt} = \vec{F}_{net,ext} \text{, or in terms of impulse, } \Delta\vec{P}_{tot} = \vec{F}_{net,ext}\Delta t$$

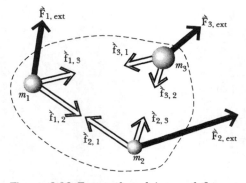

Figure 2.12 External and internal forces acting on a system of three particles.

The importance of this equation is that reciprocity has eliminated all of the internal forces (the forces that the particles in the system exert on each other); internal forces cannot affect the motion of the system as a whole. All that matters in determining the rate of change of (total) momentum is the net external force. The equation has exactly the same form as the momentum principle for a single particle.

Moreover, if the object is a sphere whose density is only a function of radius, the object exerts a gravitational force on other objects as though the sphere were a point particle. We can predict the motion of a star or a planet or an asteroid as though it were a single point particle of large mass.

2.6 Orbits

As we see within our own Solar System, it is common for objects to travel in circular or elliptical orbits around other objects, as the Earth and other planets travel around the Sun, and the Moon and other, human-made satellites travel around the Earth. Given only the momentum principle and the gravitational force law, we should be able to model how gravitational interactions lead to circular or elliptical orbits.

2.6.1 The orbit of a planet around a star

Let's try to model the motion of a planet around a star. We first need to list all the objects that interact significantly with the planet.

? Make a list of all the objects whose interactions with the planet you would need to consider in a detailed model.

One might list the nearby massive objects that interact gravitationally with the planet: the star, its moons if it has them, other planets, etc. Other objects, such as distant stars, do exert forces on the planet, but they are usually sufficiently far away to neglect safely, unless the planet is orbiting a double star system.

Simplifying the model

To make a complete model of the planet's motion, we would have to include at least the star, any moons, and other planets. However, if we begin with a complex situation it will be more difficult to identify the effects of particular factors, such as choice of initial conditions. Additionally, it will be much more difficult to find errors in our model or in our computer implementation of it. We will start by considering a drastically simplified model. We suppose that the planet is the only planet orbiting a single star, and that there are no other large objects near enough to have a significant influence.

Further simplifications

? Compare the magnitude of the velocity change that the planet experiences in each time interval Δt with the magnitude of the velocity change that the star experiences in this same time interval.

Thanks to reciprocity (Newton's third law of motion, due to the symmetrical nature of the gravitational force law), the force that the planet exerts on the massive star is equal in magnitude (and opposite in direction) to the force that the star exerts on the planet. Same force, same magnitude of change of momentum for both bodies.

But because the star is so much more massive, its velocity change $\Delta \vec{v} = \Delta \vec{p}/m$ is negligible. For the purpose of determining the orbit of the planet, we can safely make the approximation that the star doesn't move at all and carry out a calculation just for the planet acted upon by a stationary star. The scheme we will use is this:

- Determine the gravitational force on the planet.
- Use the force to update the momentum.
- Use the velocity to update the position.
- Repeat.

The next sections deal with the details of how to organize this computation.

2.6.2 Representing the gravitational force as a vector

The momentum principle, $\Delta \vec{p} = \vec{F}_{net}\Delta t$, is expressed in terms of a vector net force, so we need to be able express the gravitational force as a vector, with x, y, and z components. Take the case of the gravitational force on a planet by a star. At the instant shown in Figure 2.13, the planet and star are at known positions relative to some origin. From these positions we can construct a vector $\vec{r}$ that points from the planet to the star, in the direction of the gravitational force acting on the planet. We use "p" for planet and "s" for star:

$$\vec{r}_p = <x_p, y_p, z_p>$$

$$\vec{r}_s = <x_s, y_s, z_s>$$

$$\vec{r} = \vec{r}_s - \vec{r}_p = <x_s - x_p, y_s - y_p, z_s - z_p>$$

Sometimes it is confusing whether to use $\vec{r}_s - \vec{r}_p$ or $\vec{r}_p - \vec{r}_s$. Make a diagram like Figure 2.13 and ask yourself a simple numerical question, such as "If x_p is 3 and x_s is 8, like the diagram, then we would have $x_s - x_p = +5$, which makes sense for the x component of $\vec{r}$ on the diagram, whereas $x_p - x_s = -5$ doesn't make sense."

The gravitational force is proportional to $1/r^2$, and the distance r can be calculated from the 3D version of the Pythagorean theorem:

$$|\vec{r}| = r = \sqrt{(x_s - x_p)^2 + (y_s - y_p)^2 + (z_s - z_p)^2}$$

We use this distance to calculate the magnitude of the gravitational force acting on the planet by the star:

$$F_{p,s} = \frac{GM_sM_p}{r^2}$$

The remaining question is how to express the direction of the force as a vector. What we need is a vector $\hat{r}$ (pronounced "r-hat") that points in the direction of $\vec{r}$ but has a magnitude of 1:

UNIT VECTOR

$$\hat{r} = \frac{\vec{r}}{r}, \text{ with magnitude } \frac{|\vec{r}|}{r} = \frac{r}{r} = 1\frac{\text{meter}}{\text{meter}} = 1$$

Because its magnitude is 1, $\hat{r}$ is called a "unit vector." It is dimensionless because it has the units of meter/meter. To be absolutely clear about how to determine this unit vector, let's expand its definition:

$$\hat{r} = \frac{\vec{r}}{r} = \frac{\vec{r}_s - \vec{r}_p}{|\vec{r}_s - \vec{r}_p|} = \frac{<(x_s - x_p), (y_s - y_p), (z_s - z_p)>}{\sqrt{(x_s - x_p)^2 + (y_s - y_p)^2 + (z_s - z_p)^2}}$$

The specific example and exercises at the end of this section are intended to make this concrete. While we will occasionally calculate such quantities by hand, for the repetitive calculations required to predict orbits we will let a computer do the calculations for us.

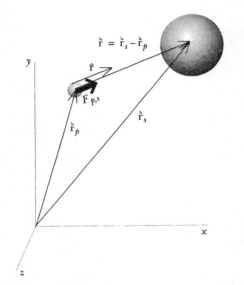

Figure 2.13 Developing a vector expression for the gravitational force on the planet by the star.

Using this unit vector we can now write the gravitational force on the Earth as a vector:

$$\vec{F}_{p,s} = \frac{GM_sM_p}{r^2}\hat{r}$$

This has the correct magnitude (because the magnitude of $\hat{r}$ is 1), and it has the correct direction—from the planet toward the star, which exerts an attractive gravitational force on the planet.

? How would you calculate the gravitational force exerted on the star by the planet?

This requires very little effort—it is simply the negative of the force you already calculated:

$$\vec{F}_{s,p} = -\vec{F}_{p,s}$$

This is an example of reciprocity. The force exerted on the star by the planet is equal in magnitude but opposite in direction to the force exerted on the planet by the star.

Component form of the gravitational force

For concreteness, here are the x, y, and z components of the gravitational force, where $r = \sqrt{(x_s - x_p)^2 + (y_s - y_p)^2 + (z_s - z_p)^2}$:

$$F_x = \frac{GM_sM_p}{r^2}\frac{(x_s - x_p)}{r}$$

$$F_y = \frac{GM_sM_p}{r^2}\frac{(y_s - y_p)}{r}$$

$$F_z = \frac{GM_sM_p}{r^2}\frac{(z_s - z_p)}{r}$$

Watch out!

? The gravitational force is proportional to $1/r^2$, so why isn't the x component of the force simply proportional to $1/(x_s - x_p)^2$?

This is a common error. Consider the situation shown in Figure 2.14, where the planet is directly below the star, so that $(x_s - x_p)$ is zero, or nearly zero. What is the value of $1/(x_s - x_p)^2$ in this case? Huge! Dividing by zero, or a very small number, gives a number that is enormous. Yet the true x component of the gravitational force in this case is not enormous but zero, or nearly zero.

The moral of the story is that we cannot take components of scalar quantities such as r^2. Nor can we take components in the denominator. We must calculate the magnitude of the gravitational force as a separate quantity, in the form GM_sM_p/r^2, and multiply by the unit vector $\hat{r} = \vec{r}/r$ to get the correct magnitude and direction of the vector force.

Example: A proton is at position <20, 15, –40> m, and an electron is at position <25, 12, –34> m. Calculate the unit vector that points from the proton toward the electron.

Solution:

$$\hat{r} = \frac{\vec{r}}{r} = \frac{<5, -3, 6> \text{ m}}{\sqrt{5^2 + 3^2 + 6^2} \text{ m}} = \frac{<5, -3, 6> \text{ m}}{8.37 \text{ m}} = <0.598, -0.359, 0.717>$$

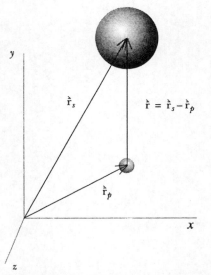

Figure 2.14 If $x_s - x_p$ is zero or nearly zero, what is the gravitational force?

Check: $\sqrt{0.598^2 + 0.359^2 + 0.717^2} = 1.00$. The vector points from the proton toward the electron and has magnitude 1.

Ex. 2.7 With respect to the example just discussed, write the unit vector that points from the electron toward the proton.

Ex. 2.8 A cube has one corner at the origin and is aligned with the axes. Each edge of the cube has length d. Write a unit vector that points from the origin to the farthest corner of the cube. Hint: First write the vector $\vec{r}$ before constructing the unit vector $\hat{r}$.

Ex. 2.9 In the previous exercise, suppose there is a mass m_1 at the origin and a mass m_2 at the farthest corner of the cube. Write a vector expression for the gravitational force acting on mass m_1.

2.6.3 Step by step evaluation of the momentum principle

We can now use the momentum principle one step at a time to calculate numerically the motion of a planet around a star. The basic idea is to take very short time steps, calculating what happens in each step. For now, make the simplifying assumption that the massive star hardly moves. Start the planet with known initial momentum, at a known initial position, and do this:

▶ Find the net force acting on the planet using the gravitational force law, which requires recalculating $\hat{r} = \hat{r}_s - \hat{r}_p$ based on current positions:

$$\vec{F}_{p,s} = \frac{GM_s M_p}{r^2}\hat{r}$$

Update the planet's momentum:

$$\vec{p}_p \Leftarrow \vec{p}_p + \vec{F}_{p,s}\Delta t$$

Update the planet's position:.

$$\hat{r}_p \Leftarrow \hat{r}_p + \frac{\vec{p}_p}{M_p}\Delta t$$

At the planet's new position, ─────┐
                                  do it all again.

Here are the details of this scheme.

Calculate force: At time t_0 you know the initial positions of the planet and star, so you can calculate the vector $\hat{r}_0$ from the position of the planet toward the position of the star;

$$\hat{r}_0 = \hat{r}_{s,0} - \hat{r}_{p,0}$$

Use $\hat{r}_0$ to calculate the force on the planet from Newton's law of gravitation:

$$\vec{F}_{p,s} = \frac{GM_s M_p}{r_0^2}\hat{r}_0$$

Update momentum: The initial momentum of the planet is $\vec{p}_{p,0}$. In the first short time interval Δt, according to the momentum principle the approximate change of the momentum will be $\Delta\vec{p}_p \approx \vec{F}_{p,s}\Delta t$. At the end of the time interval Δt the new momentum $\vec{p}_{p,1} = \vec{p}_{p,0} + \Delta\vec{p}_p$ is approximately this:

$$\vec{p}_{p,1} \approx \vec{p}_{p,0} + \vec{F}_{p,s}\Delta t$$

This is only approximate, because during even a short time interval the position will change, so the force is not constant. However, if we use a short enough time interval Δt this approximation may not introduce much error.

Update position: The initial position of the planet is $\vec{r}_{p,0}$. In the first short time interval Δt, we can calculate the approximate change of position $\Delta \vec{r}_p \approx \vec{v}_1 \Delta t$, where $\vec{v}_1$ is the velocity of the planet at the end of the first time interval. At the end of the first time interval Δt the new position $\vec{r}_{p,1} = \vec{r}_{p,0} + \Delta \vec{r}_p$ is approximately this:

$$\vec{r}_{p,1} = \vec{r}_{p,0} + \frac{\vec{p}_{p,1}}{M_p} \Delta t \text{ (assuming } v \ll c)$$

This is only approximate (even for low speeds), because the velocity is changing from $\vec{v}_0$ to $\vec{v}_1$ during the time interval Δt. However, if we use a short enough time interval Δt it is possible that this approximation will not introduce much error.

After making these calculations (calculate force, update momentum, and update position), we now know at the new time $t_1 = t_0 + \Delta t$ the new position $\vec{r}_{p,1}$ and the new momentum $\vec{p}_{p,1}$. Starting from these new values, we can repeat the same calculations for the next time interval, and the next, and the next. We can calculate the position and momentum of the planet, one step at a time. This process is called "numerical integration."

Let's summarize the computational steps to be performed for each step of the numerical integration, dropping the step subscripts (0, 1, etc.):

$$\vec{r} = \vec{r}_s - \vec{r}_p \text{ (vector from planet toward star)}$$

$$r = \sqrt{(x_s - x_p)^2 + (y_s - y_p)^2 + (z_s - z_p)^2}$$

$$\hat{r} = \frac{\vec{r}}{r} \text{ (unit vector from planet toward star)}$$

$$\vec{F}_{p,s} = \frac{GM_s M_p}{r^2} \hat{r} \text{ (gravitational force law)}$$

$$\vec{p}_p \Leftarrow \vec{p}_p + \vec{F}_{p,s} \Delta t \text{ (the momentum principle)}$$

$$\vec{r}_p \Leftarrow \vec{r}_p + \frac{\vec{p}_p}{M_p} \Delta t \text{ (update position; } v \ll c)$$

What is meant by the arrow in $\vec{p}_p \Leftarrow \vec{p}_p + \vec{F}_{p,s} \Delta t$ is "calculate the quantity $\vec{p}_p + \vec{F}_{p,s} \Delta t$ and replace the old value of $\vec{p}_p$ with this new value." In many computer programming languages a simple "=" is used to mean "replace by"; in a few programming languages this is written as ":=".

After executing the statements listed above, the old values of the planet's position and momentum have been replaced by new values, and we're ready to execute the statements again, and again, and again... The result is to predict the motion of the planet into the future, one step at a time.

Make sure that you understand the significance of each part of the computation shown above, and ask questions about any aspects that are unclear.

Vector component computations

Some computational environments include facilities for dealing with vector quantities directly, as vectors. In other environments you have to expand each vector calculation into individual x, y, and z calculations. For example, here are component calculations for updating the momentum:

$$p_x \Leftarrow p_x + F_x \Delta t, \text{ where } F_x = \frac{GM_sM_p(x_s - x_p)}{r^2} \frac{}{r}$$

$$p_y \Leftarrow p_y + F_y \Delta t, \text{ where } F_y = \frac{GM_sM_p(y_s - y_p)}{r^2} \frac{}{r}$$

$$p_z \Leftarrow p_z + F_z \Delta t, \text{ where } F_z = \frac{GM_sM_p(z_s - z_p)}{r^2} \frac{}{r}$$

All of the other vector calculations expand into three component calculations in a similar way (or two if the motion is confined to the *xy* plane).

Check by trying a smaller Δ*t*

Because the numerical integration process is only approximate, before trusting a result you should try a smaller Δ*t* (say by cutting it in half), and make sure that the physical results change very little. If there is a major change, try cutting Δ*t* in half again. Keep cutting the time step until the physical results don't change with further cuts.

Carrying out the computations

The step-by-step calculations can be done conveniently in a computer programming language, on a graphing calculator, in a spreadsheet, or in a math package. Your instructor will assist you in choosing and using an appropriate tool.

In organizing the repetitive computations, it is wise to use symbols rather than numeric values for the important quantities, such as "m1" for a mass, "x0" for an initial position, or "dt" for the step size Δ*t*. For example, if a mass is 5 kg, and you write "5" rather than "m1" in many places in the computation, you would have to make many changes to make the mass be 10 kg. But if you set "m1 = 5" at the start of the computation, and always write "m1" rather than "5" everywhere, you can easily make the change "m1 = 10" in just one place and have it affect the entire computation.

2.6.4 Initial Conditions

In order to compute a trajectory for the planet, we need to specify its initial position and its initial momentum.

?   What would happen if we make its initial momentum zero?

The planet would simply begin to fall into the star, speeding up as it goes. Eventually it would plunge into the star, where it would burn up.

More interesting trajectories are possible if we give the planet a nonzero initial momentum. To get a trajectory most different from a straight line toward the star, let's give it an initial momentum (and velocity) at a right angle to the line between the planet and the star (Figure 2.15).

In your model, it is instructive to start with parameters similar to those for our own Earth and Sun:

mass of Earth 6×10^{24} kg   mass of Sun 2×10^{30} kg

distance from Earth to Sun 1.5×10^{11} m

radius of Sun 7×10^8 m   radius of Earth 6.4×10^6 m

$G = 6.7 \times 10^{-11} \text{N} \cdot \text{m}^2/\text{kg}^2$

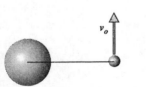

Figure 2.15 A possible set of initial conditions, with the planet's momentum at a right angle to the line between the planet and the star.

For planetary orbits, it is convenient to choose coordinate axes oriented so that the star, the planet, and the initial momentum of the planet all lie in the *xy* plane. Then the force exerted by the star lies in this same *xy* plane, and the planet will move at all times in the plane defined by $z = 0$.

Problem 2.1 Planetary orbits

In this problem you will study the motion of a planet around a star. To start with a somewhat familiar situation, you will begin by modeling the motion of our Earth around our Sun. Write answers to questions either as comments in your program or on paper, as specified by your instructor.

Planning

(a) The Earth goes around the Sun in a nearly circular orbit, taking one year to go around. Using data on page 50, what initial speed should you give the Earth in a computer model, so that a circular orbit should result? (If your computer model does produce a circular orbit, you have a strong indication that your program is working properly.)

(b) Estimate an appropriate value for Δt to use in your computer model. Remember that t is in seconds, and consider your answer to part (a) in making this estimate. If your Δt is too small, your calculation will require many tedious steps. Explain briefly how you decided on an appropriate step size.

Circular orbit

(c) Starting with speed calculated in (a), and with the initial velocity perpendicular to a line connecting the Sun and the Earth, calculate and display the trajectory of the Earth. Display the whole trail, so you can see whether you have a closed orbit (that is, whether the Earth returns to its starting point each time around). Include the Sun's position $\vec{r}_s = <x_s, y_s, z_s>$ in your computations, even if you set its coordinates to zero. This will make it easier to modify the computation to let the Sun move (Problem 2.2 Binary stars).

Computational accuracy

(d) As a check on your computation, is the orbit a circle as expected? Run the calculation until the accumulated time t is the length of a year: does this take the Earth around the orbit once, as expected? What is the largest step size you can use that still gives a circular orbit and the correct length of a year? What happens if you use a much larger step size?

Noncircular trajectories

(e) Set the initial speed to 1.2 times the Earth's actual speed. Make sure the step size is small enough that the orbit is nearly unaffected if you cut the step size. What kind of orbit do you get? What step size do you need to use to get results you can trust? What happens if you use a much larger step size? Produce at least one other qualitatively different noncircular trajectory for the Earth. What difficulties would humans have in surviving on the Earth if it had a highly noncircular orbit?

Force and momentum

(f) Choose initial conditions that give a noncircular orbit. Continuously display a vector showing the momentum of the Earth (with its tail at the Earth's position), and a different colored vector showing the force on the Earth by the Sun (with its tail at the Earth's position). You must scale the vectors appropriately to fit into the scene. A way to figure out what scale factor to use is to print the numerical values of momentum and force, and compare with the scale of the scene. For example, if the width of the scene is W meters, and the force vector has a typical magnitude F (in newtons), you might scale the force vector by a factor $0.1W/F$, which would make the length of the vector be one-tenth the width of the scene. Is the force in the same direction as the momentum? How does the momentum depend on the distance from the star?

Problem 2.2 Binary stars

About a half of the visible "stars" are actually systems consisting of two stars orbiting each other, called "binary stars." In your computer model of the Earth and Sun (Problem 2.1 Planetary orbits), replace the Earth with a star whose mass is half the mass of our Sun, and take into account the gravitational effects that the second star has on the Sun.

(a) Give the second star the speed of the actual Earth, and give the Sun zero initial momentum. What happens? Try a variety of other initial conditions. What kinds of orbits do you find?

(b) One of the things to try is to give the Sun the same magnitude of momentum as the second star, but in the opposite direction. If the initial momentum of the second star is in the +y direction, give the Sun the same magnitude of momentum but in the −y direction, so that the total momentum of the binary star system is initially zero, but the stars are not headed toward each other. What is special about the motion you observe in this case?

2.6.5 Improving the accuracy

You saw that you could improve the accuracy of your calculations by cutting the step size. But of course you pay a penalty in having to do more calculations, so there is a trade-off between accuracy and computation time. It isn't useful to use a step size smaller than one that gives you adequate accuracy.

People have developed various numerical integration schemes that give better accuracy for the *same* step size. This may seem like getting something for nothing, but it is really "working smarter rather than working harder." One of the most commonly used schemes goes by the name of "4th-order Runge-Kutta," which is a method that improves the averaging within a time interval. There is a vast lore on how to do numerical integration with high speed and high accuracy. An excellent book on the entire subject is *An Introduction to Computer Simulation Methods: Applications to Physical Systems* by Harvey Gould and Jan Tobochnik (Addison-Wesley 1996).

For the computer modeling that we will do in this course, computers are now fast enough to permit using a small step size to achieve adequate accuracy. Our simple numerical integration scheme is adequate as long as you are careful to check that cutting the step size does not change the results significantly.

2.7 Physical models

Our model of the motion of the Earth around the Sun is good enough for many purposes. However, we have left out important aspects of the actual motion. For example, we have ignored the gravitational effects of the Moon, which are sizable, and of the other planets, which are small but real.

Making and using models is an activity central to physics, and it is important to reflect on the criteria for a "good" physical model, which will depend on how we intend to use the model. In fact, one of the most important problems a scientist or engineer faces is deciding what interactions must be included in a model of a real physical, chemical, or biological system, and what interactions can reasonably be ignored.

Simplified (or "idealized") models

If we neglect the gravitational forces exerted on the Earth by the Moon or Venus, we say that we are constructing a simplified model of the situation. A useful model should omit extraneous detail but retain the main features of the real-world situation. We hope that the main results of our analysis of the simplified model will apply adequately to the complex real-world situation.

Of course if we do a poor job of modeling and make inappropriate approximations or neglect effects that are actually sizable, we may get a rather inaccurate result (though perhaps adequate for some purposes). There is therefore a certain art to making a good model. One goal of this course is to help you develop skill in formulating simplified but meaningful physical models of complex situations.

Another common way of describing a model is to say that it is an "idealization," by which we mean a simple, clean, stripped-down representation, free of messy complexities. "Ideally," a ball will roll forever on a level floor (but a real ball rolling on a real floor eventually comes to a stop). An "ideal" gas is a fictitious gas in which the molecules don't interact with each other (as opposed to a real gas whose molecules do interact, but only when they come close to each other).

The analysis of a simplified model gives an approximate result that differs somewhat from what actually happens in the real world. Alternatively, you can consider your result to be exact for a *different* real-world situation, one where the neglected influences are not actually present. For example, the calculations we make when we neglect the effect of objects other than the Earth would be closer to reality in a different solar system in which there is only one planet and no moons.

An important aspect of physical modeling that will engage us throughout this course is making appropriate approximations to simplify the messy, real-world situation enough to permit (approximate) analysis using Newton's laws. However, using Newton's laws is itself an example of modeling and making approximations, because we are neglecting the effects of quantum mechanics and of general relativity (Einstein's treatment of gravitation). Newton's laws are actually only an approximation to the way the world works (though frequently an extremely good one).

2.8 Circular motion at constant speed

Using your computational model of orbital motion, it is possible to produce trajectories that are shaped like circles, ellipses, portions of parabolas, portions of hyperbolas, or straight lines. Motion in a circle at constant speed (and constant magnitude of momentum) is a special case that we can treat "analytically" (that is, without a step-by-step numerical integration). Even though it is a special case we can learn some things of general interest from this analysis.

For concreteness, consider a planet in circular motion around a star (Figure 2.16). We need to calculate the rate of change $d\vec{p}/dt$ of the momentum due to the changing direction; the magnitude $|\vec{p}|$ of the momentum isn't changing. Before tackling this directly, we will do a similar calculation for a more familiar aspect of the situation.

In Figure 2.16 the position vector that points from the center of the star to the planet rotates, taking a time T (called the "period") to go around once. We will calculate the rate of change $d\vec{r}/dt$ of the position vector due to the changing direction; the magnitude $|\vec{r}|$ of the position vector isn't changing.

Derivative of a rotating vector

We will be dealing with derivatives of vectors. If position changes by an amount $\Delta\vec{r} = \vec{r}_2 - \vec{r}_1$ in a time Δt, the average velocity is the ratio:

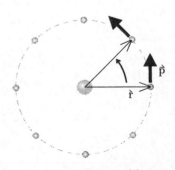

Figure 2.16 A rotating position vector describes the location of a planet orbiting a star.

$$\vec{v}_{\text{avg}} = \frac{\Delta \vec{r}}{\Delta t}; \quad v_{\text{avg,x}} = \frac{\Delta x}{\Delta t} \text{ and } v_{\text{avg,y}} = \frac{\Delta y}{\Delta t}$$

If the time interval Δt is very short, this ratio is approximately the derivative:

$$\frac{\Delta \vec{r}}{\Delta t} \approx \frac{d \vec{r}}{dt}; \quad \frac{\Delta x}{\Delta t} \approx \frac{dx}{dt} \text{ and } \frac{\Delta y}{\Delta t} \approx \frac{dy}{dt}$$

So what we mean by $d\vec{r}/dt$ is simply the displacement in a short time, divided by that short time (in the limit of *very* short time). The magnitude of $d\vec{r}/dt$ is written as

$$\left| \frac{d \vec{r}}{dt} \right|$$

and represents the speed of the object at a particular instant.

Note carefully that

$$\left| \frac{d \vec{r}}{dt} \right| \text{ and } \frac{d |\vec{r}|}{dt} \text{ are not the same things!}$$

For a rotating vector $d|\vec{r}|/dt = 0$ because the magnitude of the vector $\vec{r}$ isn't changing. However $|d\vec{r}/dt|$ is nonzero; it represents the magnitude of the velocity (the speed) and corresponds to the fact that the vector $\vec{r}$ is changing direction.

In Figure 2.17, when the planet moves along a circular path from location 1 to location 2 during a short time interval Δt, the position vector rotates from $\vec{r}_1$ to $\vec{r}_2$, rotating through an angle $\Delta \theta$. We're going to calculate the magnitude $|\Delta \vec{r}|$ of the change in position $\Delta \vec{r} = \vec{r}_2 - \vec{r}_1$, where the magnitude of the position vector is constant: $|\vec{r}| = |\vec{r}_1| = |\vec{r}_2|$. This will lead to a general procedure for finding the rate of change of any rotating vector, one whose magnitude is constant but whose direction is changing.

It is a fact of geometry that the length of an arc of a circle is $R\Delta \theta$ when the angle is measured in radians. For example, when the angle is 2π radians, $R\Delta \theta = 2\pi R$, the circumference of a circle (Figure 2.18). Looking at Figure 2.17, we see that we can write $|\Delta \vec{r}| \approx |\vec{r}|\Delta \theta$. This is approximately correct as long as $\Delta \theta$ is small, because in that case the short portion of the triangle is almost the same as the arc of the circle.

Divide by the time interval Δt:

$$\left| \frac{\Delta \vec{r}}{\Delta t} \right| \approx |\vec{r}| \frac{\Delta \theta}{\Delta t}$$

In the limit of letting Δt approach zero, we have

$$\left| \frac{d \vec{r}}{dt} \right| = |\vec{r}| \frac{d\theta}{dt} \text{ when } |\vec{r}| \text{ is constant}$$

This is the magnitude of the rate of change of the position vector. The direction of this rate of change is perpendicular to the position vector, in the direction of motion of the planet. What we have calculated is the orbital speed v of the planet around the star (Figure 2.19).

$\left| \dfrac{d \vec{r}}{dt} \right|$ is the magnitude of the rate of change of position, $|\vec{v}|$ (or just v).

$\dfrac{d\theta}{dt}$ is called the "angular speed" and is denoted by the symbol ω.

The angular speed ω (lowercase Greek omega) has units of radians per second. For example, an angular speed of 2π radians per second means that we go around the circle once (2π radians, 360°) in one second. If the time to go around once is T (the period), we have $\omega = 2\pi / T$.

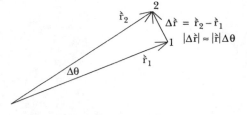

Figure 2.17 Change in position vector for an object moving along a circular path.

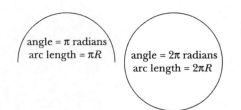

Figure 2.18 Arc length is radius times angle measured in radians.

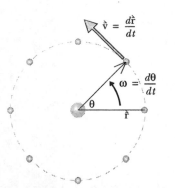

Figure 2.19 The time derivative of a rotating position vector is the velocity: $v = \omega r$.

We have obtained this useful result:

$$\left|\frac{d\hat{\vec{r}}}{dt}\right| = \omega|\hat{\vec{r}}|, \text{ and the direction of } \frac{d\hat{\vec{r}}}{dt} \text{ is perpendicular to } \hat{\vec{r}}$$

This reduces to the result $v = \omega r$ for circular motion at constant speed, but what we have done has much wider applicability. In particular, we can now calculate the rate of change of any rotating vector whose magnitude is constant.

A rotating momentum vector

During one period T the position vector rotates through 2π radians, and during that same period T the momentum vector also rotates through 2π radians. If you place the tail of the momentum vector at the origin (the center of the star), the rotating momentum vector behaves just like the rotating position vector (Figure 2.20). Therefore the same line of argument starting from the diagram in Figure 2.21 leads to the following result:

$$\left|\frac{d\vec{p}}{dt}\right| = \omega|\vec{p}|, \text{ direction of } \frac{d\vec{p}}{dt} \text{ perpendicular to } \vec{p}.$$

This direction, perpendicular to the momentum $\vec{p}$, points from the planet toward the star.

? What is the physical significance of the fact that the direction of $d\vec{p}/dt$ points from the planet toward the star?

The rate of change of the momentum of the planet is equal to the net force acting on the planet, which is the attractive gravitational force exerted by the star. So it makes sense that $d\vec{p}/dt$ points from the planet toward the star.

A rotating velocity vector

The same line of argument for the rotating velocity vector (Figure 2.22) leads to the following result:

$$\left|\frac{d\vec{v}}{dt}\right| = \omega|\vec{v}|, \text{ and the direction of } \frac{d\vec{v}}{dt} \text{ is perpendicular to } \vec{v}.$$

This says that the acceleration (rate of change of velocity) in circular motion at constant speed is $\omega v = \omega(\omega r) = \omega^2 r$. Because $v = \omega r$, we can also write $\omega v = (v/r)v = v^2/r$. The acceleration is perpendicular to the velocity, so it is directed toward the center of the circle, in the same direction as the inward-pointing force that is changing the velocity direction.

From the pattern of these examples we see a general result:

RATE OF CHANGE OF A ROTATING VECTOR

If $\hat{\vec{x}}$ is any rotating vector whose direction changes but not its magnitude (position, velocity, momentum, etc.), then

$$\left|\frac{d\hat{\vec{x}}}{dt}\right| = \omega|\hat{\vec{x}}| \text{ (direction of } \frac{d\hat{\vec{x}}}{dt} \text{ is perpendicular to } \hat{\vec{x}})$$

$$\omega = \frac{d\theta}{dt} = \frac{2\pi}{T} \text{ (}T \text{ is the time to go around once)}$$

We'll mention that when we study angular momentum in Chapter 9 we'll show that angular velocity can usefully be considered to be a vector $\vec{\omega}$ pointing along the axis of rotation, in which case the rate of change of a rotating vector quantity such as momentum $\vec{p}$ can be expressed as a "cross product": $d\vec{p}/dt = \vec{\omega} \times \vec{p}$.

Figure 2.20 The momentum vector rotates through 2π radians in a time T. (The tails of the vectors have been moved to the origin for clarity.)

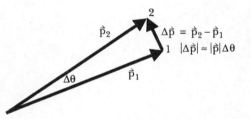

Figure 2.21 Change in momentum during circular motion.

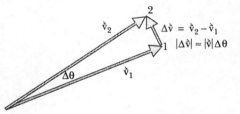

Figure 2.22 Change in velocity during circular motion.

The general case of the rate of change of a vector

A rotating vector is a special case, because its magnitude isn't changing. More generally, both the magnitude and the direction of a vector change with time. In the general case it is useful to think of the rate of change of a vector $\vec{x}$ in terms of two components, radial (perpendicular to the vector, representing a rate of change of direction as we have calculated here) and tangential (parallel to the vector, representing a rate of change of magnitude, as in the case where the speed is changing).

Ex. 2.10 A merry-go-round takes 10 seconds to go around once. What is the angular speed ω? What are the units of ω?

Ex. 2.11 If the radius of this merry-go-round is 8 meters, what is the speed at the outer rim, based on $|d\vec{r}/dt| = \omega|\vec{r}|$? What is the direction of the velocity?

Ex. 2.12 What is the magnitude of the acceleration at the outer rim of this merry-go-round, based on $|d\vec{v}/dt| = \omega|\vec{v}|$? What is the direction of the acceleration?

2.8.1 Circular planetary orbits

For a planet of mass m moving around a star of mass M in a circular orbit at constant speed, we can write

$$\left|\frac{d\vec{p}}{dt}\right| = |\vec{F}_{net}| = G\frac{Mm}{r^2}$$

$$\left|\frac{d\vec{p}}{dt}\right| = \omega p = G\frac{Mm}{r^2}, \text{ where } \omega = \frac{d\theta}{dt} = \frac{2\pi}{T}$$

The period T is the time it takes the planet to go around its star once, a distance of $2\pi r$. So we have $vT = 2\pi r$ which makes it possible to predict a relationship between the period T of a planet and its distance r from the Sun. This relationship was first discovered empirically by Johannes Kepler, who lived before Newton. Kepler produced his formula by fitting mathematical curves to accurate observations of planetary orbits made by Tycho Brahe (without telescopes, which hadn't been invented yet). The later explanation of Kepler's equation in terms of dynamical laws was a triumph of the Newtonian approach and is the subject of the following problem.

Problem 2.3 Relating radius to period for circular orbits

The planets in our Solar System have orbits around the Sun that are nearly circular, and $v \ll c$. Calculate the period T (a "year"—the time required to go around the Sun once) for a planet whose orbit radius is r. This is the relationship discovered by Kepler and explained by Newton. (It can be shown by advanced techniques that this result also applies to elliptical orbits if you replace r by the "semimajor axis," which is half the longer, "major" axis of the ellipse.) Use this analytical solution for circular motion to predict the Earth's orbital speed, using the data for Sun and Earth on page 50.

Newtons laws nodal ways dra

2.9 Predicting the future of gravitating systems

We can use the Newtonian method and numerical integration to try to predict the future of a group of objects that interact with each other mainly gravitationally, such as the Solar System consisting of the Sun, planets, moons, asteroids, comets, etc. There are other, non-gravitational interactions present. Radiation pressure from sunlight makes a comet's tail sweep away from the sun. Streams of charged particles (the "solar wind") from the Sun hit the Earth and contribute to the Northern and Southern Lights (auroras). But the main interactions within the Solar System are gravitational.

The Solar System as a group is orbiting around the center of our Milky Way galaxy, and our galaxy is interacting gravitationally with other galaxies in a local cluster of galaxies (Figure 2.23), but changes in these speeds and directions take place very slowly compared to those within the solar system. To a very good approximation we can neglect these interactions while studying the inner workings of the Solar System itself.

We have modeled the motion of planets and stars using the momentum principle and the universal law of gravitation. We can use the same approach to model the gravitational interactions of more complex systems involving three, four, or any number of gravitationally interacting objects. All that is necessary is to include the forces associated with all pairs of objects.

In principle we could use these techniques to predict the future of our entire Solar System. Of course our prediction would be only approximate, because we take finite time steps which introduce numerical errors, and we are neglecting various small interactions. We would need to investigate whether the prediction of our model is a reasonable approximation to the real motion of our real Solar System.

Our Solar System contains a huge number of objects. Most of the nine planets have moons, and Jupiter and Saturn have many moons. There are thousands of asteroids and comets, not to mention an uncounted number of tiny specks of dust. This raises a practical limitation: the fastest computers would take a very long time to carry out even one time step for all the pieces of the Solar System.

2.9.1 The three-body problem

As we have seen, in order to carry out a numerical integration we need to know the position and velocity of each object at some time t, from which we can calculate the forces on each object at that time. This enables us to calculate the position and momentum for each object at a slightly later time $t + \Delta t$. We need the net force on an object, which we get from Newton's universal law of gravitation and the superposition principle.

Suppose that there are three gravitating objects, as shown in Figure 2.24. The force on object m_2, which is acted upon gravitationally by objects m_1 and m_3, is a natural extension of what we did earlier for the analysis of the star and planet. We simply use the superposition principle to include the interaction with m_3 as well as the interaction with m_1:

$$\vec{F}_{\text{on } m_2} = \frac{Gm_2m_1}{r_{21}^2}\hat{r}_{21} + \frac{Gm_2m_3}{r_{23}^2}\hat{r}_{23}$$

We use this force to update the momentum of object m_2. Similar computations apply to the other masses, m_1 and m_3. These calculations are tedious, but let the computer evaluate them! We can carry out a numerical integration for each mass, because given the current locations of all the masses we can calculate the vector net force exerted on each mass at a time t, and then we can calculate the new values of momentum and position at a slightly later time $t + \Delta t$.

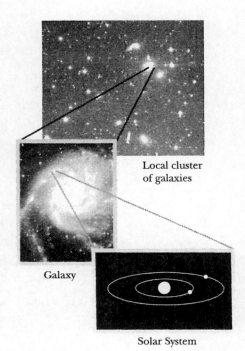

Local cluster
of galaxies

Galaxy

Solar System

Figure 2.23 Our Solar System orbits around the center of our galaxy, which interacts with other galaxies.

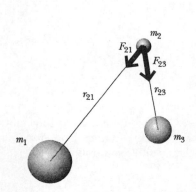

Figure 2.24 In the "three body problem" each of the three objects interacts with two other objects.

While there exist analytical (non-numerical) solutions for two-body gravitational motion, except for some very special cases the "three-body problem" has not been solved analytically. However, in a numerical integration,
the computer doesn't mind that there are lots of quantities and a lot of repetitive calculations to be done. You yourself can solve the three-body problem numerically, by adding additional calculations to your two-body
computations (see Problem 2.6 The three-body problem).

You have seen that for a single object orbiting a very massive object, or for
two objects orbiting each other (with zero net momentum), the possible trajectories are only a circle, ellipse, parabola, hyperbola, or straight line. The
motion of a three-body system can be vastly more complex and diverse. Figure 2.25 shows the numerical integration of a complicated three-body trajectory for one particular set of initial conditions. Adding just one more
object opens up a vast range of complex behaviors. Imagine how complex
the motion of a galaxy can be!

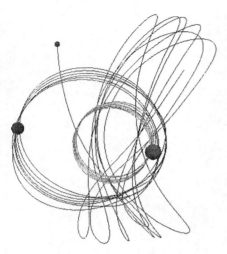

Figure 2.25 One example of three bodies
interacting gravitationally.

2.9.2 Sensitivity to initial conditions

For a two-body system, slight changes in the initial conditions make only
slight changes in the orbit, such as changing a circle into an ellipse, or an
ellipse into a slightly different ellipse, but you never get anything other than
a simple trajectory.

The situation is very different in systems with three or more interacting
objects. Consider a low-mass object orbiting two massive objects which we
imagine somehow to be nailed down so that they cannot move (Figure
2.26). This could also represent two positive charges that are fixed in position, with a low-mass negative charge orbiting around them.

Figure 2.26 A low mass object orbiting two
massive objects.

In Figure 2.27 are two very different orbits (one shown in black, the other
in gray), starting from only slightly different initial conditions. In both of
these cases, the low-mass object was released from the rest, but at slightly different initial locations. The trajectories are wildly different! (The mass of
the object on the left is twice the mass of the object on the right in this computation.)

This sensitivity to initial conditions generally becomes more extreme as
you add more objects, and the Solar System contains lots of objects, big and
small. Also, we can anticipate that if small errors in specifying the initial conditions can have large effects, so too it is likely that failing to take into account the tiny force exerted by a small asteroid might make a big difference
after a long integration time.

The Solar System is actually fairly predictable, because there is one giant
mass (the Sun), and the other, much smaller masses are very far apart. This
is unlike the example shown here, which deliberately emphasizes the sensitivity observed when there are large masses near each other.

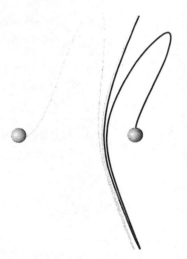

Figure 2.27 Two different orbits (one gray,
the other black), starting from rest but
from slightly different positions.

2.10 Determinism

If you know the net force on an object as a function of its position and momentum (or velocity), you can predict the future motion of the object by
simple step-by-step calculations based on the momentum principle. Newton's demonstration of the power of this approach induced many seven-
teenth and eighteenth century philosophers and scientists to adopt the view
already proposed by Descartes, Boyle, and others, that the Universe is a giant clockwork device, whose future is completely determined by the present
positions and momenta of all the macroscopic and microscopic objects in
it. Just turn the crank, and predict the future! This point of view is called
"determinism," and taken to its extreme it raises the question of whether humans actually have any free will, or whether all of our actions are predeter-

mined. Scientific and technological advances in the twentieth century, however, have led us to see that although the Newtonian approach can be used to predict the long-term future of a simple system or the short-term future of a complicated system, there are both practical and theoretical limitations on predictions in some systems.

2.10.1 Practical limitations

One reason we may be unable to predict the long-term future of even a simple system is a practical limitation in our ability to measure its initial conditions with sufficient accuracy. Over time, even small inaccuracies in initial conditions can lead to large cumulative errors in calculations. This is a practical limitation rather than a theoretical one, because in principle if we could measure initial conditions more accurately, we could perform more accurate calculations.

Another practical limitation is our inability to account for all interactions in our model. Every object in the Universe interacts with every other object. In constructing simplified models, we ignore interactions whose magnitude is extremely small. However, over time, even very small interactions can lead to significant effects. Even the tiny "radiation pressure" exerted by sunlight has been shown to affect noticeably the motion of an asteroid over a long time. With larger and faster computers, we can include more and more interactions in our models, but our models can never completely reflect the complexity of the real world.

A branch of current research that focuses on how the detailed interactions of a large number of atoms or molecules lead to the bulk properties of matter is called "molecular dynamics." For example, one way to test our understanding of the nature of the interactions of water molecules is to create a computational model in which the forces between molecules and the initial state of a large number of molecules—several million—are described, and then to see if running the model produces a virtual fluid that actually behaves like water. However, even with an accurate force law for the (electric) interactions between atoms or molecules, it is not feasible to compute and track the motions of the 10^{25} molecules in a glass of water. Significant efforts are underway to build faster and faster computers, and to develop more efficient computer algorithms, to permit realistic numerical integration of more complex systems.

2.10.2 Chaos

A second kind of limitation occurs in systems that display an extreme sensitivity to initial conditions. We saw something like this in our experimentation with the three-body problem in gravitating systems—small changes in initial conditions produced large changes in behavior. In recent years scientists have discovered systems in which a change in the initial conditions, no matter how small (infinitesimal), can lead to complete loss of predictability: the difference in the two possible future motions of the system diverges exponentially with time. Such systems are called "chaotic." It is thought that over a long time period the weather may be literally unpredictable in this sense.

An interesting popular science book about this new field of research is *Chaos: Making a New Science,* by James Gleick (Penguin 1988). The issue of small changes in initial conditions is a perennial concern of time travellers in science fiction stories. Perhaps the most famous such story is Ray Bradbury's "The Sound of Thunder," in which a time traveller steps on a butterfly during the Jurassic era, and returns to an eerily changed present time.

2.10.3 Breakdown of Newton's laws

There are other situations of high interest where we cannot usefully apply Newton's laws of motion, because these classical laws do not adequately describe the behavior of physical systems. To model systems composed of very small particles such as protons and electrons, quarks, and photons, it is necessary to use the laws of quantum mechanics. To model in detail the gravitational interactions between massive objects, it is necessary to apply the principles of general relativity. Given these principles, one might assume that it would be possible to follow a procedure similar to the one we have followed in this chapter, and predict in detail the future of these systems. However, there appear to be more fundamental limitations on what we can know about the future.

2.10.4 Probability and uncertainty

Our understanding of the atomic world of quantum mechanics suggests that there are fundamental limits to our ability to predict the future, because at the atomic and subatomic levels the Universe itself is non-deterministic. We cannot know exactly what will happen at a given time, but only the probability that certain events will occur within a given time frame.

A simple example may make this clear. Earlier we mentioned that a free neutron (one not bound into a nucleus) is unstable and eventually decays into a proton, electron, and anti-neutrino:

$$n \rightarrow p^+ + e^- + \bar{\nu}$$

Figure 2.28 The electron may continue to travel in a straight line, but we cannot be certain that it will, because we do not know when the neutron will decay.

The average lifetime of the free neutron is about 15 minutes. Some neutrons survive longer than this, and some last a shorter amount of time. All of our experiments and all of our theory are consistent with the notion that *it is not possible to predict when a particular neutron will decay*. There is only a (known) probability that the neutron will decay in the next microsecond, or the next, or the next.

If this is really how the Universe works, then there is an irreducible lack of predictability and determinism of the Universe itself. As far as our own predictions using the momentum principle are concerned, consider the following simple scenario. An electron is traveling with constant velocity through nearly empty space, but there is a free neutron in the vicinity (Figure 2.28).

We might predict that the electron will move in a straight line, and some fraction of the time we will turn out to be right (the electrically uncharged neutron exerts no electric force on the electron, there is a magnetic interaction but it is quite small, and the gravitational interaction is tiny). But if the neutron happens (probabilistically) to decay just as our electron passes nearby, suddenly our electron is subjected to large electric forces due to the decay proton and electron, and will deviate from a straight path. Moreover, the directions in which the decay products move are also only probabilistic, so we can't even predict whether the proton or the electron will come closest to our electron (Figure 2.29).

This is a simple but dramatic example of how quantum indeterminacy can lead to indeterminacy even in the context of the momentum principle.

The Heisenberg uncertainty principle

The Heisenberg uncertainty principle states that there are actual theoretical limits to our knowledge of the state of physical systems. One quantitative formulation states that the position and the momentum of a particle cannot both be simultaneously measured exactly:

$$\Delta x \Delta p \geq h$$

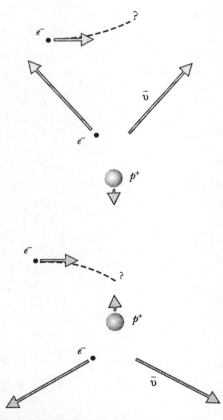

Figure 2.29 If the neutron does decay when the electron is near it, the trajectory of the electron will depend on the directions of the momenta of the decay products.

This relation says that the product of the uncertainty in position Δx and the uncertainty in momentum Δp is equal to a constant h, called Planck's constant. Planck's constant is tiny ($h = 6.6 \times 10^{-34}$ kg·m^2/s), so this limitation is not noticeable for macroscopic systems, but for objects small enough to require a quantum mechanical description, the uncertainty principle puts fundamental limits on how accurately we can know the initial conditions, and therefore how well we can predict the future.

2.11 Why do we call them "Newton's" laws?

You might ask why physicists refer to "Newton's" laws of motion. One could argue that these physical laws are embedded in the structure of the Universe and are not "owned" by anyone, not even Newton, even if he was the first to write them down. However, Newton devised a particular explanatory scheme in which the analysis is divided into two distinct parts:

1) Quantify the interaction in terms of a concept called "force." A specific example is Newton's universal law of gravitation.
2) Quantify the change of motion in terms of the change in a quantity called "momentum." The change in the momentum is equal to the force times Δt.

This scheme, called the "Newtonian synthesis," has turned out to be extraordinarily successful in explaining a huge variety of diverse physical phenomena, from the fall of an apple to the orbiting of the Moon. Yet we have no way of asking whether the Universe "really" works this way. It seems unlikely that the Universe actually uses the human concepts of "force" and "momentum" in the unfolding motion of an apple or the Moon.

In fact, people have invented alternative conceptual frameworks. In the twentieth century Albert Einstein developed a radically different way of analyzing the motion of the apple and the Moon. In Einstein's general theory of relativity, the Earth's huge mass alters space itself in the neighborhood of the Earth in such a way that the apple and the Moon follow "natural" paths, without any Newtonian "forces."

For many gravitational phenomena the analyses of Newton and Einstein predict nearly the same results, and Newton's formulation is easier to use where it is adequately accurate for the purpose. However, Einstein's theory correctly predicts some tiny corrections to Newtonian analyses of certain gravitational phenomena (for example, the orbit of the planet Mercury around the Sun), and it also correctly predicts some bizarre phenomena undreamed of in Newton's conception, such as "black holes," which are collapsed stars whose gravitational field is so strong that light cannot escape from them (see Problem 2.17).

We refer to "Newton's laws" both to identify and honor the particular, highly successful analysis scheme introduced by Newton, but also to remind ourselves that this is not the only possible way to view and analyze the universe. Although the range of validity of Newton's laws is large, in many interesting and important situations involving very small objects or very large masses, Newton's laws are not valid.

2.12 Summary

Fundamental Physical Principles

- The gravitational force law for "point" masses:

$$F_g = G\frac{m_1 m_2}{r^2}$$

where $G = 6.7 \times 10^{-11} \text{ N·m}^2/\text{kg}^2$

- Reciprocity (Newton's third law of motion): $\vec{F}_{\text{on 1 by 2}} = -\vec{F}_{\text{on 2 by 1}}$

- The superposition principle: the effective force on an object is the "net" force, the vector sum of all forces acting on the object. Each individual interaction is unaffected by the presence of other interacting objects.

- For multiparticle systems $\dfrac{d\vec{P}_{\text{tot}}}{dt} = \vec{F}_{\text{net,ext}}$

- Momentum is conserved: if one object gains momentum, other objects lose momentum.

New concepts

Computation of trajectories of moving objects by repeated application of the momentum principle:

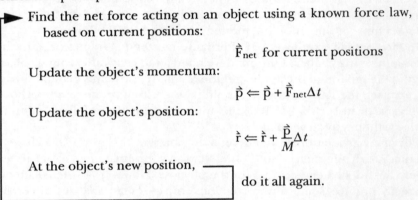

Find the net force acting on an object using a known force law, based on current positions:

$$\vec{F}_{\text{net}} \text{ for current positions}$$

Update the object's momentum:

$$\vec{p} \Leftarrow \vec{p} + \vec{F}_{\text{net}}\Delta t$$

Update the object's position:

$$\vec{r} \Leftarrow \vec{r} + \frac{\vec{p}}{M}\Delta t$$

At the object's new position, ——————

do it all again.

Initial conditions: starting values of quantities involved in a numerical calculation, such as position and momentum.

A physics "model": a simplified, approximate version of the complex real-world situation, suitable for analysis to obtain useful though approximate results.

Issues

Is the Universe deterministic? There are practical and theoretical limitations to our ability to model the Universe. Predicting the future is limited by

- neglecting small effects (imperfect model of the situation)
- computational errors due to finite Δt in a numerical integration
- inability to know initial conditions exactly
- sensitivity in complex situations to initial conditions
- quantum mechanics (probabilistic behavior)

Results

Numerical integration of a planet around a stationary star yields trajectories that are a circle, ellipse, parabola, hyperbola, or straight line.

RATE OF CHANGE OF A ROTATING VECTOR

If $\vec{x}$ is any rotating vector (position, velocity, momentum, etc.), then

$$\left|\frac{d\vec{x}}{dt}\right| = \omega|\vec{x}| \quad \text{(direction of } \frac{d\vec{x}}{dt} \text{ is perpendicular to } \vec{x}\text{)}$$

$$\omega = \frac{d\theta}{dt} = \frac{2\pi}{T} \quad (T \text{ is the time to go around once})$$

Examples:

$$\left|\frac{d\vec{r}}{dt}\right| = \omega|\vec{r}|, \text{ or } v = \omega r$$

$$\left|\frac{d\vec{v}}{dt}\right| = \omega|\vec{v}|, \text{ or } a = \omega v$$

$$\left|\frac{d\vec{p}}{dt}\right| = \omega|\vec{p}|$$

Useful data (also on inside back cover of this volume)

mass of Sun = 2×10^{30} kg   Earth-Sun distance = 1.5×10^{11} m

mass of Earth $\approx 6\times10^{24}$ kg   radius of Earth = 6.4×10^{6} m = 6400 km

mass of Moon $\approx 7\times10^{22}$ kg   radius of Moon = 1.75×10^{6} m

distance from Earth to Moon $\approx 4\times10^{8}$ m (400,000 km, center to center)

2.13 Example problem: A proton and an electric dipole

An electric "dipole" consists of equal and opposite charges very close to each other. Electric force is proportional to $1/r^2$, like gravitational force, but because of the near cancellation of the effects of the plus and minus charges, the electric force that the dipole exerts on a distant single charge is to a good approximation proportional to $1/r^3$ rather than $1/r^2$.

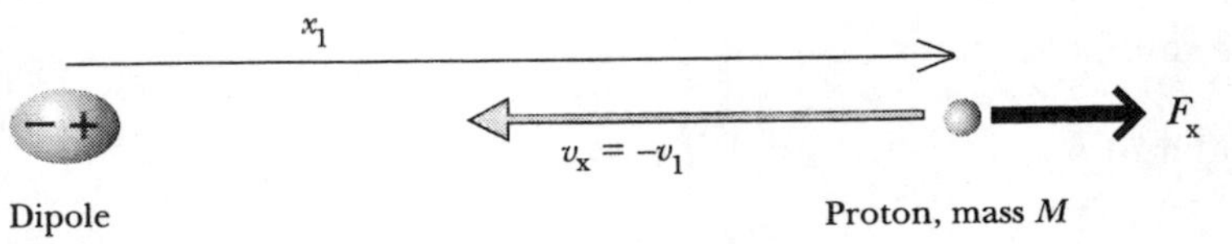

At a particular instant, as shown in Figure 2.30, a proton of mass M is located at a position x_1 to the right of a dipole and is moving to the left toward the dipole with x component of velocity $v_x = -v_1$, where v_1 is a positive quantity. The electric force on the proton is to a good approximation $F_x = K/x^3$ to the right, where K is a known positive constant.

Figure 2.30 A proton interacts with an electric dipole.

(a) After a short time Δt, short enough that the proton has moved only a small distance, what is the approximate x component of velocity? What is the approximate position of the proton?

(b) On a similar diagram, draw the proton at its new position. Show its new velocity, and the net force acting on it. Use a correct relative scale for the vectors, so that if the magnitude of a vector has increased, the new vector should be longer.

Solution

Start from a fundamental principle:

(a) The fundamental principle here is the momentum principle:

$$p_{1x} = -Mv_1 \qquad \text{initial } x \text{ component of momentum}$$

$$F_x = K/x_1^3 \qquad \text{initial force on proton}$$

$$p_{2x} = p_{1x} + F_x\Delta t \qquad \text{momentum principle; update } p_x$$

$$p_{2x} = -Mv_1 + (K/x_1^3)\Delta t \qquad \text{assemble the pieces}$$

$$v_{2x} = \frac{p_{2x}}{M} = -v_1 + \left(\frac{K}{Mx_1^3}\right)\Delta t \qquad \text{assuming } v \ll c$$

$$x_2 = x_1 + v_{2x}\Delta t = x_1 + \left[-v_1 + \left(\frac{K}{Mx_1^3}\right)\Delta t\right]\Delta t \qquad \text{update } x$$

These results are approximate, because the force and momentum (and velocity) are changing slightly during the short time interval. Although there are numerical techniques for making better approximations for the same time step Δt, in our computer modeling it will be sufficient simply to use very small values of Δt.

(b) In a very short time Δt the proton will have moved slightly to the left, its speed will have decreased slightly, and the force will have increased slightly:

We could have used full vector notation for the analysis. For example, the momentum update looks like this:

$$\langle p_{2x}, 0, 0\rangle = \langle -Mv_1, 0, 0\rangle + \langle(K/x_1^3), 0, 0\rangle\Delta t$$

2.14 Review questions

These questions are intended to help you *review* the most important concepts in this chapter *after* you have studied the material thoroughly. For maximum usefulness in learning the material, do not flip through the chapter for answers but use only the information on the preceding Summary pages.

The gravitational force law

RQ 2.1 Calculate the approximate gravitational force exerted by the Earth on a human standing on the Earth's surface. Compare with the approximate gravitational force of a human on another human at a distance of 3 meters. What approximations or simplifying assumptions must you make? (The mass of the Earth is 6×10^{24} kg and its radius is 6400 km.)

Reciprocity (Newton's third law of motion)

RQ 2.2 The windshield of a speeding car hits a hovering insect. Compare the magnitude of the force that the car exerts on the bug to the force that the bug exerts on the car. Which is bigger? Compare the magnitude of the change of momentum of the bug to that of the car. Which is bigger? Compare the magnitude of the change of velocity of the bug to that of the car. Which is bigger? Explain briefly. (Note: the interatomic forces between bug and windshield are electric forces.)

Voyage to the Moon

RQ 2.3 Newspaper reports on the space program sometimes talk about "getting free of the Earth's gravity." What's wrong with this description?

The superposition principle

RQ 2.4 In order to pull a sled across a level field at constant velocity you have to exert a constant force. Doesn't this violate Newton's first and second laws of motion, which imply that no force is required to maintain a constant velocity? Explain this seeming contradiction.

Orbits

RQ 2.5 Without looking up the relevant derivation, write the x and y components of the gravitational force exerted by the Earth on the Moon.

Circular motion

RQ 2.6 A 30 kg child rides on a playground merry-go-round, 2 m from the center. The merry-go-round makes one complete revolution every 4 seconds. How large is the net force on the child? In what direction does the net force act?

Determinism

RQ 2.7 List as many limitations as you can on our ability to predict the future.

2.15 Problems

2.15.1 In-line problems from this chapter

Problem 2.1 (page 51) Planetary orbits
In this problem you will study the motion of a planet around a star. To start with a somewhat familiar situation, you will begin by modeling the motion of our Earth around our Sun. Write answers to questions either as comments in your program or on paper, as specified by your instructor.

Planning

(a) The Earth goes around the Sun in a nearly circular orbit, taking one year to go around. Using data on page 50, what initial speed should you give the Earth in a computer model, so that a circular orbit should result? (If your computer model does produce a circular orbit, you have a strong indication that your program is working properly.)

(b) Estimate an appropriate value for Δt to use in your computer model. Remember that t is in seconds, and consider your answer to part (a) in making this estimate. If your Δt is too small, your calculation will require many tedious steps. Explain briefly how you decided on an appropriate step size.

Circular orbit

(c) Starting with speed calculated in (a), and with the initial velocity perpendicular to a line connecting the Sun and the Earth, calculate and display the trajectory of the Earth. Display the whole trail, so you can see whether you have a closed orbit (that is, whether the Earth returns to its starting point each time around). Include the Sun's position $\vec{r}_s = <x_s, y_s, z_s>$ in your computations, even if you set its coordinates to zero. This will make it easier to modify the computation to let the Sun move (Problem 2.2 Binary stars).

Computational accuracy

(d) As a check on your computation, is the orbit a circle as expected? Run the calculation until the accumulated time t is the length of a year: does this take the Earth around the orbit once, as expected? What is the largest step size you can use that still gives a circular orbit and the correct length of a year? What happens if you use a much larger step size?

Noncircular trajectories

(e) Set the initial speed to 1.2 times the Earth's actual speed. Make sure the step size is small enough that the orbit is nearly unaffected if you cut the step size. What kind of orbit do you get? What step size do you need to use to get results you can trust? What happens if you use a much larger step size? Produce at least one other qualitatively different noncircular trajectory for the Earth. What difficulties would humans have in surviving on the Earth if it had a highly noncircular orbit?

Force and momentum

(f) Choose initial conditions that give a noncircular orbit. Continuously display a vector showing the momentum of the Earth (with its tail at the Earth's position), and a different colored vector showing the force on the Earth by the Sun (with its tail at the Earth's position). You must scale the vectors appropriately to fit into the scene. A way to figure out what scale factor to use is to print the numerical values of momentum and force, and compare with the scale of the scene. For example, if the width of the scene is W meters, and the force vector has a typical magnitude F (in newtons), you might scale the force vector by a factor $0.1W/F$, which would make the length of the vector be one-tenth the width of the scene. Is the force in the same direction as the momentum? How does the momentum depend on the distance from the star?

Problem 2.2 (page 52) Binary stars

About a half of the visible "stars" are actually systems consisting of two stars orbiting each other, called "binary stars." In your computer model of the Earth and Sun (Problem 2.1 Planetary orbits), replace the Earth with a star whose mass is half the mass of our Sun, and take into account the gravitational effects that the second star has on the Sun.

(a) Give the second star the speed of the actual Earth, and give the Sun zero initial momentum. What happens? Try a variety of other initial conditions. What kinds of orbits do you find?

(b) One of the things to try is to give the Sun the same magnitude of momentum as the second star, but in the opposite direction. If the initial momentum of the second star is in the $+y$ direction, give the Sun the same magnitude of momentum but in the $-y$ direction, so that the total momentum of the binary star system is initially zero, but the stars are not headed toward each other. What is special about the motion you observe in this case?

Problem 2.3 (page 56) Relating radius to period for circular orbits

The planets in our Solar System have orbits around the Sun that are nearly circular, and $v \ll c$. Calculate the period T (a "year"—the time required to go around the Sun once) for a planet whose orbit radius is r. This is the relationship discovered by Kepler and explained by Newton. (It can be shown by advanced techniques that this result also applies to elliptical orbits if you replace r by the "semimajor axis," which is half the longer, "major" axis of the ellipse.) Use this analytical solution for circular motion to predict the Earth's orbital speed, using the data for Sun and Earth on page 50.

2.15.2 Additional problems

Problem 2.4 The effect of the Moon and Venus on the Earth

In Problem 2.1 you analyzed a simple model of the Earth orbiting the Sun, in which there were no other planets or moons. Venus and the Earth have similar size and mass. At its closest approach to the Earth, Venus is about 40 million kilometers away (4×10^{10} m). The Moon's mass is about 7×10^{22} kg, and the distance from Earth to Moon is about 4×10^8 m (400,000 km, center to center).

(a) Calculate the ratio of the gravitational forces on the Earth exerted by Venus and the Sun. Is it a good approximation to ignore the effect of Venus when modeling the motion of the Earth around the Sun?

(b) Calculate the ratio of the gravitational forces on the Earth exerted by the Moon and the Sun. Is it a good approximation to ignore the effect of the Moon when modeling the motion of the Earth around the Sun?

Problem 2.5 Other force laws

Modify your orbit computation to use a different force law, such as a force that is proportional to $1/r$ or $1/r^3$, or a constant force, or a force proportional to r (this represents the force of a spring whose relaxed length is nearly zero). How do orbits with these force laws differ from the circles and ellipses that result from a $1/r^2$ law? If you want to keep the magnitude of the force roughly the same as before, you will need to adjust the force constant G.

Problem 2.6 The three-body problem

Carry out a numerical integration of the motion of a three-body gravitational system and plot the trajectory, leaving trails behind the objects. Calculate all of the forces before using these forces to update the momenta and positions of the objects. Otherwise the calculations of gravitational forces would mix positions corresponding to different times.

Try different initial positions and initial momenta. Find at least one set of initial conditions that produces a long-lasting orbit, one set of initial conditions that results in a collision with a massive object, and one set of initial conditions that allows one of the objects to wander off without returning. Report the masses and initial conditions that you used.

Problem 2.7 The Ranger 7 mission to the Moon

The first U.S. spacecraft to photograph the Moon close up was the unmanned "Ranger 7" photographic mission in 1964. The spacecraft, shown in the illustration at the right (NASA photograph), contained television cameras that transmitted close-up pictures of the Moon back to Earth as the spacecraft approached the Moon. The spacecraft did not have retro-rockets to slow itself down, and it eventually simply crashed onto the Moon's surface, transmitting its last photos immediately before impact.

Figure 2.31 is the first image of the Moon taken by a U.S. spacecraft, Ranger 7, on July 31, 1964, about 17 minutes before impacting the lunar surface. The large crater at center right is Alphonsus (108 km diameter); above it (and to the right) is Ptolemaeus and below it is Arzachel. The Ranger 7 impact site is off the frame, to the left of the upper left corner.

You can find out more about the actual Ranger lunar missions at
http://nssdc.gsfc.nasa.gov/planetary/lunar/ranger.html

Figure 2.31 The first Ranger 7 photo of the Moon (NASA photograph).

To send a spacecraft to the Moon, we put it on top of a large rocket containing lots of rocket fuel and fire it upward. At first the huge ship moves quite slowly, but the speed increases rapidly. When the "first-stage" portion of the rocket has exhausted its fuel and is empty, it is discarded and falls back to Earth. By discarding an empty rocket stage we decrease the amount of mass that must be accelerated to even higher speeds. There may be several stages that operate for a while and then are discarded before the spacecraft has risen above most of Earth's atmosphere (about 50 km, say, above the Earth), and has acquired a high speed. At that point all the fuel available for this mission has been used up, and the spacecraft simply coasts toward the Moon through the vacuum of space.

We will model the Ranger 7 mission. Starting 50 km above the Earth's surface ($5{\times}10^4$ m), a spacecraft coasts toward the Moon with an initial speed of about 10^4 m/s. Here are data we will need:

mass of spacecraft = 173 kg   mass of Earth $\approx 6{\times}10^{24}$ kg

mass of Moon $\approx 7{\times}10^{22}$ kg   radius of Moon = $1.75{\times}10^6$ m
distance from Earth to Moon $\approx 4{\times}10^8$ m (400,000 km, center to center)

We're going to ignore the Sun in a simplified model even though it exerts a sizable gravitational force. We're expecting the Moon mission to take only a few days, during which time the Earth (and Moon) move in a nearly straight line with respect to the Sun, because it takes 365 days to go all the way around the Sun. We take a reference frame fixed to the Earth as representing (approximately) an inertial frame of reference with respect to which we can use the momentum principle.

For a simple model, make the Earth and Moon be fixed in space during the mission. Factors that would certainly influence the path of the spacecraft include the motion of the Moon around the Earth, and the motion of the Earth around the Sun. In addition, the Sun and other planets exert gravitational forces on the spacecraft. As a separate project you might like to include some of these additional factors.

(a) Compute the path of the spacecraft, and display it either with a graph or with an animated image. Here, and in remaining parts of the problem, report the step size Δt that gives accurate results (that is, cutting this step size has little effect on the results).

(b) By trying various initial speeds, determine the approximate *minimum* launch speed needed to reach the Moon, to two significant figures (this is the speed that the spacecraft obtains from the multistage rocket, at the time of release above the Earth's atmosphere). What happens if the launch speed is less than this minimum value? (Be sure to check the step size issue.)

(c) Use a launch speed 10% larger than the approximate minimum value found in part (b). How long does it take to go to the Moon, in hours or days? (Be sure to check the step size issue.)

(d) What is the "impact speed" of the spacecraft (its speed just before it hits the Moon's surface)? Make sure that your spacecraft crashes on the surface of the Moon, not at the Moon's center! (Be sure to check the step size issue.)

You may have noticed that you don't actually need to know the mass m of the spacecraft in order to carry out the computation. The gravitational force is proportional to m, and the momentum is also proportional to m, so m cancels. However, nongravitational forces such as electric forces are not proportional to the mass, and there is no cancellation in that case. We kept the mass m in the analysis in order to illustrate a general technique for predicting motion, no matter what kind of force, gravitational or not.

Problem 2.8 The effect of Venus on the Moon voyage

In the Moon voyage analysis, you used a simplified model in which you neglected among other things the effect of Venus. An important aspect of physical modeling is making estimates of how large the neglected effects might be. If we take Venus into account, make a rough estimate of whether the spacecraft will miss the Moon entirely. How large a sideways deflection of the crash site will there be? Explain your reasoning and approximations.

Venus and the Earth have similar size and mass. At its closest approach to the Earth, Venus is about 40 million kilometers away (4×10^{10} m). In the real world, Venus would attract the Earth and the Moon as well as the spacecraft, but to get an idea of the size of the effects, pretend that the Earth, the Moon, and Venus are all fixed in position (Figure 2.32) and just investigate Venus's effect on the spacecraft.

Venus ⊛   Moon · ⊕ Earth

Figure 2.32 The relative positions of Venus, Earth, and Moon in Problem 2.8.

Problem 2.9 Book and table

Consider an ordinary book (perhaps a physics textbook) lying on an ordinary table. Calculate a very rough, approximate value for the gravitational force that the table and the book exert on each other. Explicitly list all quantities that you had to estimate, and all simplifications and approximations you had to make to do this calculation. Compare your result to the gravitational force on the book by the Earth.

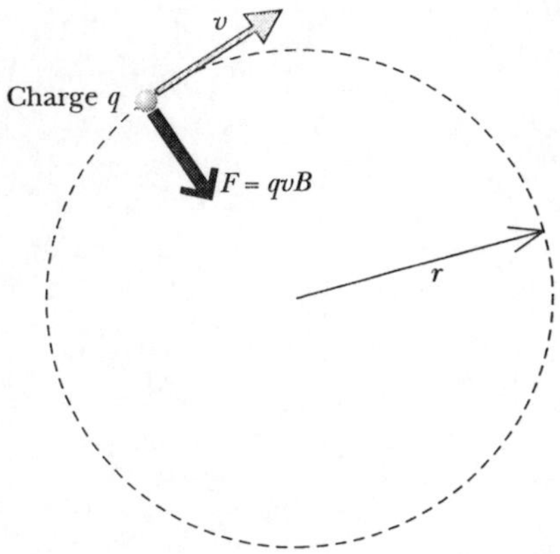

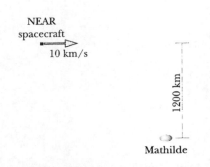

Figure 2.33 A charge moving in a circle due to a magnetic force (Problem 2.10).

Problem 2.10 Circular motion in a magnetic field

When a particle with electric charge q moves with speed v in a plane perpendicular to a magnetic field B, there is a magnetic force at right angles to the motion with magnitude qvB, and the particle moves in a circle of radius r (Figure 2.33). This formula for the magnetic force is correct even if the speed is comparable to the speed of light. Show that

$$p = \frac{mv}{\sqrt{1 - v^2/c^2}} = qBr$$

even if v is comparable to c. Remember that $\vec{F} = m\vec{a}$ is *not* valid for high speeds.

This result is used to measure relativistic momentum: if the charge q is known, we can determine the momentum of a particle by observing the radius of a circular trajectory in a known magnetic field.

Problem 2.11 Determining the mass of an asteroid

In June 1997 the NEAR spacecraft ("Near Earth Asteroid Rendezvous"; see http://near.jhuapl.edu/), on its way to photograph the asteroid Eros, passed within 1200 km of asteroid Mathilde at a speed of 10 km/s relative to the asteroid (Figure 2.34). From photos transmitted by the 805 kg spacecraft, Mathilde's size was known to be about 70 km by 50 km by 50 km. It is presumably made of rock. Rocks on Earth have a density of about 3000 kg/m^3 (3 grams/cm^3).

(a) Make a rough diagram to show qualitatively the effect on the spacecraft of this encounter with Mathilde. Explain your reasoning.

(b) Make a very rough estimate of the change in momentum of the spacecraft that would result from encountering Mathilde. Explain how you made your estimate.

(c) Using your result from part (b), make a rough estimate of how far off course the spacecraft would be, one day after the encounter.

(d) From actual observations of the location of the spacecraft one day after encountering Mathilde, scientists concluded that Mathilde is a loose arrangement of rocks, with lots of empty space inside. What was it about the observations that must have led them to this conclusion?

Experimental background: The position was tracked by very accurate measurements of the time that it takes for a radio signal to go from Earth to the spacecraft followed immediately by a radio response from the spacecraft being sent back to Earth. Radio signals, like light, travel at a speed of 3×10^8 m/s, so the time measurements had to be accurate to a few nanoseconds (1 ns $= 10^{-9}$ s).

Problem 2.12 The SLAC two-mile accelerator

SLAC, the Stanford Linear Accelerator Center, located at Stanford University in Palo Alto, California, accelerates electrons through a vacuum tube two miles long (it can be seen from an overpass of the Junipero Serra freeway that goes right over the accelerator). Electrons which are initially at rest are subjected to a continuous force of 2×10^{-12} newton along the entire length of two miles (one mile is 1.6 kilometers) and reach speeds very near the speed of light. Carry out a numerical integration to determine the final speed of an electron, and the time required to go the two-mile distance. Give the final speed in terms of v/c, and also in terms of $1 - (v/c)$, which is a small difference.

Note that in the numerical integration you cannot use a velocity of p/m but must instead determine the velocity v in terms of the relativistic momentum p, as explained in Chapter 1.

Figure 2.34 The NEAR spacecraft passes the asteroid Mathilde (Problem 2.11).

Problem 2.13 Dark matter in the Universe

In the 1970's the astronomer Vera Rubin made observations of distant galaxies that she interpreted as indicating that perhaps 90% of the mass in a galaxy is invisible to us ("dark matter"). She measured the speed with which stars orbit the center of a galaxy, as a function of the distance of the stars from the center. The orbital speed was determined by measuring the "Doppler shift" of the light from the stars, an effect which makes light shift toward the red end of the spectrum ("red shift") if the star has a velocity component away from us, and makes light shift toward the blue end of the spectrum if the star has a velocity component toward us. She found that for stars farther out from the center of the galaxy, the orbital speed of the star hardly changes with distance from the center of the galaxy, as is indicated in the diagram below. The visible components of the galaxy (stars, and illuminated clouds of dust) are most dense at the center of the galaxy and thin out rapidly as you move away from the center, so most of the visible mass is near the center.

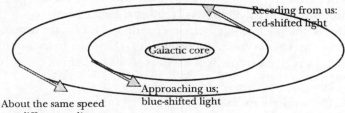

Receding from us: red-shifted light

Galactic core

Approaching us; blue-shifted light

About the same speed at a different radius

(a) Predict the speed v of a star going around the center of a galaxy in a circular orbit, as a function of the star's distance r from the center of the galaxy, assuming that almost all of the galaxy's mass M is concentrated at the center.

(b) Construct a logical argument as to why Rubin concluded that much of the mass of a galaxy is not visible to us. Reason from principles discussed in this chapter, and your analysis of part (a). Explain your reasoning. You need to address the following issues:

• Rubin's observations are not consistent with your prediction in (a).

• Most of the *visible* matter is in the center of the galaxy.

• Your prediction in (a) assumed that most of the mass is at the center.

This issue has not yet been resolved, and is still a current topic of astrophysics research. Here is a reference to the original work: "Dark Matter in Spiral Galaxies" by Vera C. Rubin, *Scientific American*, June 1983 (96-108). You can find several graphs of the rotation curves for spiral galaxies on page 101 of this article.

Problem 2.14 Orbital periods

(a) Many communication satellites are placed in a circular orbit around the Earth at a radius where the period (the time to go around the Earth once) is 24 hours. If the satellite is above some point on the equator, it stays above that point as the Earth rotates, so that as viewed from the rotating Earth the satellite appears to be motionless. That is why you see dish antennas pointing at a "fixed" point in space. Calculate the radius of the orbit of such a "synchronous" satellite. Explain your calculation in detail.

(b) Electromagnetic radiation including light and radio waves travels at a speed of 3×10^8 m/s. If a phone call is routed through a synchronous satellite, what is the minimum delay between saying something and getting a response? Explain. Include in your explanation a diagram of the situation.

(c) Some human-made satellites are placed in "near-Earth" orbit, just high enough to be above almost all of the atmosphere. Calculate how long it takes for such a satellite to go around the Earth once, and explain any approximations you make.

(d) Calculate the orbital speed for a near-Earth orbit, which must be provided by the launch rocket. (Recently large numbers of near-Earth communications satellites have been launched. Their advantages include making the signal delay unnoticeable, but with the disadvantage of having to track the satellites actively and having to use many satellites to ensure that at least one is always visible over a particular region.)

(e) When the first two astronauts landed on the Moon, a third astronaut remained in an orbiter in circular orbit near the Moon's surface. During half of every complete orbit, the orbiter was behind the Moon and out of radio contact with the Earth. On each orbit, how long was the time when radio contact was lost?

Problem 2.15 Analytical 3-body orbit

There is no general analytical solution for the motion of a 3-body gravitational system. However, there do exist analytical solutions for very special initial conditions. Figure 2.35 shows three stars, each of mass m, which move in the plane of the page along a circle of radius r. Calculate how long this system takes to make one complete revolution. (In many cases 3-body orbits are not stable: any slight perturbation leads to a break-up of the orbit.)

Problem 2.16 Newton's apple

Newton suspected that the Moon falls toward the Earth due to the gravitational attraction of the Earth, just as an apple falls toward the Earth for the same reason. An apple falls from a tree with an acceleration that is measured to be 9.8 m/s^2 (effects of air resistance are negligible for short falls). The Moon goes around the Earth once in about a month (actually 27 days). The Moon's orbit as seen from the Earth is actually an ellipse, but this ellipse is very close to being a circle. The radius of the Earth is 6400 km. The distance between the centers of the Earth and Moon is 3.8×10^5 km (this distance had been measured by Newton's time, although not very accurately). Show what Newton showed, that the same ("universal") law of gravitation applies to the falling apple and to the "falling" Moon, in that the accelerations of both objects are consistent with a single force law.

(You might well wonder whether this analysis makes much sense, since the Earth-Moon system is itself in orbit around the Sun. In one month the Earth-Moon system moves around the Sun in a very gentle arc which can be approximately considered to be motion at constant velocity, because the direction doesn't change much, about one-twelfth of a circle, since there are 12 months in a year. The principle of relativity says that we should be able to do this analysis from a frame of reference in uniform motion with respect to the distant stars, and we choose a reference system moving with the Earth, which is close enough to a uniform-motion reference frame for our analysis to be valid.)

Problem 2.17 The center of our galaxy

Remarkable data indicate the presence of a massive black hole at the center of our Milky Way galaxy. One of the two W. M. Keck 10 meter diameter telescopes in Hawaii was used by Andrea Ghez and her colleagues to observe infrared light coming directly through the dust surrounding the central region of our galaxy (visible light is multiply scattered by the dust, blocking a direct view). Stars were observed for several consecutive years to determine their orbits near a motionless center that is completely invisible ("black") in the infrared but whose precise location is known due to its

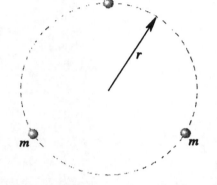

Figure 2.35 Three stars in a circular orbit (Problem 2.15).

strong output of radio waves, which are observed by radio telescopes. The data were used to show that the object at the center must have a mass that is huge compared to the mass of our own Sun, whose mass is a "mere" 2×10^{30} kg.

(a) The speeds of the stars as a function of distance from the center were consistent with Newtonian predictions for orbits around a massive central object. How should the speed of each star depend on its distance from the center, and the mass of the central object?

(b) The star nearest the center of our galaxy was observed to be about 10^{14} m from the center, with a period of about 15 years. What is the speed of this star in m/s? Also express the speed as a fraction of the speed of light. (The actual orbit is an ellipse as reproduced in Figure 2.36, but for purposes of this analysis you may approximate it as being circular.)

(c) This is an extraordinarily high speed for a star. Is it reasonable to approximate the star's momentum as mv?

(d) Calculate the mass of the massive black hole about which this star is orbiting.

(e) How many of our Suns does this represent?

It is thought that all galaxies may have such a black hole at their centers, as a result of long periods of mass accumulation. When many bodies orbit each other, sometimes an object happens in an interaction to acquire enough speed to escape from the group, and the remaining objects are left closer together. Some simulations show that over time, as much as half the mass may be ejected, with agglomeration of the remaining mass. This could be part of the mechanism for forming massive black holes.

For more information, see the home page of Andrea Ghez, http://www.astro.ucla.edu/faculty/ghez.htm. The papers available there refer to "arcseconds," which is an angular measure of how far apart objects appear on the sky, and to "parsecs," which is a distance equal to 3.3 light-years (a light-year is the distance light goes in one year).

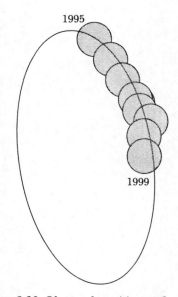

Figure 2.36 Observed positions of a star from 1995 to 1999 near the center of our Milky Way Galaxy, and an orbit fit to the data (Problem 2.17).

Problem 2.18 In the space shuttle

A space shuttle is in a circular orbit near the Earth. An astronaut floats in the middle of the shuttle, not touching the walls. On a diagram, draw and label (a) the momentum $\vec{p}_1$ of the astronaut at this instant; (b) all of the forces (if any) acting on the astronaut at this instant; (c) the momentum $\vec{p}_2$ of the astronaut a short time Δt later; (d) the momentum change (if any) $\Delta \vec{p}$ in this time interval. Why does the astronaut seem to "float" in the shuttle?

2.16 Answers to exercises

2.1 (page 39) 2.5×10^{-10} N in both cases

2.2 (page 39) $3F/16$

2.3 (page 39) 6×10^{24} kg

2.5 (page 40)   63 km. The height of Earth's atmosphere is roughly 50 km.

2.6 (page 43)   1: No, because a force acted on it.

2: No, because a force acted on it.

Total momentum was constant, because the net force on the combined system was zero. Momentum change of 1 is equal and opposite the momentum change of 2.

2.7 (page 48) $\hat{r} = \langle -0.598, 0.359, -0.717 \rangle$

2.8 (page 48) $\hat{r} = \langle \dfrac{1}{\sqrt{3}}, \dfrac{1}{\sqrt{3}}, \dfrac{1}{\sqrt{3}} \rangle$

2.9 (page 48) $\dfrac{G m_1 m_2}{3 d^2} \left(\langle \dfrac{1}{\sqrt{3}}, \dfrac{1}{\sqrt{3}}, \dfrac{1}{\sqrt{3}} \rangle \right)$

2.10 (page 56)   0.63 rad/s

2.11 (page 56)   5 m/s, tangential

2.12 (page 56) 3 m/s^2, toward the center

Chapter 3

The Atomic Nature of Matter

Chapter 3

The Atomic Nature of Matter

"If, in some cataclysm, all of scientific knowledge were to be destroyed, and only one sentence passed on to the next generations of creatures, what statement would contain the most information in the fewest words? I believe it is the *atomic hypothesis* (or the atomic *fact*, or whatever you wish to call it) that *all things are made of atoms— little particles that move around in perpetual motion, attracting each other when they are a little distance apart, but repelling upon being squeezed into one another.* In that one sentence, you will see, there is an *enormous* amount of information about the world, if just a little imagination and thinking are applied."

Richard Feynman, in *The Feynman Lectures on Physics*
by R. P. Feynman, R. B. Leighton, & M. Sands, 1965;
Palo Alto: Addison-Wesley.

Much of physics was developed before it was known that matter was composed of atoms. This older physics is called "classical" physics. Although the classical physics principles allow us to explain a wide variety of situations, there are important aspects of matter and interactions that we cannot understand unless we include the atomic nature of matter in our models.

In this chapter we will use a simple atomic model of a solid to begin to understand some of the important properties of solids. We will emphasize the need for making simplified "models" of the real-world situations, in order to obtain approximate but useful results. We will emphasize modeling interactions at the atomic level, and making connections between microscopic and macroscopic analyses. We will also consider some aspects of how to model gases, a topic to which we'll return in a later chapter. As we explained in Chapter 1, we will say little in this course about liquids, which are much more difficult to analyze.

3.1 A model of a solid: Balls connected by springs

In the quote given above, Feynman has summarized the basic properties of atoms and of interatomic forces. The main properties of atoms that will be important to us in this course are these:

- All matter consists of atoms, whose typical radius is about 10^{-10} meter.
- Atoms attract each other when they are close to each other but not too close.
- Atoms repel each other when they get too close to each other.
- Atoms move around in perpetual motion.

These properties were established during a century of intense study of atoms by physicists and chemists, using a wide variety of experimental techniques. You can see for yourself some of the evidence for these properties. For example, consider a solid block of aluminum, which we have good reason to believe is made up of a large number of tiny aluminum atoms.

? What properties of this block of aluminum, observable without special equipment, support the claims that "atoms attract each other when they are not too close to each other" and "atoms repel each other when they get too close to each other"?

Since the block of metal does not spontaneously fall apart or evaporate, the atoms in the block must be attracting each other. It is very difficult to compress the block, so the atoms must resist an attempt to push them closer together. Evidently the atoms in a block of aluminum are normally at just the right distance from each other, not too close and not too far away. This just-right distance is called the "equilibrium" distance.

Given these properties of solids, a very useful model for a block of aluminum is an array of tiny balls in constant motion, connected by springs (Figure 3.1). If two balls (models of atoms) are pulled apart and then released, the spring connecting the balls will pull them back toward their original equilibrium positions. But if the balls are moved closer together, the spring will push them back toward the equilibrium positions. In general, the balls will continually oscillate short distances away from and back toward their equilibrium positions.

In a solid at room temperature, the atoms are in motion, continually oscillating around their equilibrium positions. If we heat the solid, these atomic oscillations become more vigorous. One of the most important outcomes of research on the atomic nature of matter was the realization that for many materials, the "temperature" of an object, as measured by an ordinary thermometer, is just an indicator of the average energy of the atoms; the higher the temperature, the more vigorous the atomic motion. During this course we will have much more to say about the relationship between temperature and energy. Figure 3.1 is a "snapshot" of the atoms in a model solid at a particular instant.

One can think of a ball in Figure 3.1 as representing the atomic nucleus plus the inner electrons. Although almost all of the mass of an atom is concentrated in the nucleus, the nucleus is only about 1/100000 as large as the electron cloud surrounding it, so nuclei would not be visible on the scale of this picture. If the balls represented the entire electron cloud of the atoms, including the outer electrons, the balls would of course touch each other. A spring in the picture can be thought of as representing the interatomic forces associated with the outer electrons, which are involved in the bonds between atoms. Compare the ball-and-spring model with the actual scanning tunneling microscope (STM) images we first showed in Chapter 1, where the electron clouds of the atoms on the surface of solid silicon touch each other (Figure 3.2).

Limitations of the ball-and-spring model

The simple ball-and-spring model of a block of aluminum is not perfect. Some properties of a block of aluminum, such as its thermal conductivity, electrical conductivity, and color can be fully explained only with quantum mechanics. This branch of physics is called "condensed matter physics" (as opposed to the study of individual atoms or subatomic particles). Nevertheless, many properties can be explained fully or at least approximately in terms of the simple model. Even professional condensed matter physicists often use this simplified model in their day-to-day work, both to get an approximate idea of what will happen in a particular situation, and to check approximately the results of full quantum-mechanical calculations.

A solid vs. a single atom

A block of aluminum has very different properties than a single atom of aluminum. A block of aluminum looks metallic, melts at high temperature, can be bent or dented, is a good conductor of electric current, etc. A single atom of aluminum has none of these properties! When atoms bind to each other to form a solid, macroscopic object, that object has special properties arising from the interactions of the individual atoms. Aggregates of mole-

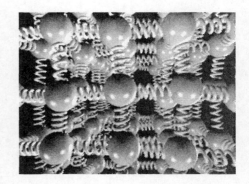

Figure 3.1 A simple model of a solid: tiny balls in constant motion, connected by springs.

Figure 3.2 STM images of atoms on two different surfaces of crystalline silicon.

cules also have different properties from those of individual molecules; a single H_2O molecule is not "wet."

Ex. 3.1 One mole of aluminum (6×10^{23} atoms) has a mass of 27 grams. The density of aluminum is 2.7 grams/cm^3. What is the approximate diameter of an aluminum atom? Make the simplifying assumption that the atoms are arranged in a "cubic" array, so that an atom of diameter d occupies an amount of space d^3 in the solid (the atomic arrangement in aluminum is not actually cubic).

Ex. 3.2 One mole of lead has a mass of 207 grams. The density of lead is 11.4 grams/cm^3. What is the approximate diameter of a lead atom? You may be surprised by the comparison between these very different atoms, aluminum and lead.

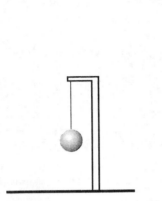

Figure 3.3 A thin wire is stretched by a heavy ball.

3.2 Static properties of a solid: The interatomic "spring"

We begin our study of a solid by looking at its "static" properties, when the solid is not moving. We will use our simple ball-and-spring model of a solid to analyze how much the interatomic "springs" inside a thin wire are stretched when we hang a heavy ball from the end of the wire (Figure 3.3). The upper end of the wire is fastened to a support whose base is firmly embedded in the ground.

We need a quantitative description of springs in general and a quantitative description of the gravitational forces acting on the ball and the wire. Then we will be able to determine the strength of the interatomic "spring."

3.2.1 Force exerted by a spring

We need a quantitative formulation of the force exerted by a spring on an object touching it. As we discussed in Chapter 1, the magnitude of the force required to stretch or compress a spring is observed to be proportional to the magnitude of the stretch or compression s. For example, if we double the force F acting on the spring, we get double the stretch or compression s (Figure 3.4 and Figure 3.5). We write this:

$$F = k_s s, \text{ where } s = L - L_{\text{relaxed}}$$

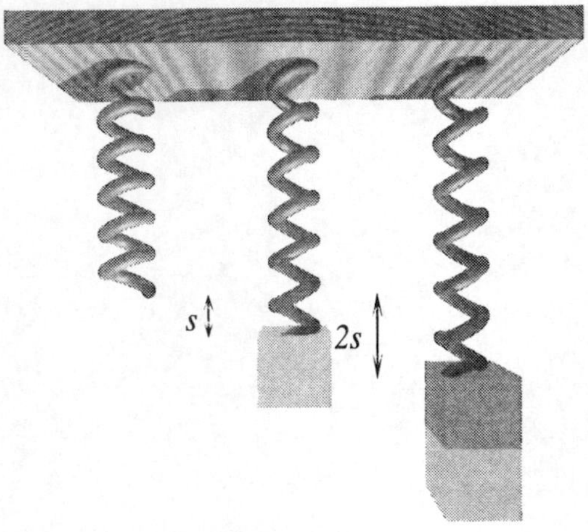

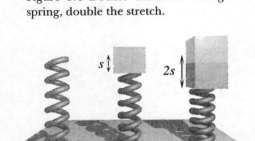

Figure 3.4 Double the force acting on a spring, double the stretch.

The constant k_s is called the "spring stiffness" or "spring constant." This is an approximate relationship, because it is true only over a limited range. If you stretch a spring so much that it turns into a straight wire, it suddenly becomes extremely difficult to stretch it any further. If you compress a spring so much that its coils come into contact, it suddenly becomes extremely difficult to compress the spring any further.

Ex. 3.3 A spring has a length of 8 cm when relaxed, and a length of 10 cm when stretched by a force of 60 N. What is the stiffness k_s of this spring?

Ex. 3.4 This same spring is then compressed by an unknown force so that the length of the spring becomes 7 cm. What is the magnitude of this force?

Ex. 3.5 A different spring of length L has stiffness k_s. It is cut into two springs, each of length $L/2$. What is the stiffness of one of these

Figure 3.5 Double the force acting on a spring, double the compression.

shorter springs? (Note that when you stretch the full spring, each half of the spring stretches half the total amount of stretch.)

3.2.2 A microscopic view: Interactions within the wire

For a moment consider a wire that supports only its own weight; nothing is hanging from it. For simplicity, Figure 3.6 shows a segment near the bottom of a wire that is only one atom wide. The atom at the very bottom of the wire experiences a gravitational force mg due to its interaction with the Earth. Since its momentum is not changing, the net force on the atom must be zero ($d\vec{p}/dt = \vec{F}_{net}$), so the "spring" between the atoms must exert an upward force $k_s s_1$. The second atom feels a gravitational force mg, and also a downward force mg due to the first spring (since the force exerted by a spring is proportional to its stretch, the magnitude of the force it exerts on both atoms connected to it must be the same).

Since the second atom's momentum is not changing, the next spring must exert an upward force of $k_s(2s_1) = 2mg$, so it must be stretched twice as much as the bottom "spring," the next spring must be stretched three times as much, and so on. Clearly the interatomic bonds at the top of the wire are stretched more than the bonds at the bottom of the wire, so the tension in the wire is not actually uniform. You can see a similar effect if you hold up a long spring such as a Slinky. The coils at the top of the spring are stretched considerably more than the coils at the bottom.

In Figure 3.6 the amount of stretch of the interatomic springs is greatly exaggerated. In a thin wire supporting only its own weight, even the interatomic bond at the top is stretched very little, because the mass of all the atoms in the wire itself is small.

Wire supporting a ball

When a large ball is hung from the wire, it will have a very large effect on the stretch of the interatomic springs (Figure 3.7). Now the interatomic bond at the bottom of the wire must support the large ball. The mass of the ball is much greater than the mass of one of the atoms, so the interatomic "spring" stretch will be much greater than it was without the ball. As before, the "springs" at the top of the wire will be stretched slightly more than the "springs" at the bottom of the wire, due to the accumulated weight of the atoms in the wire. However, if the mass of the wire is very small compared to the mass of the ball, this small increase in stretch will be very small compared to the stretch caused by the ball, and all the stretches will be similar.

Note that for a wire whose mass is very small, it is a good approximation to neglect the effect of the mass of the wire and attribute the interatomic stretch entirely to the ball. However if the wire's mass is comparable to that of the ball, this is a very poor approximation.

You may have seen the phrase "a massless wire" or a "massless spring" in textbook physics problems. Of course there is no such thing as a wire or a spring with no mass. This phrase refers to a frequently used model or "idealization" of a thin wire, a string, or a rope. An "ideal" string has no mass at all and does not stretch at all when under tension. A "massless, inextensible string" is a useful model of a real string or wire in situations where the mass of the string is very small and the effective spring constant of the string is very large (that is, the "spring" is very stiff), so the string lengthens very little.

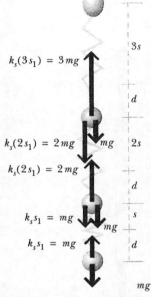

$k_s(3s_1) = 3mg$

$k_s(2s_1) = 2mg$

$k_s(2s_1) = 2mg$

$k_s s_1 = mg$

$k_s s_1 = mg$

mg

$3s$

d

$2s$

d

s

d

Figure 3.6 The bottom of a hanging wire, with nothing hanging from the wire. d is the equilibrium interatomic distance (for a wire lying on a table). The stretch of the interatomic bonds is greatly exaggerated in this figure.

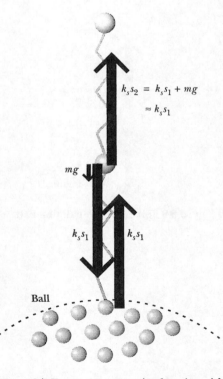
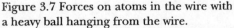

$k_s s_2 = k_s s_1 + mg$

$\approx k_s s_1$

mg

$k_s s_1$ $k_s s_1$

Ball

Figure 3.7 Forces on atoms in the wire with a heavy ball hanging from the wire.

3.2.3 A macroscopic view: Tension

Instead of taking an atomic viewpoint, suppose we consider the wire as a macroscopic object with mass m_w, and we mostly ignore its internal (atomic) structure. The Earth exerts a downward gravitational force on the wire, $m_w g$, which is the sum of the gravitational forces on the individual atoms in the wire (Figure 3.8). Although the wire interacts gravitationally with other objects such as the support, the ball, and the Moon, we will neglect these forces because they are very small.

The hanging ball exerts a downward force on the bottom of the wire, of magnitude $F_{T,\text{bottom}}$. This is an interatomic electric force. The "T" is for "tension," which is what we call the force internal to the wire that is associated with the stretch of the interatomic "springs" in our simple ball-and-spring model of a solid.

The support exerts a force $F_{T,\text{top}}$ on the top end of the wire. This is an interatomic electric force.

The wire is at rest, so according to the momentum principle (Newton's second law of motion) the net force should be zero. In the y direction, we have

$$\frac{dp_y}{dt} = 0 = F_{T,\text{top}} - F_{T,\text{bottom}} - m_w g.$$

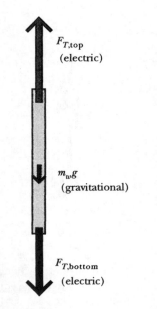

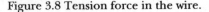

Figure 3.8 Tension force in the wire.

We can solve this equation to get $F_{T,\text{top}} = F_{T,\text{bottom}} + m_w g$. In other words, the "tension" in the top of the wire is equal to the total downward force acting on the wire, which is the (electric) force exerted by the ball on the bottom of the wire plus the gravitational force of the Earth on the wire. If the mass of the thin wire is very small compared to that of the hanging mass, we can use the "massless, inextensible wire" idealization, and say that the tension $F_{T,\text{top}}$ in the topmost portion of the wire is approximately equal to the tension $F_{T,\text{bottom}}$ in the bottom of the wire. We then do not have to worry about the different values for the tension along the length of the wire, and we can use the approximate value $F_T \approx F_{T,\text{top}} \approx F_{T,\text{bottom}}$ for the tension everywhere in the wire.

Now consider the equilibrium of the ball, of mass M (Figure 3.9). According to the principle of reciprocity (Newton's third law), the wire must exert an upward force with the same magnitude $F_{T,\text{bottom}}$ on the ball that the ball exerts on the wire. This force must equal the gravitational force Mg exerted by the Earth on the ball. If the ball has much more mass than the wire ($M \gg m_w$), we have $F_T \approx F_{T,\text{top}} \approx F_{T,\text{bottom}} \approx Mg$; the tension F_T in the wire is approximately equal to the weight Mg of the ball, everywhere throughout the wire. (By "weight" we mean the gravitational force on an object.)

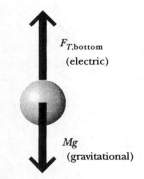

Figure 3.9 Forces on the hanging ball.

3.2.4 How much does the wire stretch?

In the real, non-ideal world the thin wire will stretch somewhat due to the tension F_T. We are now in a position to discuss quantitatively how much the wire will stretch, and then deduce the effective stiffness of the interatomic "springs" in our ball-and-spring model.

If we knew the spring constant k_s for the wire we could calculate the stretch s from the relation $F_T = k_s s$. Usually a different relationship is used to describe the stretching of straight elements such as a wire. If the wire has a length L, we call the amount of stretch of the wire ΔL (a small increment in the length, what we have called s in a spring), and the fractional stretch $\Delta L/L$ is called the "strain." As long as the strain isn't too big, the strain is proportional to the tension per unit area, F_T/A, where A is the cross-sectional area of the wire (Figure 3.10). The tension per unit area is called the "stress."

$$\text{strain} = \frac{\Delta L}{L}; \text{ stress} = \frac{F_T}{A}$$

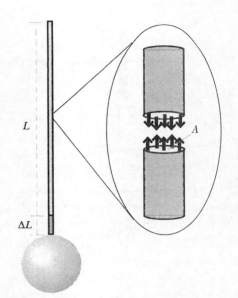

At the atomic level, the stress F_T/A can be related to the force that neighboring atoms exert on each other, and the strain $\Delta L/L$ can be related to the stretch of the interatomic bond. The ratio of stress to strain is a property of the material and differs for steel, aluminum, etc. A quantitative form of the relation between stress and strain in a wire is written as follows:

$$\frac{F_T}{A} = Y\frac{\Delta L}{L}$$

The parameter Y is called "Young's modulus" and is a property of the material. The stiffer the material, the larger is Young's modulus. Note the similarity to the spring formula, $F_T = k_s s \ (= k_s \Delta L)$.

Figure 3.10 The heavy ball stretches the wire an amount ΔL. We are interested in the stretch of the interatomic "springs." A is the cross-sectional area—the area of a slice through the wire, perpendicular to its length.

Limit of applicability of Young's modulus

If you apply too large a stress, the wire "yields" (stretches a great deal) or breaks, and the proportionality of stress and strain is no longer even approximately true. Figure 3.11 shows a graph of the strain $\Delta L/L$ that results as a function of the applied stress F_T/A for a particular aluminum alloy.

You see that for moderate stress the resulting strain is proportional to the applied stress: double the stress, double the strain. But once you reach the yield stress, a further slight increase in the stress leads to a very large increase in the length of the metal. This can be quite dramatic: as you add more weights to the end of a wire the wire gets slightly longer, then suddenly it starts to lengthen a lot, very rapidly, and then the wire breaks. (In materials reference manuals stress is usually plotted against strain, so that the slope of the line is equal to Young's modulus. We plot strain vs. stress to emphasize that stress is the *cause* and strain the *effect*.)

Ex. 3.6 In Figure 3.11, what is the approximate value of Young's modulus for this aluminum alloy?

Ex. 3.7 Suppose we hang a heavy ball with a mass of 10 kg (about 22 pounds) from a steel wire 3 m long that is 3 mm in diameter. Steel is very stiff, and Young's modulus for steel is unusually large, 2×10^{11} N/m^2. Calculate the stretch ΔL of the steel wire. This calculation shows why in many cases it is a very good approximation to pretend that the wire doesn't stretch at all ("ideal non-extensible wire").

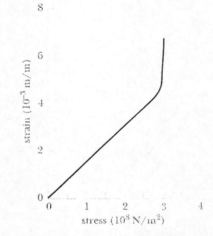

Figure 3.11 Strain vs. stress for a particular aluminum alloy.

Problem 3.1 Experiment—study of strain vs. stress

Measure and graph the strain vs. stress curve for a long metal wire by hanging weights from the end of the wire and carefully measuring the stretch as a function of the hanging weight. If possible, go beyond the "yield" stress and observe the rapid, large increase in the strain. BE CAREFUL! Once you exceed the yield stress, the wire may break and dump the weights on whatever is underneath.

3.2.5 Effective interatomic spring stiffness

From macroscopic measurements of Young's modulus we can calculate an approximate spring stiffness k_s for the interatomic force. Consider a slice through a wire that is being stretched, with a force uniformly distributed

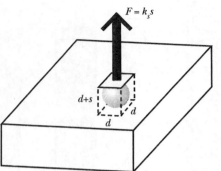

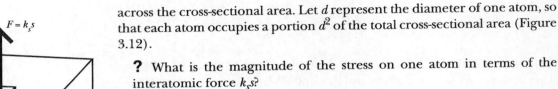

across the cross-sectional area. Let *d* represent the diameter of one atom, so that each atom occupies a portion d^2 of the total cross-sectional area (Figure 3.12).

? What is the magnitude of the stress on one atom in terms of the interatomic force $k_s s$?

The atomic stress (force per unit area) is $k_s s/d^2$.

? In terms of the stretch *s* of the interatomic "spring," what is the strain for one atom (change of height of the atom, divided by its normal height)?

The strain for one atom is s/d, because its height was originally *d*, and the change in its height is *s*. Therefore we can express Young's modulus in terms of atomic quantities:

$$Y = \frac{(k_s s/d^2)}{(s/d)} = \frac{k_s}{d}$$

Figure 3.12 One atom occupies a portion d^2 of the total cross-sectional area, and the force acting on that portion is $k_s s$.

This is a good example of an important theme in this course, that of relating macroscopic properties to microscopic (atomic-level) properties.

The following short problem is quite important because we will use these results in various contexts at later times in the course.

Problem 3.2 Interatomic spring constant

Young's modulus for aluminum is 6.2×10^{10} N/m^2. The density of aluminum is 2.7 grams/cm^3, and the mass of one mole is 27 grams. If we model the interactions of neighboring aluminum atoms as though they were connected by springs, determine the approximate spring constant of such a spring. Repeat this analysis for lead: Young's modulus for lead is 1.6×10^{10} N/m^2, the density of lead is 11.4 grams/cm^3, and the mass of one mole is 207 grams. Make a note of these results, which we will use for various purposes later on. Note that aluminum is a rather stiff material, whereas lead is quite soft.

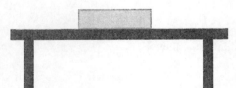

3.3 Brick lying on a table: Compression

A closely related situation is that of a lead brick lying at rest on a table (Figure 3.13), except that the brick is compressed while the wire was stretched.

Take the lead brick as the system of interest (Figure 3.14). The Earth pulls down on the brick of mass *M*, and the magnitude of the gravitational force is *Mg*. The table pushes up on the brick with a force whose magnitude we'll call F_N, a "normal" force—that is, a force that is normal (perpendicular) to the table. What kind of a force is this, exactly?

Figure 3.13 A brick lies on a table.

If the lead brick were extremely heavy, we might not be surprised to observe the table sagging under its weight. Even if the table does not sag visibly, the molecules in the top layer of the table are pushed down by molecules in the bottom layer of the brick. Both objects deform somewhat (their interatomic "springs" are compressed).

Figure 3.14 Forces acting on the brick.

Atoms in the table and the brick are squeezed closer than their normal equilibrium positions, like compressing a spring. Figure 3.15 shows a tiny contact region. The bottom layers of atoms in the brick are compressed the most. Higher layers support less of the total weight of the brick. Young's modulus relates stress and strain for compression as well as tension.

It would be appropriate to call the upward force of the table F_C for "compression force," but it is common practice to label this kind of force as a "normal" force. The word is used in the mathematics sense, meaning "perpendicular to." The table exerts a "normal" force—that is, perpendicular to its surface.

Note: Neither "normal" force F_N nor "tension" force F_T
is a different kind of fundamental force!

These are merely names for interatomic electric interactions.

Since the brick is at rest, the vertical component of the net force must be zero:

$$\frac{dp_y}{dt} = 0 = +F_N - Mg$$

We conclude that the normal force F_N exerted by the table is equal to the weight Mg of the brick, an unsurprising result. This neglects the effects of the air on the brick, which we will discuss later in this chapter.

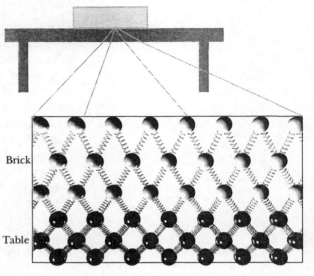

Figure 3.15 Interatomic "springs" in a tiny contact region are compressed by the weight of atoms above.

3.4 Dynamic properties of a solid: A spring-mass system

We have modeled a solid as a collection of atoms connected to their neighbors by spring-like interatomic forces, and we have studied some "static" aspects of these forces. Next we will study the "dynamics" of masses and springs.

We have commented that the atoms in a solid are in constant motion around their equilibrium positions, and that there is more such motion at higher temperatures. In the following sections we'll study the motion of a mass on a spring. This will help us visualize the behavior of atoms in a solid, and this model will be important later when we attempt to understand the thermal properties of solids. Many of the concepts associated with the motion of a mass on a spring have remarkably broad application to other kinds of phenomena, including phenomena as disparate as the behavior of certain kinds of electric circuits, and the vibrational motion of a diatomic molecule such as oxygen, O_2.

3.4.1 Experimental observation of a spring-mass system

To begin our study of the dynamic properties of a spring-mass system, you will observe and measure such a vertical spring-mass system (Figure 3.16). You will use these measurements later to create a computer model of the system.

Figure 3.16 A vertical spring-mass system.

Problem 3.3 Experiment—a vertical spring-mass system
You will be given a spring and one or more masses with which to study the motion of a mass hanging from a spring.

(a) Measure the value of the spring stiffness k_s of the spring in N/m. The magnitude of a spring force is $k_s s$, where s is the stretch of the spring (change from the unstretched length). Explain briefly how you measured k_s. Include in your report the unstretched length of the spring.

(b) Measure the period (the round-trip time) of a mass hanging from the vertical spring. Report the mass that you use, and the amplitude of the oscillation. Amplitude is the maximum displacement, plus or minus, from the equilibrium position (the position where the mass can hang motionless).

(c) With twice the amplitude that you reported in part (b), measure the period again. Since the mass has to move twice as far, one might expect the period to lengthen. Does it?

How to measure the period accurately

There are unavoidable fluctuations in starting and stopping the timing, but you can minimize the error this contributes by timing many complete cycles so that the starting and stopping fluctuations are a small fraction of the total time measured.

It is good practice to count out loud *starting from zero, not one*. To count five cycles, say out loud "Zero, one, two, three, four, five." If on the other hand you say "One, two, three, four, five," you have actually counted only four cycles, not five.

Since the motion is periodic, you can start (say "zero") at any point in the motion. It is best to start and stop when the mass is moving fast past some marker near the equilibrium point, because it is difficult to estimate the exact time when the mass reaches the very top or the very bottom, because it is moving slowly at those turnaround points. Be sure to measure full round-trip cycles, not the half-cycles between returns to the equilibrium point (but going in the opposite direction).

3.4.2 Applying the momentum principle to a spring-mass system

Consider a block of mass m connected to a spring whose stiffness is k_s (that is, the force exerted by the spring is $k_s s$ when the spring is stretched or compressed an amount s). The block slides with almost no friction on an air table, where it is supported on a cushion of air. The other end of the spring is attached to the wall (Figure 3.17). The horizontal motion of this horizontal spring-mass system is very similar to the motion of the vertical spring-mass system you studied experimentally in Problem 3.3.

Identifying interactions

We will apply the momentum principle to the block. To apply this principle, we must identify the interactions the block has with its surroundings. Interactions always involve pairs of *objects* (two stars, a table and a book, a proton and an electron). To be systematic in identifying all forces on a system, list all *objects* interacting with the system. This allows you to make sure you have neither overlooked real forces, nor included mythical ones. Note that each interaction we list must have a mechanism. Either an object must touch the system (a contact interaction involving interatomic forces), or it must interact at a distance through a specific long-range interaction (gravitational or electric).

Non-contact interactions

? What objects exert non-contact forces on the block?

The only kinds of distant interactions we know are gravitational interactions and electric interactions. In this case there are no non-contact electrical interactions, because none of the objects is charged. All of the objects in the Universe exert gravitational attractions on the block (universal law of gravitation), but the Earth's contribution is so much larger than the others' that we can safely make the approximation that the other gravitational interactions are negligibly small compared to the interaction with the Earth. We emphasize that all of these gravitational interactions (with the Moon, Mars, the Sun, the wall, etc.) are actually present in the real world, but they have only a tiny effect on the behavior of the ball. We are deliberately making a simplified "model" of the situation in order to be able to carry out an adequately accurate but not exact analysis of the more complicated real-world situation.

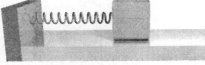

Figure 3.17 A block connected by a spring to a wall slides back and forth with little friction.

method

Having decided in the context of our model that only the Earth exerts a significant distant (non-contact) force on the block, we make the important step of recording our decision on a sketch of the system (Figure 3.18). Drawing a physics diagram is a quick and compact way of keeping track of all the forces on the system. Artistic talent is not needed to draw physics diagrams; the simpler the diagram is, the easier it is to understand. We draw the system (the block), and we draw a force vector whose tail is positioned at the center of the block and which points in the direction of the applied force (that is, downward). We also write the known magnitude of this force beside the vector. We will use the approximation of a constant gravitational field (constant force per unit mass) near the Earth's surface which we discussed on page 40.

(Diagram incomplete)

$F_g = mg$

Figure 3.18 The only significant non-contact force is the gravitational force exerted by the Earth.

Contact interactions

? What objects touch the block? Which of these exert significant forces on the block? Include these forces on the physics diagram.

You should have listed the spring and the air (including the air cushion underneath the block). Because the block is supported on a cushion of air which is blown out of the air table, the table does not actually touch the block. Since the block will be moving, there will be friction between the block and the air. To begin with as simple a model as possible, we will assume that the friction with the air is small compared to the force exerted by the spring. We may wish to check this assumption later by performing actual experiments.

In Figure 3.19, we add the contact forces to the physics diagram, where F_N is the upward force of the air cushion supporting the block. Note that although we have drawn the force on the block by the spring pointing to the left, this force could be to the right, to the left, or zero, depending on the position of the block.

We are neglecting other contact forces. Specifically, we neglect air resistance (opposing the motion of the block through the air) and a tiny upward "buoyancy" force (related to slight differences in air pressure between the top and bottom of the block). We will study these effects later in the course.

$k_s s$ F_N

mg

Figure 3.19 Add the contact forces to the physics diagram.

Choose an origin, and write an expression for the spring force

In order to write the equations corresponding to the momentum principle we need to select an origin with respect to which we will measure the position of the block. We can simplify our calculations by choosing to measure the position of the block with respect to an origin chosen to be at the equilibrium position, the location of the block when the spring is relaxed, so that the stretch s of the spring is equal to the position x of the block.

In the middle of Figure 3.20, the spring is relaxed, at the equilibrium position, with zero stretch ($x = L - L_{\text{relaxed}} = 0$). Therefore the force that the spring exerts on the block is zero. At the top of Figure 3.20, the spring has been stretched, and the position of the block is $x > 0$. The length of the spring is more than its equilibrium length by an amount x, and the quantity $k_s x$ is positive. The x component of the force that the spring exerts on the block, F_x, is given by $F_x = -k_s x$, which is a negative quantity and correctly predicts an x component of force on the block to the left, as we would expect.

At the bottom of Figure 3.20, the spring is compressed, and the position of the block is $x < 0$. The length of the spring is less than its equilibrium length by an amount x, and the quantity $k_s x$ is negative. The x component of the force that the spring exerts on the ball, F_x, is given by $F_x = -k_s x$, which is a positive quantity and correctly predicts an x component of force on the block to the right, as we would expect.

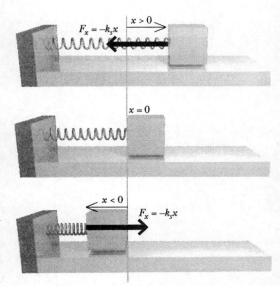

$F_x = -k_s x$ $x > 0$

$x = 0$

$x < 0$ $F_x = -k_s x$

Figure 3.20 The x component of the force that the spring exerts on the block is equal to $-k_s x$.

Therefore the equation $F_x = -k_s x$ describes the component of the force exerted on the block by the spring no matter whether the spring is stretched or compressed. The direction of the force is opposite to the direction of the displacement from the equilibrium position.

? What would be the formula for the spring force in terms of the position of the block if the origin were located at the wall, and the relaxed length of the spring is L?

In that case we would write $F_x = -k_s(x - L)$, because what matters in this formula is the stretch (change in length of the spring), which is $(x - L)$. A check on this is the fact that when the block is at location $x = L$, the stretch $(x - L)$ is zero.

? Given the physics diagram we constructed, write equations for the momentum principle, in a form suitable for numerical computation. Also record any information you have about the motion of the system.

Reading forces off the physics diagram, the momentum principle in vector form is this:

$$\Delta \vec{p} = \; <(-k_s x), (F_N - mg), 0> \Delta t$$

We also write down motion information, which includes anything we know about the motion. In this case, we chose to analyze only simple horizontal motion, with no bouncing up and down, so we know this:

$$p_y = 0 \text{ at all times, so } \frac{dp_y}{dt} = 0$$

3.4.3 Determining forces when we don't know the force law

The motion information that $dp_y/dt = 0$ for the block in horizontal motion tells us that $(F_N - mg)$ must be zero in this restricted situation. Therefore the air cushion not surprisingly supports the weight of the block with a force $F_N = mg$.

This is an example of being able to determine a force even when you don't know the force law. In Chapter 2 you were able to predict the motion of a binary star into the future because you knew the gravitational force law that determined the motion. Often the task is to predict motion from known force laws.

But we can also turn the momentum principle around and determine unknown forces from known motions. If we know that the block is moving horizontally, with no vertical motion, the y component of momentum isn't changing, so the y component of the net force must be zero: $(F_N - mg) = 0$.

Evidently there are two main kinds of analyses:

Known force laws: predict unknown motion

Known motion: determine unknown forces

The horizontal spring-mass system involves both kinds of analyses. In the x direction we have a known force law (the spring force $F_x = -k_s x$). In the y direction we don't have a force law for the complicated interaction between the block and the air track. The upward force on the block is due to air molecules striking the bottom of the block and depends on the speed of the air stream. Nevertheless we can determine this upward force because we know $p_y = 0$ at all times (no motion is a special case of known motion). Knowing that $dp_y/dt = 0$ means that we can conclude that $(F_N - mg) = 0$. (It is also important that we have a force law mg for the gravitational force near the Earth's surface.)

3.4.4 Predicting the motion of a spring-mass system

In order to predict the horizontal motion, we can apply the same method to the block that we applied to computing the motion of the binary star in Chapter 2. At each step in the numerical integration we can evaluate the known force law $-k_s x$ in terms of the current position x. (Later we will also discuss a way to obtain an analytical solution.)

Suppose that at some instant in time you know the position of the block and its momentum exactly. We call these data the "initial conditions" x_0 and p_0 at time $t = 0$. Using the momentum principle, we will predict what the position x and momentum p of the block will be at some future time t.

Problem 3.4 Numerical integration of a spring-mass system without friction

Carry out a numerical integration of the motion of a horizontal spring-mass system. Use the spring stiffness k_s and mass m that you measured experimentally for a vertical spring-mass system. Predict the motion of this spring-mass system sufficiently far into the future to observe several oscillations.

(a) Initial conditions: Make the initial position of the block be such that the stretch of the spring is equal to the amplitude of the oscillations in your experiment, and release the block with zero initial momentum. Display the motion in two ways:

(1) Plot a graph of the position of the block as a function of time. Label the scales on both axes, so your numerical results are clear.

(2) If possible using your computer tool, make an animation of the motion of the block. (You don't need to draw a real spring; it is sufficient to draw a line connecting the block to the wall.)

Very briefly, what are the most important aspects of the predicted motion?

(b) Vary the time step and report the maximum step size that gives reasonable results, in the sense that a smaller step size has little effect.

(c) Read the "period" (the round-trip time) of the oscillation off the graph. How does this compare with the period you measured in your experiment?

(d) Repeat the calculation with double the initial stretch. What happens to the period? Is this what you observed in your experiment?

(e) Temporarily change the force law to be $-k_s x^3$, and observe the period with the amplitude of part (a) and with double this amplitude. What happens to the period when you change the amplitude?

(f) How long would the spring-mass system theoretically continue to oscillate? Would a real block connected to a spring behave that way? What *would* happen?

Your calculations should have shown that the block oscillates back and forth along the x axis, producing a graph of the position of the block *vs.* time that should look something like Figure 3.21.

Oscillating back and forth is familiar behavior for a spring. The graph looks suspiciously like a cosine function, which is somewhat intriguing, since we didn't use any trigonometry in our calculation. Later we will show that the function is indeed a cosine, but for now the most important point is not what "standard" math function this graph may represent, but rather that we were able to predict the future one step at a time. This illustrates the great power of Newtonian analysis; if you know the net force on an object as a function of its position and velocity, you can predict the motion of the object far into the future by simple step-by-step calculations based on the momentum principle.

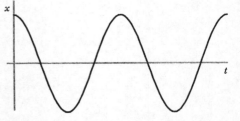

Figure 3.21 A graph of the position of the block as a function of time.

3.4.5 Harmonic oscillators

? How did the period change when you started out with an initial stretch twice as big?

Doubling the initial stretch apparently just doubles the size of the oscillations without changing the period of the oscillations.

A system that oscillates sinusoidally is called a "harmonic oscillator," and a spring-mass system is an example of a harmonic oscillator. Other kinds of oscillators are sometimes called "anharmonic" oscillators. A simple example is a bouncing ball, whose motion is repetitive but not sinusoidal; its height as a function of time involves t^2, not a sine or cosine of the time.

A very special property of harmonic oscillators is that the period does not depend on the amplitude. If you double the amplitude, the round-trip time doesn't change despite having to go twice as far. The bigger force due to the larger stretch makes the mass go just enough faster to compensate for the longer distance. This makes the harmonic oscillator extremely important in measuring time, because even if the amplitude changes somewhat (for example, decreases slowly due to friction), the period stays the same. This is not true for other kinds of oscillators, as you saw in part (e) of Problem 3.4.

Here is another example of an anharmonic oscillator:

Ex. 3.8 A bouncing ball is an example of an anharmonic oscillator. If you double the maximum height, what happens to the period? (Assume that the ball keeps returning almost to the same height.)

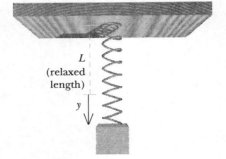

Figure 3.22 A vertical spring-mass system minimizes sliding friction.

3.4.6 A vertical spring-mass system

To reduce friction, it was easier to experiment with a mass oscillating vertically up and down on a spring (Figure 3.22) rather than sliding on an air table. We will show that we expect the same period whether the spring is vertical or horizontal.

We choose a coordinate system with the positive y axis pointing downward (Figure 3.22). We'll measure y downward from the location where the end of the relaxed spring would be (the relaxed length of the spring is L).

Neglecting air resistance and gravitational interactions with objects other than the Earth, the physics diagram of the forces acting on the hanging block is that shown in Figure 3.23. We then have the following:

$$\frac{dp_y}{dt} = -k_s y + mg$$

We choose to analyze only simple vertical motion, with no bouncing side to side, so we know this:

$$p_x = 0 \text{ at all times, so } \frac{dp_x}{dt} = 0$$

We can reduce the p_y equation to a familiar form if we measure a new coordinate, w, from the equilibrium position, when the block is hanging motionless, rather than from the unstretched spring position (Figure 3.24). When the mass hangs motionless from the spring, the spring is stretched an amount $k_s s_0 = mg$, so $s_0 = mg/k_s$. We'll measure w downward from the end of the spring, so

$$y = (s_0 + w) = \frac{mg}{k_s} + w, \text{ and therefore}$$

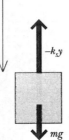

Figure 3.23 Physics diagram showing the forces acting on the hanging block.

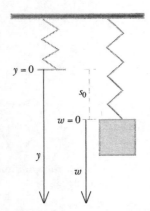

Figure 3.24 Measure w from the place where the block can hang motionless.

$$\frac{dp_y}{dt} = -k_s\left(\frac{mg}{k_s} + w\right) + mg = -k_s w$$

This says that if we measure from the end of the spring at its equilibrium location, mg/k_s below the bottom of the unstretched spring, the equation for the motion of the mass is the same as in a horizontal spring-mass system, whose motion we already know. The moving mass will oscillate up and down around this equilibrium position, as you observed in your experiment.

3.4.7 Discussion of modeling

At this point it is useful to have a discussion about the validity and accuracy of our calculational attempt to predict the future motion of an actual spring-mass system. Think about the following issues and be prepared to discuss them in other students:

1) Does your actual spring-mass system behave *exactly* as predicted? How does the observed motion differ from the prediction?

2) A wide range of observations are consistent with the momentum principle (at least in the domain where we need not invoke quantum mechanics or general relativity), and we expect the momentum principle to be valid for predicting the motion of a spring-mass system. Therefore it is prudent to look at failings of our *model* of the spring-mass system. What assumptions or approximations did we make that aren't entirely appropriate for your own mass oscillating on your own spring?

3) What could be done to make your own spring-mass system correspond more closely to the simple model? What could be done to make your own spring-mass system behave even more differently from the simple model?

4) On the other hand, how could we make the *model* more realistic? That is, could we remove some of the simplifying approximations, and then apply our step-by-step analysis to this more realistic model?

5) Is it conceivable that you could model your spring-mass system *completely*, taking into account *all* interactions, no matter how small?

6) Can we know the initial conditions *exactly*? If we don't know them exactly, can we predict the future exactly?

3.4.8 An analytical prediction for a spring-mass system

Even if one uses a high-accuracy numerical integration scheme, and takes very short time steps (with correspondingly lots of calculations), there is some unavoidable error associated with finite time steps in a numerical integration. Presumably Nature uses *infinitesimal* time steps! In some cases we can use calculus to predict the future using infinitesimal time steps, and eliminate entirely the errors introduced by a numerical integration. Even so, our prediction of the future is still constrained by the inevitable failings of our physics model to encompass all of the messiness of the real world.

The spring-mass model is one of the (few!) systems for which we can use calculus to obtain an "analytical" prediction for our model, with the prediction in the form of a symbolic expression rather than a set of numbers. As mentioned already, the graph of our numerical results on page 87 looks suspiciously like a cosine function. In a moment we will show that a cosine function is indeed compatible with the momentum principle, and that an analytical solution to the motion is given by this:

AN ANALYTICAL SOLUTION FOR SPRING-MASS

$$x = A\cos(\omega t), \text{ with } \omega = \sqrt{\frac{k_s}{m}}$$

A (the "amplitude") is a constant that depends on the initial conditions.

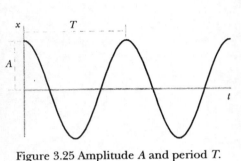

Figure 3.25 Amplitude A and period T.

Angular frequency, period, and frequency

The constant ω in $x = A\cos(\omega t)$ is called the "angular frequency" and is measured in radians per second (ω is lower-case Greek omega). When the argument ωt of the cosine increases by 2π radians, the motion repeats (one complete cycle), so if we call the "period" T (the time required for one complete cycle, Figure 3.25), we have $\omega T = 2\pi$.

One other quantity often used to describe oscillating systems is the "frequency" f, which is the number of complete cycles per second: $f = (1 \text{ cycle})$ per T seconds. Cycles per second are also called "hertz."

$$\text{Angular frequency } \omega = \sqrt{\frac{k_s}{m}}, \text{ radians per second}$$

$$\text{Period } T = \frac{2\pi}{\omega}, \text{ seconds}$$

$$\text{Frequency } f = \frac{1}{T} = \frac{\omega}{2\pi}, \text{ cycles per second or "hertz"}$$

In Chapter 2 you calculated the period of a planet orbiting the Sun. Both the oscillation of a mass on a spring and the orbital motion of a planet are called "periodic" motion. If you project the motion of a body in a circular orbit onto a line (for example, by shining a light edge-on to the orbital plane and observing the resulting shadow), this component of the circular motion will look like the motion of a harmonic oscillator.

Ex. 3.9 The constant A is called the "amplitude." What was the amplitude in your numerical integration? What are the minimum and maximum values of x during a cycle?

Ex. 3.10 Calculate the period T from the analytical solution and compare with the result obtained from your numerical integration.

3.4.9 Checking the analytical solution

In this section we verify that the cosine function given above does indeed satisfy the momentum principle for the spring-mass system. If you have not yet learned in calculus the derivatives of sines and cosines, or have not yet learned what is usually called the "chain rule," consult the appendix on basic calculus at the end of this volume before studying this section.

If we think we have an expression for $x(t)$, we can check it by differentiating it twice to see if it is consistent with our equation for the momentum principle, because nonrelativistically the time derivative of momentum that appears in the momentum principle is actually the mass times the second derivative of position with respect to time:

$$\frac{dp_x}{dt} \approx \frac{d(mv_x)}{dt} = m\frac{d}{dt}\left[\frac{dx}{dt}\right]$$

Based on our observations of a spring-mass system, we make the guess that

$$x = A\cos(\omega t),$$

where A and ω are initially unknown constants, and we check whether this function works when we plug it into the derivative form of the momentum principle, $dp_x/dt = F_{net}$. Since we need the time derivative of the momentum p_x, we start by differentiating x to get v_x:

? $v_x = \dfrac{dx}{dt} = \dfrac{d[A\cos(\omega t)]}{dt} = ?$

Using the chain rule of differentiation, we have the following:

$$\frac{d[A\cos(\omega t)]}{dt} = -A\sin(\omega t)\frac{d(\omega t)}{dt} = -A\sin(\omega t)\omega$$

So $v_x = -A\sin(\omega t)\omega$. Next we take the derivative of v_x:

? $\dfrac{dv_x}{dt} = \dfrac{d[-A\sin(\omega t)\omega]}{dt} = ?$

Apply the chain rule a second time:

$$\frac{d[-A\sin(\omega t)\omega]}{dt} = -A\cos(\omega t)\omega\frac{d(\omega t)}{dt} = -A\cos(\omega t)\omega^2$$

Therefore $\dfrac{dv_x}{dt} = -A\cos(\omega t)\omega^2$. We know from the momentum principle that

$$\frac{dp_x}{dt} = m\frac{dv_x}{dt} = -k_s x$$

Therefore for *any* time t that you care to choose, the following equation must be correct for our guessed solution, $x = A\cos(\omega t)$, to be valid:

$$m[-A\cos(\omega t)\omega^2] = -k_s A\cos(\omega t)$$

We can indeed satisfy this equation for any and all time t if (and only if) we choose ω^2 to be equal to k_s/m. In that case we have proven that

$$x = A\cos(\omega t)$$

satisfies the differential equation (the momentum principle) *for all values of the time t*. We have an "analytical" solution of the prediction problem. If someone gives you a time t, you can predict what the stretch x will be at that time, without having to carry out a numerical integration step by step from the present time to the specified time.

3.4.10 Real springs have mass

Real springs have mass, but we have ignored this fact of the real world. It is not easy to include the effect of the mass of the spring, either in a numerical integration or in an analytical treatment. One way to model a real spring with mass is with a chain of point masses connected by massless springs.

? Under what circumstances is it likely to be a good approximation to neglect the mass of the spring?

In many cases the spring has much less mass than the objects that are connected to it, in which case the effect of the spring's mass is likely to be small. A contrasting example is an oscillating Slinky, in which the only mass is the mass of the spring.

3.4.11 Existence of analytical solutions

It is fortunate that we can obtain an analytical solution—a prediction for all time of the motion of our model spring-mass system not subject to the additional inaccuracies introduced by a finite time step. Unfortunately, for many systems the differential equation corresponding to the momentum principle has a mathematical form that cannot be solved analytically and must be solved numerically. The power of numerical integration applied to the Newtonian scheme is that it can be used even in those situations where no analytical solution seems possible.

? As a simple example, show that $x = A\cos(\omega t)$ *cannot* solve the following "nonlinear" differential equation for any value of ω:

$$\frac{dp_x}{dt} = m\frac{d}{dt}\left[\frac{dx}{dt}\right] = -k_s x^3$$

There may or may not exist an analytical solution of this differential equation in terms of well-known mathematical functions, but *any* differential equation can be solved numerically, just as we did for the spring-mass equation. Given x and p at some instant in time, we can calculate the force ($-k_s x^3$) and determine the new values of x and p at a time Δt later, then repeat as often as necessary to reach the time of interest. In fact, you did just that in part (e) of Problem 3.4.

It is important to keep clearly in mind that either an analytical or a numerical prediction of the future applies only to our *model* of the actual real-world situation. If our model is a good approximation to the real world, our prediction will be a good approximation to what will actually happen, but even an analytical prediction cannot be exact because of the practical impossibility of knowing exactly the initial conditions and the net force due to all the other objects in the Universe.

One might think that analytical solutions are better than numerical ones, but note that one of the ways to calculate a table of cosine values is to do a numerical integration of the equation that applies to spring-mass systems! In general, a fruitful way to describe a mathematical function is by specifying the differential equation for which it is a solution, because then a numerical integration of that equation yields numerical values of the function.

3.4.12 General solution for the spring-mass system

It is easy to show that the function $x = A\cos(\omega t + \phi)$ is a general solution for a spring-mass system with arbitrary k_s, m, and initial conditions, where $\omega = \sqrt{k_s/m}$, and where the amplitude A and the "phase shift" ϕ (measured in radians and denoted by the Greek letter phi) are constants determined by the initial conditions, as is explained below. The "phase shift" essentially shifts the starting point of the oscillation. For example, if $\phi = \pi/2$, the cosine function actually represents a negative sine function (Figure 3.26):

$$x = A\cos\left(\omega t + \frac{\pi}{2}\right) = -A\sin(\omega t)$$

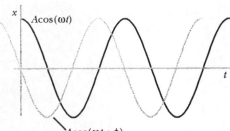

Figure 3.26 Phase shift ϕ.

Determining the amplitude and phase shift

The amplitude A and phase shift ϕ can be determined from the initial values at time t_0 of position x_0 and velocity v_0:

$$x_0 = A\cos\left(\sqrt{\frac{k_s}{m}}t_0 + \phi\right)$$

$$v_0 = -\sqrt{\frac{k_s}{m}}A\sin\left(\sqrt{\frac{k_s}{m}}t_0 + \phi\right)$$

Using the important trigonometric identity $\sin^2\theta + \cos^2\theta = 1$ (which is the Pythagorean theorem applied to a triangle whose hypotenuse has length 1), we can obtain an equation that can be solved for the amplitude A:

$$x_0^2 + \frac{m}{k_s}v_0^2 = A^2\text{, so we have } A = \sqrt{x_0^2 + \frac{m}{k_s}v_0^2}$$

For example, if the initial velocity is zero, the amplitude is simply equal to the initial stretch.

The value for A can be plugged back into either the x_0 equation or the v_0 equation to obtain the phase shift ϕ. Alternatively, by dividing the x_0 equation by the v_0 equation we obtain an equation that can be solved for the phase shift ϕ:

$$\frac{v_0}{x_0} = -\sqrt{\frac{k_s}{m}} \tan\left(\sqrt{\frac{k_s}{m}} t_0 + \phi\right)$$

$$\sqrt{\frac{k_s}{m}} t_0 + \phi = \arctan\left(-\sqrt{\frac{m}{k_s}} \frac{v_0}{x_0}\right)$$

$$\phi = \arctan\left(-\sqrt{\frac{m}{k_s}} \frac{v_0}{x_0}\right) - \sqrt{\frac{k_s}{m}} t_0$$

Ex. 3.11 Suppose the system is not oscillating and you strike it with a hammer, applying a large force F for a *very* short time Δt. What is the initial speed v_0? Explain briefly. What will be the amplitude (the maximum displacement away from the equilibrium position)?

Ex. 3.12 In the case you just analyzed (striking with a hammer at time $t_0 = 0$), what is the phase shift ϕ? (Does this make sense?)

Comment: Two constants for a second-order differential equation

Note that whether we specify the starting conditions in terms of x_0 and v_0 or in terms of the amplitude A and the phase shift ϕ, two constants are required. You can see why two constants are needed by looking at the start of the numerical integration. You needed the initial position x_0 in order to be able to calculate the initial force (which lets you step the velocity), and you also needed the initial momentum p_0 or velocity v_0 in order to be able to step the position.

We have been working with what is called a "second-order differential equation." That is, the highest derivative in the equation is a second derivative—the acceleration $dv/dt = d(dx/dt)/dt$. A second-order differential equation normally requires two constants to specify the starting conditions completely.

3.4.13 *Linear systems

We saw that doubling the amplitude of a spring-mass system simply multiplies the solution by a factor of two without changing the time dependence of the motion. This is an extremely important property of systems described by "linear" differential equations (equations in which the variables appear only to the first power, not squared, for example). More formally, let x be replaced by the quantity Rx, where R is a constant:

$$m\frac{d}{dt}\left(\frac{d(Rx)}{dt}\right) = -k_s(Rx)$$

The R's cancel, leaving us with the original equation. Therefore starting with a different initial stretch will simply scale up the vertical axis of the graph of x vs. t. Also, since the motion is periodic, a change in the initial velocity corresponds to starting the motion at a different time. For example, instead of starting with an initial stretch and zero velocity, you could start with zero stretch and some initial velocity, and this would correspond to a point on the original graph where the function crosses the time axis.

In contrast, consider a differential equation in which there is an x^3, and try scaling by R:

$$m\frac{d}{dt}\left(\frac{d(Rx)}{dt}\right) = -b(Rx)^3 = -bR^3x^3$$

In this case the R's do *not* cancel. The solutions to this "nonlinear" equation do not scale up in the simple way that solutions to a linear equation do.

3.5 The speed of sound in a solid

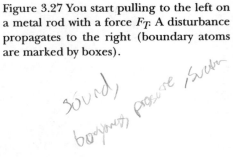

Figure 3.27 You start pulling to the left on a metal rod with a force F_T. A disturbance propagates to the right (boundary atoms are marked by boxes).

Our ball-and-spring model of a solid should allow us to predict real properties of matter—otherwise the model is of no value. One of the properties we might try to predict is the speed of propagation of a disturbance in a solid.

When you first start pulling to the left on the end of a metal rod, you very slightly lengthen the interatomic bonds between neighboring atoms in the nearby section of the rod. As a result of their leftward displacement, these atoms stretch the bonds of their neighbors to the right, and very quickly this new interatomic bond-stretching propagates to the right, all the way to the other end of the rod, and the whole rod is then in tension.

On Figure 3.27 we have marked atoms on the boundary between the already stretched region of the rod and the newly stretched region. It is this boundary that moves to the right, along the length of the rod. Individual atoms move extremely short distances, though for clarity in the diagram we have enormously exaggerated the interatomic stretches and the distance any individual atom moves.

This process is very fast, but it is not instantaneous. The boundary (or the the place where there is a change in the average length of the interatomic bond) propagates at a rate that we will try to determine, using our ball-and-spring model and our understanding of spring-mass motion.

The rate of propagation of the disturbance is called the "speed of sound" in the material, and we should explain what the propagation has to do with sound. When you clap your hands, a pulse of sound propagates through the air to your ear (Figure 3.28). This pulse starts as a momentary compression of the air (an increase in the air density) made by your hands. This compression is passed on to neighboring regions of the air, and the air near the hands relapses to its original density.

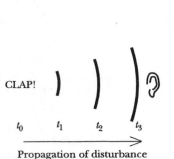

Figure 3.28 Clap your hands—sound travels to your ear as a pulse of air compression.

Eventually your eardrum is hit by the pulse and moves in response to the sudden increase in density (and pressure). The motion of the eardrum is detected by your inner ear and passed on to the brain, which interprets the signal as meaning that you clapped your hands. Note that no individual air molecule moves from near your hands to your ear. Rather it is the disturbance that moves from one place to another.

Similarly, a burst of sound hitting a solid surface compresses the atoms at the surface of the solid very slightly, and this compression propagates into the solid at a rate called the "speed of sound" in the solid. The sudden change in tension caused by your pulling on one end of the rod also propagated at this speed.

Dimensional analysis

? How would you expect the speed of sound in a solid to depend on the interatomic spring constant k_s? On the mass m of one of the atoms?

Our ball-and-spring model of a solid correctly suggests that the larger the interatomic spring stiffness k_s, the faster the propagation, due to quicker response of a neighboring atom. Also, the larger the mass m of the atoms, the slower the propagation (slower response of a neighboring atom). It is reasonable to guess that the speed of sound is proportional to the angular frequency $\omega = \sqrt{k_s/m}$ of atomic oscillations.

The angular frequency ω has units of $1/s$. If we multiply by some atomic distance to get units of m/s, we might have a rough prediction for the speed of sound. A plausible guess is to say that ω is related to the rate at which one atom hits its neighbor, propagating a change through a distance of one interatomic spacing. Therefore multiplying by the interatomic spacing d (the distance between nuclei) might give us the speed of sound:

$$v = \omega d \ ?$$

This informal reasoning is an example of "dimensional analysis," in which we identify meaningful quantities in the situation and combine them to guess a formula for some other quantity of interest. While this can be a fruitful procedure, we shouldn't be surprised if we find that the speed of sound is actually given by $v = 2\omega d$, or $v = \omega d/\pi$, for example. Moreover, note that atomic oscillations in the x direction involve two atomic "springs" on each side of an atom, not one, so we can't even be sure that $\omega = \sqrt{k_s/m}$ is the right way to estimate ω. Nevertheless, we can guess that $(\sqrt{k_s/m})d$ is proportional to the actual speed of sound, with the correct dependence on k_s and m. As it happens, it can be shown that this *is* the correct formula. We omit the difficult derivation.

Speed of sound in different materials

In aluminum, which is very stiff (large k_s and small atomic mass m), the speed of sound is very high, about 5000 m/s. If a wire is half a meter long, it takes about

$$(0.5 \text{ m})/(5000 \text{ m/s}) = 10^{-4} \text{ s} \text{ (a tenth of a millisecond)}$$

from the time you pull on one end for the other end to become tense. On the other hand, lead is quite soft (small k_s and large atomic mass m), and the speed of sound in lead is only about 1200 m/s, much lower than in aluminum. (The speed of sound in solids is much higher than the speed of sound in air, which is about 340 m/s. Because air molecules are usually far from one another, the speed of sound in air is determined by the average speed of the air molecules, not by interatomic forces.)

If possible, it is interesting to measure the speed of sound in a metal bar directly, by hitting one end of a long bar with a short pulse and observing on an oscilloscope when sensors at two locations along the bar detect the sound.

Ex. 3.13 Using your results from Problem 3.2 (page 82), predict the speed of sound in aluminum and compare with the measured value (about 5000 m/s). Also predict the speed of sound in lead and compare with the measured value (about 1200 m/s).

3.5.1 Computing and visualizing propagation

Problem 3.6 involves modeling the propagation of a disturbance in a metal. From the computer modeling you have already done, you have most of the tools required to study the speed of sound. We discuss two important aspects of the modeling that may be less familiar.

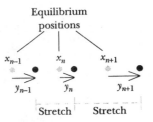

Figure 3.29 Figuring out the interatomic stretches to the left and the right of the *nth* atom, in terms of the displacements y_n away from the equilibrium positions.

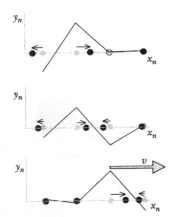

Figure 3.30 Visualizing the propagation by plotting the displacements y_n away from the equilibrium positions x_n.

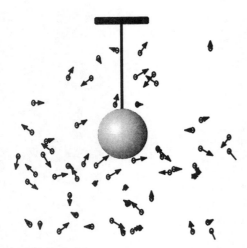

Figure 3.31 Air molecules in motion near a hanging ball.

For every atom in a long chain of atoms, you need to take into account interatomic spring forces exerted by atoms both to the left and to the right of the atom. Consider the *nth* atom in the chain, located at equilibrium position x_n (Figure 3.29). When there is no disturbance, the distance between the centers of the atoms is d, and the interatomic spring forces are zero. If atoms *n*–1, *n*, and *n*+1 are displaced from their equilibrium positions by amounts y_{n-1}, y_n, and y_{n+1}, the interatomic forces are computed from the stretches:

$$k_s(y_{n+1} - y_n) - k_s(y_n - y_{n-1})$$

Knowing the current displacements of all the atoms (y_1, y_2, etc.), you can compute the forces on each atom and determine how they will move in the next time step.

Visualizing the propagation of the disturbance can be done in various ways. A particularly useful visualization is to plot the horizontal displacements (y_n) vertically, at the locations of the horizontal equilibrium locations (x_n). An example is shown in Figure 3.30, where you see that a displacement to the left is plotted below the axis, and a displacement to the right is plotted above the axis. For further details, see Problem 3.6 (page 109) Computer model of the propagation of sounds in a metal.

3.6 Buoyancy

Until now we have neglected the force that the air exerts on objects. Consider the surrounding air that is in contact with a ball hanging from a wire, as shown in Figure 3.31. You might think that the air has no effect on the ball, especially since the ball is not moving, so we don't need to worry about air resistance. To be certain, let's consider this issue at the microscopic level.

Air is a gas, composed of about 80% nitrogen and 20% oxygen, with very small amounts of other gases. The molecules in the air are continuously in motion, moving in random directions. Some of the molecules in the air near the ball will occasionally bump into the ball, exerting a small force on it. If there is no wind, the motion of the gas molecules is random, so in a short time period there should be just as many molecules colliding with the right hand side of the ball as with the left hand side, and there should on average be no sideways force on the ball.

However, the situation is not quite the same for collisions with the top and bottom of the ball. The directions of molecular velocities are still random, but there are more air molecules per cubic meter below the ball than there are above it. The density of the Earth's atmosphere decreases as one moves upward away from the Earth's surface, until finally at a sufficient distance (approximately 50 km) there is essentially no air left. You may have observed this variation yourself if you have ever gone from sea level to a location several thousand meters higher; you may feel light-headed if you exercise at high altitude because the density of oxygen is significantly lower.

This variation in air density occurs over kilometers; can it possibly be important over a distance of a few centimeters? Interestingly enough, the answer is yes. There is actually a "buoyant force" upward on the ball, because the number of air molecules hitting the bottom of the ball per second is slightly greater than the number of molecules hitting the top of the ball per second.

A macroscopic view of buoyancy

How can we estimate the magnitude of the buoyant force on the ball due to the air? Let's consider a macroscopic viewpoint. The following discussion, based on the momentum principle, applies equally well to the buoyant force exerted by any fluid (air, water, other liquids, other gases) on a floating object. The result we will obtain is called the Archimedes principle, for the Greek thinker who first understood it.

Imagine a box filled with air (or water). Consider a ball-shaped region that a ball will eventually occupy, at a moment when the ball is not yet there, so the ball-shaped region is filled with the fluid. Let's choose this ball-shaped mass of fluid as our system, and consider what forces must be acting on it (Figure 3.32). This may seem odd, but it is a perfectly legitimate choice of system.

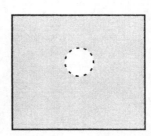

Figure 3.32 A system consisting of a ball-shaped spherical region of fluid, surrounded by more fluid.

? Is the momentum of the indicated sphere of fluid changing? What must the net force on this sphere be?

The momentum of the indicated sphere of fluid is not changing. The Earth exerts a downward gravitational force on it, but the sphere does not sink. Because its momentum isn't changing, the net force on the sphere must be zero, so there must be an upward force that balances the gravitational force.

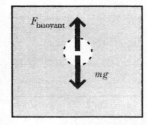

Figure 3.33 Forces on the sphere of fluid. Here m is the mass of the fluid in the sphere.

? What exerts this upward force on the sphere?

The only object in contact with this sphere is the rest of the fluid, so it must be that the rest of the fluid exerts an upward force F_b ("buoyant force") whose magnitude is mg, where m is the mass of the sphere of fluid (Figure 3.33).

Now remove the sphere of fluid and replace it with the ball. It must be that initially the rest of the fluid still exerts the same upward force F_b of magnitude mg (Figure 3.34). This force may not be large enough to counteract the downward gravitational force on the ball; in the case of a ball hanging in air it clearly is not, since if the ball were not supported by a wire it would begin accelerating downward.

The upward force is called a "buoyant" force but is simply an interatomic contact force due to fluid molecules striking atoms in the surface of the ball. The key point is that despite the complexity of the interatomic interactions, the net effect is simply an upward force whose magnitude is the mass of an equivalent volume of fluid, times g. Often it is reasonable to neglect buoyant forces in air (though usually not in water). To see why, let's compare the buoyant force to the gravitational force F_g on a ball hanging in air, where we write V for the volume of the ball:

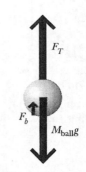

Figure 3.34 The tension force is actually less than the weight of the ball, because the buoyant force F_b due to the air is also in the upward direction.

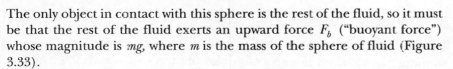

$$\frac{F_b}{F_g} = \frac{m_{air}g}{M_{ball}g} = \frac{m_{air}}{M_{ball}} = \frac{m_{air}/V}{M_{ball}/V} = \frac{\rho_{air}}{\rho_{ball}}$$

The ratio of the forces is equal to the ratio of ρ_{air} (the density of air) to ρ_{ball} (the density of the ball; ρ is Greek lower-case letter rho). The density of air at STP ("standard temperature and pressure," 0°C and 1 atmosphere) can be calculated if we remember from chemistry that one mole of a gas at STP occupies 22.4 liters. The molecular mass of N_2 is 28 and that of O_2 is 32, which gives an average molecular mass of about 29 for air (which is about 80% nitrogen and 20% oxygen):

$$\rho_{air} \approx \frac{29 \text{ grams/mole}}{22.4 \times 10^3 \text{ cm}^3 / \text{mole}} \approx 1.3 \times 10^{-3} \text{ grams/cm}^3$$

What is the density of the ball? We can estimate it by noting that the density of water is 1 gram/cm^3, and that most solids are more dense than water. Using 1 g/cm^3 as the density of the ball, we find that $F_b/F_g \approx 1.3 \times 10^{-3}/1$.

The buoyant force $m_{air}g$ of air on a solid object is therefore very small compared to the gravitational force mg. In many cases we can safely choose to neglect it. Since the upward force exerted by the air on the ball is small compared to the gravitational force or the tension force exerted by the wire, we can choose to neglect it in an analysis of the forces on a hanging ball. However, it is not always appropriate to neglect the buoyant force; in accurate analytical chemistry it is necessary to correct for buoyancy when weighing a sample of liquid or low-density solid on an analytical balance.

A final note: the buoyant force being equal to $m_{air}g$ is really a time average, because the buoyant force is due to random collisions with air molecules and therefore has an intermittent character. However, the collision rate is so high that the force seems nearly continuous and constant. If, however, the ball is of microscopic size, it may not be possible to ignore the intermittent nature of the collisions. In a microscope it is possible to observe small particles being jostled about by the random collisions with the water molecules. This effect is called "Brownian motion."

Ex. 3.14 Calculate the buoyant force in air on a kilogram of iron (whose density is about 8 grams per cubic centimeter). Compare with the weight mg of this much iron.

3.7 Pressure and suction

There is an interesting case where we cannot neglect the effect of the air. Suppose that an object lies on a table and the surfaces of the table and object are so smooth that upon squeezing them together all air is expelled. Such a situation occurs when a suction cup is pressed down onto a smooth surface. Now there are no longer any air molecules striking the bottom surface of the suction cup, while air molecules continue to strike the top surface. The force of the air on the top surface is quite large, about 10^5 newtons on each square meter at sea level! This large force per unit area is called the "pressure" P of the air. The pressure can be thought of as due to the weight of a one-square-meter column of the entire atmosphere, extending upward many kilometers (although with diminishing density), as illustrated in Figure 3.35.

$$\text{Pressure } P = F/A \text{ (force per unit area)}$$

If the atmosphere had a constant density ρ like that near the surface (1.3 grams per cubic centimeter), we could calculate the height h of the atmosphere in the following way. The volume of a column of air of height h and cross-sectional area A is Ah, and its mass is $M_{air} = \rho(Ah)$. The pressure, which is about 10^5 newtons per square meter, is the weight $M_{air}g$ divided by the cross-sectional area: $P = M_{air}g/A = \rho gh$.

The height of a constant-density atmosphere, $h = P/(\rho g)$, would be about 8000 meters (about 5 miles), about the height of Mount Everest. In reality air density is not constant but decreases as you go higher, so there is still some low-density air at much higher altitudes than this.

A physics diagram for a suction cup looks like Figure 3.36, where PA represents the force of the air on the area A of the top surface of the suction cup. The normal force exerted upward by the table and the downward force of the air are huge compared to the weight of the suction cup mg.

The normal force $F_N = PA + mg \approx PA$, since the force of the air is so much larger than the weight mg of the suction cup. For example, a rubber suction cup with a diameter of 4 cm has an area of $\pi(0.02\text{ m})^2 = 1.3\times10^{-3}$ m^2, so the force of the air on the suction cup is

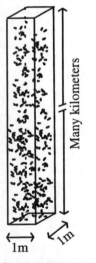

Figure 3.35 A one-square-meter column of air extending upward through the entire atmosphere.

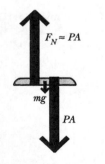

Figure 3.36 The weight of the suction cup is very small compared to the downward force by the air or the normal force by the table.

$$PA = (10^5 \text{ N/m}^2)(1.3 \times 10^{-3} \text{ m}^2) = 130 \text{ newtons}$$

whereas its own weight mg might be about 0.01 kg (10 grams) times 9.8 N/kg, or only about 0.1 newton. Note that a force of 130 newtons is the weight of an object whose mass is about 13 kg, or about 29 pounds, which is why it isn't easy to pull even a small suction cup off a flat surface.

Every solid object in air, such as a hanging ball, is subjected to large compression forces by the pressure of the surrounding air (Figure 3.37). The upward buoyant force is the tiny difference between these huge forces. A hollow object must be very strong to support this compression, if air is removed from the inside. There is no problem however if air can flow into and out of the container, because then it is just the solid walls that are compressed, since the air exerts comparable pressure on the inner and outer surfaces.

All of this is even more relevant under water, because the density of water is about 1000 times greater than the density of air, and the compression forces are about 1000 times as strong. A hollow submersible vehicle that dives very deep must have an extremely strong hull. When a scuba diver descends to greater depth where the water pressure is larger, the breathing apparatus automatically increases the pressure of the air supplied to the diver to equal the increased pressure of the outside water at the new depth, in order to prevent the lungs from being crushed.

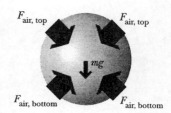

Figure 3.37 A ball hanging in air is subject to large compression forces due to the surrounding air.

Ex. 3.15 It is hard to imagine that there can be enough air between a book and a table so that there is a net upward (buoyant) force on the book despite the large downward force on the top of the book. About how many air molecules are there between a textbook and a table, if there is an average distance of about 0.01 mm between the uneven surfaces of the book and table?

3.8 Analysis with unknown force laws

In the case of a planet orbiting a star and the horizontal motion of a mass on a spring, you knew the forces at each step in the numerical integration, because these forces were completely determined by the positions of the objects (the gravitational force depends on the distances to other objects; the spring force depends on the stretch). Without known formulas for the gravitational force or the spring force you wouldn't have been able to predict the motion.

However, for the vertical forces acting on the horizontal spring-mass oscillator, we didn't have a force law for the upward force F_N exerted by the air table. We were able to deduce this force from the motion: $F_N = mg$.

What makes such deductions possible is knowing something about the motion. For example, if for some reason you know that there is no motion in the z direction, then you know that the z component of momentum isn't changing, and that the z component of the net force must be zero, no matter what force laws apply.

A circular pendulum

As another, more complex example of an analysis involving unknown force laws, we will analyze a ball hanging at the end of a string, but with the ball in motion. There are many possible motions of this system. The ball could swing back and forth in a plane; this is the motion of a "simple pendulum" (see Problem 3.15). The ball could go around in a path something like an ellipse, in which case the ball's height would vary, just as with simple pendu-

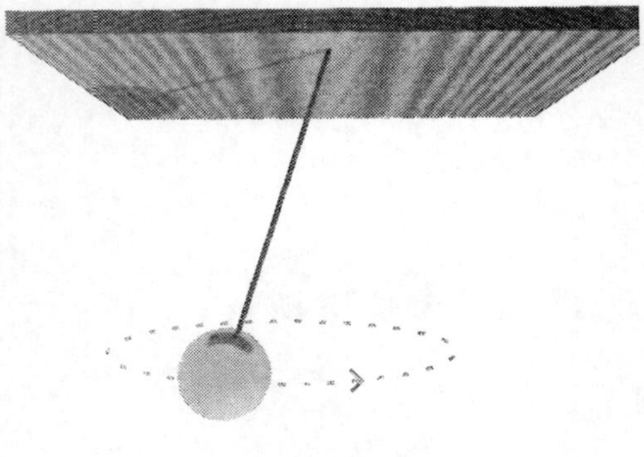

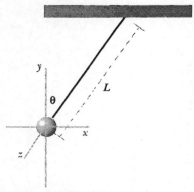

Figure 3.38 A circular pendulum.

lum motion. The ball could move so violently that the string goes slack for part of the time, with abrupt changes of momentum every time the string suddenly goes taut. If the string can stretch noticeably, there can be a sizable oscillation superimposed on the swinging. All of these motions are difficult to analyze. It is even difficult to do a numerical integration, because it is difficult to express the force of the string in terms of the position of the ball.

There is one particular motion of the ball that is simple enough to model and to analyze without a lot of mathematics—circular motion with a fixed stretch of the string. This is a "circular pendulum" (Figure 3.38). As you discovered in Chapter 2 when studying planetary orbits, circular motion doesn't just happen. To get the ball moving in a circle you have to start it moving with just the right initial conditions. We tackle a problem that we can solve, by imposing the requirement that we only consider the case of circular motion.

A large circular pendulum was used by Newton to measure *g*, the strength of the gravitational field, with considerable accuracy. You will be able to observe and measure the motion of a circular pendulum yourself, and your analysis of the motion will let you deduce a value for *g*, which you can compare with the accepted value.

Unknown forces and known motion

The ball interacts with the string and with the Earth. We know the force exerted by the Earth. In principle, we could predict the general motion of the ball hanging from the string if we had a formula for the tension force in the string, as a function of its length. Essentially, we would need the effective spring stiffness for this stiff "spring." However, the string may be so stiff that even a tiny stretch implies a huge tension (a nearly "inextensible" string). The observable length of the string is nearly constant, but the tension force that the string applies to the ball can vary a great deal. In a practical sense, the tension force is unknown.

As we will see, what makes an analysis possible is knowing that there is circular motion with constant length of the string. This motion information compensates for our lack of a force law describing the effect of the string on the ball.

A physics diagram (Figure 3.39) shows a snapshot of a side view of the circular pendulum, at an instant when the string (of length *L* to the center of the ball) lies in the *xy* plane (at an angle θ to the vertical), and the ball is at the leftmost point in its circular path. Let's assume that the ball is moving counter-clockwise as seen from above.

? What is the direction of the ball's momentum at this instant?

At this instant the ball's velocity is momentum along the +*z* axis (out of the page).

Identify forces

? List all the objects that exert forces on the ball. Be sure to consider both distant interactions and contact interactions. Draw and label the significant forces on the physics diagram. If you don't have a force law for a particular force, just give the force a name and consider the force to be unknown for now.

Non-contact interactions

Only the Earth exerts a significant non-contact force on the ball.

Figure 3.39 A diagram of the circular pendulum, showing the main geometrical aspects of the situation.

Contact interactions

Only the (bottom of the) string and the air exert contact forces on the ball. In particular, note that the support at the upper end of the string doesn't touch the system of interest. The force of the string on the ball is shown in a physics diagram (Figure 3.40).

We label the string force F_T because it is equal to the tension in the string, but its magnitude is unknown because we don't know the force law for the string. An important property of a flexible string is that it cannot exert a sideways force, or a compression force; it can only exert a tension force in the direction of the string.

? What approximations and simplifying assumptions are appropriate, in order to be able to analyze the motion approximately?

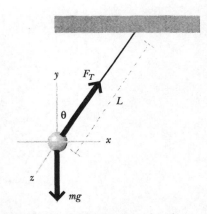

Figure 3.40 We add known and unknown forces to create a useful physics diagram.

Approximations and assumptions

Since the mass is moving through the air, we must consider air resistance. To start with a simple case, we make a *simplifying assumption* that air resistance is negligible, although we don't have a good way to estimate how big it is actually likely to be. This simplifying assumption is presumably valid if the mass is sufficiently dense. If on the other hand the mass is just a slightly crumpled piece of paper, air resistance will be comparable to the other forces. It will be easy to tell whether air resistance is significant for our own circular pendulum; if so, the mass will quickly come to a stop. Later we could refine our model and include the effects of air resistance.

In this analysis we don't care how much the string has stretched and whether or not the tension is uniform throughout the string, because we are focusing on the ball's motion. We do not need to make any assumptions about the string except that its stretch does not change, and we intend to make the ball move in such a way that this will be true.

Write the momentum principle and motion information

? Look at the system diagram (with axes and forces shown, both known and unknown), and write the components for the momentum principle and for the motion information. Important: This is *not* a case of static equilibrium. The ball is *not* sitting still. It not only is moving, but its momentum is changing, even though the magnitude of the momentum is constant. Think carefully about what information you have about the motion. The information about the motion can be used to determine the unknown forces.

Here is the vector form of the momentum principle at the instant for which we drew the physics diagram (Figure 3.40), given our assumptions and approximations:

$$\frac{d\vec{p}}{dt} = \vec{F}_{net} = <F_T\sin\theta, (F_T\cos\theta - mg), 0>$$

There are two pieces of motion information. The chosen motion is circular motion, and there is no change in height y.:

$$\frac{dp_x}{dt} = \omega p_x, \text{ where } \omega = \frac{v}{r}$$

$$p_y = 0 \text{ at all times, so } \frac{dp_y}{dt} = 0$$

As the ball travels in a circle, the direction of its momentum is changing, and as we saw in Chapter 2, this rate of change $d\vec{p}/dt$ is toward the center of the circle. Our x axis points in this direction, so the x component of

$d\vec{p}/dt$ is nonzero. In Chapter 2 we found that the magnitude of $d\vec{p}/dt$ for a mass moving at constant angular speed ω in a circle of radius r is ωp, so at the instant of the snapshot we have $dp_x/dt = \omega p_x$.

The y position of the mass is constant (it is not going up or down); both p_y and dp_y/dt are zero. This known information about the motion compensates for our lack of knowledge of a force law for the string.

At this instant there is no component of force in the z direction (tangent to the circle), which implies that dp_z/dt must be zero. Since the z direction is the direction of the ball's momentum at this instant, this means that the magnitude of the momentum cannot change; only the *direction* of the momentum can change.

? We assumed that the ball was moving counter-clockwise, as seen from above. Would the motion information change if we assumed instead that the ball was moving clockwise?

No, the key information is basically the same.

Solve the equations

? Solve the equations in terms of the quantities m, g, L, v, and θ (eliminate the unknown string tension F_T, and also eliminate the path radius r, which is simply $L\sin\theta$).

The result is $v^2 = gL\sin\theta\tan\theta$. Now you have an equation relating the speed v and the gravitational field strength g. It was possible to do this analysis despite not knowing a force law for the string, thanks to the motion information that $dp_y/dt = 0$. Given a value for g, you can predict how long it will take the pendulum to go around once. Alternatively, you can measure the time to go around and calculate a value for g.

? Your equation relates speed and g. How could you use this equation to predict the time T it takes to go around once?

The speed is the circumference of the circle divided by the time it takes the pendulum to go around once.

No "centrifugal force"

We point out that in this analysis there is no role for a radially outward "centrifugal force" (Figure 3.41). There is no interaction associated with such a fictitious force. Moreover, if the net radial force were zero, Newton's first law says that the ball would have to move in a straight line!

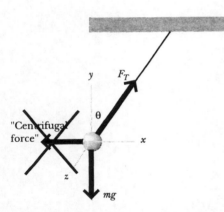

Figure 3.41 There is no "centrifugal force." There is no interaction associated with such a fictitious force. Moreover, if the net radial force were zero, the ball would have to move in a straight line!

Problem 3.5 Experiment—determine g with a circular pendulum
Use a circular pendulum to determine g. You can increase the accuracy of the time it takes to go around once by timing N revolutions and then dividing by N. This minimizes errors contributed by inaccuracies in starting and stopping the clock. It is wise to start counting from zero (0, 1, 2, 3, 4, 5) rather than starting from 1 (1, 2, 3, 4, 5 represents only 4 revolutions, not 5). It also improves accuracy if you start and stop timing at a well-defined event, such as when the mass crosses in front of an easily visible mark.

This was the method used by Newton to get an accurate value of g. Newton was not only a brilliant theorist but also an excellent experimentalist. For a circular pendulum he built a large triangular wooden frame mounted on a vertical shaft, and he pushed this around and around while making sure that the string of the circular pendulum stayed parallel to the slanting side of the triangle.

3.8.1 Comment: Forward reasoning

A common technique for solving problems is to start by considering what you want to know (or what you have been asked to determine), and to work backward from that to the solution. We have approached the analysis of physical systems in the opposite way. We have begun by organizing our knowledge about the system according to a fundamental principle—the momentum principle— and systematically recording our knowledge in equations and diagrams. After we have recorded all the information we have, whatever is missing is what we need to figure out. At this point it is usually quite clear how to figure out this missing information.

This way of working problems, sometimes called "forward reasoning," is typical of the way expert physicists solve problems of this kind—starting each new problem from the fundamental principles. This approach works! For the student, however, approaching problems in this way may require a bit of faith—it is not necessarily obvious how the desired answer will emerge from the analysis. If you practice approaching problems in this way, you will develop confidence in the approach and will gain important experience in explaining complex phenomena starting from fundamental principles.

3.9 Summary

Fundamental principles

No new fundamental principles were introduced in this chapter.

New concepts

Macroscopic spring: $|\vec{\mathbf{F}}| = k_s|s|$, where $s = L - L_{\text{relaxed}}$

Forces describing interatomic interactions (contact forces):

  Tension force in a wire or a string due to stretching of interatomic bonds in the object

  "Normal" force due to compression of object

$$\frac{F_T}{A} = Y\frac{\Delta L}{L}; \text{ atomic version } Y = \frac{(k_s s/d^2)}{(s/d)} = \frac{k_s}{d}$$

Concepts of amplitude A, angular frequency ω, frequency f, and period T for oscillator.

Buoyancy force due to variation in air density.

Two classes of problems

Given the forces, predict the unknown motion.

Given the motion, determine the unknown forces.

Problem solving technique

Choose a system and make a physics diagram

Identify objects interacting with the system and record forces in diagram

  Non-contact interactions (electric or gravitational)

  Contact interactions

Write the momentum principle and motion information

Solve equations numerically or analytically

Explicitly list and discuss all assumptions/approximations made

Results

Near the Earth's surface, $F_g \approx mg$, where $g = 9.8\frac{\text{N}}{\text{kg}}$

Spring-mass motion

Numerical solution of the momentum principle $\frac{dp_x}{dt} = -k_s x$ for a horizontal mass on a spring produces an oscillatory motion. The idealized spring-mass system has an analytical prediction:

$$x = A\cos\left(\sqrt{\frac{k_s}{m}}t\right)$$

Angular frequency $\omega = \sqrt{\frac{k_s}{m}}$, Period $T = \frac{2\pi}{\omega}$, Frequency $f = \frac{1}{T} = \frac{\omega}{2\pi}$

Speed of sound in a solid

$v = d\sqrt{k_s/m}$, where d is distance between nuclei, m is atomic mass

Buoyancy and pressure

Upward buoyancy force $= m_{\text{fluid}}g$, where m_{fluid} is mass of equivalent volume of fluid.

Pressure is force per unit area; air pressure at sea level is about $10^5\,\text{N}/\text{m}^2$.

3.10 Example problems

3.10.1 Example 1: A vertical steel rod

Steel has a rather high Young's modulus, 2×10^{11} N/m^2.

(a) A thin steel rod is 4 m long with diameter 2 mm (2×10^{-3} m). You hang a 50 kg mass from the bottom of the vertical rod, and the rod does not sag catastrophically. How much does the rod stretch? Explain your work clearly.

(b) You pull the 50 kg mass down very slightly and release it. In the resulting vertical oscillations of the large mass, how long does it take for the mass to move from its lowest position to its highest position? Explain your work, including a brief explanation of the model of the rod-mass system you used.

Solution

Start from a fundamental principle: the momentum principle. This is a case of known motion (no motion) and unknown forces.

(a) The rod pulls up on the 50 kg mass with a force F_T, the tension in the rod. Since the momentum of the 50 kg mass isn't changing, the net force on it must be zero, so

$$F_T - mg = 0 \text{, and the tension } F_T = mg$$

Since the rod doesn't sag catastrophically, we can assume that the stress-strain relation in terms of Young's modulus is valid:

$$\frac{F_T}{A} = Y\frac{\Delta L}{L}$$

$$\Delta L = \frac{F_T L}{AY} = \frac{(50 \text{ kg})(9.8 \text{ N/kg})(4 \text{ m})}{\pi(10^{-3} \text{ m})^2(2\times10^{11} \text{ N/m}^2)} = 3.1\times10^{-3} \text{ m} = 3.1 \text{ mm}$$

(b) We can consider the rod to be a very stiff spring, and model the system as a mass on a spring, since the stretch of the rod, $s = \Delta L$, is proportional to the force $F_T = mg$ that the 50 kg mass exerts on the rod. (By the principle of reciprocity, the interatomic force that the mass exerts on the bottom of the rod is equal and opposite to the interatomic force the rod exerts on the top of the mass.)

$$F_T = \left(\frac{AY}{L}\right)\Delta L = k_s s$$

Therefore the effective spring stiffness of the rod is this:

$$k_s = \frac{AY}{L}$$

We have a known solution for the motion of a mass acted on by a force proportional to displacement, $y = A\cos(\omega t)$. Since $\omega = \sqrt{k_s/m} = 2\pi/T$, we find that the period is

$$T = 2\pi\sqrt{\frac{m}{k_s}} = 2\pi\sqrt{\frac{mL}{AY}} = 2\pi\sqrt{\frac{(50 \text{ kg})(4 \text{ m})}{\pi(10^{-3} \text{ m})^2(2\times10^{11} \text{ N/m}^2)}} = 0.112 \text{ s}$$

Moving from the lowest position to the highest position takes half a period, which is

$$\frac{T}{2} = 0.056 \text{ s}$$

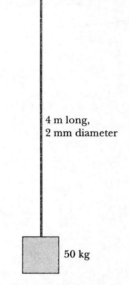

4 m long,
2 mm diameter

50 kg

Figure 3.42 A thin steel rod supports a 50 kg mass.

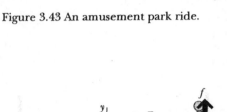

Figure 3.43 An amusement park ride.

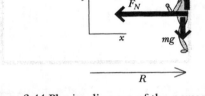

Figure 3.44 Physics diagram of the person. At this instant the person is moving in the –z direction.

3.10.2 Example 2: An amusement park ride

There is an amusement park ride that some people love and others hate in which a bunch of people stand against the wall of a cylindrical room of radius R, and the room starts to rotate at higher and higher angular speed ω (Figure 3.43). When a certain critical angular speed is reached, the floor drops away, leaving the people stuck against the whirling wall.

Explain why the people stick to the wall without falling down. Include a carefully labeled force diagram of a person, and discuss how the person's momentum changes, and why.

Solution

Start from a fundamental principle: the momentum principle. This is a case of known motion with unknown forces.

We draw a physics diagram for a person of mass m at the moment when the person is at the far right, moving in the –z direction (Figure 3.44).

Non-contact forces

The Earth exerts a known force mg downward.

Contact forces

The wall exerts an unknown force which must have a y component $+f$ (because the person isn't falling) and an x component $-F_N$ normal to the wall (because the person's momentum is changing direction). Note that there is no outward-going "centrifugal" force! There is a momentum change inward; if the net force were zero, the person would move in a straight line.

We can deduce something about the unknown wall force from the known motion, which is circular motion with no change in y. Here are the momentum principle and the motion information:

$$\frac{d\vec{p}}{dt} = <-F_N, (f - mg), 0>$$

$$p_y = 0 \text{ at all times, so } \frac{dp_y}{dt} = 0$$

$$\left|\frac{d\vec{p}}{dt}\right| = \omega p, \text{ circular motion with constant } p = |\vec{p}|$$

Combining the momentum principle with the motion information we have

$$F_N = \omega p \text{ and } f = mg$$

We can also write $p = mv = m\left|\dfrac{d\vec{R}}{dt}\right| = m\omega R$, so the normal component of the wall force is

$$F_N = \omega p = m\omega^2 R$$

The vertical component f of the wall force is a frictional force. If the wall has friction that is too low, the person won't be supported. In Chapter 5 we will see that $f \le \mu F_N$, where the "coefficient of friction" μ has a value for many materials between about 0.1 and 1.0. The angular speed has to be large enough that

$$mg \le \mu(m\omega^2 R)$$

$$\omega^2 \ge \frac{g}{\mu R}$$

The smaller the friction, the higher the angular speed needed.

3.11 Review questions

The nature of solids

RQ 3.1 Suppose you attempt to pick up a very heavy object. Before you tried to pick it up, the object was sitting still—its momentum was not changing. You pull very hard, but do not succeed in moving the object. Isn't this a violation of the momentum principle? How can you be exerting a large force on the object without causing a change in its momentum? What does change when you apply this force?

Spring-mass system without friction

For the following review questions, assume that friction is negligible.

RQ 3.2 Write the momentum principle for a horizontal spring-mass system with spring stiffness k_s and mass m, and write an analytical solution.

RQ 3.3 In terms of k_s and m, what is the angular frequency? The frequency? The period? The amplitude?

RQ 3.4 If you double the amplitude, by what factor does the period change? If you double the mass, by what factor does the period change? If you double the spring stiffness, by what factor does the period change?

RQ 3.5 How should you start the system going at $t = 0$ in order for the motion to be $A\cos(\omega t)$? How should you start the system going at $t = 0$ in order for the motion to be $A\sin(\omega t)$?

Other oscillators

RQ 3.6 Describe two examples of oscillating systems that are *not* harmonic oscillators.

The atmosphere

RQ 3.7 Calculate the height that the atmosphere would have if the air density were constant.

Pressure

RQ 3.8 Air pressure at the surface of a fresh water lake near sea level is about $10^5 \, \text{N/m}^2$. At approximately what depth below the surface does a diver experience a pressure of $2 \times 10^5 \, \text{N/m}^2$? How would this be different in sea water, which has higher density than fresh water?

Circular motion

RQ 3.9 You are driving an American car, sitting on the left side of the front seat. You make a sharp right turn. You feel yourself "thrown to the left" and your left side hits the left door. Is there a force that pushes you to the left? What object exerts that force? What really happens? Draw a diagram to illustrate and clarify your analysis.

3.12 Problems

3.12.1 In-line problems from this chapter

Problem 3.1 (page 81) Experiment—study of strain vs. stress
Measure and graph the strain vs. stress curve for a long metal wire by hanging weights from the end of the wire and carefully measuring the stretch as a function of the hanging weight. If possible, go beyond the "yield" stress and observe the rapid, large increase in the strain. BE CAREFUL! Once you exceed the yield stress, the wire may break and dump the weights on whatever is underneath.

Problem 3.2 (page 82) Interatomic spring constant
Young's modulus for aluminum is 6.2×10^{10} N/m^2. The density of aluminum is 2.7 grams/cm^3, and the mass of one mole is 27 grams. If we model the interactions of neighboring aluminum atoms as though they were connected by springs, determine the approximate spring constant of such a spring. Repeat this analysis for lead: Young's modulus for lead is 1.6×10^{10} N/m^2, the density of lead is 11.4 grams/cm^3, and the mass of one mole is 207 grams. Make a note of these results, which we will use for various purposes later on. Note that aluminum is a rather stiff material, whereas lead is quite soft.

Problem 3.3 (page 83) Experiment—a vertical spring-mass system
You will be given a spring and one or more masses with which to study the motion of a mass hanging from a spring.

(a) Measure the value of the spring stiffness k_s of the spring in N/m. The magnitude of a spring force is $k_s s$, where s is the stretch of the spring (change from the unstretched length). Explain briefly how you measured k_s. Include in your report the unstretched length of the spring.

(b) Measure the period (the round-trip time) of a mass hanging from the vertical spring. Report the mass that you use, and the amplitude of the oscillation. Amplitude is the maximum displacement, plus or minus, from the equilibrium position (the position where the mass can hang motionless).

(c) With twice the amplitude that you reported in part (b), measure the period again. Since the mass has to move twice as far, one might expect the period to lengthen. Does it?

How to measure the period accurately

There are unavoidable fluctuations in starting and stopping the timing, but you can minimize the error this contributes by timing many complete cycles so that the starting and stopping fluctuations are a small fraction of the total time measured.

It is good practice to count out loud *starting from zero, not one*. To count five cycles, say out loud "Zero, one, two, three, four, five." If on the other hand you say "One, two, three, four, five," you have actually counted only four cycles, not five.

Since the motion is periodic, you can start (say "zero") at any point in the motion. It is best to start and stop when the mass is moving fast past some marker near the equilibrium point, because it is difficult to estimate the exact time when the mass reaches the very top or the very bottom, because it is moving slowly at those turnaround points. Be sure to measure full round-trip cycles, not the half-cycles between returns to the equilibrium point (but going in the opposite direction).

Problem 3.4 (page 87) Numerical integration of a spring-mass system without friction

Carry out a numerical integration of the motion of a horizontal spring-mass system. Use the spring stiffness k_s and mass m that you measured experimentally for a vertical spring-mass system. Predict the motion of this spring-mass system sufficiently far into the future to observe several oscillations.

(a) Initial conditions: Make the initial position of the block be such that the stretch of the spring is equal to the amplitude of the oscillations in your experiment, and release the block with zero initial momentum. Display the motion in two ways:

(1) Plot a graph of the position of the block as a function of time. Label the scales on both axes, so your numerical results are clear.

(2) If possible using your computer tool, make an animation of the motion of the block. (You don't need to draw a real spring; it is sufficient to draw a line connecting the block to the wall.)

Very briefly, what are the most important aspects of the predicted motion?

(b) Vary the time step and report the maximum step size that gives reasonable results, in the sense that a smaller step size has little effect.

(c) Read the "period" (the round-trip time) of the oscillation off the graph. How does this compare with the period you measured in your experiment?

(d) Repeat the calculation with double the initial stretch. What happens to the period? Is this what you observed in your experiment?

(e) Temporarily change the force law to be $-k_s x^3$, and observe the period with the amplitude of part (a) and with double this amplitude. What happens to the period when you change the amplitude?

(f) How long would the spring-mass system theoretically continue to oscillate? Would a real block connected to a spring behave that way? What *would* happen?

Problem 3.5 (page 102) Experiment—determine g with a circular pendulum

Use a circular pendulum to determine g. You can increase the accuracy of the time it takes to go around once by timing N revolutions and then dividing by N. This minimizes errors contributed by inaccuracies in starting and stopping the clock. It is wise to start counting from zero (0, 1, 2, 3, 4, 5) rather than starting from 1 (1, 2, 3, 4, 5 represents only 4 revolutions, not 5). It also improves accuracy if you start and stop timing at a well-defined event, such as when the mass crosses in front of an easily visible mark.

This was the method used by Newton to get an accurate value of g. Newton was not only a brilliant theorist but also an excellent experimentalist. For a circular pendulum he built a large triangular wooden frame mounted on a vertical shaft, and he pushed this around and around while making sure that the string of the circular pendulum stayed parallel to the slanting side of the triangle.

3.12.2 Additional problems

Problem 3.6 Computer model of the propagation of sounds in a metal

Create a computer model of a long metal bar as many atomic masses connected by interatomic springs, all in a straight line. Give the first atom a sudden push and compute the motions of all the atoms, one time step after another. Calculate all of the forces before using these forces to update the momenta and positions of the objects. Otherwise the calculations of the forces would mix positions corresponding to different times.

After each time step, display the displacements of all the atoms away from their equilibrium positions, and determine the speed of sound (the distance between the first and last atoms, divided by the amount of time it takes for the pulse to reach the last atom). Compare with the prediction that $v = (\sqrt{k_s/m})d$.

Problem 3.7 The interatomic spring stiffness for copper

A hanging copper wire with diameter 2 mm (2×10^{-3} m) is initially 3 m long. When a 5 kg mass is hung from it, the wire stretches an amount 0.425 mm, and when a 10 kg mass is hung from it, the wire stretches an amount 0.85 mm. A mole of copper has a mass of 63 grams, and its density is 9 grams/cm^3. Find the approximate value of the effective spring stiffness of the interatomic force, and explain your analysis.

Problem 3.8 Observing spring-mass systems

Using masses and springs similar to those you used, it was found that a 20 gram mass hanging from a spring had an oscillation period of 1.2 seconds.

(a) When two 20 gram masses are hung from this spring, what would you predict for the period in seconds? Explain briefly.

(b) When one 20 gram mass is supported by two of these vertical, parallel springs (Figure 3.45), what would you predict for the period in seconds? Explain briefly.

(c) Suppose you cut one spring into two equal lengths, and you hang one 20 gram mass from this half spring. What would you predict for the period in seconds? Explain briefly.

(d) Suppose you take a single (full-length) spring and a single 20 gram mass to the Moon and watch the system oscillate vertically there. Will the period you observe on the Moon be longer, shorter, or the same as the period you measured on Earth? (The gravitational field strength on the Moon is about one-sixth that on the Earth.) Explain briefly.

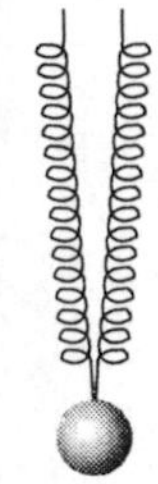

Figure 3.45 Hang the mass from two springs (Problem 3.8).

Problem 3.9 Vibration of diatomic molecules

In Problem 3.2 (page 82) we found the effective spring constant corresponding to the interatomic force for aluminum and lead. Let's assume for the moment that, *very* roughly, other atoms have similar values.

(a) What is the (very) approximate frequency f for the vibration of H$_2$, a hydrogen molecule?

(b) What is the (very) approximate frequency f for the vibration of O$_2$, an oxygen molecule?

(c) What is the approximate vibration frequency f of D$_2$, a molecule both of whose atoms are deuterium atoms (that is, each nucleus has one proton and one neutron)?

(d) Explain why the *ratio* of the deuterium frequency to the hydrogen frequency is quite accurate, even though you have estimated each of these quantities very approximately, and the effective spring constant is normally expected to be significantly different for different atoms. (Hint: what interaction is modeled by the effective "spring"?)

Problem 3.10 Speed of sound in uranium

Uranium-238 (U^{238}) has three more neutrons than uranium-235 (U^{235}). Compared to the speed of sound in a bar of U^{235}, is the speed of sound in a bar of U^{238} higher, lower, or the same? Explain your choice, including justification for assumptions you make.

Problem 3.11 Speed of sound in nickel

One mole of nickel (6×10^{23} atoms) has a mass of 59 grams, and its density is 8.9 grams per cubic centimeter. You have a bar of nickel 2.5 m long, with a square cross section, 2 mm on a side. You hang the rod vertically and attach a 40 kg mass to the bottom, and you observe that the bar becomes 1.2 mm longer.

Next you remove the 40 kg mass, place the rod horizontally, and strike one end with a hammer.

How much time T will elapse before a microphone at the other end of the bar will detect a disturbance? Be sure to show clearly the steps in your analysis.

Problem 3.12 A spring-like device

A certain spring-like device with uneven windings has the property that to stretch it an amount s from its relaxed length requires a force that is given by $F = bs^3$. You suspend this device vertically, and its unstretched length is 20 cm.

(a) You hang a mass of 15 grams from the device, and you observe that the length is now 24 cm. What is b, including units?

(b) You hold the 15 gram mass and throw it downward, releasing it when the length of the spring-like device is 27 cm and the speed of the mass is 4 m/s. One millisecond later (10^{-3} s), what is the stretch of the device, and what is the speed of the mass?

Problem 3.13 Falling through the Earth

In the approximation that the Earth is a sphere of uniform density, it can be shown that the gravitational force it exerts on a mass m inside the Earth at a distance r from the center is $mg(r/R)$, where R is the radius of the Earth. (Note that at the surface, the force is indeed mg). Suppose that there were a hole drilled along a diameter straight through the Earth, and the air were pumped out of the hole. If an object is released from one end of the hole, how long will it take to reach the other side of the Earth? Include a numerical result.

Problem 3.14 Floating objects

(a) A block of wood 20 cm long by 10 cm wide by 6 cm high has a density of 0.7 grams/cm^3 and floats in water (density 1.0 grams/cm^3). How far below the surface of the water is the bottom of the block? Explain your reasoning.

(b) An advertising blimp consists of a gas bag, the supporting structure for the gas bag, and the gondola hung from the bottom, with its cabin, engines, and propellers. The gas bag is about 30 meters long with a diameter of about 10 meters, and it is filled with helium (density about 4 grams per 22.4 liters under these conditions). Estimate the total mass of the blimp, including the helium.

Problem 3.15 The simple pendulum

A "simple" pendulum (Figure 3.46; next page) consists of a small mass of mass m swinging at the end of a low-mass string of length L (in contrast to a pendulum whose mass is distributed, such as a rod swinging from one end).

(a) Show that the momentum of the small mass obeys the following equation, where s is the arc length $L\theta$:

$$\frac{dp}{dt} = -mg\sin\theta = -mg\sin\left(\frac{s}{L}\right)$$

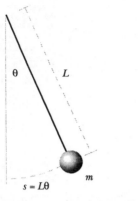

Figure 3.46 A simple pendulum (Problem 3.15).

(b) What form does this equation take for small-amplitude swings? (Note that if θ is measured in radians, $\sin\theta \approx \theta$ for small angles. For example, even for an angle as large as $\theta = 30°$, $\theta = \pi/6 = 0.524$ radians, which is close to $\sin(30°) = 0.5$, and the approximation gets better for smaller angles.)

(c) Compare the form of the approximate equation you obtained in part (b) with the form of the momentum principle for a spring-mass system. By comparing these equations for different systems, determine the period of a simple pendulum for small-amplitude swings. (Note that any time you can approximate the momentum principle for some system by an equation that looks like the equation for a mass on a spring, you know that the system will move approximately sinusoidally.)

(d) Make a simple pendulum and predict its period, then measure the period. Do this for a long pendulum and for a short pendulum. Report your experimental data and results, with comparisons to your theoretical predictions.

(e) For large-amplitude swings, the small-angle approximation is not valid, but you can use numerical integration to calculate the motion. For a mass on a string, the largest possible amplitude is 90°, but if the string is replaced by a lightweight rod, the amplitude can be as large as 180° (standing upside-down, with the mass at the top). It turns out that such a pendulum can be modeled by the same equation as the equation derived in part (a), but the concepts of "torque" and "angular momentum" are required to prove it. Assume that the equation of part (a) is valid, and use numerical integration to plot the position and velocity as a function of time for a pendulum whose amplitude is nearly 180°.

Note how different these plots are from sines or cosines, because the system is not well approximated by a "spring-mass" equation. Also note that for this anharmonic oscillator the period increases with increasing amplitude, unlike the situation with a harmonic oscillator.

Unknown force laws

The following problems deal with analyses involving unknown force laws. They offer you additional practice in analyzing such situations using the momentum principle.

Problem 3.16 Pushing a rod through space

(a) In outer space, a rod is pushed to the right by a constant force F. Describe the pattern of interatomic distances along the rod. Include a specific comparison of the situation at locations A, B, and C. Explain briefly in terms of fundamental principles. Hint: Consider the motion of an individual atom inside the rod, and various locations along the rod.

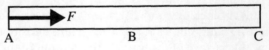

(b) After the rod in part (a) reaches a speed v, the object that had been exerting the force on the rod is removed. Describe the subsequent motion of the rod, and the pattern of interatomic distances inside the rod. Include a specific comparison of the situation at locations A, B, and C. Explain briefly.

Problem 3.17 Strings and wall

A mass M hangs from a string (Figure 3.47). You tie another string to the middle of the first string and pull horizontally. What force F must you apply in order for the upper part of the first string to be at an angle of θ to the vertical? What is the tension in the upper and lower parts of the first string? What approximations or simplifying assumptions did you make?

Figure 3.47 A string attached to a wall holds up a mass (Problem 3.17).

Problem 3.18 Mass on a spring with circular motion

A small block of mass m is attached to a spring with stiffness k_s and relaxed length L. The other end of the spring is fastened to a fixed point on a low-friction table. The block slides on the table in a circular path of radius $R > L$. How long does it take for the block to go around once?

Problem 3.19 Pushing on a block

Two blocks of mass m_1 and m_3, connected by a rod of mass m_2, are sitting on a low-friction surface, and you push to the left on the right block (mass m_1) with a constant force (Figure 3.48).

(a) What is the acceleration dv/dt of the blocks?

(b) What is the compression force in the rod (mass m_2) near its right end? Near its left end?

(c) How would these results change if you *pull* to the left on the *left* block (mass m_3) with the same force, instead of pushing the right block?

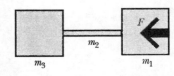

Figure 3.48 Push on a block (Problem 3.19).

Problem 3.20 Looping roller coaster

What is the minimum speed v that a roller coaster car must have in order to make it around an inside loop and just barely lose contact with the track at the top of the loop (Figure 3.49)? The center of the car moves along a circular arc of radius R. Include a carefully labeled force diagram. State briefly what approximations you make. Design a plausible roller coaster loop, including numerical values for v and R.

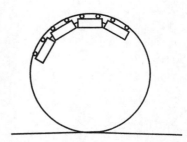

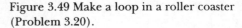

Figure 3.49 Make a loop in a roller coaster (Problem 3.20).

Problem 3.21 Ferris wheel

A Ferris wheel is a vertical, circular amusement ride with radius R. Riders sit on seats that swivel to remain horizontal. The Ferris wheel rotates at a constant rate, going around once in a time T. At the bottom of the ride, what are the magnitude and direction of the force exerted by the seat on a rider of mass m? Include a diagram of the forces on the rider.

Problem 3.22 Weightlessness

By "weight" we usually mean the gravitational force exerted on an object by the Earth. However, when you sit in a chair your own perception of your own "weight" is based on the contact force the chair exerts upward on your rear end rather than on the gravitational force. The smaller this contact force is, the less "weight" you perceive, and if the contact force is zero, you feel peculiar and "weightless" (an odd word to describe a situation when the only force acting on you is the gravitational force exerted by the Earth!). Also, in this condition your internal organs no longer press on each other, which presumably contributes to the odd sensation in your stomach.

(a) How fast must a roller coaster car go over the top of a circular arc for you to feel "weightless"? The center of the car moves along a circular arc of radius R (Figure 3.50). Include a carefully labeled force diagram.

(b) How fast must a roller coaster car go through a circular dip for you to feel three times as "heavy" as usual, due to the upward force of the seat on your bottom being three times as large as usual? The center of the car moves along a circular arc of radius R (Figure 3.51). Include a carefully labeled force diagram.

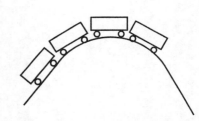

Figure 3.50 Go over the top in a roller coaster (Problem 3.22a).

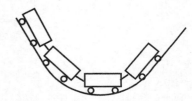

Figure 3.51 Go through a dip in a roller coaster (Problem 3.22b).

Problem 3.23 Your weight on a bathroom scale

If you stand on a bathroom scale at the North Pole, the scale shows your "weight" as an amount Mg (actually, it shows the force F_N that the scale exerts on your feet). At the North Pole you are 6357 km from the center of the Earth. If instead you stand on the scale at the equator, the scale reads a different value due to two effects: 1) The Earth bulges out at the equator (due to its rotation), and you are 6378 km from the center of the Earth. 2) You are moving in a circular path due to the rotation of the Earth (one rotation every 24 hours). What does the scale read at the equator?

3.13 Answers to exercises

3.1 (page 78)	2.6×10^{-10} m
3.2 (page 78)	3.1×10^{-10} m
3.3 (page 78)	3000 N/m
3.4 (page 78)	30 N
3.5 (page 78)	$2k_s$
3.6 (page 81)	$Y \approx 6 \times 10^{10}$ N/m^2
3.7 (page 81)	$\Delta L \approx 0.2$ millimeter
3.8 (page 88)	Period increases by a factor of $\sqrt{2}$
3.9 (page 90)	$A = 0.05$ m. Minimum and maximum values of x are -0.05 m and $+0.05$ m.
3.10 (page 90)	$T = 2\pi\sqrt{m/k}$
3.11 (page 93)	$v_0 = \dfrac{F}{m}\Delta t; \ A = v_0\sqrt{\dfrac{m}{k}}$
3.12 (page 93)	90° ($\pi/2$ radians), corresponding to a sine rather than a cosine (start at x = 0)
3.14 (page 98)	$F_{\text{buoyant}} = 1.6 \times 10^{-3}$ N
	$\dfrac{F_{\text{buoyant}}}{F_{\text{weight of iron}}} = 1.6 \times 10^{-4}$
3.15 (page 99)	For a book the size of this book, there are approximately 10^{19} air molecules between the book and the table.

Chapter 4

Conservation of Energy: Limits on the Possible

Chapter 4

Conservation of Energy: Limits on the Possible

If your numerical integration of a spring-mass system showed the maximum displacement (the amplitude) growing with time, why would you know that there is something wrong with your calculation? If your computer model of the Earth going around the Sun showed the Earth spiraling outward with greater and greater speed, why would you suspect something is wrong with your calculations? How would you predict how far a car could travel by burning one gallon of gasoline?

These are questions that take us up to and beyond the boundaries of what we can analyze using only the momentum principle. The concept of "energy" greatly expands the scope of our models of physical systems. Conservation of energy is a powerful principle that allows us to decide what is possible and what is impossible in the physical world. Although we will introduce the concept of energy conservation in the context of Newtonian mechanics, we will see later that energy conservation is a very general principle, which applies even in situations where Newton's laws do not apply.

4.1 Work: Mechanical energy transfer

We already have a good feel for some aspects of energy from our everyday experience. It is clear that energy is associated with motion—the faster an object is moving, the more energy it has. It is also clear that energy must be expended to get the object moving rapidly—as the old saying goes, there ain't no such thing as a free lunch. Suppose you apply a force on an object and increase its speed. The object gains "kinetic energy" (energy associated with motion). You have lost "chemical" energy, because you have expended some of the energy stored in your body. The net effect is that energy has been transferred from you to the object.

To be quantitative, suppose you apply a constant force F in a constant direction to an object of mass m that was initially at rest, pulling it through a displacement r (Figure 4.1). The object moves faster and faster, acquiring some kinetic energy. If you were to pull twice as hard ($2F$) through the same displacement r, the object would presumably acquire twice as much kinetic energy (and you would expend twice as much chemical energy).

If you lift a heavy weight to a height h above the ground, you will have expended a certain amount of chemical energy (Figure 4.2). If you lift the weight twice as high (a height $2h$ above the ground), you will presumably have had to expend twice as much energy.

Because the amount of energy expended appears to be proportional to the magnitude of the force exerted, and to the distance through which an object is moved, let's tentatively define the amount of energy transfer between you and the object as the product of force times displacement, *Fr*. This type of mechanical energy transfer is called "work." Here we use the word "work" as a technical term in the context of physics. This technical meaning is only loosely connected with everyday uses of the word.

Contrast between distance and time

Doubling the displacement is not the same as doubling the time. If you pull the block on the air track through twice the displacement, you do twice as

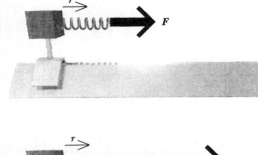

Figure 4.1 Pulling on a spring connected to an object.

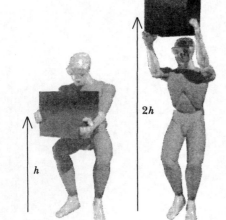

Figure 4.2 You expend more energy when lifting the weight higher.

much work, and the block acquires twice as much kinetic energy. But because the block is moving faster in the second half of the doubled displacement, the time interval is less than doubled (and the impulse is less than doubled), so the momentum change is not doubled.

Note carefully: work and kinetic energy change are proportional to displacement, while impulse and momentum change are proportional to time interval. In symbols: work $F\Delta r$ is not the same as impulse $F\Delta t$.

Force at an angle

Let's consider the motion of a block pulled through a small displacement Δr by a spring stretched a constant amount s, held at an angle θ to the horizontal, as shown in Figure 4.3. We resolve the spring force into vertical and horizontal components, as shown in Figure 4.4. We pull at an angle such that the block does not move in the vertical direction.

? In calculating the amount of work done to pull the block, should we use the magnitude of the force ($k_s s$), or just the horizontal component of the force ($k_s s \cos\theta$)?

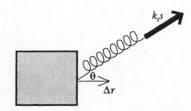

Figure 4.3 Exerting a force at an angle to the motion.

Because there is no vertical displacement, the vertical component of the force does no work:

$$F_y \Delta r_y = F_y \cdot 0 = 0$$

Only the component $k_s s \cos\theta$ in the direction of the motion does work. If the block moves to the right through a displacement r, the amount of work done by the spring force is:

$$F_x \Delta r_x = (k_s s \cos\theta)\Delta r$$

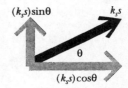

Figure 4.4 x and y components of the force.

Because only the component of a force parallel to the direction of motion does any work, the work done is given by the product of the component of the force parallel to the displacement times the small displacement:

$$\text{work} \equiv F_{\parallel}\Delta r$$

The unit of energy, 1 newton · meter , is called a joule, abbreviated as J. Mechanical energy transfer ("work") is also measured in joules.

Work can be positive or negative. If work is positive, the object gains energy and the surroundings lose energy. If work is negative, the object loses energy and the surroundings gain energy.

Example: You throw a 250 gram ball straight up, and it rises 8 meters into the air. Neglecting air resistance, how much work is done by the gravitational force?

Solution: $F_{\parallel} = mg\cos\theta = -mg$, because $\theta = 180°$.

work $\equiv F_{\parallel}\Delta r = [-(0.25 \text{ kg})(9.8 \text{ N/kg})](8 \text{ m}) = -19.6 \text{ joule}$

Negative work is done on the ball, and the ball slows down, losing some energy.

Ex. 4.1 You pull your little sister across a flat snowy field on a sled. Your sister plus the sled have a mass of 20 kg. The rope is at an angle of 35 degrees to the ground. You pull a distance of 50 m with a force of 30 N. How much work do you do?

Ex. 4.2 About how much mechanical work is done by the Sun on the Earth during six months? (A separate kind of energy transfer is solar radiation, which warms up the Earth.)

Ex. 4.3 Contrast work and impulse. If the magnitude of the Earth's momentum is P, what is the magnitude of the impulse that the Sun applies to the Earth during six months?

4.1.1 The dot product

As we saw above, the component $F_{\parallel}$ of the force vector $\vec{F}$ in the direction of the displacement $\vec{r}$ is equal to $F\cos\theta$, where θ is the angle between the two vectors $\vec{F}$ and $\vec{r}$. The work done by this force is therefore $(F\cos\theta)\Delta r$, or $F\Delta r\cos\theta$.

A more compact notation involves the "dot product." This mathematical operation is defined for any two vectors in terms of their magnitudes and the cosine of the angle between them, as shown in Figure 4.5.

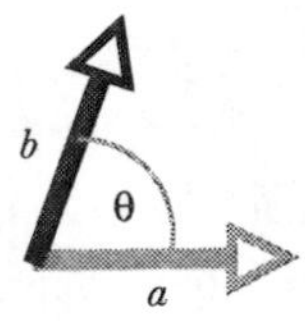

DOT PRODUCT: FORM USING ANGLE θ

$$\vec{a} \bullet \vec{b} \equiv ab\cos\theta \quad \text{(a scalar quantity)}$$

Note that the result of the dot product is a *scalar* quantity, a single number, and not a vector involving components.

Using this definition, we can write the following compact expression for the work W done by a force $\vec{F}$ acting through a small displacement $\Delta\vec{r}$:

$$W = \vec{F} \bullet \Delta\vec{r}$$

? Using the definition of the dot product, what is the work done by a force of magnitude F acting through a small displacement of magnitude Δr if the angle is 0°? 90°? 180°?

If the force is in the same direction as the displacement (0°), the work is $F\Delta r\cos(0°) = F\Delta r$; if the force is perpendicular to the displacement (90°), the work is $F\Delta r\cos(90°) = 0$; and if the force is opposite to the displacement (180°), the work is $F\Delta r\cos(180°) = -Fr$.

The dot product can also be expressed as the sum of products involving the individual components. This form is often more useful in computations:

DOT PRODUCT: FORM USING COMPONENTS

$$\vec{F} \bullet \Delta\vec{r} = F\Delta r\cos\theta = F_x\Delta r_x + F_y\Delta r_y + F_z\Delta r_z \quad \text{(a scalar quantity)}$$

To confirm this for yourself, write each vector using unit vectors (vectors of unit length along the x, y, and z axes):

$$\vec{F} = F_x\hat{\imath} + F_y\hat{\jmath} + F_z\hat{k} \quad \text{and} \quad \Delta\vec{r} = \Delta r_x\hat{\imath} + \Delta r_y\hat{\jmath} + \Delta r_z\hat{k}$$

When you multiply out the product $\vec{F} \bullet \Delta\vec{r}$, note that $\hat{\imath} \bullet \hat{\imath} = 1$ (because $\cos 0° = 1$), and $\hat{\imath} \bullet \hat{\jmath} = 0$ (because $\cos 90° = 0$).

? Writing each vector in terms of its components, again calculate the work done by a force of magnitude F acting through a displacement of magnitude Δr if the angle is 0°, 90°, or 180°. (Choose axes so that $\Delta\vec{r}$ points in the x direction.)

If the force is in the x direction, the work is

$$F_x\Delta r_x + F_y\Delta r_y + F_z\Delta r_z = F\Delta r + (0)(0) + (0)(0) = F\Delta r$$

If the force is in the y direction, the work is

$$F_x\Delta r_x + F_y\Delta r_y + F_z\Delta r_z = (0)\Delta r + F(0) + (0)(0) = 0$$

Figure 4.5 θ is the angle between two vectors whose magnitudes are a and b.

And if the force is in the $-x$ direction (angle is 180°), the work in that case is

$$F_x \Delta r_x + F_y \Delta r_y + F_z \Delta r_z = (-F)\Delta r + (0)(0) + (0)(0) = -F\Delta r$$

Example: You pull a block 1.5 m across a table in the $-x$ direction, while exerting a force of $(-0.3 \text{ N})\hat{\imath} + (0.25 \text{ N})\hat{\jmath}$. How much work do you do on the block?

Solution: $W = \vec{F} \bullet \Delta \vec{r} = [(-0.3 \text{ N})\hat{\imath} + (0.25 \text{ N})\hat{\jmath}] \bullet [(-1.5 \text{ m})\hat{\imath}]$

$W = (-0.3 \text{ N})(-1.5 \text{ m}) = 0.45 \text{ J}$

Ex. 4.4 A 2 kg ball rolls off a 30 m high cliff, and lands 25 m from the base of the cliff. Express the displacement and the gravitational force in terms of unit vectors and calculate the work done by the gravitational force. Note that the gravitational force is $-mg\hat{\jmath}$, where g is a *positive* number (+9.8 N/kg).

4.1.2 Non-constant force

If the force changes in magnitude or direction during the process, we can't simply multiply a constant force times the net displacement. The nonconstant force acts on the object along a path, and we split the path into small increments $\Delta \vec{r}$ (Figure 4.6). This is similar to the way in which we took small time steps Δt when computing the trajectories of objects.

If our increments along the path are sufficiently small, the force is approximately constant in magnitude and direction within one increment. We can write the total work done as a sum along the path:

$$W = \vec{F}_1 \bullet \Delta \vec{r}_1 + \vec{F}_2 \bullet \Delta \vec{r}_2 + \vec{F}_3 \bullet \Delta \vec{r}_3 + \dots \text{ (small steps)}$$

More compactly, we can use the symbol Σ (Greek capital sigma) to mean "sum":

WORK

$$W = \sum \vec{F} \bullet \Delta \vec{r} \text{ (joules; a sum of scalar quantities)}$$

It is worth noting that this technical definition of "work" by itself is merely a definition, with no significance for physics unless we can show that it is related to something else. However we will soon show that, as we suspected from the beginning, work is equal to the amount of energy transferred from the surroundings to an object, through the actions of forces.

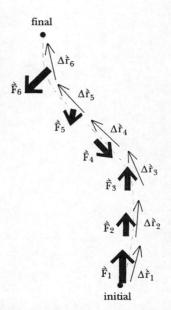

Figure 4.6 A varying force is exerted on an object moving along a curved path.

Ex. 4.5 You push a heavy crate out of a carpeted room and down a hallway with a waxed linoleum floor. While pushing the crate 2.3 m out of the room you exert a force of 30 N; while pushing it 8 m down the hallway you exert a force of 15 N. How much work do you do in all?

Ex. 4.6 You pull on a spring whose spring constant is 20 N/m, and stretch it from its equilibrium length of 0.2 m to a length of 0.3 m. Estimate the work done by dividing the stretching process into two stages and using the average force you exert to calculate work done during each stage.

4.1.3 Comment: Work as an integral

If we go to the limit where the increments are infinitesimal ($\Delta \vec{r} \to d\vec{r}$), the sum used for calculating work turns into a sum of an infinite number of infinitesimal contributions. Such a summation is called a "definite integral." It is written as a distorted "S" for "sum," with an indication of the initial position "i" and the final position "f":

WORK

$$W = \int_i^f \vec{F} \bullet d\vec{r} \text{ (joules)}$$

If you have already studied integration in a calculus course, you already know some techniques for evaluating definite integrals such as this; if not, you will learn them soon in calculus. In many cases it is not possible to find an analytical form for evaluating an integral, in which case one approximates the infinite sum by a finite sum, with finite increments $\Delta \vec{r}$. This is what you did in the computer analyses to calculate the sums of $\Delta \vec{r}$ and $\Delta \vec{p}$ to find $\vec{r}_f$ and $\vec{p}_f$ from $\vec{r}_i$ and $\vec{p}_i$, and the procedure is called "numerical integration." For example, you calculated the net change in the momentum like this:

$$\vec{p}_{\text{final}} - \vec{p}_{\text{initial}} = \vec{F}_1 \Delta t_1 + \vec{F}_2 \Delta t_2 + \vec{F}_3 \Delta t_3 + \ldots\ldots$$

And you calculated the net change in the position like this:

$$\vec{r}_{\text{final}} - \vec{r}_{\text{initial}} = (\vec{p}_1 / m)\Delta t_1 + (\vec{p}_2 / m)\Delta t_2 + (\vec{p}_3 / m)\Delta t_3 + \ldots\ldots$$

Much of our work in this course will involve finite sums such as these, rather than integrals, because many integrals cannot be evaluated analytically.

4.2 Energy of a particle

Remember that we began our study of work in the hopes of understanding something about energy: when you do work on an object, its energy changes. The question we will now address is, "How much will the speed of a single particle change if we do a specific amount of work on it?"

What is a "particle"?

It is important to be clear about what we mean by a "particle." A particle is an object that does not undergo any internal changes during the process of interest. Some examples may be helpful. A proton may often be treated as a particle, as long as there is no change in the configuration of the quarks that make up the proton. A single water molecule may often be treated as a particle, but not if it rotates and the rotation rate changes, nor if the molecule vibrates and the mode of vibration changes. A baseball or even the whole Earth may often be treated as a particle, in situations where there is no significant change in shape or rotation or vibration or temperature of the object.

Sometimes the notion of particle is emphasized by speaking of a "point particle," which is a particle so small that its size is negligible compared to other distances in the situation. On the scale of our galaxy, the Sun can often be treated as though it were a "point particle."

The best example of a point particle is the electron. The most precise measurements are consistent with the radius of the electron being zero! Of course even more precise measurements might discover a nonzero radius, but we have good reasons to suspect that the radius really is zero. And yet the electron also behaves in some ways like a spinning ball, despite its zero size.

Looking for the right formula for particle energy

We expect the energy of a particle to depend on its speed v and its mass m. For a definition of energy E to be consistent with our definition of work, the following should be true for a particle:

$$\Delta E = W$$

What we mean by the notation ΔE is a *change* in particle energy, which is the final energy minus the initial energy, $E_{\text{final}} - E_{\text{initial}}$. If the particle energy increases from 5 joules to 7 joules, the *change* is +2 joules (final minus initial). If the particle energy decreases from 9 joules to 8 joules, the *change* is –1 joule (final minus initial).

Consider a single point-like particle of mass m. For convenience in the reasoning we are about to do, choose coordinate axes so that the x axis is in the direction of the motion at this instant, so the displacement in the next short time interval is simply Δx (Figure 4.7). There is a net force $\vec{F}$ (magnitude F) acting at an angle θ to the displacement. The physical results we will obtain do not depend on our choice of coordinate axes; we are free to choose axes that simplify our calculations.

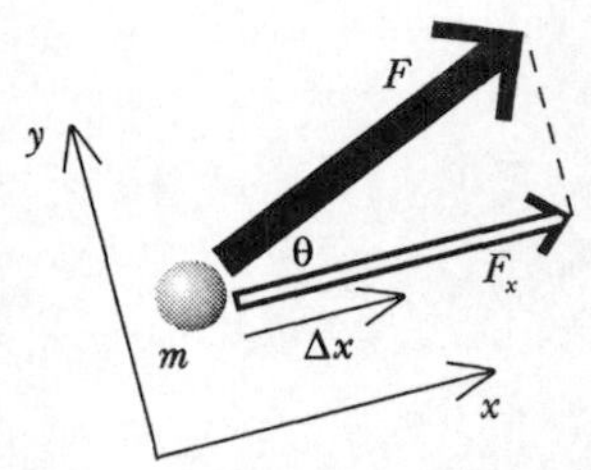

The component of the net force that is parallel to the displacement, $F_x = F\cos\theta$, affects the speed and hence the kinetic energy of the particle. The perpendicular component F_y merely changes the direction of the momentum.

In the very short displacement Δx, the energy E of the particle changes by a small amount:

$$\Delta E = \vec{F} \bullet \Delta\vec{r} = F_x\Delta x$$

We can rewrite the product $F_x\Delta x$ using the momentum principle:

$$\Delta E = F_x\Delta x = \left(\frac{\Delta p_x}{\Delta t}\right)\Delta x$$

Dividing by Δx, we have this:

$$\frac{\Delta E}{\Delta x} = \frac{\Delta p_x}{\Delta t}$$

In the limit as $\Delta t \to 0$, this relates derivatives in the direction of the motion:

$$\frac{dE}{dx} = \frac{dp_x}{dt}$$

In words, this says that the change in energy of the particle per unit distance (in the direction of the motion) is equal to the change in the particle's momentum in that direction per unit time. What we are looking for is a formula for the energy E that satisfies this relationship between the two derivatives. In a section at the end of this chapter we show that the following formula has the right properties:

ENERGY OF A PARTICLE MOVING AT SPEED *v*

$$E = \frac{mc^2}{\sqrt{1 - v^2/c^2}} \quad \text{(single point particle with rest mass } m\text{)}$$

As usual, c is the speed of light (3×10^8 m/s). Like the formula for momentum, this was predicted by Einstein in his special theory of relativity, and it has been confirmed in a very wide range of experiments with high-speed particles.

Note the similarities with and differences from the formula for momentum, $\vec{p} = m\vec{v}/\sqrt{1 - v^2/c^2}$. Both have the same square root in the denomi-

Figure 4.7 A particle of mass m. We choose the x axis to be in the direction of the motion. The net force acting on the particle has a component F_x in the direction of motion.

nator, and both are proportional to the rest mass m. The energy formula has a scalar value c^2 where the momentum formula has the vector value $\vec{v}$.

Interpreting the energy formula—low speed

Let's see what the energy formula reduces to when the speed v is small compared to the speed of light.

? If a particle is at rest, what is its energy?

Setting $v = 0$ in the energy formula we find the famous formula $E = mc^2$. This has the striking interpretation that a particle has energy even when it is sitting still. We will return to this important point in a moment.

? If the particle is moving with speed v, is its energy greater than mc^2?

If $v > 0$, $\sqrt{1 - v^2/c^2}$ is less than 1, so $1/\sqrt{1 - v^2/c^2}$ is greater than 1, and the particle energy $E = mc^2/\sqrt{1 - v^2/c^2}$ increases with increasing speed. As a result of its motion, the particle has additional, "kinetic" energy K:

KINETIC ENERGY K OF A PARTICLE

$$E = \frac{mc^2}{\sqrt{1 - v^2/c^2}} = mc^2 + K$$

$$K = E - mc^2 = \frac{mc^2}{\sqrt{1 - v^2/c^2}} - mc^2$$

If a particle is at rest, the particle energy is just its "rest energy" mc^2. If the particle moves, it has not only its rest energy mc^2 but also an additional amount of energy which we call the kinetic (motional) energy K.

A speed limit

Doing more and more work on a particle increases its energy more and more, and there is no theoretical limit on how much energy a particle can acquire. As the speed approaches the speed of light, the denominator $\sqrt{1 - v^2/c^2}$ becomes very small, and the energy becomes very large.

However, even though you can increase the energy, it becomes very difficult to make large increases in the speed v, because a sizable speed increase would mean a huge energy increase. There is a kind of speed limit in effect: it would require an infinite amount of work to raise the speed v all the way to the speed of light c.

The nonrelativistic approximation

Next consider the case of a particle moving with a speed v that is small compared to the speed of light ($v \ll c$). An approximate formula for the energy at low speeds can be obtained through use of the mathematics of the "binomial expansion," as is shown in a section at the end of this chapter:

$$\frac{mc^2}{\sqrt{1 - v^2/c^2}} = mc^2\left[1 + \frac{1}{2}\frac{v^2}{c^2} - \frac{3}{8}\left(\frac{v^2}{c^2}\right)^2 - \frac{5}{16}\left(\frac{v^2}{c^2}\right)^3 + \ldots\right]$$

$$\frac{mc^2}{\sqrt{1 - v^2/c^2}} \approx mc^2 + \frac{1}{2}mv^2 \text{ if } v/c \ll 1$$

$E = mc^2 + K$, so at low speeds the kinetic energy is approximately this:

NONRELATIVISTIC KINETIC ENERGY K OF A PARTICLE

$$K \approx \frac{1}{2}mv^2 = \frac{p^2}{2m} \text{ (if } v \ll c)$$

Example: A proton has a mass of 1.7×10^{-27} kg. What is its rest energy in joules? If its total energy is 3 times its rest energy, what is its kinetic energy K in joules and its speed v in m/s?

Solution: $mc^2 = (1.7 \times 10^{-27} \text{ kg})(3 \times 10^8 \text{ m/s})^2 = 1.5 \times 10^{-10}$ J

$$E = mc^2 + K = 3mc^2, \text{ so } K = 2mc^2 = 3 \times 10^{-10} \text{ J}$$

Because $E = mc^2 / \sqrt{1 - v^2/c^2}$, $1 - v^2/c^2 = (mc^2/E)^2 = 1/9$

$v/c = \sqrt{8/9}$, and $v = (0.943)(3 \times 10^8 \text{ m/s}) = 2.82 \times 10^8$ m/s

Note that we cannot use $K \approx \frac{1}{2} mv^2$, because v is large.

Ex. 4.7 It is not very difficult to accelerate an electron to a speed that is 99% of the speed of light, because it has such a very small mass. What is the ratio of the kinetic energy K to the rest energy mc^2 in this case? In the definition of what we mean by kinetic energy K, $E = mc^2 + K$, you must use the full relativistic formula for E, because v/c is not small compared to 1.

Ex. 4.8 A pitcher can throw a baseball at about 100 miles/hour (about 44 m/s). What is the ratio of the kinetic energy to the rest energy mc^2? (Can you use $K \approx \frac{1}{2} mv^2$?)

4.2.1 Relativistic energy and momentum

Because both relativistic energy and momentum have the factor

$$\sqrt{1 - v^2/c^2}$$

in the denominator, it seems likely that they can be related in some way. As you can show in Ex. 4.9 on page 124, the relationship turns out to be:

RELATIVISTIC ENERGY AND MOMENTUM OF A PARTICLE

$$E^2 - (pc)^2 = (mc^2)^2 \text{ (rest mass } m)$$

This relation is true in all reference frames. Both energy and momentum will change if a phenomenon is viewed in a different reference frame, but this relation will still be true. The quantity $E^2 - (pc)^2$ is therefore called an "invariant" quantity—a quantity that does not vary when the reference frame is changed.

This relation turns out to be $E^2 - (pc)^2 = 0$ for those particles such as photons (and probably neutrinos) which always travel at the speed of light. Even though they can never be at rest, it is as though they had zero rest mass. The relation for so-called "massless" particles reduces to $E^2 = (pc)^2$, or $E = pc$. This relation is also approximately true for any particle that is traveling at very high speed, because in that case mc^2 is small compared to E.

Example: A proton has a mass of 1.7×10^{-27} kg. If its total energy is 3 times its rest energy, what is its momentum?

Solution: $mc^2 = (1.7 \times 10^{-27} \text{ kg})(3 \times 10^8 \text{ m/s})^2 = 1.5 \times 10^{-10}$ J

$$E = mc^2 + K = 3mc^2 \text{ and } (pc)^2 = E^2 - (mc^2)^2$$

$pc = \sqrt{E^2 - (mc^2)^2} = \sqrt{(3mc^2)^2 - (mc^2)^2} = \sqrt{8(mc^2)^2} = \sqrt{8}(mc^2)$

$p = \sqrt{8}(mc) = \sqrt{8}(1.7 \times 10^{-27} \text{ kg})(3 \times 10^8 \text{ m/s}) = 1.4 \times 10^{-18}$ kg $\cdot$ m/s

Ex. 4.9 Show the validity of the relation $E^2 - (pc)^2 = (mc^2)^2$ in the case where $m \neq 0$, by making the substitutions:

$$E = \frac{mc^2}{\sqrt{1 - v^2/c^2}} \text{ and } p = \frac{mv}{\sqrt{1 - v^2/c^2}}$$

4.3 The work-energy relation

So far we have discussed the work done by a force acting through a small displacement, and the energy of a particle. We now return to the connection between work and energy. Here again is the small energy change ΔE that results from a small amount of work:

$$\Delta E = F_\parallel \Delta r = \vec{F} \bullet \Delta \vec{r}$$

If we add up lots of small amounts of work done on a particle as it moves from some initial position to a final position (Figure 4.8), we have this:

$$\Delta E_1 + \Delta E_2 + \Delta E_3 + \ldots = \vec{F}_1 \bullet \Delta \vec{r}_1 + \vec{F}_2 \bullet \Delta \vec{r}_2 + \vec{F}_3 \bullet \Delta \vec{r}_3 + \ldots$$

But the left side of this equation, the sum of all the energy changes, is simply the difference between the initial and final values of the particle energy. Therefore we have this result:

$$\Delta E = \vec{F}_1 \bullet \Delta \vec{r}_1 + \vec{F}_2 \bullet \Delta \vec{r}_2 + \vec{F}_3 \bullet \Delta \vec{r}_3 + \ldots$$

Using the summation symbol, we have the following important result, that the total work W done on a particle is equal to the change in its energy:

WORK-ENERGY RELATION FOR A PARTICLE

$$\Delta E = \Delta\left(\frac{mc^2}{\sqrt{1 - v^2/c^2}}\right) = \sum \vec{F}_{net} \bullet \Delta \vec{r} \text{ (for a particle)}$$

If there is no change in the identity of the particle, the mass doesn't change, in which case $\Delta E = \Delta(mc^2 + K)$ is the same as ΔK, the change in the kinetic energy, so we can write $\Delta K = W$.

The work-energy relation for a particle is true for any net force, and does not depend on the net force being constant. Even if the force varies (requiring us to add up the various contributions in a sum), the change in the kinetic energy of a particle is equal to the work done on the particle.

The work done by the net force on the particle can be calculated as a finite sum of small quantities, or as an infinite sum of infinitesimal quantities:

$$W = \sum \vec{F}_{net} \bullet \Delta \vec{r} \rightarrow \int_i^f \vec{F}_{net} \bullet d\vec{r} \text{ (for a particle)}$$

The nonrelativistic limit

At low speeds we have the following approximate form:

NONRELATIVISTIC WORK-ENERGY RELATION FOR A PARTICLE

$$\Delta K = \Delta\left(\frac{p^2}{2m}\right) = \frac{p_f^2}{2m} - \frac{p_i^2}{2m} = \sum \vec{F}_{net} \bullet \Delta \vec{r} \text{ } (v \ll c, \text{ no change in identity})$$

We have developed the relationship between work and particle energy for the general, relativistic case, in order to establish the right foundation. Much of what we will do next will deal with the nonrelativistic approximation, but we will return to relativistic energy later in this chapter.

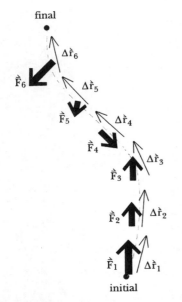

final

$\Delta \vec{r}_6$

$\vec{F}_6$ $\Delta \vec{r}_5$

$\vec{F}_5$ $\Delta \vec{r}_4$

$\vec{F}_4$ $\Delta \vec{r}_3$

$\vec{F}_3$

$\vec{F}_2$ $\Delta \vec{r}_2$

$\vec{F}_1$ $\Delta \vec{r}_1$

initial

Figure 4.8 Each small amount of work adds some energy ΔE_i to the particle.

Example: If you pull on a 3 kg block and change its speed from 4 m/s to 5 m/s, what is the minimum change of chemical (food) energy in you? Give magnitude, sign, and units. (Your temperature rises when you do work like this, involving an increase in thermal energy, so there is more actual chemical cost than the minimum amount you calculate.)

Solution: The speed is tiny compared to *c*, so the change in kinetic energy of the block is

$$\Delta K = \tfrac{1}{2}m(v_f^2 - v_i^2) = \tfrac{1}{2}(3 \text{ kg})[(5 \text{ m/s})^2 - (4 \text{ m/s})^2]$$

$$\Delta K = 13.5 \text{ J}$$

Because $\Delta K = W$, you did an amount of work equal to +13.5 J, which means your store of chemical (food) energy has decreased at least this much; the change is at least −13.5 J.

Ex. 4.10 Show that the units of $\dfrac{p^2}{2m}$ and mc^2 are indeed joules. (Note that 1 N is 1 kg · m/s².)

Ex. 4.11 An automobile traveling on a highway has an average kinetic energy of 1.1×10^5 J. Its mass is 1.5×10^3 kg; what is its average speed? Convert your answer to miles per hour to see if it makes sense.

Ex. 4.12 If you could use all of the mc^2 rest energy of some fuel to provide the car in the previous exercise with its kinetic energy of 1.1×10^5 J, how much mass of fuel would you need?

Problem 4.1 Work done in Moon shot

If you have already done Problem 2.7 The Ranger 7 mission to the Moon, you can use the program you wrote for that problem, and go directly to part (b). Otherwise:

(a) Write a program to model the journey of a spacecraft coasting from the Earth to the Moon. Start the spacecraft at a height of 50 km above the Earth's surface with a speed of 1.3×10^4 m/s. This is approximately the speed it would have after all the rockets have fired, and is high enough to be above most of the Earth's atmosphere. Include the spacecraft's interactions with both the Earth and the Moon. Use a *dt* of 5 seconds. Make sure you stop the program when the spacecraft reaches the Moon's surface (not its center!). The data you need may be found on page 68, or on the inside back cover of this book. If you are interested in more of the details of this trip, see Problem 2.7 The Ranger 7 mission to the Moon, on page 68.

(b) Add a calculation of the work done by the gravitational forces of the Earth and the Moon to your analysis of sending a spacecraft to the Moon. You need to approximate the work by adding up the amount of work done by gravitational forces along each step of the path:

$$W = \sum \vec{F} \bullet \Delta \vec{r} = \vec{F}_1 \bullet \Delta \vec{r}_1 + \vec{F}_2 \bullet \Delta \vec{r}_2 + \vec{F}_3 \bullet \Delta \vec{r}_3 + \ldots$$

Compare the numerical value of the work with the change in the kinetic energy (final kinetic energy just before crashing on the Moon, minus initial kinetic energy when released above the Earth's atmosphere). Try shorter time steps to make sure that the errors introduced by a finite time step are not significant. Report your results.

4.4 Particle interactions: Potential energy

For isolated, individual point particles, there is only one kind of energy, the relativistic particle energy:

$$E = \frac{mc^2}{\sqrt{1 - v^2/c^2}} \quad \text{(single point particle, rest mass } m\text{)}$$

It is often convenient to split this into two parts, $E = mc^2 + K$, though this is only a convenience. We call mc^2 the rest energy, and we call K the kinetic energy (the additional energy the particle has if it is moving).

In systems of interacting particles, such as galaxies of stars interacting gravitationally, or atoms in which the protons and electrons interact electrically, there is energy associated with the interactions. This interaction energy is in addition to the energies of the individual particles. As you might expect, changes in interaction energy are associated with changes in shape of the multiparticle system.

4.4.1 Choosing a system

The word "system" has a special scientific meaning. When we talk about a "system of three particles," as in Figure 4.9, we are implicitly dividing the Universe into two parts: the objects within the system, and everything in the rest of the Universe. Having created this division, it makes sense to talk about interactions that are internal to the system (that is, interactions between the particles within the system) and interactions with objects external to the system.

When considering a problem, we are free to define any collection of objects we wish as the "system" of interest. Once we have done this, however, we must stick strictly to this definition in our analysis. If we decide to include an object in a system, it must be considered this way throughout our analysis.

4.4.2 Interaction energy in a multiparticle system

First we will consider a generic multiparticle system to see how to calculate interaction energy in general. Then we'll be able to calculate the interaction energy for several specific systems of interest to us.

Consider a system of three particles that exert forces on each other, and which are also acted upon by objects outside this system, as shown in Figure 4.9. For example, these could be three charged particles that exert electric forces on each other but also have springs attached to them that exert additional forces. The forces that are exerted by other particles in the system are called "internal" forces. The forces exerted by objects outside the system are called "external" forces.

The work done by these internal and external forces on each particle changes the particle energy E of that particle. We can write the work-energy relation for each particle:

$$\Delta E_1 = (\vec{f}_{1,2} + \vec{f}_{1,3} + \vec{F}_{1,\text{ext}}) \bullet \Delta \vec{r}_1 = W_{1,\text{int. forces}} + W_{1,\text{ext. forces}}$$

$$\Delta E_2 = (\vec{f}_{2,1} + \vec{f}_{2,3} + \vec{F}_{2,\text{ext}}) \bullet \Delta \vec{r}_2 = W_{2,\text{int. forces}} + W_{2,\text{ext. forces}}$$

$$\Delta E_3 = (\vec{f}_{3,1} + \vec{f}_{3,2} + \vec{F}_{3,\text{ext}}) \bullet \Delta \vec{r}_3 = W_{3,\text{int. forces}} + W_{3,\text{ext. forces}}$$

If we add up the changes in the particle energy for each of the three particles (the left sides of the equations) and the work done on each particle (the right sides of the equations) we can write this:

$$\Delta(E_1 + E_2 + E_3) = W_{\text{int. forces}} + W_{\text{ext. forces}}$$

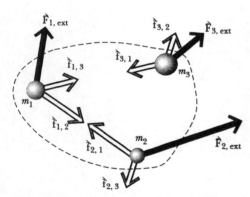
Figure 4.9 A system consisting of three particles. They exert "internal" forces on each other, and additional "external" forces are exerted by objects outside the system.

where $W_{\text{int. forces}}$ is the sum of the work done by internal forces on each of the particles, and $W_{\text{ext. forces}}$ is the sum of the work done by external forces on each of the particles. We emphasize that this represents the sum of the changes in the individual particle energies. Note that the work done by each individual external force involves the displacement of that force (equivalently, the displacement of the particle on which that particular force acts).

We now make a simple algebraic rearrangement that has major conceptual consequences. We move the work done on the particles by the internal forces, $W_{\text{int. forces}}$, to the other side of the equation:

$$\Delta(E_1 + E_2 + E_3) + (-W_{\text{int. forces}}) = W_{\text{ext. forces}}$$

We relabel the term $(-W_{\text{int. forces}})$ as a change of what is called "potential" energy U, energy associated with the interactions of the particles with each other:

$$\Delta(E_1 + E_2 + E_3) + \Delta U = W_{\text{ext. forces}}$$

We define the total energy of the three-particle system, E_{system}, to be the sum of all the particle energies in the system, plus the potential energy (interaction energy):

$$E_{\text{system}} \equiv (E_1 + E_2 + E_3) + U$$

With this definition, we can write an extremely important equation for how the energy of a multiparticle system changes as a result of work done by external forces (Figure 4.10):

THE ENERGY PRINCIPLE FOR A MULTIPARTICLE SYSTEM

$$\Delta E_{\text{system}} = W_{\text{ext. forces}}$$

$$E_{\text{system}} = (E_1 + E_2 + \dots) + U$$

E_1, E_2, etc. are the particles energies in the system

U is the potential energy of the interacting particles in the system

The importance of this way of describing energy changes is that to the left of the equal sign are quantities that are internal to the system, while to the right of the equal sign are the external effects on the system. The interpretation is, "if you do work on a system, you change its energy." The crucial difference between the single-particle work-energy relation and this broader energy principle for multiparticle systems is the role of the potential energy U associated with interactions inside the system.

You get to choose what system you want to analyze, but once you have specified what portion of the Universe you're considering to be the system of interest, you must be consistent about what effects act on the system (external work done by external forces), versus what energy changes occur inside your system (possible changes in particle energies or potential energy).

It is customary to drop the subscript and write $W_{\text{ext. forces}}$ simply as W, with the understanding that the work W is mechanical energy transfer from the surroundings to the system, with a resulting change in the energy of the system:

$$\Delta E_{\text{system}} = W$$

Remember that W can be positive or negative, resulting in either an energy gain or an energy loss in the system of interest.

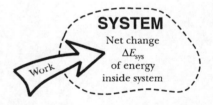

Figure 4.10 Change of system energy is equal to work. Work can be either positive or negative.

Ex. 4.13 Consider an isolated system, by which we mean a system free of external forces. If the sum of the particle energies in the system increases by 50 joules, what must have been the change ΔU in the system's potential energy?

4.4.3 Conservation of energy

A *conserved* quantity is defined to be a quantity that is constant for a system plus its surroundings, where by "surroundings" we mean everything but the system of interest. An example of a conserved quantity is momentum: If an object gains momentum due to forces acting on it, the surroundings lose the same amount of momentum. Thanks to the reciprocity principle, electric or gravitational forces on the object and the surroundings are equal in magnitude and opposite in direction, so the impulses are equal and opposite.

In the 1800's it was discovered experimentally that energy is a conserved quantity. If a system gains energy, the surroundings lose the same amount of energy. If a system loses energy, the surroundings gain the same amount of energy. When new forms of energy such as nuclear energy were discovered in the 1900's, it was found experimentally that energy conservation was still valid. Here is a statement of the conservation of energy:

CONSERVATION OF ENERGY

$$\Delta E_{\text{system}} + \Delta E_{\text{surroundings}} = 0$$

Many systems of particles of practical interest are nearly isolated from their surroundings and can be considered to be nearly free of external influences. For isolated systems we can say that the energy of the system stays constant with time, and the energy is called "a constant of the motion."

In an isolated system the total energy doesn't change, but energy can be swapped back and forth among particle energies and potential energy. For example, in an isolated binary star system the kinetic energies of the stars can increase if there is a corresponding decrease in potential energy. In collisions in an isolated container of helium, some helium atoms gain particle energy and others lose, but the total energy of the gas doesn't change.

A system can exchange energy with its surroundings, and one form of energy exchange is mechanical energy transfer, called work W. In that case we have this:

$$\Delta E_{\text{system}} = W$$

$$\Delta E_{\text{surroundings}} = -W$$

$$\Delta E_{\text{system}} + \Delta E_{\text{surroundings}} = 0$$

4.4.4 The potential energy of pairs of particles

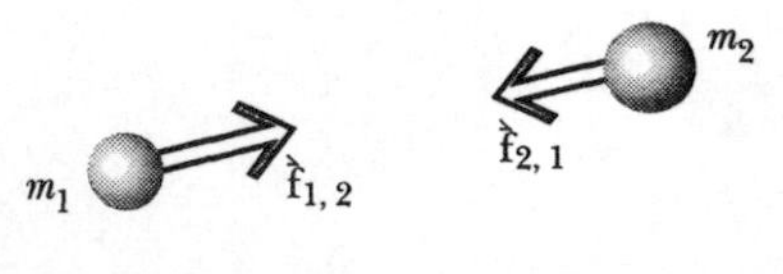

The reciprocity of gravitational forces and electric forces makes it useful to consider pair-wise contributions to potential energy. Consider the sum of the work done by internal gravitational forces in a short time interval on the individual stars in a system consisting of two stars (Figure 4.11):

$$\Delta U = -W_{\text{int. forces}} = -(\vec{f}_{1,2} \bullet \Delta \vec{r}_1 + \vec{f}_{2,1} \bullet \Delta \vec{r}_2)$$

Because $\vec{f}_{12} = -\vec{f}_{21}$, we have

$$\Delta U = -(-\vec{f}_{2,1} \bullet \Delta \vec{r}_1 + \vec{f}_{2,1} \bullet \Delta \vec{r}_2) = -\vec{f}_{2,1} \bullet (\Delta \vec{r}_2 - \Delta \vec{r}_1)$$

Figure 4.11 A system consisting of two stars interacting gravitationally.

Because $\Delta \vec{r}_2 - \Delta \vec{r}_1 = \Delta(\vec{r}_2 - \vec{r}_1)$, we can rewrite the position changes:

$$\Delta U = -\vec{f}_{2,1} \bullet \Delta(\vec{r}_2 - \vec{r}_1)$$

But $(\vec{r}_2 - \vec{r}_1)$ is the position of star 2 relative to star 1 (Figure 4.12). Therefore we can calculate the amount of change of potential energy ΔU simply in terms of the relative displacements between pairs of particles. We will express the energy of a multiparticle system in terms of the particle energies plus the pair-wise potential energies. We write U_{12} for the potential energy of the pair consisting of particles 1 and 2, U_{23} for pair 2 and 3, etc.

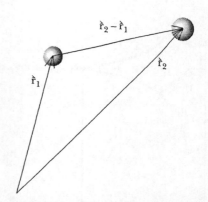

Figure 4.12 The difference in the position vectors of the two stars is the position of star 2 relative to star 1.

Ex. 4.14 Write the energy of a system containing four particles, including the relativistic particle energies E_1 etc. plus the potential energy pairs U_{12} etc. How many potential energy pairs are there?

4.4.5 Potential energy change involves a change of shape or size

We found that in a short time interval, $\Delta U_{12} = -\vec{f}_{2,1} \bullet \Delta \vec{r}_{21}$, where $\Delta \vec{r}_{21}$ is the change in the position of particle 2 relative to particle 1.

? Suppose that both particles move together, with the same velocity (Figure 4.13). In that case, what is $\Delta \vec{r}_{21}$? What is ΔU_{12}?

In this case the relative positions don't change, $\Delta \vec{r}_{21}$ is zero, and there is no change in the pair-wise potential energy.

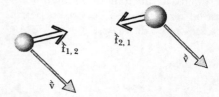

Figure 4.13 The two particles move together, with the same velocity.

? Suppose instead that the two particles rotate around each other (Figure 4.14). What can you say about ΔU_{12}?

This is a bit trickier, because if there is rotation, $\Delta \vec{r}_{21}$ isn't zero, because $\vec{r}_{21}$ is now a rotating vector. But in rotation, $\Delta \vec{r}_{21}$ is perpendicular to $\vec{r}_{21}$, and to the force $\vec{f}_{2,1}$, so the dot product $\vec{f}_{2,1} \bullet \Delta \vec{r}_{21}$ is zero. So again there is no change in the pair-wise potential energy.

? What can you conclude about potential energy for a rigid system?

Evidently U is constant for a rigid system, one whose shape doesn't change. For the potential energy to change, there must be a change of shape or size. Conversely, a change of shape or size is an indicator that U may have changed.

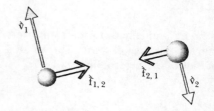

Figure 4.14 The two particles rotate around each other.

4.4.6 The sign of potential energy

The total energy of a multiparticle system is $E_{\text{system}} = (E_1 + E_2 + \dots) + U$, where $E_1 = m_1 c^2 / \sqrt{1 - v_1^2/c^2}$ is the particle energy of particle 1, etc. When the particles are very far apart, we must define U to be zero, so that the total energy of these far-apart particles is simply the sum of the particle energies:

$U = 0$ WHEN THE PARTICLES ARE VERY FAR APART

This is necessary for consistency with particle energy.

If the particles start to come closer together, we find that U starts to become negative if the interactions are attractive, while U starts to become positive if the interactions are repulsive. To see this, first consider an isolated system consisting of two stars that are very far apart, initially at rest, so the initial energy of the system is just the sum of the rest energies of the two stars. At first slowly, then more and more rapidly, their kinetic energies increase due to their mutual gravitational attractions (Figure 4.15).

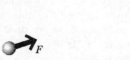

Figure 4.15 Initially two stars are very far apart, at rest, but they attract each other.

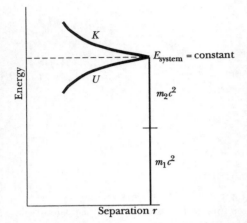

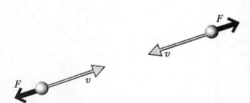

Figure 4.16 A graph of the individual energies of two stars, and the potential energy, as a function of separation. U must be zero at large r. As r decreases, K increases and U decreases.

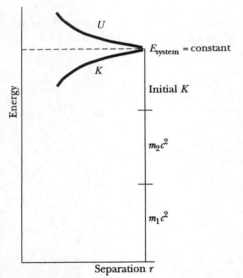

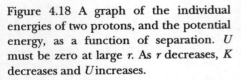

Figure 4.17 Initially two protons are very far apart, heading inward. The mutual repulsions slow the protons down.

Figure 4.18 A graph of the individual energies of two protons, and the potential energy, as a function of separation. U must be zero at large r. As r decreases, K decreases and U increases.

? U was initially zero, so what must be the sign of U when the stars have acquired significant kinetic energies?

If the system is isolated, there are no external forces and hence no external work, so E_{system} does not change. If the relativistic particle energies increase (so the total kinetic energy increases), the potential energy must decrease. Because U was defined to be zero initially, U must become negative. A graph of the energies is shown in Figure 4.16. (In this graph the kinetic and potential energies are greatly exaggerated for clarity. We will soon see that K and U are normally extremely small compared to the rest energies.)

Does this negative sign make sense in terms of the definition of potential energy change, $\Delta U = -W_{\text{int. forces}}$? The attractive gravitational forces that the stars exert on each other certainly do a positive amount of work on each star, so the potential energy change must indeed be negative.

Next consider an isolated system consisting of two protons that are very far apart, and suppose that initially they have high speeds, heading inward (Figure 4.17). Their mutual electric repulsions make the protons slow down.

? U was initially zero, so what must be the sign of U when the protons have slowed down somewhat?

Again, there is no external work done on this isolated multiparticle system, so E_{system} does not change. The relativistic particle energies decrease, so the potential energy must increase; U becomes positive (Figure 4.18). Also, the repulsive electric forces do negative work on each other, so $\Delta U = -W_{\text{int. forces}}$ is positive. (Again, in Figure 4.18 we have greatly exaggerated the kinetic and potential energies in comparison with the rest energies.)

More precisely, we will soon show that it is the *change* of U, positive or negative, that indicates whether the interaction is attractive or repulsive. But we wanted to give you a qualitative feel for the issues right away, and in particular to emphasize that there is nothing wrong with a negative value of U; this is expected for an attractive force.

Note that the change in potential energy for the two stars or the two protons was associated with a change in shape of the two-particle system. If there is no change in shape or size, there is no change in potential energy.

4.4.7 A specific example: Gravitational potential energy

We have investigated some general properties of potential energy. Next we will show how to find a specific formula for potential energy, starting from a force law. We will apply this method to find a formula for gravitational potential energy.

Remember that we found that in a short time interval the change in potential energy depends on the relative displacement $\Delta \vec{r}_{21}$ of object 2 relative to object 1:

$$\Delta U_{12} = -\vec{f}_{2,1} \bullet \Delta \vec{r}_{21}$$

It is the relative displacement that matters, so we can pretend that object 1 remains stationary, and just calculate the work done on object 2. The force is in a line with $\vec{r}_{21}$, so we can evaluate the dot product and write the following, where $F_{g,r}$ is the component of the force on object 2 in the direction of $\vec{r}_{21}$, and r is the distance from object 1 to object 2:

$$\Delta U_g = -F_{g,r}\Delta r$$

Therefore if we already knew the formula for potential energy, we could calculate the associated force like this: $F_{g, r} \approx -\Delta U_g / \Delta r$. If we let Δr become very small, we have the following:

FORCE IS NEGATIVE GRADIENT OF POTENTIAL ENERGY

$$F_{g, r} = -\frac{dU_g}{dr}$$

A rate of change of a quantity with respect to position is called a "gradient." For example, if a hill is said to have "a 7% grade," this means that it rises 7 m vertically for every 100 meters horizontally—the tangent of the angle of the hill (the slope) is 0.07 (Figure 4.19).

All that remains to be done to find a formula for gravitational energy is to think of an expression whose (negative) derivative is the component of the gravitational force acting on m_2 in the direction of $\hat{r}_{21}$. From Figure 4.20 we see that this component of the gravitational force is negative: $F_{g, r} = -G(m_1 m_2 / r^2)$. The minus sign reflects the fact that the attractive gravitational force on object 2 points back toward object 1.

? Can you think of a function of r whose (negative) derivative with respect to r has this value of $F_{g, r}$? (If necessary, review the appendix on basic calculus at the end of this book.)

The derivative of r^n with respect to r is nr^{n-1}. We have a r^{-2} factor in the gravitational force, so $n = -1$, and the "antiderivative" of r^{-2} (the function whose derivative gives r^{-2}) must be $-r^{-1}$.

? Therefore, what is the formula for gravitational energy?

Being careful about the signs, we find that $U_g = -Gm_1 m_2 / r$

? Check that the negative gradient of this gravitational energy is indeed the force, by differentiating with respect to r.

What about the minus sign on U_g? Is it correct? Yes, because the potential energy decreases as the particles get closer together. When stars fall toward each other, their kinetic energies increase and the pair-wise potential energies must decrease to more and more negative values. In summary,

GRAVITATIONAL POTENTIAL ENERGY U_g

$$U_g = -G\frac{m_1 m_2}{r}, \text{ where } r \text{ is the separation of } m_1 \text{ and } m_2$$

A possible constant?

It is true that the gradient of $U_g = (-Gm_1 m_2 / r + \text{constant})$ would also give us the correct gravitational force, because the derivative of a constant is zero. However, remember that when the objects are very far apart the total energy of the system must be equal just to the sum of the particle energies, and the potential energy must be zero.

? When the particle separation r is very large, what is the value of $-Gm_1 m_2 / r$?

When r is very large, $-Gm_1 m_2 / r$ is nearly zero. Therefore, the value of the constant in $U_g = (-Gm_1 m_2 / r + \text{constant})$ must be zero, in order for U_g to be zero at large separations. We have shown that the pair-wise gravitational potential energy must be simply $U_g = -Gm_1 m_2 / r$, with no added constant.

The minus sign

It may still seem odd that the expression for gravitational potential energy has a minus sign in it. Figure 4.21 shows that although U_g is negative, it does

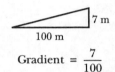

Gradient $= \dfrac{7}{100}$

Figure 4.19 The gradient is the rate of change of y with respect to x.

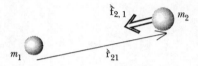

Figure 4.20 The component of the gravitational force, $F_{g,r}$

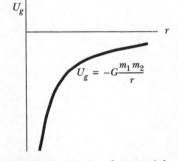

Figure 4.21 Gravitational potential energy is negative. It increases with increasing separation r.

increase with increasing separation r, which means that you have to do work to move two gravitationally attracting objects farther apart.

Here are two little exercises that will help you see why the minus sign is appropriate:

Ex. 4.15 If a block falls toward the Earth, the kinetic energy increases and the gravitational potential energy must decrease. In this process, what is the sign of $\Delta(-GMm/r)$ (the change in the potential energy), in which the final separation r_f is less than the initial separation r_i?

Ex. 4.16 If a block rises away from the Earth, the kinetic energy decreases and the gravitational potential energy must increase. In this process, what is the sign of $\Delta(-GMm/r)$ (the change in the potential energy), in which the final separation r_f is greater than the initial separation r_i?

4.4.8 More about force as gradient of potential energy

Because the (negative) gradient of U is equal to the force, the force is the (negative) slope of a graph of potential energy. This means that you can mentally read the force right off a potential energy graph.

? Where the curve on a potential energy curve is steep, is the force large or small? How can you determine the direction of the force?

Where the curve is steep, the force is large (large gradient), and where the curve is shallow, the force is small (Figure 4.22). Where the slope of the curve is horizontal, the force is zero, and this represents a possible equilibrium position for the system. The component of the force is given by the negative gradient, so a negative slope corresponds to a force to the right, while a positive slope corresponds to a force to the left.

Actually, in determining the x component of the force, we differentiate with respect to x while holding y and z constant. This kind of derivative is called a "partial derivative" and has a special symbol to emphasize that everything else is held constant:

$$F_x = -\frac{\partial U}{\partial x}$$

To get all three components of the force we take partial derivatives of the potential energy with respect to x, y, and z:

FORCE IS NEGATIVE GRADIENT OF POTENTIAL ENERGY

$$F_x = -\frac{\partial U}{\partial x}, \; F_y = -\frac{\partial U}{\partial y}, \; F_z = -\frac{\partial U}{\partial z}$$

$$\vec{F} = <-\frac{\partial U}{\partial x}, -\frac{\partial U}{\partial y}, -\frac{\partial U}{\partial z}>$$

(Another standard mathematical notation for this relation is $\vec{F} = -\vec{\nabla} U$.)

Ex. 4.17 With y pointing upward from the center of the Earth, find the y component of the gravitational force from $-\partial U/\partial y$, where $U = -GMm/y$. Find the x component of the force from $-\partial U/\partial x$.

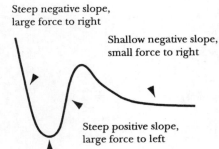

Steep negative slope, large force to right

Shallow negative slope, small force to right

Steep positive slope, large force to left

Zero slope, equilibrium position

Figure 4.22 Force is (negative) gradient of potential energy, so force is the (negative) slope of the potential energy graph.

4.4.9 Energy diagrams

A graphical representation of energy can be exceptionally helpful in reasoning about a process. Consider an isolated system of a star of mass M and a planet of mass m. As a function of star-planet separation r, we plot a graph of the potential energy $-GMm/r$ (Figure 4.23). We omit the rest energies, which are not changing during the motion.

We add to the graph a horizontal line whose height above or below the axis represents the total kinetic plus potential energy of the two-particle system. The line for $K+U_g$ equal to the value A is horizontal because no external work is done on this isolated system, and there is no change of the rest masses (no change of identity), so $K+U_g$ is constant during the motion.

? Is A positive or negative for the system shown in (Figure 4.23)?

The horizontal line representing $K+U_g = A$ is below the axis representing $U_g = 0$, so A is negative. Go down to the curve by an amount U_g, then up by an amount K, to arrive at the horizontal line that represents the fact that $K+U_g$ is a constant (A). By mentally drawing a vertical line from the potential energy curve up to the horizontal total energy line, you can visualize how the total kinetic energy of the star and planet changes during the motion.

The gray area in (Figure 4.23) represents a forbidden region, where the kinetic energy K would be negative, no matter what the value of $K+U_g$ (more about this in the next section).

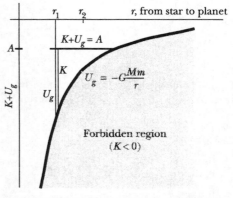

Figure 4.23 $K+U_g$ vs. the separation distance between a star and a planet.

Ex. 4.18 The star-planet separation increases from r_1 to r_2 (Figure 4.23). Does the kinetic energy of the star-planet system increase, decrease, or remain constant?

4.4.10 Limits on possible motion

The energy graph shows at a glance the limits on the motion for a given energy. In Figure 4.24, a star and planet with $K+U_g < 0$ cannot get farther away from each other than r_3, a bound that is set by the potential-energy curve. At a larger separation r_4, the kinetic energy would have to be negative, which is "classically forbidden" (that is, impossible according to the laws of classical, prequantum mechanics). Physically, because kinetic energy is the energy a particle has as a result of being in motion (in addition to its rest energy), kinetic energy cannot be negative.

If the system has the total $K+U_g$ shown ($A < 0$), the separation of the star and planet can never be larger than r_3. An energy graph shows at a glance the range of possible positions achievable during the motion, for a given total $K+U$. We should mention, however, that in the world of atoms where a full analysis requires quantum mechanics, in some cases a system can "tunnel through" a region that is classically forbidden!

An energy graph can show at a glance whether a system is "bound" or "unbound." An example of a bound system is a planet in circular or elliptical orbit around a star: the planet cannot escape. Another example is an electron bound to an atom.

An example of an unbound system is the Earth plus a spacecraft whose initial speed is great enough that the spacecraft will get away from the Earth. For another example, if an amount of energy greater than or equal to the ionization energy is supplied to an atom, an electron can become unbound and escape from the atom. An unbound system has total $K + U_g \geq 0$ (which is easy to see on an energy graph). Only when $K + U_g \geq 0$ can the system be-

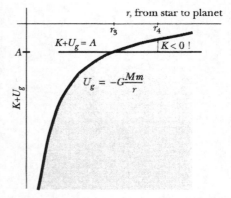

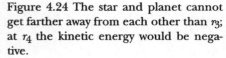

Figure 4.24 The star and planet cannot get farther away from each other than r_3; at r_4 the kinetic energy would be negative.

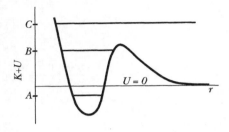

Figure 4.25 An energy graph for some system. Which states are bound? (Exercise 4.19)

come separated by large distances, because K is never negative, and U_g goes to zero at large separations.

> *Ex. 4.19* In this energy graph for some system (Figure 4.25), consider the various energy states indicated. Which of these values for the energy $K+U$ (A, B, or C) represent a bound state? Which represent an unbound state (the particle can escape)? Which represents a trapped state with enough energy to be unbound but with a barrier that (classically) prevents escape?

4.5 Applications of gravitational potential energy

In the next sections we apply the energy principle to the analysis of a variety of situations involving gravitational energy. We will frequently use energy diagrams as a powerful tool.

4.5.1 Application: Energy graphs and types of orbits

In Figure 4.26 are plotted graphs of potential energy, and potential plus kinetic energy, for three different orbits of a planet and star. The following exercise asks you to interpret the meaning of these graphs.

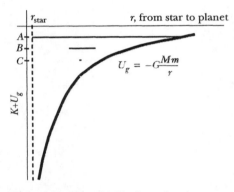

Figure 4.26 What kind of motion is represented by each of these situations? (Exercise 4.20).

> *Ex. 4.20* In Figure 4.26, what kind of motion is represented by the situation with $K+U_g = A$? B? C? Think about the range of r in each situation. For example, C represents a circular orbit (constant r).

4.5.2 Application: Escape speed

An analysis based on the energy principle will allow us to answer the following question: With what minimum speed v must a spacecraft leave the surface of an airless planet (no air resistance) of mass M and radius R in order that it can coast away without ever coming back (Figure 4.27)? Assume there are no other objects near by.

If you did the Moon voyage problem (Problem 2.7 The Ranger 7 mission to the Moon) you found that there was a minimum initial speed required to reach the Moon. With an initial speed smaller than this, the spacecraft fell back to Earth without reaching the Moon. Your determination of the minimum initial speed was done by trial and error. However, we can now use energy relationships to calculate such minimum speeds directly.

As is often the case, an energy diagram can be very helpful. Consider Figure 4.28 (page 135). Motions with $K+U_g =$ A, B, C, or D all start from the surface of an airless planet, with a separation R between the center of the planet and the spacecraft. (Remember that the gravitational force outside a uniform sphere is exactly the same as it would be if all the mass collapsed to the center of the sphere.)

? Which of these motions starts with the largest kinetic energy K? Which starts with the least K?

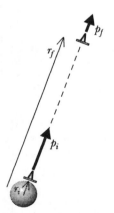

Figure 4.27 A spacecraft is launched with momentum p_i from near an airless planet's surface; it has momentum p_f when it has traveled far away.

Each of the horizontal lines represents motion with $K+U_g$ constant. Evidently A is the highest value of $K+U_g$, and therefore involves the largest initial kinetic energy. D is the lowest value of $K+U_g$ and starts with the least kinetic energy.

? For which of these motions does the spacecraft "escape" and never come back?

For motions with $K+U_g = A$, B, or C, no matter how far from the planet the spacecraft gets, there is still some kinetic energy, so the spacecraft will never return. This is another example of an unbound state being associated with a positive value of $K+U_g$.

In contrast, with motion $K+U_g = D$ the spacecraft will reach a maximum separation from the planet, at which point its kinetic energy has fallen to zero. The spacecraft will fall back to the planet.

? How would you describe the motion that requires the least possible initial kinetic energy to achieve escape?

Evidently the least costly escape is to have $K = 0$ when the spacecraft has reached a distance very far from the planet. At a large separation, $U_g = 0$, so $K+U_g = 0$. This corresponds on the diagram to a horizontal line lying on the axis. Because $K+U_g$ doesn't change (no external work on the planet-spacecraft system), it must also be true that $K+U_g = 0$ at the start of the motion:

MINIMAL CONDITION FOR ESCAPE

$$K+U_g = 0$$

The minimal initial kinetic plus potential energy of the system composed of planet plus spacecraft is this, assuming that the kinetic energy of the planet is negligible:

$$K_i + U_i = \frac{p^2}{2m} + \left[-G\frac{Mm}{r}\right]_{r=R} = 0 \text{ (for } v \ll c)$$

This lets us calculate the minimum speed, which is called the "escape speed" v_{esc}, from a planet with mass M and radius R.

$$\frac{p_{esc}^2}{2m} = \frac{1}{2}mv_{esc}^2 = G\frac{Mm}{R}, \text{ or } v_{esc} = \sqrt{\frac{2GM}{R}}$$

Though we speak of "escape speed," it is wrong to say that the spacecraft has "escaped from gravity." As long as the spacecraft is a finite distance from the planet, it feels a finite gravitational attraction to the planet.

If you give the spacecraft more than the minimum kinetic energy, it not only escapes but has nonzero kinetic energy when it is far away (for example, see motions A and B in Figure 4.28).

Don't memorize the formula for escape speed!

You should not memorize the formula for the escape speed. Instead, you should study the analysis leading to this result, especially the graphical analysis, and be able to rederive the formula whenever you need it. Memorization of such easily derivable formulas turns them into dead knowledge, with no context, and carries the very real danger that the formulas will be misapplied outside their range of validity. For example, the formula above does not apply if the spacecraft has some nonzero speed when it is far from the planet.

Acquiring the ability to obtain such results starting from fundamental principles makes the result come alive, gives you practice in reasoning from fundamental principles, and helps you avoid using the formula in situations where it isn't valid.

Initial direction

It is interesting to note that in this analysis, the initial direction of motion doesn't matter. An energy analysis is insensitive to direction. If the initial velocity points directly away from the planet, the path is a straight line. If the initial velocity points in some other direction, the path can be shown to be

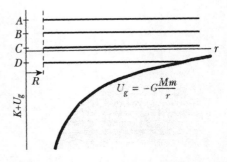

Figure 4.28 What is the minimum K required to escape?

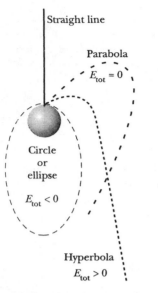

Figure 4.29 Possible orbital trajectories.

a parabola if the initial speed is exactly equal to escape speed, and a hyperbola for higher speeds.

If the speed is below escape speed, there is no escape. The various possible orbits (Figure 4.29) can be classified on the basis of their total kinetic plus potential energy, with initial speed v_i:

$$\tfrac{1}{2}mv_i^2 + \left(-G\frac{Mm}{r}\right) > 0 : \text{escape with } v_\infty > 0 \text{ ; straight line or hyperbola}$$

$$\tfrac{1}{2}mv_i^2 + \left(-G\frac{Mm}{r}\right) = 0 : \text{escape with } v_\infty = 0 \text{ ; straight line or parabola}$$

$$\tfrac{1}{2}mv_i^2 + \left(-G\frac{Mm}{r}\right) < 0 : \text{no escape; straight line, circular, or elliptical}$$

Bound vs. unbound states

This is a specific example of an important general principle. If the kinetic plus potential energy is negative, the system is "bound": the objects cannot become widely separated, because that would require that the kinetic energy become negative, which is impossible (classically). If the kinetic plus potential energy is positive, the system is "unbound" or "free": the objects can become widely separated, with net kinetic energy (and zero potential energy).

BOUND AND UNBOUND STATES

If $K + U < 0$, the system is in a bound state.

If $K + U \geq 0$, the system is unbound (free).

Example: What is escape speed from Earth?

Solution: Consider the system of Earth + spacecraft.

Minimum K for escape means $K + U_g = 0$:

$$\tfrac{1}{2}mv_{esc}^2 + (-GMm/r) = 0 \text{ , so } v_{esc} = \sqrt{2GM/r}$$

$$v_{esc} = \sqrt{\frac{2(6.7\times10^{-11} \text{ N} \cdot \text{m}^2/\text{kg}^2)(6\times10^{24} \text{ kg})}{(6.4\times10^6 \text{ m})}} = 1.12\times10^4 \text{ m/s}$$

Ex. 4.21 This escape speed is slightly larger than the launch speed you found was necessary to reach the Moon in Problem 2.7 The Ranger 7 mission to the Moon. Explain why.

Ex. 4.22 Turn the argument around. If an object falls to Earth starting from rest a great distance away, what is the speed with which it will hit the upper atmosphere? (Actually, a comet or asteroid coming from a long distance away might well have an even larger speed, due to its interaction with the Sun.) Small objects vaporize as they plunge through the atmosphere, but a very large object can penetrate and hit the ground at very high speed. Such a massive impact is thought to have killed off the dinosaurs (see *T. Rex and the Crater of Doom*, Walter Alvarez, Princeton University Press, 1997).

Ex. 4.23 The radius of the Moon is 1750 km, and its mass is 7×10^{22} kg. What is escape speed from the Moon? Why was a small rocket adequate to lift the lunar astronauts back up from the surface of the Moon?

4.5.3 Application: The Moon voyage

We have been applying energy considerations to two-particle systems. In our simple model for the Moon voyage (Problem 2.7 The Ranger 7 mission to the Moon), there were three "particles": the spacecraft with mass m, the Earth with mass M_{Earth}, and the Moon with mass M_{Moon}.

? How many interaction pairs are there in this three-particle system?

In a three-particle system there are three interaction pairs: U_{12}, U_{13}, and U_{23}. We can write the gravitational potential energy like this:

$$U_g = \left[-G\frac{M_{Earth}\,m}{r_{\text{to Earth}}}\right] + \left[-G\frac{M_{Moon}\,m}{r_{\text{to Moon}}}\right] + \left[-G\frac{M_{Earth}\,M_{Moon}}{r_{\text{Earth to Moon}}}\right]$$

In our simple model, the Earth and Moon were fixed in space, so the Earth-Moon term of this expression doesn't change. Also, there is no change of identity of the three particles, so their rest masses do not change. Therefore the energy equation for the system can be written like this, where p is the momentum of the spacecraft, the only moving particle in the model system (and $v \ll c$):

$$\Delta\left(\frac{p^2}{2m}\right) + \Delta\left[-G\frac{M_{Earth}\,m}{r_{\text{to Earth}}}\right] + \Delta\left[-G\frac{M_{Moon}\,m}{r_{\text{to Moon}}}\right] = 0$$

The change in the energy of the system is zero because no external work is done on the system (we're ignoring the effects of other objects, including the Sun). We can rearrange the terms and write the energy equation in terms of initial and final energies, which must be equal:

$$\frac{p_f^2}{2m} + \left[-G\frac{M_{Earth}\,m}{r_{\text{to Earth},\,f}}\right] + \left[-G\frac{M_{Moon}\,m}{r_{\text{to Moon},\,f}}\right] = \frac{p_i^2}{2m} + \left[-G\frac{M_{Earth}\,m}{r_{\text{to Earth},\,i}}\right] + \left[-G\frac{M_{Moon}\,m}{r_{\text{to Moon},\,i}}\right]$$

That is, the final (kinetic plus potential) energy of the system is equal to the initial (kinetic plus potential) energy of the system, because no external work is done on the system (and the rest masses don't change; no change of identity).

Problem 4.2 Energy in the Moon shot
Energy conservation is a powerful check on the accuracy of a numerical integration. Plot a graph of kinetic energy, of gravitational potential energy, and of the sum of the kinetic energy and the gravitational potential energy, *vs.* position during the Moon voyage, Problem 4.1 (page 125). Does the kinetic plus potential energy remain constant? What if you vary the step size (which varies the accuracy of the numerical integration)? Vary the launch speed, and explain the effect that this has on your graphs.

Problem 4.3 Calculating the impact speed
Use energy conservation to calculate analytically (that is, without doing a numerical integration) the final speed of the spacecraft just before it hits the Moon. Include the gravitational effect of the Moon. Use a launch speed of 1.3×10^4 m/s. Modify your program for Problem 4.2 to print out the speed of the spacecraft when it hits the surface of the Moon, and compare this value to your analytical result. What questions that could be addressed in the numerical integration are you *not* able to answer by doing this energy calculation?

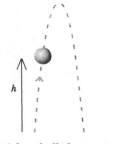

Figure 4.30 A baseball thrown upward has reached a height h.

4.5.4 Gravitational potential energy near the Earth's surface

An important special case of gravitational potential energy arises in projectile motion near the surface of the Earth. Suppose you throw a baseball upward (Figure 4.30). When it has reached a height h above you, what is the change in gravitational energy? Let M be the mass of the Earth and m the mass of the ball:

$$\Delta U_g = \Delta\left[-G\frac{Mm}{r}\right] = \left[-G\frac{Mm}{R+h}\right] - \left[-G\frac{Mm}{R}\right] = GMm\left[\frac{(R+h)-R}{R(R+h)}\right]$$

Because h is very small compared to R, the product $R(R+h)$ is nearly the same as R^2, so

$$\Delta U_g \approx GMm\left[\frac{h}{R^2}\right] = \left(G\frac{Mm}{R^2}\right)h$$

Note that while we can neglect h in the denominator, we cannot neglect it in the numerator, where it is the largest (indeed the only) term. Remembering that $GMm/R^2 \approx mg$, and writing $h = \Delta y$, we can rewrite the change in the gravitational potential energy like this:

CHANGE IN GRAVITATIONAL POTENTIAL ENERGY NEAR THE SURFACE OF THE EARTH

$$\Delta U_g \approx \Delta(mgy)$$

Notice that there is no minus sign in this (approximate) expression for the change in gravitational potential energy. As you go up (farther from the center of the Earth), the change $\Delta(mgy)$ is positive as expected; if you go down, it is negative as expected. In the full formula there is a $(-1/r)$ because moving to larger distance r means an increase (gravitational potential energy *less negative*), as can be seen in Figure 4.31.

In the formula $\Delta U_g = \Delta(mgy)$, Δy must be small compared to the radius of the Earth (6400 kilometers), because the gravitational force gets smaller as you go farther away from the Earth, inversely proportional to the square of the distance from the center of the Earth. But if y doesn't change much, neither does gravity. If you go up 0.1 km (100 meters, the length of a football field), the gravitational force decreases by a tiny factor:

$$\left(\frac{6400\ \text{km}}{6400.1\ \text{km}}\right)^2 \approx 0.99997$$

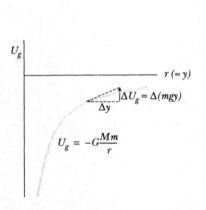

Figure 4.31 A small increase Δy in height y corresponds to a small increase in the (negative) gravitational energy.

Example: At a certain instant a ball is falling with a speed of 6 meters/second, and its position is $y = 35$ meters above the Earth's surface. How fast is the ball falling when it has fallen to a position $y = 20$ meters above the Earth's surface, assuming that we can neglect air resistance?

Solution: Choose (ball + Earth) as the system. Neglect external forces on the system, such as air resistance. There are no changes in rest mass. Then

$$\Delta K + \Delta U = 0$$

$$(K_f - K_i) + (U_f - U_i) = 0, \text{ so } K_f = K_i - (U_f - U_i)$$

$$\frac{1}{2}mv_f^2 = \frac{1}{2}mv_i^2 - (mgy_f - mgy_i)$$

$$v_f^2 = v_i^2 - 2g(y_f - y_i)$$

$$v_f = \sqrt{v_i^2 - 2g(y_f - y_i)}$$

$$v_f = \sqrt{(6 \text{ m/s})^2 - 2(9.8 \text{ N/kg})[(20 \text{ m}) - (35 \text{ m})]} = 18.2 \text{ m/s}$$

We did not have to say whether the ball was initially heading upward or downward. In either case, if we neglect air resistance, the speed will be 18.2 m/s when the ball reaches a height of 20 m above the ground. (If it was initially headed upward, it will again have a speed of 6 m/s when it returns to a height of 35 m.)

Ex. 4.24 Suppose in this situation we measure y from 25 meters above the surface, so that $y_{initial} = +10$ meters, and $y_{final} = -5$ meters. Why do we get the same value for the final speed?

Ex. 4.25 Use energy conservation to find the approximate final speed of a basketball dropped from a height of 2 meters (roughly the height of a professional basketball player). Why don't you need to know the mass of the basketball?

Ex. 4.26 Suppose a pitcher can throw a ball straight up at 100 miles per hour (about 45 m/s). Use energy conservation to calculate how high the baseball goes. (See the Answers section for a hint, if necessary.) Explain your work. Actually, a pitcher probably can't attain this high a speed when throwing straight up, so your result will be an overestimate of what a human can do; air resistance also reduces the achievable height.

4.5.5 Including horizontal motion

Moving vertically up or down involves a change in gravitational energy approximately equal to $\Delta(mgy)$. What about sideways, horizontal movements? There is negligible change in the gravitational energy of the Universe when an object is moved a short distance horizontally (that is, tangent to the Earth's surface). Consider moving a block along a surface with it rolling very easily on low-friction wheels, or a hockey puck sliding easily along the ice. You hardly have to use any energy at all to move the block or puck horizontally, as long as there is little friction.

What if a block is moved horizontally, but at a constant height of 3 meters above the surface? You could support the block with a tall cart, and again there would be little effort involved in moving the block horizontally. Or you could pile up dirt 3 meters deep and place a slick surface on top, and then you could easily move the block horizontally. We conclude that horizontal position doesn't affect the gravitational energy. (Again, we're considering movements over distances small compared with the radius of the Earth, so we can ignore the Earth's curvature.

Suppose you throw a ball at an angle to the horizontal, and just after it leaves your hand at a height y_i it has a speed v_i (Figure 4.32). At the top of its trajectory, at the instant that it is momentarily traveling horizontally, it has a speed v_f

Ex. 4.27 What is the height y_f at the top of the trajectory, in terms of the other known quantities, assuming that we can neglect air resistance? Start by picking a system. Choose an initial and a final state, and apply the energy principle. If necessary, see the Answers section for a hint.

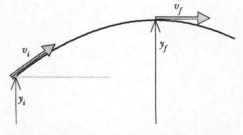

Figure 4.32 Path of a ball thrown at an angle to the horizontal.

4.6 Electric potential energy

In this section we will study the potential energy associated with the electric interactions of charged particles, and the potential energy associated with the electric interactions of neutral atoms, which contain charged particles.

4.6.1 Electric potential energy involving charged particles

The electric force law for two charged particles is similar to the gravitational force law:

THE ELECTRIC FORCE

$$F_{\text{electric}} = \frac{1}{4\pi\varepsilon_0} \frac{|q_1 q_2|}{r^2} \text{ , where } \frac{1}{4\pi\varepsilon_0} = 9\times10^9 \frac{\text{N}\cdot\text{m}^2}{\text{coulomb}^2}$$

The constant $1/(4\pi\varepsilon_0)$ is read as "one over four pi epsilon-zero". The quantities q_1 and q_2 represent the amount of charge on the two particles (measured in coulombs), and r is the distance between the two particles. Because this is an inverse square force, like gravitation, the electric potential energy has a formula very similar to that for gravitational potential energy:

ELECTRIC POTENTIAL ENERGY U_e

$$U_e = \frac{1}{4\pi\varepsilon_0} \frac{q_1 q_2}{r}$$

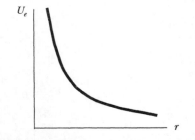

Figure 4.33 Electric potential energy is positive when like charged objects repel each other.

There is no minus sign in the formula for electric potential energy, because when the electric interaction is attractive, like gravitation, q_1 and q_2 have different signs, and the minus sign shows up automatically. However, if the two charges have the same sign, the interaction is repulsive, and there is a plus sign in the formula for the electric energy (Figure 4.33). If the two charges have opposite signs, the interaction is attractive, and there is a minus sign in the formula for electric energy (Figure 4.34).

A certain amount of energy is required to remove an outer electron from a neutral atom, leaving behind a positive "ion" that is missing one electron and whose charge is $+e$, where $e = 1.6\times10^{-19}$ coulomb. (The electron's charge is of course $-e$.) The minimum energy to remove one outer electron is called the "ionization" energy.

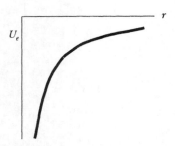

Figure 4.34 Electric potential energy is negative (like gravitational potential energy) when objects attract.

Example: A typical atom has a radius of about 10^{-10} m . Calculate a typical ionization energy in joules. Then convert your result to "electron volts," a unit of energy equal to 1.6×10^{-19} joules. This small energy unit is frequently used in discussions of atomic phenomena.

Solution: As the system we choose two particles: the outer electron with charge $-e$, and the remaining part of the atom, the ion with charge $+e$. Assume that no external forces act on this isolated system. In the initial state, the electron is bound to the atom, at a distance r_{atom} from the center; in the final state the two particles are very far from each other, at rest.

Make the rough approximation that initially the electron has negligible kinetic energy, so $K_f = K_i = 0$. Then, applying the energy principle we find the amount of work required to ionize the atom:

$$\Delta E_{\text{sys}} = W$$

$$\Delta U_e + \Delta K = \Delta U_e + 0 = W$$

$$\Delta U_e = \frac{1}{4\pi\varepsilon_0}\frac{q_1 q_2}{r_\infty} - \frac{1}{4\pi\varepsilon_0}\frac{q_1 q_2}{r_{\text{atom}}} = -\frac{1}{4\pi\varepsilon_0}\frac{(+e)(-e)}{r_{\text{atom}}}$$

$$\Delta U_e \approx \left(9\times10^9\frac{\text{N}\cdot\text{m}^2}{\text{C}^2}\right)\frac{(1.6\times10^{-19}\ \text{C})^2}{(10^{-10}\ \text{m})} = 2.3\times10^{-18}\ \text{J}$$

$$\Delta U_e = (2.3\times10^{-18}\ \text{J})\left(\frac{1\ \text{eV}}{1.6\times10^{-19}\text{J}}\right) = 14\ \text{eV}$$

This is a rough estimate, but typical ionization energies are indeed a few electron volts.

Ex. 4.28 Two protons are hurled straight at each other, each with a kinetic energy of 0.1 MeV, where 1 mega electron volt is equal to $10^6\times(1.6\times10^{-19})$ joules. What is the separation of the protons from each other when they momentarily come to a stop?

4.6.2 Electric potential energy involving neutral atoms

In the preceding section we obtained the formula for electric potential energy for pairs of charged particles. In Chapter 3 we studied the spring-like forces between electrically *neutral* atoms. These interatomic forces are the superposition of electric attractions and repulsions among the positively charged protons and negatively charged electrons of which the atoms are made.

When two electrically neutral atoms come very near each other (Figure 4.35), they attract each other. If you try to push the two atoms closer together, you eventually encounter a rapidly increasing repulsion. Both of these properties are due to the superposition of the electric forces of the various protons and electrons in the atoms.

In Chapter 3 we found that a good approximation to the net force over a limited range of separations is $F_s = -k_s s$; that is, the component of the force in the direction of the interatomic bond is opposite to the stretch s. When s is positive (longer bond), the force is attractive and pulls the atoms toward each other (negative component). When the atoms are forced closer together (s negative), the force is repulsive (positive component).

Remember that the (negative) gradient of potential energy is equal to the associated force. So we're looking for a function whose derivative is $+k_s s$.

? Can you think of a function of s whose derivative with respect to s is $+k_s s$? (You might review the Appendix on basic calculus at the end of this volume.)

The derivative with respect to s of $\frac{1}{2}k_s s^2$ is $+k_s s$, and the negative gradient of $\frac{1}{2}k_s s^2$ is the force $-k_s s$:

CHANGE IN INTERATOMIC POTENTIAL ENERGY

$$\Delta U_s = \Delta[\tfrac{1}{2}k_s s^2]$$

Note that the stretch s appears squared in the formula for spring energy. This is due to the fact that either a lengthening or a compression of the interatomic bond involves increased potential energy.

As usual, change of the interatomic potential energy involves a change of shape of the two-atom system.

What about large separations?

Our analysis is fine as far as it goes, but there's a serious problem. The potential energy U_s of the two atoms should go to zero when they are far apart,

Figure 4.35 Two atoms attract each other at intermediate distances, but repel each other at very short distances.

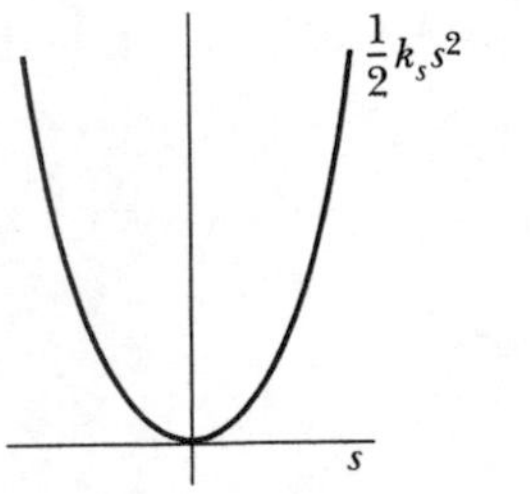

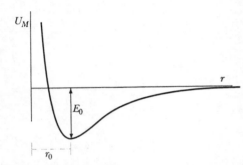

Figure 4.36 The predicted form of the potential energy becomes *infinite!*

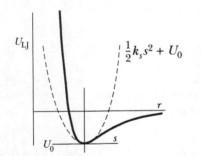

Figure 4.37 The Morse potential energy function U_M represents the potential energy of two neutral atoms.

so that the total energy is just the sum of their particle energies, but $\frac{1}{2}k_s s^2$ becomes *infinite!* (See Figure 4.36.) We need to go deeper into the details of the interatomic force.

Figure 4.37 shows a function called the Morse potential energy U_M that approximately represents the potential energy of two neutral atoms, as a function of the distance r between their nuclei:

$$U_M = E_0[1 - e^{-\alpha(r - r_{eq})}]^2 - E_0$$

The Morse approximation is designed to model the basic behavior of the interatomic force, and is in approximate agreement with measurements of the interatomic force. The parameter E_0 gives the depth of the potential energy where it is a minimum, and the parameter r_{eq} (on the order of 10^{-10} m) is the distance at which the potential energy curve is a minimum. The parameter α is adjusted to make the curve have the right width. When the interatomic separation r is very large, $U_M = 0$, as it should.

The depth of the potential well, E_0, is typically a few electron volts, where 1 eV is equal to 1.6×10^{-19} joule, a tiny energy unit often used in describing the world of atoms. For example, the potential energy well is about 5 eV deep for oxygen, O_2. (As you saw in the example on page 140, a typical ionization energy is also in the range of electron volts). However, for the noble gases such as helium and argon the well is so shallow that these atoms don't form stable diatomic molecules at room temperature, because collisions easily provide enough energy to break the molecules apart.

? What is the physical significance of the minimum of the curve?

At the minimum of the potential energy curve, the slope is zero, so the force between the two atoms is zero. This is the position of stable equilibrium.

? What is the direction of the force on an atom at a distance smaller than the equilibrium distance?

The left side of the curve has a large negative slope, so the sign of the force will be positive. This corresponds to a force to the right (in the $+r$ direction), away from the other atom. This makes sense; when the atoms are too close together they push each other apart.

Likewise, if the atom moves to the right (farther away) there should be an attractive restoring force to the left. The slope of this region of the curve is positive, so the sign of the force is negative, meaning its direction is to the left, as expected.

As indicated by the dashed curve in Figure 4.38, the bottom of the interatomic potential energy curve can be approximated like this:

THE "SPRING" APPROXIMATION FOR INTERATOMIC *U*

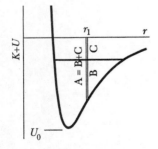

Figure 4.38 The bottom of the interatomic potential energy curve may be approximated by a harmonic oscillator energy curve.

$\frac{1}{2}k_s s^2 + U_0$ (U_0 is a negative number; s measured from equilibrium)

For the Morse potential energy formula, $U_0 = -E_0$

Because we can make this approximate fit, we know that for small oscillations around the equilibrium point the two atoms will act as though they are connected by a spring with a $-k_s s$ force. This is the rationale for modeling a solid as a collection of balls connected by springs.

Bound and unbound states with interatomic potential energy

In Figure 4.39, a horizontal line represents motion in which the sum of the kinetic energies and the potential energy of the two-atom system does not change. A horizontal line corresponds to a possible state of the system when it is isolated from external forces.

Figure 4.39 A horizontal line represents constant kinetic plus potential energy.

Ex. 4.29 At separation r_1 in Figure 4.39, what is the physical significance of the quantity A? Of the quantity B? Of the quantity C?

Ex. 4.30 In Figure 4.40, which of the states are bound states? Which are unbound states?

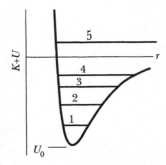

Figure 4.40 Which are bound states? Which are unbound states?

4.7 Rest energy in a multiparticle system

If the individual particles in a multiparticle system don't change their identity, both the initial and the final energies of the system include mc^2 rest-energy terms for each particle, and these cancel out in the energy equation:

$$m_1 c^2 + m_2 c^2 + K_i + U_{s,i} + U_{g,i} = m_1 c^2 + m_2 c^2 + K_f + U_{s,f} + U_{g,f}$$

$$K_i + U_{s,i} + U_{g,i} = K_f + U_{s,f} + U_{g,f}$$

This is an important special case. Next we consider the interesting effects observed when the rest masses of the particles can change due to change of identity.

Neutron decay is a good example to illustrate the issues. A free neutron (one not bound into a nucleus) is unstable, with an average lifetime of about 15 minutes. As shown in Figure 4.41, it decays into a proton, electron, and a (nearly) massless antineutrino, which travels at (nearly) the speed of light. All three of these particles have kinetic energy. That is, they have energy above and beyond their rest energy; the energy of the antineutrino is (nearly) all kinetic, because it has (almost) no rest energy. Here is a situation with a change of particle identity, and a change in particle rest mass.

The rest energy of a proton is 938.3 MeV (1 MeV is a million electron volts, where 1 electron volt is 1.6×10^{-19} joule). One mole of atomic hydrogen has a mass of 10^{-3} kg, so we have approximately

$$n \rightarrow p^+ + e^- + \bar{\nu}$$

$$m_n c^2 = (m_p c^2 + K_p) + (m_e c^2 + K_e) + E_\nu$$

Figure 4.41 A stationary neutron decays into a proton, an electron, and a massless antineutrino (Exercise 4.31).

$$mc^2 \approx \frac{\left(\dfrac{10^{-3} \text{ kg}}{6 \times 10^{23} \text{ protons}}\right)(3 \times 10^8 \text{ m/s})^2}{1.6 \times 10^{-19} \text{ J/eV}} \text{ ; accurate value is } 938.3 \times 10^6 \text{ eV}$$

We may use the unit MeV/c^2 as a unit of mass, because dividing energy by c^2 gives mass. The rest mass of a proton is 938.3 MeV/c^2. The rest mass of a neutron is 939.6 MeV/c^2. The rest mass of the electron is 0.511 MeV/c^2. The antineutrino is nearly massless.

Ex. 4.31 In the decay of the free neutron (Figure 4.41), how much kinetic energy (in electron volts) do the decay products (proton, electron, and antineutrino) share? Why can't a proton decay into a neutron?

4.7.1 The mass of a multiparticle system

A surprising prediction of Einstein's special theory of relativity (1905) was that mass and energy are actually the same thing. This prediction has been confirmed in many different kinds of experiments. The most dramatic example of mass-energy equivalence is the annihilation reaction that occurs between matter and anti-matter, as in the case of the annihilation of an electron and a positron (a positive electron) into two photons (high-energy quanta of light denoted by lower-case Greek gamma, γ):

$$e^- + e^+ \rightarrow \gamma + \gamma$$

Before the reaction, an electron and a positron can be almost at rest far from each other, a situation we would naturally describe as one where there is mass but seemingly no energy. After the reaction there is light (in the form of two photons), a situation which we would naturally describe as pure energy but seemingly no mass. But from the point of view of special relativity, the mass of the electron-positron system *is* energy, and the energy of the two-photon system *is* mass.

The equivalence of mass and energy is of course visible in Einstein's formula for the energy of a particle:

$$E = \frac{mc^2}{\sqrt{1 - v^2/c^2}} \equiv mc^2 + K \quad (\text{\textit{definition} of kinetic energy } K)$$

In this equation m is called the rest mass of the particle, and is a constant. The particle has "rest energy" mc^2 even if it is not moving. If it is moving, it has additional, "kinetic" energy K and additional mass. A valid way of thinking about a fast-moving particle is that it has a larger mass, E/c^2, than its rest mass m, because mass is equivalent to energy.

What is the mass of a multiparticle system consisting of interacting particles? Consider the simple case where the multiparticle system is overall "at rest." The individual particles are moving around, but the system as a whole is not moving (technically, as we will study in Chapter 7, the total momentum of the system is zero, and its center of mass is at rest). An example would be a box containing air molecules that are moving around rapidly in random directions, but the box isn't going anywhere. Another example is a diatomic molecule such as O_2 which is vibrating but has no overall translational motion.

It would seem natural to suppose that the total mass of such a system is just the sum of the rest masses of the particles: $M = m_1 + m_2 + ...$ However, this is not quite right. Rather, as predicted by the theory of relativity, the energy of the multiparticle system is Mc^2, and therefore its mass is this:

$$M = \frac{E_{\text{system}}}{c^2} = \frac{m_1 c^2 + K_1 + m_2 c^2 + K_2 + ... + U}{c^2}$$

$$M = (m_1 + m_2 + ...) + \left(\frac{K_1 + K_2 + ... + U}{c^2} \right)$$

The mass M of the system may be greater or less than the sum of the individual particle rest masses, because the final term in this equation may be positive or negative.

In most everyday situations the kinetic energies of the individual particles are very small compared to their rest energies (that is, the speeds of the particles are small compared to the speed of light), and the interaction energy U (the sum of all the pair-wise interactions) is also small compared to the rest energies. In that case, the mass of the multiparticle system is very nearly equal to the sum of the individual rest masses. However, this is really only an approximation, and it is often not a good approximation in nuclear and particle physics.

4.7.2 The total energy of a multiparticle system

Consider an O_2 molecule at rest. Instead of plotting only the kinetic and potential energy for this two-atom system, let's include the rest energy of the atoms, too, as we attempt to show in Figure 4.42. We can't draw the full vertical scale, because the rest energies are enormous compared to the kinetic and potential energies. In order to show the rest energy to the same scale as *K* and *U*, we would need a page stretching from the Earth to the Moon!

The total energy of the system (indicated by the horizontal line inside the potential energy well) is the sum of the rest energies of the two atoms, the potential energy *U* (which is negative), and *K*, the kinetic energy of the two atoms. For the bound state drawn on the graph you can see that the rest energy of the two-atom O_2 system is very slightly less than the sum of the individual atomic rest energies. When the diatomic molecule O_2 forms, about 5 eV of energy are given off (for example, by emitting photons, which carry energy). Correspondingly, it takes an input of about 5 eV of energy to break O_2 apart, if the molecule's energy is near the bottom of the well. This is called the "binding energy."

Two ways of thinking about the energy of the system

Given these aspects of particle energy, there are two rather different ways to think about the energy of a multiparticle system:

- The energy of a multiparticle system consists of the individual particle energies of the particles that make up the system, plus their pair-wise interaction energies.
- A multiparticle system itself has energy, rather like a particle, and if the system is at rest its energy *E* is simply Mc^2, where *M* is the mass of the system.

The mass of the system

The mass *M* may be greater or smaller than the sum of the rest masses of the individual particles, depending on whether the total *K+U* of the particles is positive or negative. A vertical line on the left side of the graph indicates that in this case the quantity Mc^2 (where *M* is the mass of the whole molecule) is less than $2\,mc^2$. From the graph we can see that:

$$Mc^2 = 2\,mc^2 + K + U$$

Mass of a hot object

This way of thinking about mass leads us to conclude that a hot block has a greater mass than a cold block. As we'll discuss in more detail in the next chapter, a hot metal block has more thermal energy than a cold metal block, with more kinetic energy of the atoms and more potential energy in the interactions with neighboring atoms (represented as springs in our simple model of a solid). Although the effect is much too small to measure directly, the equivalence of mass and energy embodied in the special theory of relativity, $M = E/c^2$, leads to the conclusion that a hot block is actually more massive than a cold block! The same impulse $\vec{F}\Delta t$ applied to the two blocks should change the momentum of the cold block very slightly more than the momentum of the hot block.

> *Ex. 4.32* Calculate the approximate ratio of the binding energy of O_2 (about 5 eV) to the rest energy of O_2. Most oxygen nuclei contain 8 protons and 8 neutrons, and the rest energy of a proton or neutron is about 940 MeV. Do you think you could use a laboratory scale to detect the difference in mass between a mole of molecular oxygen (O_2) and two moles of atomic oxygen?

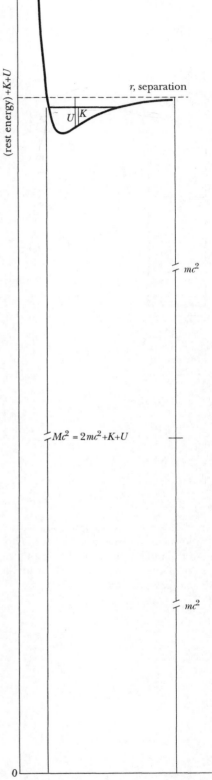

Figure 4.42 A plot of rest energy, kinetic energy, and potential energy for O_2.

4.7.3 Application: Nuclear binding

In chemical reactions it is nearly impossible to measure the change of rest mass directly, but in nuclear interactions the changes are easily measurable. Consider the nuclear binding energy of an iron nucleus, which contains 26 protons and 30 neutrons, a total of 56 nucleons. When the protons and neutrons bind to each other through the "strong" force, the resulting iron nucleus has a mass that is about 1% smaller than the total mass of its individual constituents when separated. There is a sizable negative potential energy U associated with the strong force, which makes the multiparticle system's rest energy measurably smaller than the sum of the individual rest energies.

The average binding energy per nucleon is largest in iron. Lighter nuclei (hydrogen through manganese) are less tightly bound and can in principle fuse together to form the more tightly bound iron, with the release of energy. In fact, fusion reactions in stars build up from hydrogen to heavier and heavier nuclei, stopping at iron (heavier elements are made in supernova explosions).

Heavier nuclei (cobalt through uranium and beyond) are also less tightly bound than iron and can in principle fission to form the more tightly bound iron. However, only a few nuclei such as uranium and plutonium do actually spontaneously fission.

Figure 4.43 shows an energy diagram (not to scale) for one of the protons in an iron nucleus, subjected to the forces of all the other protons and neutrons. There is a deep potential well corresponding to the strong or nuclear force, which is essentially the same for protons and neutrons. This potential well has a very steep side corresponding to the very strong nuclear force (force is negative gradient of potential energy). The well is so steep it can be described approximately as a "square well." This deep well extends only out to the radius R of the nucleus, because the nuclear force has a very short range. A proton (but not a neutron) in addition experiences the repulsive electric force of the other protons (the electric U curve outside the nucleus). The total rest mass of the nucleus is about 1% less than the sum of the rest masses of the individual nucleons (protons and neutrons).

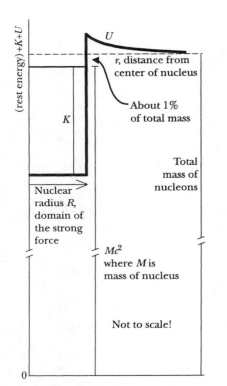

Figure 4.43 An energy diagram for a proton in an iron nucleus, not to scale. The mass of the nucleus is about 1% less than the sum of the masses of the individual nucleons (protons and neutrons).

Example: The rest energy of an iron nucleus (26 protons and 30 neutrons) is 52107 MeV. Show that the average binding energy per nucleon is about 1% of the rest energy of a nucleon, which is easily measurable. The rest energy of a proton is 938.3 MeV; of a neutron, 939.6 MeV.

Solution: As shown in Figure 4.42, the binding energy is essentially the difference between the rest energy of the multiparticle system (the nucleus) and the sum of the rest energies of its constituent particles.

binding energy = ((26)(938) MeV + (30)(939) MeV) − 52107 MeV

= 481 MeV

$$\text{energy per nucleon} = \frac{481 \text{ MeV}}{26 + 30} = 8.6 \text{ MeV}$$

This is about 1% of the rest energy of one nucleon.

Ex. 4.33 The deuteron, the nucleus of the deuterium atom ("heavy" hydrogen), consists of a proton and a neutron. It is observed experimentally that a high-energy photon ("gamma ray") with a minimum energy of 2.2 MeV can break up the deuteron into a free proton and a free neutron; this process is called "photodissociation." About what fraction of the deuteron rest energy corresponds to its binding energy? The result shows that the deuteron is very lightly bound compared to the iron nucleus.

4.7.4 Application: Nuclear fission

The nuclear fission of uranium also leads to an easily measured change in rest masses. The average binding energy per nucleon in uranium is about 7.8 MeV, so the uranium nucleus is less tightly bound than an iron nucleus, whose binding is 8.5 MeV per nucleon. This suggests the possibility of obtaining energy from the fission of a uranium nucleus, corresponding to roughly 0.1% (0.7 MeV per nucleon) of its rest energy (about 1000 MeV per nucleon).

In other words, the "mass deficiency" of a medium-mass nucleus is about 1%, with a variation of about a tenth of that. Even though there is only a small difference in average binding energy between iron and uranium, the energy involved is vast compared to the energy of a chemical reaction, which only involves the outer electrons of an atom, not the nucleus.

A uranium-235 nucleus contains 92 protons and 143 neutrons. The 92 protons exert strong electric repulsions on each other, but the large number of neutrons spreads the protons throughout a larger volume and reduces the electric repulsions. The neutrons and the protons also attract neighboring protons with strong nuclear forces. As discussed in Chapter 1, uranium-235 has an unusual property that few other nuclei have. If you hit a U-235 nucleus with a slowly moving neutron, the nucleus may deform into a dumbbell-like form and then split apart because the electric repulsion of the protons can now overcome the short-range attraction of the nucleons for each other (Figure 4.44). (In addition to the two main fission fragments there are typically one or more free neutrons in the final state.)

The fission fragments quickly acquire very high speed, and they leave the original atom's 92 electrons behind. These highly charged nuclei storm through the block of uranium metal, tearing up the material and violently heating the metal. In a nuclear power reactor the reaction is controlled in such a way that the metal merely gets hot enough to boil water. With many uranium-235 nuclei fissioning, the metal can boil water in a nuclear power plant to produce steam and power a turbine to drive an electric generator.

The rest masses of the two fission-fragment nuclei total less than the rest mass of the uranium-235 nucleus. The rest mass difference Δm is only about one-tenth of one percent of the original rest mass, but this rest mass change is easily measurable, and c^2 times this rest mass change is indeed found to be equal to the observed kinetic energy of the fission fragments. This 0.1% change is about ten million times as big an effect as the rest mass changes that occur in chemical reactions, which only involve rearrangements of the outer electrons in the atoms, and do not affect the nucleus.

Figure 4.44 A U-235 nucleus in the process of fissioning.

4.8 Reflection: Why energy?

It is appropriate at this point to reflect on why we introduced the topic of energy. Energy methods often make it possible to analyze some aspect of a phenomenon with much less effort than is required by direct application of the momentum principle. A striking example is the ability to calculate escape speed, or the impact speed of the Ranger spacecraft on the Moon, without carrying out a lengthy numerical integration.

Limits on the possible

Another powerful capability of energy analyses is that they can easily predict whether some process can occur or not. For example, if there is insufficient

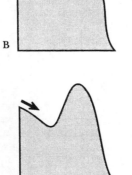

energy, the rocket can't escape from the planet, or the atom cannot be ionized, or the biochemical reaction cannot proceed.

Less detail

However, energy analyses give less detailed information than is given by a step-by-step numerical integration of the momentum principle. In particular, energy analyses tell us nothing about how fast a process will proceed; time does not appear in the statement of conservation of energy.

Here is an example that illustrates the strengths and weaknesses of energy analyses. Consider snowboarding down the three hills shown in Figure 4.45, starting from rest. If we can neglect friction, the final speed at the bottom should be the same in all three cases, because the final kinetic energy should be equal to *mgh*, the amount by which the gravitational energy decreases.

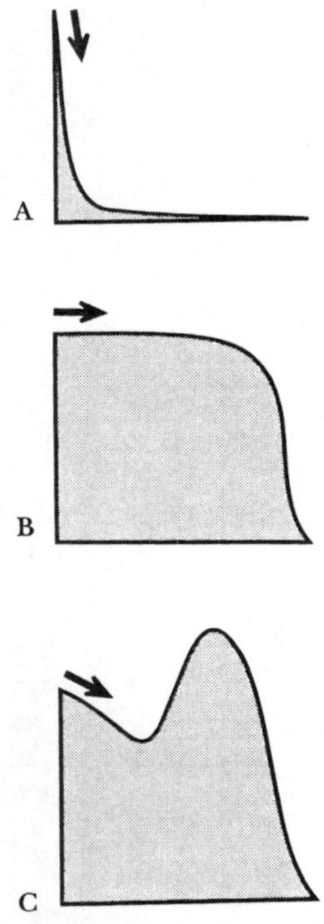

? But how will the amount of *time* compare for the ride down the hill?

On the first hill you get going very fast very quickly, and the trip is short. On the second hill you move very slowly for a long time before dropping to the bottom, and the total trip takes a long time. On the third hill you never get to the bottom, despite the fact that the change in potential energy would be negative, because you would have to pass through a region that is energetically forbidden. So energy methods are powerful, but at the cost of losing some information, including the amount of time required for a process.

Figure 4.45 The final speed of the snowboarder will be the same at the bottom of each hill.

4.9 *A puzzle

The principle of conservation of energy states that when a system's energy increases, the energy of the surroundings decreases by the same amount. For example, take a block as the system of interest. If you push the block away from you and raise its kinetic energy, your store of chemical energy decreases by the same amount. The flow of energy from you (the "surroundings") to the block takes the form of mechanical energy transfer we call "work." We can demonstrate that there is a puzzle concerning the energy in the surroundings of a star whose kinetic energy increases due to an interaction with another star.

Consider two stars of equal mass initially at rest, far apart (Figure 4.46). The two stars do work on each other:

$$\Delta E_1 = W_{\text{on }1} \quad \text{star 1 as system}$$

$$\Delta E_2 = W_{\text{on }2} \quad \text{star 2 as system}$$

If we add these two energy equations for the two stars, we have this:

$$\Delta E_1 + \Delta E_2 = W_{\text{on }1} + W_{\text{on }2} \quad \text{sum of two energy equations}$$

Figure 4.46 Initially the stars were very far apart, at rest, but they attract each other.

Moving $(W_{\text{on }1} + W_{\text{on }2})$ to the left side of the equation and renaming the negative of this quantity ΔU, we have this familiar result:

$$\Delta E_1 + \Delta E_2 + \Delta U = 0 \quad \text{energy equation for two stars}$$

Our interpretation of this equation is that the gravitational potential energy U of the two-star system (calculated from the work on the individual particles) decreases as the particle energies increase. This point of view is valid, yields physically correct results, and is the view that we will take in this introductory course.

However, it is not difficult to show that there is a puzzle concerning energy flow between one of the stars and its surroundings. Consider star 1 as the system. Since it is a single object, there is no potential energy U—no pairwise interaction energy. The energy of star 1 increases due to work done on it: $\Delta E_1 = W_{\text{on }1}$. There are very strong reasons to believe in energy conservation, so we expect that when star 1 gains an amount of energy ΔE_1 the surroundings of star 1 should lose that much energy.

? Part of the surroundings is star 2. Does star 2 lose energy?

No! Star 2 doesn't lose energy, it *gains* energy, and of the same amount $\Delta E_2 = \Delta E_1$, due to the symmetry of the situation, with the two stars having equal mass.

What else is there in the surroundings that could experience a change in energy? The only other matter is star 2, but its energy change is inconsistent with the principle of energy conservation. Evidently our model of the world is incomplete. To be able to analyze this situation fully, we need to introduce the abstract idea of a "field." Fields, which are the subject of Volume II of this textbook, are associated with objects having mass (gravitational fields) or charge (electric and magnetic fields), and they extend throughout all space. It is energy stored in the gravitational field that accounts for the energy in the surroundings of star 1. Fields can also have momentum, as we will see in Volume II.

Despite the fact that we haven't yet defined what a "field" is, we can determine indirectly how much the field energy must change. Since the energy in the surroundings of star 1 changes by an amount $-\Delta E_1$, and the energy of star 2 increases by an amount $+\Delta E_1$, it must be that the energy stored in the gravitational field changes by an amount $-2\Delta E_1$. (It is an indication of the power of the principle of conservation of energy that we can calculate the energy change of an entity that we know nothing about!) This energy change is precisely the amount of change in the potential energy of the two-star system: $\Delta U = -(\Delta E_1 + \Delta E_2) = -2\Delta E_1$.

Because the potential energy U correctly accounts for the energy changes, we don't have to consider field energy in our calculations. Our analyses in terms of U that focus on particles and their interactions, not their surroundings, are completely correct. However, the difficulty in accounting for the energy in the surroundings of a star suggests that eventually we will need to include fields in our models of the world.

4.10 *Integrals and antiderivatives

In determining the formula for gravitational potential energy, we started from these complementary formulas:

$$\Delta U_g = -F_r \Delta r \text{ and } F_r = -\frac{dU_g}{dr}$$

We calculated the change in potential energy from the negative of the internal work:

$$U_{s,f} - U_{s,i} = \Delta\left(-G\frac{m_1 m_2}{r}\right) = -\Sigma F_r \Delta r$$

But if we had let Δr approach zero, the finite sum would have turned into an integral:

$$\Delta\left(-G\frac{m_1 m_2}{r}\right) = -\int_i^f F_r dr = \int_i^f G\frac{m_1 m_2}{r^2} dr$$

You can see that a definite integral (the infinitesimal version of a finite sum) can be evaluated in terms of the antiderivative. The antiderivative of $Gm_1 m_2 / r^2$ is $-Gm_1 m_2 / r$, and the change in the value of the antiderivative turns out to be equal to the definite integral (the infinite sum of infinitesimal elements).

The reason why this works this way can be traced back to the fact that the force is the derivative of the energy, so the energy must be the antiderivative of the force, yet the energy is also the sum of lots of $F_i\Delta r$'s. Hence the sum is given by the change in the antiderivative.

When the quantity to be integrated (here, the force) is a simple function, it may be easy to find the antiderivative. However, in many practical cases there is no known simple function that is the antiderivative, in which case an integral must be evaluated by numerical integration, as a finite sum.

4.11 *Approximation for particle energy

An approximation for the kinetic energy of a particle at low speeds can be obtained through use of the mathematics of the "binomial expansion":

$$(1 + \varepsilon)^n = 1 + n\varepsilon + \frac{n(n-1)}{2}\varepsilon^2 + \frac{n(n-1)(n-2)}{3 \cdot 2}\varepsilon^3 + \ldots$$

This formula comes from working through what happens when you multiply the quantity $(1 + \varepsilon)$ times itself n times. The first term is 1 multiplied by itself n times (1^n); the next term is made up of all those products, n of them, in which ε appears only once [$n\varepsilon$]; etc.

Now for the important point that makes the binomial expansion useful in obtaining approximate results. If ε is a number that is very small compared to 1, each term in the expansion is much smaller than the preceding term:

$n\varepsilon$ is much smaller than 1, $\frac{n(n-1)}{2}\varepsilon^2$ is much smaller than $n\varepsilon$, etc.

Therefore we can get a good approximation if we keep just the first few terms and ignore the rest.

Although we have developed the binomial expansion for the case where n is an integer, it can be shown that the formula is valid even for non-integer values of n. In particular,

$$\frac{1}{\sqrt{1-\varepsilon}} = (1 - \varepsilon)^{-1/2} = 1 + (-\tfrac{1}{2})(-\varepsilon) + \frac{(-\tfrac{1}{2})(-\tfrac{3}{2})}{2}\varepsilon^2 + \frac{(-\tfrac{1}{2})(-\tfrac{3}{2})(-\tfrac{5}{2})}{6}\varepsilon^3 + \ldots$$

If v/c is small compared to 1, then v^2/c^2 is even smaller compared to 1. For example, $1/10$ is small compared to one, but $(1/10)^2 = 1/100$ is very small indeed. Therefore we can write this for low speeds:

$$\frac{mc^2}{\sqrt{1 - v^2/c^2}} = mc^2\left[1 + \frac{1}{2}\frac{v^2}{c^2} - \frac{3}{8}\left(\frac{v^2}{c^2}\right)^2 - \frac{5}{16}\left(\frac{v^2}{c^2}\right)^3 + \ldots\right] \approx mc^2 + \frac{1}{2}mv^2$$

This last element can be rewritten in terms of momentum $p \approx mv$:

$$\frac{1}{2}mv^2 = \frac{1}{2}\frac{(mv)^2}{m} = \frac{p^2}{2m}$$

4.12 *Determining the formula for particle energy

If we choose our axes so that x is in the direction of the motion at this instant (Figure 4.47), we have

$$\Delta E = F_x\Delta x = \left(\frac{\Delta p_x}{\Delta t}\right)\Delta x$$

for the change in energy of a point particle acted on by a net force. The energy E should satisfy the following relationship between two derivatives:

$$\frac{dE}{dx} = \frac{dp_x}{dt}$$

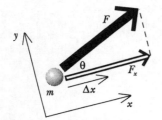

Figure 4.47 A particle of mass m. We choose the x axis to be in the direction of the motion. The net force acting on the particle has a component F_x in the direction of motion.

We will show that if we choose the following function for E, the relationship between the two derivatives will be satisfied:

$$E = \frac{mc^2}{\sqrt{1 - v^2/c^2}}$$

The proof proceeds by calculating dE/dx, calculating dp_x/dt, and showing that the two derivatives are equal to each other. First, here is dE/dx:

$$\frac{dE}{dx} = \frac{d}{dx}\left(\frac{mc^2}{\sqrt{1 - v^2/c^2}}\right) = \frac{mc^2}{(1 - v^2/c^2)^{3/2}}\left(-\frac{1}{2}\right)\left(-\frac{2v}{c^2}\right)\frac{dv}{dx}$$

$$\frac{dE}{dx} = \frac{mv}{(1 - v^2/c^2)^{3/2}}\frac{dv}{dx}$$

Next we calculate dp_x/dt, keeping in mind that $v_x = v$, because we chose our axes so that x is in the direction of the motion at this instant:

$$\frac{dp_x}{dt} = \frac{d}{dt}\left(\frac{mv}{\sqrt{1 - v^2/c^2}}\right) = \frac{m}{\sqrt{1 - v^2/c^2}}\frac{dv}{dt} + \frac{mv}{(1 - v^2/c^2)^{3/2}}\left(-\frac{1}{2}\right)\left(-\frac{2v}{c^2}\right)\frac{dv}{dt}$$

$$\frac{dp_x}{dt} = \frac{m}{(1 - v^2/c^2)^{3/2}}\left[\left(1 - \frac{v^2}{c^2}\right) + \frac{v^2}{c^2}\right]\frac{dv}{dt} = \frac{m}{(1 - v^2/c^2)^{3/2}}\frac{dv}{dt}$$

This is almost in the same form as the derivative dE/dx. Now note this:

$$\frac{dv}{dt} = \left(\frac{dv}{dt}\right)\left(\frac{dx}{dx}\right) = \left(\frac{dv}{dx}\right)\left(\frac{dx}{dt}\right) = \left(\frac{dv}{dx}\right)v, \text{ because } v = \frac{dx}{dt}$$

Substituting this result for dv/dt into the derivative dp_x/dt, we find that this derivative is indeed equal to the derivative dE/dx if E is defined as

$$E = \frac{mc^2}{\sqrt{1 - v^2/c^2}}$$

Hence the relativistic formula for energy does follow from the relativistic formula for momentum.

It might seem that we could define the particle energy with an additive constant, which would drop out when we took the derivative dE/dx. However, when one considers a variety of particle reactions, it is found that to account for energy in a consistent way requires that the additive constant be zero.

This approach to solving a differential equation (proposing a solution and showing that it satisfies the equation) is a legitimate technique of solving differential equations, and in fact is used quite frequently.

4.13 Summary

Fundamental Principles

The energy principle:

$$\Delta E_{\text{system}} = W \text{ where } E_{\text{system}} = (E_1 + E_2 + \dots) + U$$

New Concepts

Work W_F done by a single force $\vec{F}$:

$$W_F = \vec{F} \bullet \Delta \vec{r} = F\Delta r \cos\theta = F_x\Delta r_x + F_y\Delta r_y + F_z\Delta r_z \text{ (constant force)}$$

$$W_F = \sum \vec{F} \bullet \Delta \vec{r} \text{ (variable force)}$$

Energy of a particle with rest mass m is $E = \dfrac{mc^2}{\sqrt{1 - v^2/c^2}} = mc^2 + K$

Kinetic energy $K = \dfrac{mc^2}{\sqrt{1 - v^2/c^2}} - mc^2$; $K \approx \dfrac{1}{2}mv^2 = \dfrac{p^2}{2m}$ if $v \ll c$

Energy and momentum: $E^2 - (pc)^2 = (mc^2)^2$ (rest mass m)

For so-called "massless" photon or neutrino, $E = pc$

Potential energy: $\Delta U \equiv -W_{\text{internal}}$

Force is negative gradient of potential energy: $F_x = -\dfrac{dU}{dx}$

Results

Gravitational potential energy $U_g = -G\dfrac{Mm}{r}$

$$\Delta U_g \approx \Delta(mgy) \text{ near Earth's surface}$$

Electric potential energy $U_e = \dfrac{1}{4\pi\varepsilon_0}\dfrac{q_1 q_2}{r}$,

$$\text{where } \dfrac{1}{4\pi\varepsilon_0} = 9\times10^9 \text{ N·m}^2/\text{coulomb}^2$$

Approx. interatomic potential energy: $U_M = E_0[1 - e^{-\alpha(r - r_{\text{eq}})}]^2 - E_0$

Near bottom of well: $U_s \approx \dfrac{1}{2}k_s s^2 + U_0$, where U_0 is negative.

Nuclear binding: heavy nuclei can fission, light nuclei can fuse, with energy released in both cases.

Electron volt: 1 eV = 1.6×10^{-19} J

Problem Solving Techniques

Use energy calculations to place limits on what is possible. For example:
How high a ball can go given its initial velocity
Escape speed for a spacecraft
Whether a particle reaction can proceed

4.14 Solving energy problems

4.14.1 Organizing your solution

Problems involving energy can appear quite complex and intimidating, since they may involve many objects, many interactions, and even changes in identity. Nonetheless, the following simple way of organizing your solution will always work!

First: Explicitly choose a system. Write it down so you will not get confused later.

Second: Identify an initial state and a final state of this system. Some problems may involve more than two states. Since you can only apply the energy principle to two states, this means you will have to solve such a problem in more than one step.

Third: Write down the energy principle as it applies to your chosen system. Be careful to note whether or not there are external energy inputs to your system.

You can write the energy principle in two different ways:

$$\Delta E_{sys} = W, \text{ or}$$

$$E_{final} = E_{initial} + W$$

These are of course equivalent, but sometimes one or the other is more easily applied to a particular problem.

4.14.2 Example problem: Collision of alpha and carbon nucleus

An alpha particle (the nucleus of a helium atom, containing two protons and two neutrons) moving with momentum p_1 is shot toward a carbon-12 nucleus (containing six protons and six neutrons) that is moving with the same momentum p_1 toward the alpha particle. The speeds are nonrelativistic.

(a) What is the minimum momentum p_1 necessary to allow the alpha particle and carbon-12 nucleus to come in contact, so that a nuclear reaction can occur? (With enough energy it is possible to produce oxygen-16.) The radius of a nucleus is approximately $(1.3 \times 10^{-15} \text{ m}) A^{1/3}$, where A is the total number of protons and neutrons. Briefly explain your analysis.

(b) What is the kinetic energy of the alpha particle in part (a), in MeV (1 MeV = 10^6 eV)?

(c) There is a head-on collision of the alpha particle and the carbon-12 nucleus, and an excited state of an oxygen-16 nucleus is formed at rest. A very short time later the oxygen nucleus drops to its ground state, emitting a photon whose energy is 10.352 MeV (1 MeV = 10^6 eV). The oxygen nucleus recoils with negligible kinetic energy. What is the minimum (nonrelativistic) momentum p_2 that the alpha particle must have in order for this sequence of phenomena to occur? Masses of the ground states of selected nuclei, where 1 u = 1.66×10^{-27} kg :

He-4 (2p+2n) = 4.00040868 u
C-12 (6p+6n) = 11.99670852 u
O-16 (8p+8n) = 15.99052636 u

(d) For the process in part (a), as a function of the separation r between the alpha particle and the carbon-12 nucleus, plot K_1+K_2 (the sum of the kinetic energies of the two nuclei), plot the potential energy, and plot the sum of the kinetic energies and the potential energy. Label each of the three plots clearly.

Figure 4.48 Initial state (part a).

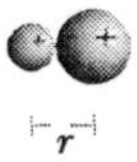

Figure 4.49 Final state (part a).

Solution

(a) *System:* alpha particle and carbon nucleus. No external forces.
Initial state: Very far apart, particles moving toward each other (Figure 4.48).
Final state: Particles at rest, touching each other (Figure 4.49).

Energy principle: $\Delta K + \Delta U_e + \Delta(mc^2) + \Delta(Mc^2) = 0$

$U_i = 0$ (particles very far apart) and $K_f = 0$ (particles at rest)

no change of identity, so $\Delta(mc^2) + \Delta(Mc^2) = 0$

so $\left[0 - \left(\dfrac{p_1^2}{2m} + \dfrac{p_1^2}{2M} \right) \right] + \left(\dfrac{1}{4\pi\varepsilon_0} \dfrac{(2e)(6e)}{r} - 0 \right) = 0$

where m is the mass of the alpha particle and M is the mass of the carbon nucleus.

$$p_1 = \sqrt{ \dfrac{\dfrac{1}{4\pi\varepsilon_0} \dfrac{12e^2}{r}}{\dfrac{1}{2}\left(\dfrac{1}{m} + \dfrac{1}{M} \right)} }$$

final separation $r = (1.3\times10^{-15}\text{ m})(4^{1/3} + 12^{1/3}) = 5.04\times10^{-15}\text{ m}$, so

$$p_1 = \sqrt{ \dfrac{\left(9\times10^9\ \dfrac{\text{N·m}^2}{\text{C}^2} \right) \dfrac{12(1.6\times10^{-19}\text{ C})^2}{(5.04\times10^{-15}\text{ m})}}{\dfrac{1}{2}\left(\dfrac{1}{4} + \dfrac{1}{12} \right)\left(\dfrac{1}{1.66\times10^{-27}\text{ kg}} \right)} } = 7.39\times10^{-20}\text{ kg·m/s}$$

(b) $K_\alpha = \dfrac{p_1^2}{2m} = \dfrac{(7.39\times10^{-20}\text{ kg·m/s})^2}{2(4\times1.66\times10^{-27}\text{ kg})}\left(\dfrac{10^{-6}\text{ MeV}}{1.6\times10^{-19}\text{ J}} \right) = 2.57\text{ MeV}$

Figure 4.50 Initial state (part c).

Figure 4.51 Final state (part c).

Photon

p_3 0

(c) *System:* alpha particle and carbon nucleus. No external forces.
Initial state: Very far apart, particles moving toward each other (Figure 4.50).
Final state: Oxygen nucleus and photon (Figure 4.51).
Energy principle: $E_{\text{initial}} = E_{\text{final}}$

since $K_{\text{oxygen}} \approx 0$,

$$\left(mc^2 + \dfrac{p_2^2}{2m} \right) + \left(Mc^2 + \dfrac{p_2^2}{2M} \right) = M_{\text{oxygen}}\, c^2 + E_\gamma$$

$$p_2 = \sqrt{ \dfrac{(M_{\text{oxygen}} - m - M)\, c^2 + E_\gamma}{\dfrac{1}{2}\left(\dfrac{1}{m} + \dfrac{1}{M} \right)} }$$

To calculate accurately the quantity $(M_{\text{oxygen}} - m - M)$ we must use very accurate values of the nuclear masses, because the differences in the rest masses are very slight:

$$M_{oxygen} - m - M = (15.99052636 - 4.00040868 - 11.99670852)u$$

$$M_{oxygen} - m - M = -0.00659084 \text{ u}$$

$$(-0.00659084 \text{ u})\left(1.66\times10^{-27} \frac{\text{kg}}{\text{u}}\right)\left(3\times10^{8} \frac{\text{m}^2}{\text{s}^2}\right) = -9.85\times10^{-13} \text{ J}$$

$$E_\gamma = (10.352 \text{ MeV})\left(\frac{1.6\times10^{-19} \text{ J}}{10^{-6} \text{ MeV}}\right) = 16.56\times10^{-13} \text{ J}$$

We don't have to be so careful in calculating the sum $\left(\dfrac{1}{m} + \dfrac{1}{M}\right)$:

$$p_2 = \sqrt{\frac{(16.56\times10^{-13} - 9.85\times10^{-13})\text{ J}}{\frac{1}{2}\left(\frac{1}{4} + \frac{1}{12}\right)\left(\frac{1}{1.66\times10^{-27} \text{ kg}}\right)}} = 8.18\times10^{-20} \text{ kg·m/s}$$

We find that the incoming momentum p_2 required to create the excited oxygen nucleus is greater than the momentum p_1 required just to overcome the electric potential energy barrier and bring the alpha particle and carbon nucleus in contact (Figure 4.52 shows the energies involved). It is necessary to bring nuclei into contact for a reaction to proceed, since the nuclear force has a very short range. However, in this reaction just making contact is insufficient. A bit more energy is needed in order to produce the larger-mass oxygen nucleus.

(d) See Figure 4.53. The sum of the kinetic and potential energies doesn't change, since no work is done on the two-particle system. The smallest separation is associated with the kinetic energy going to zero.

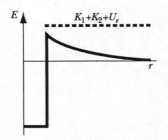
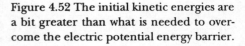

Figure 4.52 The initial kinetic energies are a bit greater than what is needed to overcome the electric potential energy barrier.

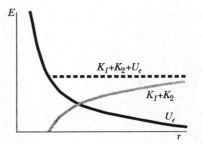

Figure 4.53 Energies as a function of separation between alpha particle and carbon nucleus.

4.15 Review questions

Work

RQ 4.1 You push a crate 3 m across the floor with a 40 N force whose direction is 30° below the horizontal. How much work do you do?

RQ 4.2 Give brief explanations for your answers to each of the following questions:

(a) You hold a 1 kg book in your hand for 1 minute. How much work do you do on the book?

(b) In a circular pendulum, how much work is done by the string on the mass in one revolution?

(c) For a mass oscillating horizontally on a spring, how much work is done by the spring on the mass in one complete cycle? In a half cycle?

Kinetic energy

RQ 4.3 Approximately what is the kinetic energy of a baseball pitcher's fast ball? Of a basketball passed from one player to another?

RQ 4.4 What is the speed of an electron whose total energy is equal to the total energy of a proton that is at rest? What is the kinetic energy of this electron?

RQ 4.5 An electron is traveling at a speed of $0.95c$. An electric force of 1.6×10^{-13} N is applied in the direction of motion while the electron travels a distance of 2 m. What is the new speed of the electron?

Work versus impulse, kinetic energy versus momentum

RQ 4.6 You pull a block of mass m across a frictionless table with a constant force. You also pull with an equal constant force a block of larger mass M. The blocks are initially at rest. If you pull the blocks through the same distance, which block has the greater kinetic energy, and which block has the greater momentum? If instead you pull the blocks for the same amount of time, which block has the greater kinetic energy, and which block has the greater momentum?

Energy conservation

RQ 4.7 You throw a ball straight up, and it reaches a height of 20 meters above your hand before falling back down. What was the speed of the ball just after it left your hand?

Gravitational energy

RQ 4.8 The escape speed from an asteroid whose radius is about 10 km is only 10 m/s. If you throw a rock away from the asteroid at a speed of 20 m/s, what will be its final speed?

Graphs of potential energy

RQ 4.9 The gravitational force inside the Earth is approximately $(GMm/R^3)r$ $= gr/R$, where r is the distance from the center and R is the radius of the Earth; see Problem 4.11 (page 160) for additional discussion. Draw the potential energy as a function of r, from $r = 0$ to very large r ($\gg R$). Note that the potential energy curve must be continuous, with no breaks.

Mass and energy

RQ 4.10 In positron-emission tomography (PET) used in medical research and diagnosis, compounds "tagged" with unstable nuclei that emit positrons are introduced into the brain, destined for a site of interest in the brain. When a positron is emitted, it goes only a short distance before coming nearly to rest, then annihilating with an electron. In the annihilation reaction two gamma rays (high-energy photons) are produced. The PET scanner detects these pairs of gamma rays, emitted at close to 180° to each other. What energy of gamma ray (in MeV, million electron volts) should each of the detectors be made sensitive to? (The mass of an electron or positron is 9×10^{-31} kg.)

RQ 4.11 One often hears the statement, "Nuclear energy production is fundamentally different from chemical energy production (such as burning of coal), because the nuclear case involves a change of mass." Critique this statement. Discuss the similarities and differences of the two kinds of energy production.

Additional exercises

Ex. 4.34 A jar of honey with a mass of 0.5 kg is knocked off the kitchen counter and falls one meter to the floor. What force acts on the jar during its fall? How much work is done by this force?

Ex. 4.35 An electron traveling through a wire in an electric circuit experiences a constant force of 5×10^{-19} N, always in the direction of its motion. How much work is done on the electron by this force as it travels through 0.5 m of the wire?

Ex. 4.36 A 0.5 kg teddy bear is knocked off a windowsill and falls 2 m to the ground. What is its kinetic energy at the instant it hits the ground? What is its speed? What assumptions or approximations did you make in this calculation?

Ex. 4.37 A 1.0 kg flowerpot is knocked off a windowsill and falls 2 m to the ground. What is its kinetic energy at the instant it hits the ground? What is its speed? How do the speed and kinetic energy compare to that of the teddy bear in the previous exercise?

Ex. 4.38 A commercial airliner flew 20,000 km nonstop from Seattle to Kuala Lumpur at an average speed of 900 km/hr. If its average mass was 140,000 kg what was its average kinetic energy? (We take an average because the mass of an airplane changes as it exhausts burnt fuel.)

4.16 Problems

4.16.1 In-line problems from this chapter

Problem 4.1 (page 125) Work done in Moon shot

If you have already done Problem 2.7 The Ranger 7 mission to the Moon, you can use the program you wrote for that problem, and go directly to part (b). Otherwise:

(a) Write a program to model the journey of a spacecraft coasting from the Earth to the Moon. Start the spacecraft at a height of 50 km above the Earth's surface with a speed of 1.3×10^4 m/s. This is approximately the speed it would have after all the rockets have fired, and is high enough to be above most of the Earth's atmosphere. Include the spacecraft's interactions with both the Earth and the Moon. Use a *dt* of 5 seconds. Make sure you stop the program when the spacecraft reaches the Moon's surface (not its center!). The data you need may be found on page 68, or on the inside back cover of this book. If you are interested in more of the details of this trip, see Problem 2.7 The Ranger 7 mission to the Moon, on page 68.

(b) Add a calculation of the work done by the gravitational forces of the Earth and the Moon to your analysis of sending a spacecraft to the Moon. You need to approximate the work by adding up the amount of work done by gravitational forces along each step of the path:

$$W = \sum \vec{F} \bullet \Delta \vec{r} = \vec{F}_1 \bullet \Delta \vec{r}_1 + \vec{F}_2 \bullet \Delta \vec{r}_2 + \vec{F}_3 \bullet \Delta \vec{r}_3 + ...$$

Compare the numerical value of the work with the change in the kinetic energy (final kinetic energy just before crashing on the Moon, minus initial kinetic energy when released above the Earth's atmosphere). Try shorter time steps to make sure that the errors introduced by a finite time step are not significant. Report your results.

Problem 4.2 (page 137) Energy in the Moon shot

Energy conservation is a powerful check on the accuracy of a numerical integration. Plot a graph of kinetic energy, of gravitational potential energy, and of the sum of the kinetic energy and the gravitational potential energy, *vs.* position during the Moon voyage, Problem 4.1 (page 125). Does the kinetic plus potential energy remain constant? What if you vary the step size (which varies the accuracy of the numerical integration)? Vary the launch speed, and explain the effect that this has on your graphs.

Problem 4.3 (page 137) Calculating the impact speed

Use energy conservation to calculate analytically (that is, without doing a numerical integration) the final speed of the spacecraft just before it hits the Moon. Include the gravitational effect of the Moon. Use a launch speed of 1.3×10^4 m/s. Modify your program for Problem 4.2 to print out the speed of the spacecraft when it hits the surface of the Moon, and compare this value to your analytical result. What questions that could be addressed in the numerical integration are you *not* able to answer by doing this energy calculation?

4.16.2 Additional problems

Problem 4.4 Boosting a satellite orbit

Calculate the speed of a satellite in an orbit near the Earth (just above the atmosphere). If the mass of the satellite is 200 kg, what is the minimum amount of energy required to move the satellite very far away from the Earth?

Problem 4.5 Comet orbit

A comet is in an elliptical orbit around the Sun. Its closest approach to the Sun is a distance of 4×10^{10} m (inside the orbit of Mercury), at which point its speed is 8.17×10^{4} m/s. Its farthest distance from the Sun is far beyond the orbit of Pluto. What is its speed when it is 6×10^{12} m from the Sun? (This is the approximate distance of Pluto from the Sun.)

Problem 4.6 An electron in an accelerator

An electron is traveling at a speed of $0.95c$ in an electron accelerator. An electric force of 1.6×10^{-13} N is applied in the direction of motion while the electron travels a distance of 2 m. What is the new speed of the electron?

Problem 4.7 Four protons

Four protons, each with mass M and charge $+e$, are initially held at the corners of a square that is d on a side. They are then released from rest. What is the speed of each proton when the protons are very far apart?

Problem 4.8 Throwing from an asteroid

You stand on a spherical asteroid of uniform density whose mass is 2×10^{16} kg and whose radius is 10 km (10^{4} m). These are typical values for small asteroids, although some asteroids have been found to have much lower average density and are thought to be loose agglomerations of shattered rocks; see Problem 2.11 (page 70). The asteroid is rotating a bit faster than once per day, so that objects on the surface have a speed of 1 m/s, including you and a rock you are holding.

(a) How fast (relative to you) do you have to throw the rock so that it never comes back to the asteroid and ends up traveling at a speed of 3 m/s when it is very far away? Explain briefly.

(b) Sketch graphs of the kinetic energy of the rock, the gravitational potential energy of the rock plus asteroid, and their sum, as a function of separation (distance from center of asteroid to rock). Label the graphs clearly.

Problem 4.9 Collision with Jupiter

This problem is closely related to the spectacular impact of the comet Shoemaker-Levy with Jupiter in July 1994 (http://www.jpl.nasa.gov/sl9/sl9.html). A rock far outside our Solar System is initially moving very slowly relative to the Sun, in the plane of Jupiter's orbit around the Sun. The rock falls toward the Sun, but on its way to the Sun it collides with Jupiter.

Calculate the rock's speed just before colliding with Jupiter. Explain your calculation and any approximations that you make.

$M_{\text{Sun}} = 2 \times 10^{30}$ kg $\qquad\qquad M_{\text{Jupiter}} = 2 \times 10^{27}$ kg

Distance, Sun to Jupiter = 8×10^{11} m $\qquad$ Radius of Jupiter = 1.4×10^{8} m

Problem 4.10 Particle-antiparticle annihilation

(a) A particle with mass M and charge $+e$ and its antiparticle (same mass M, charge $-e$) are initially at rest, far from each other. They attract each other and move toward each other. On axes like those in Figure 4.54, graph and label the various energies involved in this process, as a function of the distance r between the two particles. Be sure to include the rest energy of the particles.

(b) When the particle and antiparticle collide, they annihilate and produce a different particle with rest mass m (much smaller than M) and charge $+e$ and its antiparticle (same rest mass m, charge $-e$). When these two particles have moved far away from each other, how fast are they going? Is this speed large or small compared to c?

(c) Now take the specific case of a proton and antiproton colliding to form a positive and negative pion. Each pion has a rest mass of

Figure 4.54 Plot various energies (Problem 4.10).

2.5×10^{-28} kg . When the pions have moved far away, how fast are they going?

(d) How far apart must the two pions be (in meters) for their electric potential energy to be negligible compared to their kinetic energy? Be explicit and quantitative about your criterion and your result.

Problem 4.11 Gravitational energy inside the Earth

In the rough approximation that the density of the Earth is uniform throughout its interior, the gravitational field strength (force per unit mass) inside the Earth at a distance r from the center is gr/R, where R is the radius of the Earth. (In actual fact, the outer layers of rock have lower density than the inner core of molten iron.) Using the uniform-density approximation, calculate the amount of energy required to move a mass m from the center of the Earth to the surface. Compare with the amount of energy required to move the mass from the surface of the Earth to a great distance away.

Problem 4.12 Two neutral atoms

Figure 4.55 shows a potential energy curve for the interaction of two neutral atoms. The two-atom system is in a vibrational state indicated by the heavy solid horizontal line.

(a) At $r = r_1$, what are the approximate values of the kinetic energy K, the potential energy U, and the quantity $K+U$?

(b) What minimum energy must be supplied to cause these two atoms to separate?

(c) In some cases, when r is large, the interatomic potential energy can be expressed approximately as $U = -a/r^6$. For large r, what is the algebraic form of the magnitude of the force the two atoms exert on each other in this case?

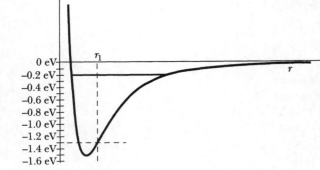

Figure 4.55 Potential energy diagram for Problem 4.12.

Problem 4.13 Alpha decay

Many heavy nuclei undergo spontaneous "alpha decay," in which the original nucleus emits an alpha particle (a helium nucleus containing two protons and two neutrons), leaving behind a "daughter" nucleus that has two fewer protons and two fewer neutrons than the original nucleus. Consider a Radium-220 nucleus that is at rest before it decays to Radon-216 by alpha decay.

The mass of the Radium-220 nucleus is 219.96274 u (unified atomic mass units) where 1 u = 1.6603×10^{-27} kg (approximately the mass of one nucleon). The mass of a Radon-216 nucleus is 215.95308 u, and the mass of an alpha particle is 4.00151 u. Radium has 88 protons, Radon 86, and an alpha particle 2.

(a) Make a diagram of the final state of the Radon-216 nucleus and the alpha particle when they are far apart, showing the momenta of each particle to the same relative scale. Explain why you drew the lengths of the momentum vectors the way you did.

(b) Calculate the final kinetic energy of the alpha particle. For the moment, assume that its speed is small compared to the speed of light.

(c) Calculate the final kinetic energy of the Radon-216 nucleus.

(d) Show that the nonrelativistic approximation was reasonable.

Problem 4.14 The SLAC two-mile accelerator

SLAC, the Stanford Linear Accelerator Center, located at Stanford University in Palo Alto, California, accelerates electrons through a vacuum tube two miles long (it can be seen from an overpass of the Junipero Serra freeway that goes right over the accelerator). Electrons which are initially at rest are subjected to a continuous force of 2×10^{-12} newton along the entire length of two miles (one mile is 1.6 kilometers) and reach speeds very near the speed of light. The analysis in Problem 2.12 (page 70) required numerical integration, but with the new techniques of this chapter you can analyze the motion analytically.

(a) Calculate the final energy, momentum, and speed of the electron.

(b) Calculate the time required to go the two-mile distance.

Problem 4.15 Nuclear fission

Uranium-235 fissions when it absorbs a slow-moving neutron. The two fission fragments can be almost any two nuclei whose charges Q_1 and Q_2 add up to $92e$ (where e is the charge on a proton, $e = 1.6 \times 10^{-19}$ coulomb), and whose nucleons add up to 236 protons and neutrons (U-236; U-235 plus a neutron). One of the possible fission modes involves nearly equal fragments, palladium nuclei with $Q_1 = Q_2 = 46e$. The rest masses of the two palladium nuclei add up to less than the rest mass of the original nucleus. (In addition to the two main fission fragments there are typically one or more free neutrons in the final state; in your analysis make the simplifying assumption that there are no free neutrons, just two palladium nuclei.)

The rest mass of the U-236 nucleus is 235.996 u (unified atomic mass units), and the rest mass of each Pd-118 nuclei is 117.894 u, where 1 u = 1.6603×10^{-27} kg (approximately the mass of one nucleon).

(a) Calculate the final speed v, when the palladium nuclei have moved far apart (due to their mutual electric repulsion). Is this speed small enough that $p^2/(2m)$ is an adequate approximation for the kinetic energy of one of the palladium nuclei? (It is all right to go ahead and make the nonrelativistic assumption first, but you then must check that the calculated v is indeed small compared to c.)

(b) Using energy considerations, calculate the distance between centers of the palladium nuclei just after fission, when they are starting from rest.

(c) A proton or neutron has a radius of roughly 10^{-15} m, and a nucleus is a tightly packed collection of nucleons. Experiments show that the radius of a nucleus containing N nucleons is approximately $(1.3 \times 10^{-15}$ m$) \times N^{1/3}$. What is the approximate radius of a palladium nucleus? Draw a sketch of the two palladium nuclei in part (b), and label the distances you calculated in parts (b) and (c).

Problem 4.16 Nuclear fusion

One of the thermonuclear or fusion reactions that takes place inside a star such as our Sun is the production of helium-3 (^{3}He, with two protons and one neutron) and a gamma ray (high-energy photon) in a collision between a proton (^{1}H) and a deuteron (^{2}H, the nucleus of "heavy" hydrogen, consisting of a proton and a neutron):

$$^1\text{H} + {}^2\text{H} \rightarrow {}^3\text{He} + \gamma$$

The rest mass of the proton is 1.0073 u (unified atomic mass unit, 1.66×10^{-27} kg), the rest mass of the deuteron is 2.0136 u, the rest mass of the helium-3 nucleus is 3.015 u, and the gamma ray is massless.

(a) What is the energy released in this reaction (the kinetic energy of the ^{3}He plus the energy of the gamma ray), in joules and in electron-volts (1 eV $= 1.6 \times 10^{-19}$ joule)?

(b) The strong interaction has a very short range and is essentially a contact interaction. For this fusion reaction to take place, the proton and deuteron have to come close enough together to touch. The approximate radius of a proton or neutron is about 10^{-15} m. What is the approximate initial total kinetic energy of the proton and deuteron required for the fusion reaction to proceed, in joules and electron volts?

Note that you do get back your investment: the energy you put in plus the energy release you calculated in part (a) will equal the kinetic energy of the ^{3}He plus the energy of the gamma ray in the new situation. So the net energy release is still the amount you calculated in part (a).

Continued on next page.

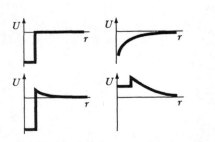

Figure 4.56 Potential energy curves for Problem 4.16 part (d).

(c) As we will study later, the average kinetic energy of a gas molecule is $\frac{3}{2}kT$, where k is the "Boltzmann constant," 1.4×10^{-23} joule/Kelvin degree (J/K), and T is the absolute or Kelvin temperature, measured from absolute zero (so that the freezing point of water is 273 K). What is the approximate temperature required for the fusion reaction to proceed? (This high temperature, required because of the electric repulsion barrier to the reaction, is the main reason why it has been so difficult to make progress toward thermonuclear power generation. Sufficiently high temperatures are found in the interior of the Sun, where fusion reactions take place.)

(d) Which of the potential energy curves in Figure 4.56 is a reasonable representation of the interaction in this fusion reaction? Why?

Problem 4.17 Determining the precise mass of the deuteron
A proton (1.6726×10^{-27} kg) and a neutron (1.6749×10^{-27} kg) at rest combine to form a deuteron, the nucleus of deuterium or "heavy hydrogen." In this process, a gamma ray (high-energy photon) is emitted, and its energy is measured to be 2.2 MeV (2.2×10^{6} eV).

(a) Keeping all five significant figures, what is the mass of the deuteron? Assume that you can neglect the small kinetic energy of the recoiling deuteron.

(b) Momentum must be conserved, so the deuteron must recoil with momentum equal and opposite to the momentum of the gamma ray. Calculate approximately the kinetic energy of the recoiling deuteron and show that it is indeed small compared to the energy of the gamma ray.

Problem 4.18 Potential energy and a pendulum
A pendulum consists of a very light but stiff rod of length L hanging from a nearly frictionless axle, with a mass m at the end of the rod (Figure 4.57).

(a) Calculate the gravitational potential energy as a function of the angle θ, measured from the vertical.

(b) Sketch the potential energy as a function of the angle θ, for angles from $-210°$ to $+210°$.

(c) Let $s = L\theta =$ the arc length away from the bottom of the arc. Calculate the tangential component of the force on the mass by taking the (negative) gradient of the energy with respect to s. Does your result make sense?

(d) Suppose you hit the stationary hanging mass so that it has an initial speed v_i. What is the minimum initial speed in order that the pendulum go over the top ($\theta = 180°$)? On your sketch of the potential energy (part b), draw and label energy levels for the case where the initial speed is less than, equal to, or greater than this critical initial speed.

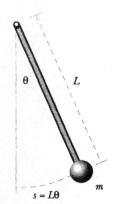

Figure 4.57 Pendulum (Problem 4.18).

4.17 Answers to exercises

4.1 (page 117)	1230 J
4.2 (page 117)	zero
4.3 (page 118)	$2P$
4.4 (page 119)	$(25\text{ m})\hat{\text{i}} - (30\text{ m})\hat{\text{j}}$, $(-19.6\text{ N})\hat{\text{j}}$, 588 J
4.5 (page 119)	189 J
4.6 (page 119)	0.1 J
4.7 (page 123)	about 6
4.8 (page 123)	about 10^{-14}
4.9 (page 124)	Make the indicated substitution; after that it's just algebra.
4.10 (page 125)	$\text{kg}\cdot\text{m}^2/\text{s}^2 = (\text{kg}\cdot\text{m}/\text{s}^2)\cdot\text{m} = \text{N}\cdot\text{m} = \text{J}$
4.11 (page 125)	12 m/s, 27 m.p.h.
4.12 (page 125)	1.2×10^{-12} kg (an extremely small amount of mass!)
4.13 (page 127)	$\Delta U = -50$ joules
4.14 (page 129)	$E_1 + E_2 + E_3 + E_4 + U_{12} + U_{13} + U_{14} + U_{23} + U_{24} + U_{34}$
4.15 (page 132)	– (more negative)
4.16 (page 132)	+ (less negative)
4.17 (page 132)	$-GMm/y^2$; 0
4.18 (page 133)	K has decreased
4.19 (page 134)	A is bound; C is unbound; B is trapped. In a quantum system in state B there is some probability of "tunneling" through the barrier and getting out of the trap!
4.20 (page 134)	A: go out in a straight line and stop, then fall back B: elliptical (motion with varying separation) C: nearly circular (nearly constant separation)
4.21 (page 136)	v to reach Moon $< v_{esc}$ because didn't go to infinity
4.22 (page 136)	1.1×10^4 m/s
4.23 (page 136)	2.4×10^3 m/s, much smaller than v_{esc} from Earth
4.24 (page 139)	same ΔU_g
4.25 (page 139)	6.3 m/s; the mass cancels from both sides of the energy equation
4.26 (page 139)	Hint: At top of trajectory the ball's vertical speed is momentarily zero. Answer on next page.
4.27 (page 139)	Hint: $E_f = E_i$. Answer on next page.
4.28 (page 141)	7.2×10^{-15} m Note: the radius of the proton is about 1×10^{-15} m.
4.29 (page 143)	A is potential energy, B is kinetic energy, C is $K+U$
4.30 (page 143)	1, 2, 3, 4 are bound states; 5 is an unbound state
4.31 (page 143)	0.789 MeV; proton mass is less than neutron mass
4.32 (page 145)	About 1.5×10^{-10}; accuracy of laboratory scale is far from what would be needed to detect the difference
4.33 (page 146)	About 0.1%

Additional exercises

4.34 (page 157)	4.9 N downward, 4.9 J Positive work makes the jar speed up.
4.35 (page 157)	2.5×10^{-19} J
4.36 (page 157)	9.8 J, 6.3 m/s, neglect air friction

4.37 (page 157)   19.6 J, 6.3 m/s; same speed, twice the kinetic energy

4.38 (page 157) 4.4×10^9 J

Additional answers

4.26 (page 139)   101 m

4.27 (page 139) $v_f = \sqrt{v_i^2 - 2g(y_f - y_i)}$

Chapter 5

Energy in Macroscopic Systems

Chapter 5

Energy in Macroscopic Systems

In this chapter we consider energy in macroscopic systems, and we make connections between microscopic and macroscopic views of energy. We also show how the choice of system affects the form of the energy equation. The last part of the chapter deals with energy dissipation in macroscopic systems when there is friction.

5.1 Potential energy of macroscopic springs

A bar of metal can be treated as a spring, since stretching or compressing the bar stretches or compresses the interatomic bonds (Figure 5.1). Double the force produces double the change in length, as long as we're in the range where the spring approximation to the interatomic force is adequate. This is usually described for a bar in terms of Young's modulus, which we studied in Chapter 3, where the tension force F_T (per unit area) is proportional to the stretch ΔL (per unit length):

$$\frac{F_T}{A} = Y\frac{\Delta L}{L}$$

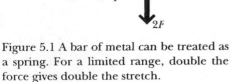

Figure 5.1 A bar of metal can be treated as a spring. For a limited range, double the force gives double the stretch.

Writing this as $F_T = (YA/L)\Delta L$, we see that if we think of the bar as a (very stiff) spring, its spring stiffness is YA/L.

If you stretch the bar too much, two things can make the bar stop behaving in a spring-like manner. You might exceed the interatomic stretch for which $k_s s$ is a good approximation to the magnitude of the interatomic force. Also, the regular array of atoms may be disrupted by dislocations of the crystal structure, leading to large-scale slippage of crystal planes. Suddenly the bar "yields" and grows very much longer with little applied force.

A straight object is not usually used as a longitudinal spring but may be used in a bending mode. For example, when a diving board bends under a load, atomic bonds in the upper part of the board are stretched, and atomic bonds in the lower part are compressed (Figure 5.2). The net effect is a spring-like behavior of the diving board, with the amount of bend proportional to the applied force.

Figure 5.2 The bending of a diving board stretches and compresses interatomic bonds.

Helical springs

We will be most interested in helical (spiral) springs. A modest force applied to a helical spring gives a substantial change in the length of the spring, even though the total length of the coiled wire hardly changes. A large stretch of the spring with a small force implies a much smaller spring stiffness than the wire would have when straightened out.

If you stretch a helical spring far enough, it straightens out into a long straight wire, and it suddenly becomes very difficult to lengthen the wire any further. If you continue to stretch the straight wire, eventually it yields and then breaks. If you compress a helical spring so far that its coils run into each other, it suddenly gets extremely hard to compress the spring any further. Therefore, the linear relation between stretch and force, $F_s = -k_s s$, is valid only over a limited region of stretch, although this linear range is much wider than for the microscopic "spring."

In many situations the entire phenomenon takes place within the valid range of the linear approximation. Or one can speak of an "infinite" well as shown in Figure 5.3, which is an idealization of the real situation. Within the linear range where $F_s = -k_s s$ is the negative gradient of $\frac{1}{2}k_s s^2$, we have this:

SPRING POTENTIAL ENERGY

$U_s = \frac{1}{2}k_s s^2 + U_0$, s measured from equilibrium point

U_0 is a negative number representing how far below $U_s = 0$ the bottom of the potential "well" is located. An amount of work $W_{external} = |U_0|$ is needed to stretch the wire out straight, and make it deform.

In Figure 5.3 (which is an idealization of the real situation) there are very steep sides of the potential well, corresponding to the large forces required to compress a spring whose coils are in contact, and the large forces required to stretch the spring after it has straightened out into a straight wire. Remember that force is the negative gradient of the potential energy, so a steep slope corresponds to a large force.

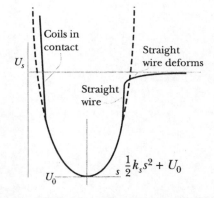

Figure 5.3 Spring potential energy, within the linear range where $F_s = -k_s s$. The dotted line represents an ideal spring, while the solid line represents a real macroscopic spring.

5.1.1 Application: Energy in an oscillating spring-mass system

Consider a block connected by a helical spring to a wall, whose oscillations we treated in Chapter 3. If you compress the spring and then let go, the block acquires kinetic energy (Figure 5.4), with a loss of spring potential energy. If you stretch the spring and then let go, the block also acquires kinetic energy, again with a loss of spring potential energy (Figure 5.5).

We'll apply the energy principle to this spring-mass system. For the system consisting of the block of mass m and the spring of stiffness k_s, the sum of the changes in the kinetic energy K of the block and the spring potential energy U_s, must be zero, because there is no external work done on this system, if we neglect friction (the rest energy mc^2 of the block doesn't change). The wall exerts a force on the left end of the spring, but this force acts through no distance and does no external work; it only affects the momentum of the system, which changes continuously. The vertical forces act at right angles to the displacement and so do no work.

$$\Delta(K + U_s) = 0 \text{ (neglecting friction)}$$

$$\Delta\left(\frac{p^2}{2m} + \frac{1}{2}k_s s^2 + U_0\right) = 0$$

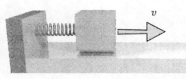

Figure 5.4 As the compressed spring expands the kinetic energy of the block increases.

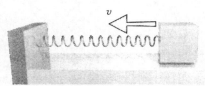

Figure 5.5 As the stretched spring contracts the kinetic energy of the block increases.

Consider a mass and spring oscillating on a horizontal air track with very little friction. We start the oscillation by stretching the spring a distance s and releasing it from rest.

? At the moment that the spring is released, what is the kinetic energy of the spring-mass system? What is the total energy of the system?

Since the block is released from rest, its kinetic energy is zero. The energy of the system is equal to the energy stored in the spring, $\frac{1}{2}k_s s^2 + U_0$.

? At what point in the oscillation is the kinetic energy of the system highest? At that moment, what is the potential energy of the system?

When the contracting spring reaches its equilibrium length, the energy stored in the spring is a minimum (U_0). The maximum possible amount of spring potential energy has now been converted to kinetic energy, so this is the instant when the kinetic energy of the system is highest.

? At what point in the oscillation will the kinetic energy of the system be lowest?

At either endpoint of the oscillation—spring fully compressed or spring fully extended—the instantaneous speed of the mass is zero. The kinetic energy of the system is lowest here, and all the energy has been momentarily converted back into spring potential energy.

We can write the energy equation for the closed system of mass plus spring to emphasize that over a time interval the total energy (kinetic plus spring) doesn't change:

$$\Delta\left(\frac{p^2}{2m} + \frac{1}{2}k_s s^2 + U_0\right) = 0$$

Or we can rearrange the terms and write the energy equation in a somewhat different form:

$$\frac{p_f^2}{2m} + \frac{1}{2}k_s s_f^2 + U_0 = \frac{p_i^2}{2m} + \frac{1}{2}k_s s_i^2 + U_0$$

That is, the final kinetic plus potential energy of the Universe is equal to the initial kinetic plus potential energy of the Universe, no matter what we decide to choose as initial and final states. Note that the potential energy U_0 at the bottom of the "well" cancels out in either view of energy conservation.

5.1.2 Shifting the zero of potential energy

For consistency with special relativity, it is essential that potential energy be zero when the particles of a system are widely separated. However, when analyzing a macroscopic spring-mass system we can conveniently ignore the negative U_0 term and pretend that the potential energy is zero when the stretch is zero: $U_s = \frac{1}{2}k_s s^2$. The reason why this works is that in a process we only care about *changes* in energy, and the change in the constant U_0 term is zero.

Example: A mass of 0.2 kg is attached to a horizontal spring whose stiffness is 12 N/m. Friction is negligible. At $t = 0$ the spring has a stretch of 3 cm and the mass has a speed of 0.5 m/s. What is the amplitude (maximum stretch) of the oscillation? What is the maximum speed?

Solution: System: mass and spring

Initial state: $s = 3$ cm, $v = 0.5$ m/s
Final state: $v = 0$
Energy principle: $E_f = E_i$ (neglecting friction)

The initial energy, ignoring changes in rest mass and in U_0, is

$$\frac{1}{2}mv_i^2 + \frac{1}{2}k_s s^2 = (\tfrac{1}{2})(0.2 \text{ kg})(0.5 \text{ m/s})^2 + (\tfrac{1}{2})(12 \text{ N/m})(0.03 \text{ m})^2$$

$$E_i = 0.03\text{J}$$

With negligible friction, the energy will be 0.03 J at all times. The stretch is maximum when the speed and kinetic energy are zero:

$$\frac{1}{2}k_s s_{max}^2 = 0.03\text{J}$$

$$s_{max} = \sqrt{2(0.03\text{J})/(12 \text{ N/m})} = 0.07 \text{ m}$$

Now consider different final state: $s = 0$

The speed is maximum when the stretch and potential energy are zero:

$$\frac{1}{2}mv_{max}^2 = 0.03\text{J}$$

$$v_{max} = \sqrt{2(0.03\text{J})/(0.2 \text{ kg})} = 0.55 \text{ m/s}$$

Ex. 5.1 About how many joules of energy can you store in a spring like the one you used in Chapter 3, starting from a relaxed spring? Assume $k_s \approx 0.6$ N/m and a stretch of about 20 cm.

Ex. 5.2 A spring with stiffness 0.5 N/m has a relaxed length of 15 cm. A mass of 20 grams is attached and the spring is stretched to a total length of 25 cm. The mass is then released from rest. What is the speed of the mass at the moment when the spring returns to its relaxed length of 15 cm?

Problem 5.1 Energy in a spring-mass system

Energy calculations provide a powerful check on the accuracy of a numerical integration. Modify your numerical integration of the motion of a mass on a spring that you did in Chapter 3, Problem 3.4 (page 87). In addition to plotting graphs of position versus time, also plot graphs of kinetic energy, of spring energy, and of the sum of the kinetic energy and the spring energy, versus time during the motion. It is up to you to choose some value for the potential energy U_0 at the bottom of the well. Label the energy graphs with numerical values and units. Make sure that the energy values are reasonable for your system.

(a) Does the sum of kinetic and potential energy remain constant as a function of time? What if you vary the step size (which varies the accuracy of the numerical integration)?

(b) Increase the energy by increasing the initial speed. How does the amplitude A change? How does the angular frequency ω change?

5.2 Path independence of potential energy

In situations describable by a potential energy, the associated interaction depends solely on position, not on speed or direction. For example, spring energy depends on the stretch s, and gravitational and electric energy depend on the separation r. It is easy to prove that the following must be true:

PATH INDEPENDENCE OF POTENTIAL ENERGY

In a round trip, the potential energy doesn't change.

Change in potential energy doesn't depend on the path taken.

Here is the proof: In Figure 5.6, if one of the particles in a multiparticle system starts at point A, moves around and comes back to the starting point, the change in potential energy U of the multiparticle system must be zero, since the initial and final potential energy depends on pair-wise distances r or s, and these return to their original values in a round trip. Similarly, in Figure 5.7 change in potential energy U of the system when one of its particles moves along some path from point A to point B depends solely on the initial and final values of r or s, so it doesn't matter what path the particle follows to get from A to B.

A simple example is projectile motion when air resistance can be neglected (Figure 5.8). You throw a ball up in the air with speed v, and when it returns to your hand it has regained the same speed v that you gave it (the gravitational energy change is zero, and the change in the kinetic energy is zero). Another example is the motion of a mass on a spring in the absence of friction: every time the mass passes a particular position, it has the same speed (same s implies same v).

In both of these examples, although the speed doesn't change, the velocity may change. The ball was headed up but is now headed down; a mass on

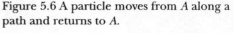

Figure 5.6 A particle moves from A along a path and returns to A.

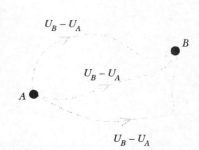

Figure 5.7 Potential energy difference is independent of the path between the initial and final locations.

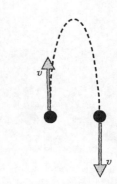

Figure 5.8 A ball goes up and down with negligible air resistance.

a spring may be going to the left and then going to the right with the same speed. The key point is this:

Potential energy depends only on position,
not on the direction of motion.

There is a connection between the fact that change of potential energy is independent of path, and that a round trip gives zero. Suppose that along some path from *A* to *B* to *C* in Figure 5.9 the change in potential energy is +5 J.

? What will the change in potential energy be along the same path in the opposite direction, from *C* to *B* to *A*?

Going backwards from *C* to *B* to *A* along this same path the change would be −5 J. Since the change is independent of the path, you could take *any* path to return from *C*, such as *C* to *D* to *A*, with a change of −5 J. Therefore any round trip from *A* to *B* to *C* to *D* and back to *A* will yield a change of zero.

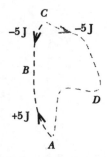

Figure 5.9 Forward and reversed paths, and a round trip.

Ex. 5.3 During one complete oscillation of a mass on a spring (one period), what is the change in potential energy of the mass + spring system, in the absence of friction?

5.3 Thermal energy

In our idealized models of spring-mass systems, the sum of kinetic plus spring potential energy is constant. In real spring-mass systems, however, the oscillations die away ("damp down") with time, which means that the kinetic plus spring potential energy must be decreasing. The decrease in the amplitude of the oscillations is mainly due to air resistance or to friction with a solid, and some energy is transferred to the surroundings, to the air or to a solid along which the mass slides. On the microscopic level, if we could look inside the solid, we would see increased atomic motion—the atoms in the solid have gained energy. A macroscopic indication of this increased atomic motion is an increase in the temperature of the surface.

We can measure "macroscopic" spring potential energy and kinetic energy with simple measurements that don't require a microscope. But when the temperature of a solid object increases, the increased energy inside the solid is not visible to the naked eye, and we face the problem of how to evaluate the increase in the "thermal" energy.

Microscopic kinetic and spring potential energy

The increased "microscopic" energy in a solid is in two forms. First, the atoms on the average are moving around faster, with increased kinetic energy $p^2/(2m)$. Second, there is increased spring potential energy $\frac{1}{2}k_s s^2 + U_0$ in the interatomic bonds ("springs" in our model of solids), where k_s is the stiffness of an interatomic spring (Figure 5.10).

In principle the microscopic, thermal energy in a solid could be evaluated by simultaneously measuring at some instant the momenta of *all* of the atoms, and the stretches or compressions of *all* of the springs (that is, changes in atomic positions away from their equilibrium positions, together with knowledge of the spring stiffness). Since there are about 10^{23} atoms (and even more "springs") in an ordinary-sized object, it is in practice impossible to evaluate the energy this way.

Perhaps we could make microscopic measurements of just one atom (and its attached "springs"), assume that this is the average energy of each atom, and then multiply by the number of atoms there are in the object to get the

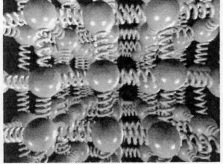

Figure 5.10 In our model solid we must account for the kinetic energy of every ball and the spring energy of every spring.

total energy. That doesn't work either, because at some instant we may find an atom that is momentarily sitting motionless at its equilibrium position (zero energy), and an instant later it is displaced away from its equilibrium position and moving rapidly. Energy keeps getting passed back and forth among the many atoms and springs, and measuring the energy of just one atom and its attached springs at one instant doesn't give us the correct average energy we would need to determine the total energy of the object.

Temperature

In the 1800's it was realized that for many systems the temperature is essentially a measure of the average energy of the atoms in the system:

> Temperature and the average energy of atoms are related

Given the impossibility of evaluating the energy of each individual atom, we can simply use a thermometer to measure thermal energy. A familiar example is the mercury thermometer (silver colored) or the alcohol thermometer (tinted red for visibility). In these thermometers, a very thin column of liquid expands when heated. By tradition, marks on the thermometer are placed to represent "degrees of temperature," but we could just as well place marks to indicate the average energy of the molecules in the thermometer, in joules (Figure 5.11). One Kelvin (or one Celsius degree of temperature) is equivalent to an average molecular energy of about 10^{-23} joules.

This is not the whole story. Later in this course we will make more precise the relationship between temperature and energy, and we will see that temperature is more directly related to a quantity called "entropy" than it is to energy. Nevertheless, for most ordinary systems at room temperature it is true that temperature is a good measure of the average energy of the atoms.

The usefulness of a thermometer lies in the important observation that when two objects are in good contact with each other, the average kinetic energies of the molecules in the two objects slowly come to be equal. Where the two objects touch each other, molecules collide with each other, and the faster molecules on average lose kinetic energy to the slower molecules in a collision. Once the average kinetic energies of molecules in both objects have come to be equal, additional collisions on average make no further change. To put it in terms of temperature (which is related to the average molecular kinetic energy), the hotter object gets cooler, and the cooler object gets hotter. Everyday examples abound: a cold drink warms up in a hot room, and warm water placed in a freezer cools down and freezes.

To measure the microscopic energy of an object, we place a thermometer in contact with the object. Then we wait for the temperatures (average molecular kinetic energies) of the object and thermometer to equilibrate, and we read the thermometer. It is of practical importance that the thermometer have a relatively small mass compared to the object of interest, so that attaching the thermometer doesn't add or subtract much energy to or from the object. Another advantage of a small thermometer is that it reaches thermal equilibrium quicker than a large thermometer does.

How was the calibration between one Kelvin (or one Celsius degree) and energy established? In one of a series of classic experiment performed by Joule in the 1840's, a paddle wheel in water was turned by a falling weight. The work done by the Earth's gravitational force on the falling weight (mgh), was accompanied by an increase in the thermal energy of the water (presumably also mgh, assuming that energy is conserved), with a measurable rise in the water temperature measured in Kelvins (after the paddle wheel had stopped turning). It was found for water that the energy required to raise the temperature of one gram of water by one Kelvin (1 K) is 4.2 joules. This is often summarized by saying that the "heat capacity" of water

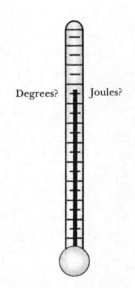

Degrees?   Joules?

Figure 5.11 A thermometer might be calibrated in joules instead of in degrees.

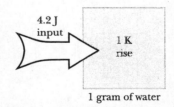

4.2 J input   1 K rise

1 gram of water

Figure 5.12 Energy input of 4.2 joules into a gram of water raises the temperature by 1 K. We say that the heat capacity of water is 4.2 J/gram/K.

is 4.2 joules per gram per Kelvin (Figure 5.12). Other materials have been measured to have different heat capacities. We will study an atomic theory of heat capacity later in the course.

Volume and temperature

An ordinary thermometer depends on the effect that higher temperature makes the material expand. Why does this happen?

For large oscillations of atoms whose interaction is described by a potential energy such as that shown in Figure 5.13, the average stretch increases. Because the actual potential energy curve is not really symmetric around the equilibrium point, with increasing energy the average interatomic bond length gets slightly longer. This is why an object typically expands at higher temperature. Higher temperature implies larger amplitude of oscillations and higher energy, and there is a slight shift in the center of the oscillations at higher energy.

In the interior of the solid the net potential curve for an atom is symmetric, not asymmetric, because the potential curve is associated with forces exerted by atoms to the left and to the right. But since the bond lengthening can start at the surface and propagate into the interior, a solid ultimately expands at higher temperatures.

Other kinds of thermometers

There are many other kinds of thermometers. All materials change in some way when they get hotter or colder, and some of these effects are the basis for useful temperature indicators. A new kind of thermometer that is increasingly used is the liquid crystal thermometer. In a liquid crystal there is some molecular order over long distances in the liquid, unlike a true liquid in which there is hardly any long-range order. In one form of liquid crystal thermometer, helical molecules remain roughly parallel to each other, and the distance between coils in the helix changes with temperature. Because the distance between coils is comparable to the wavelength of visible light, this distance affects the way light is reflected from the liquid crystal. The effect is that the material changes color with changing temperature.

Calculations using heat capacity

Joule found that it takes 4.2 J of energy input to a gram of water to raise its temperature 1 K. If such experiments are done on other materials, the temperature rise is found to be different. For example, the heat capacity of ethanol is found to be 2.4 J/gram/K, and the heat capacity of copper is only 0.4 J/gram/K. One of the many unusual and useful properties of water is that it has a very large heat capacity on a per-gram basis, which means that it is difficult to change its temperature. Here is the meaning of heat capacity on a per-gram basis, where $\Delta E_{\text{thermal}}$ is the rise in the energy of the system, in the form of increased atomic kinetic and potential energy:

HEAT CAPACITY *C* ON A PER-GRAM BASIS

$$\Delta E_{\text{thermal}} = mC\Delta T, \text{ or } C = \frac{\Delta E_{\text{thermal}}}{m\Delta T} \quad (m \text{ in grams})$$

In principle, it should be possible to calculate the heat capacity of a material if something is known about its atomic structure, since a temperature rise is a measure of increased energy at the atomic level. Comparisons of calculations with experimental heat capacity data are a good test of our understanding of atomic models. In later chapters you will learn how to calculate the heat capacity of a gas, and of a solid, based on the ball-and-spring model of a solid.

If the heat capacity of a material is known, the amount of energy transfer into the material can be determined by observing the temperature rise.

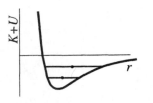

Figure 5.13 For larger oscillations the average stretch increases.

Example: You stir 12 kg of water vigorously, doing 36000 joules of work. If the water is well insulated (so that all of your energy input goes into increasing the energy of the water), what temperature rise would you expect?

Solution: $\left(\dfrac{36000 \text{ J}}{12000 \text{ grams}}\right)\left(\dfrac{1}{4.2 \text{ J/gram/K}}\right) = 0.7 \text{ K}$

Including units helps avoid making calculational mistakes.

Ex. 5.4 Niagara Falls is about 50 meters high. What is the temperature rise in Kelvins of the water from just before to just after it hits the rocks at the bottom of the falls, assuming negligible air resistance during the fall and that the water doesn't rebound but just splats onto the rock?

5.3.1 Thermal energy transfer

Work is mechanical energy transfer into or out of a system involving forces acting through macroscopic displacements. When a hot object is placed in contact with a cold object, energy is transferred from the hot object to the cold one, but there are no macroscopic forces or displacements, so we don't refer to this kind of energy transfer as work.

Instead, we speak of "thermal energy transfer" Q, a term which originated at a time before the atomic model of matter was prevalent. At the microscopic level thermal energy transfer actually does involve work; when a hot block is placed in contact with a cold block (Figure 5.14), at the interface the atoms in the two blocks collide with each other, and do work on each other. The atoms in the hot block have greater average kinetic energy than the atoms in the cold block, so in an individual collision it is likely that a fast-moving atom in the hot block loses energy to a slow-moving atom in the cold block.

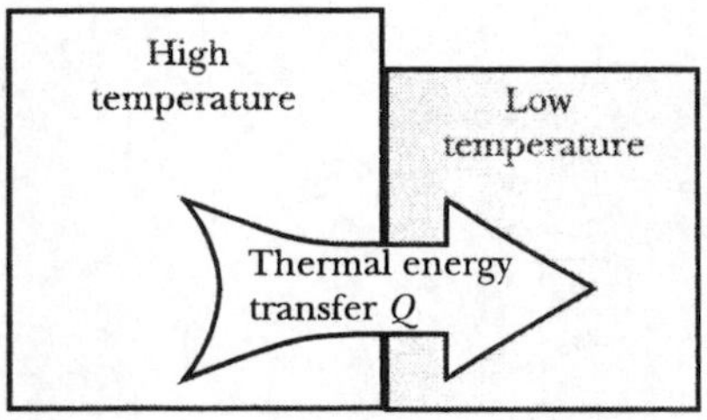

It can happen that the atom in the hot block happens to be moving slowly and gains energy when it is hit by an atom in the cold object that happens to be moving fast, but this is less likely. On average there is energy flow, "thermal energy transfer" (microscopic work), from the hot block to the cold block. These energy changes will diffuse throughout the blocks. If left to themselves, the two blocks will eventually come to the same temperature, intermediate between their two initial temperatures.

THERMAL ENERGY TRANSFER

Thermal energy transfer is the name we give to a process in which energy moves between a system and the surroundings with which it is in contact, due to a temperature difference.

If we choose just the initially cold block as our system of interest, this system gains energy (at the expense of the surroundings, the initially hot block). At the atomic level, this energy increase is the result of work done on the atoms at the surface of the cold block. Usually we are unable to observe these atomic-level interactions directly. We can infer the amount of thermal energy transfer into the system by observing the temperature rise of that system, if we know the heat capacity of the material.

It is common practice to denote thermal energy transfer by the letter Q and work (macroscopic, mechanical energy transfer) by the letter W. The energy equation for an "open" (not isolated) system is then

$$\Delta E_{\text{system}} = Q + W$$

Figure 5.14 A hot block in contact with a cold block.

That is, thermal energy transfers and mechanical energy transfers into or out of the system produce a change in the energy *of* the system. The energy of the system includes particle energy and potential energy, at the microscopic or macroscopic level.

Note that Q and/or W can be negative. If there is thermal energy transfer out of a system, we give Q a negative sign, because the change in the energy of the system is a decrease. Typically such transfer would be due to the system having a higher temperature than the surroundings. Similarly, if the system does work on its surroundings rather than the other way round, the sign of W is negative. (Some older textbooks on thermodynamics reverse the sign of W, counting energy outputs from the system as positive.)

We believe that for the Universe as a whole Q and W are zero, and the energy of the Universe does not change. But even a small system may be effectively closed, with its energy unchanging, if it is thermally well insulated (to prevent thermal energy transfer into or out of the system) and if there are no other energy transfers from the surroundings.

Terminology issues

The technical meaning of the word "heat" in science is not supposed to be the same as its everyday meaning, as is also the case with the word "work." In science the word "heat" is supposed to be reserved to refer to thermal energy transfer (microscopic work) across a system boundary. Rather than saying "there is heat in the object," one says there is energy in the object. The amount of energy in the object might increase due to "heat" (energy transfer into the system due to a temperature difference). However, even professional scientists sometimes slip in their usage of the word "heat"!

To avoid confusion, we will avoid the use of the word "heat" as a noun. Instead, we will speak of "thermal energy transfer" between system and surroundings due to a temperature difference, and change of "thermal energy" inside the system (in the form of increased atomic kinetic energy and potential energy).

Ex. 5.5 Suppose you warm up 500 grams of water (half a liter, or about a pint) on a stove, and the temperature of the water is observed to rise from 20° C to 80° C. Taking the water as the system, how much thermal energy transfer Q was there across the system boundary? What was the change in the thermal energy of the water? What was the energy change of the surroundings?

5.3.2 Other input: Electromagnetic radiation

Work and heat are very common kinds of energy transfers into or out of an open system. Also, electromagnetic radiation, including visible light and infrared, can carry energy into or out of an open system. Sunlight absorbed by the Earth raises the temperature (and the thermal energy) of the daylight side of the Earth. We will typically account for this kind of energy transfer in terms of photons (packets of energy) crossing the system boundary, which is a kind of change of identity for the system.

5.4 Reflection: Forms of energy

There are two fundamental kinds of energy in a multiparticle system:
- particle energy (expressible as rest energy plus kinetic energy)
- potential energy

All other forms of energy are examples of these basic forms. For example, chemicals in a system can react with each other and raise the temperature (and thermal energy) of the system, so we speak of "chemical energy," of which food is an example, with the molecule ATP providing energy storage in the body. The "chemical energy" is the kinetic energy and electric potential energy of molecules. Change of shape (configuration) is associated with change of potential energy, and there may also be changes of the kinetic energy of the molecules.

Nuclear energy is similar. Nucleons can be more or less tightly bound to particular nuclei, associated with different amounts of nuclear potential energy (change of shape) and different amounts of kinetic energy of the nucleons. Also, a change of identity in a nuclear reaction can be associated with a (huge) kinetic energy change associated with a change of rest mass.

5.5 Power: Energy per unit time

In technical usage, the word "power" is defined to mean "energy per unit time." From the point of view of energy usage, it makes no difference whether you take a minute or a month to climb a stairs, yet from a practical point of view you certainly do notice the *rate* of energy usage, called power. The units of power are joules per second, or watts (honoring James Watt, the developer of the first efficient steam engine).

POWER
Energy per unit time (J/s or watts)

There is a useful formula for the instantaneous power associated with the work done by a force:

INSTANTANEOUS POWER

$$\text{power} = \frac{dW}{dt} = \frac{\vec{F} \bullet d\vec{r}}{dt} = \vec{F} \bullet \frac{d\vec{r}}{dt} = \vec{F} \bullet \vec{v}$$

Example: How much energy is required to run a 100-watt light bulb for an hour?

Solution: $(100 \text{J/s})(1 \text{ hr})\left(60\frac{\text{min}}{\text{hr}}\right)\left(60\frac{\text{s}}{\text{min}}\right) = 3.6 \times 10^5 \text{ J}$

Ex. 5.6 In the Niagara Falls hydroelectric generating plant, the energy of falling water is converted into electricity. The height of the falls is about 50 meters. Assuming that the energy conversion is highly efficient, approximately how much energy is obtained from one kilogram of falling water? Therefore, approximately how many kilograms of water must go through the generators every second to produce a megawatt of power (10^6 watts)?

Ex. 5.7 A vehicle with a mass of 1000 kg has an engine whose maximum power output is 50 kilowatts (about 67 horsepower; one horsepower is 746 watts). At a speed of 20 m/s (about 45 miles/hour), what is the maximum acceleration possible?

5.6 Open and closed systems

We have already had some experience with the difference between a closed system and an open system. Because of the importance of this distinction, in this section we will go more deeply into the issues. Consider this diagram

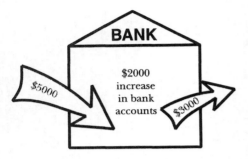

Figure 5.15 A bank is an open system.

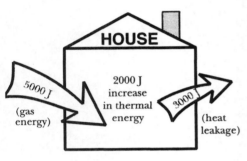

Figure 5.16 A house is an open system.

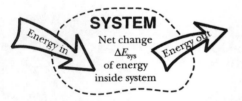

Figure 5.17 The change of energy of the system is equal to energy transferred into the system across the system boundary minus energy transferred out of the system across the system boundary.

showing transfers of money into and out of a bank during a particular time period, and the corresponding change in accounts inside the bank (Figure 5.15).

? We have shown a situation where more money is transferred into the bank than is transferred out. During these transactions, does the total amount of money inside the bank remain unchanged?

We say that the bank is an "open system"—a portion of the world open to transfers in and out, and therefore subject to changes in its internal amount of money. During times when the bank does not permit transfers, the bank is temporarily a "closed system" and its total internal amount of money is unchanged, although there may be changes in the form of money inside the bank such as transfers between checking accounts and savings accounts.

Now consider Figure 5.16 which shows energy transfers during a particular time period in the winter, into and out of a house that is heated by gas, and the corresponding change in the energy inside the house. This is a situation where more energy comes into the house than goes out, and as a result the temperature inside the house rises.

? During these energy transactions, the total amount of energy in the whole Universe is unchanged, but is the amount of energy inside the house unchanged?

The amount of energy inside the house increased by 2000 joules. We say that the house is an "open system"—a portion of the Universe open to energy transfers, and therefore subject to changes in its internal amount of energy. If the house could be perfectly insulated against heat leakage (and solar radiation), and the gas (and electricity) turned off, the house would be a "closed system" with respect to energy and its total internal amount of energy would be unchanged, although there may be changes in the form of energy inside the house such as the family dog converting chemical energy into kinetic energy when chasing its tail.

These considerations lead to the following scheme for keeping track of energy. Choose some portion of the Universe and mentally surround it by a dashed line marking the boundary (Figure 5.17). Then the energy transfer into the system across the system boundary minus the energy transfer out of the system across the system boundary is equal to the change of the energy inside the system.

The change in energy *inside* the system, ΔE_{sys}, can be either a positive or a negative amount of energy. For example, take an electric car as the system of interest:

Charge the battery: ΔE_{sys} is positive (more energy stored in the battery).
Do work on the car by pushing it: ΔE_{sys} is positive (more kinetic energy).
Run the headlights (radiate electromagnetic energy): ΔE_{sys} is negative (less energy stored in the battery).

The car is an open system whose total amount of energy changes due to energy transfers into or out of the system: the energy of an *open* system is *not* constant.

In contrast, the total energy of the Universe remains constant, because there are compensating energy changes in the surroundings of the car: the electric company lost a store of chemical energy by running its generators to charge the car, you used up some chemical energy to push the car, and the light from the headlights is absorbed by the surroundings and leads to a rise in temperature of the surroundings. The measurements that we are able to make confirm the premise that the Universe is a closed system whose total amount of energy never changes, although the form of the energy may change.

For any closed system, *inflow = outflow* = 0; $\Delta E_{sys} = 0$.

The energy of a closed system does not change.

The Universe as a whole is the most important example of a closed system, but it is often the case that a portion of the Universe can be considered to be a closed system, at least approximately. For example, put hot water and ice cubes into a very well-insulated container. During the short time that the ice melts and the water gets somewhat cooler, we can neglect the small amount of energy leakage through the walls of the insulated container. There is an increase in the energy of the ice cubes (which changed from solid to liquid), and a decrease in the energy of the hot water, but negligible net change in the energy inside the container, which is approximately a closed system whose total energy is (approximately) unchanged.

5.6.1 The choice of system affects energy accounting

What you choose to include in your chosen system affects the form of the energy equation for that system. For example, when we analyze a system consisting of an airless Earth plus a falling ball, the energy of the system does not change, because there are no energy transfers into or out of our chosen system (Figure 5.18). The ball falls a distance h, starting from rest:

$$\Delta E_{sys} = \Delta K + \Delta U_g = 0 \text{ (system is Earth + ball)}$$

$$\Delta K + (-mgh) = 0$$

On the other hand, when we analyze a system consisting of just the falling ball (Figure 5.19), the Earth (which is now outside our chosen system) does work on the ball, and the energy of the ball increases:

$$\Delta E_{sys} = \Delta K = +mgh \text{ (system is ball only)}$$

Energy terms move back and forth across the equal sign in the energy equation, depending on our choice of the system of interest. It is extremely important to be clear about the choice of system, because this affects the form of the energy equation. In fact, if one is vague about the choice of system, it is possible to make a serious mistake.

Warning: avoid double counting!

? A student said, "The Earth exerts a force mg through a distance h and does an amount of work mgh, and there is also a decrease in gravitational potential energy $-mgh$, so the kinetic energy increases by an amount $2mgh$." What is wrong with the student's statement?

This is an easy mistake to make. The problem here is double counting due to a lack of clarity about the choice of system. If the Earth is part of the system, the force mg is internal to the system and does no external work on the system; rather, there is work done by internal forces on both the Earth and the ball, and the negative of this work by internal forces is the change in potential energy. On the other hand, if the ball alone is the system, there is no shape change of the ball and no change of gravitational potential energy, but there is an external force mg which does an amount of work mgh.

In the next section we will illustrate the issue of choice of system by analyzing the same problem three times, using three different choices of system. We should find that the form of the energy principle will be different in each case, but that we should obtain the same physical results each time.

Figure 5.18 System: Earth + ball.

Figure 5.19 System: ball only.

The woman, the barbell, and the Earth

To help avoid double-counting mistakes, it is useful practice to write the energy equation for various choices of system in a slightly more complicated situation, that of a woman applying a constant force F to lift a barbell of mass m from rest through a distance h, at which point the barbell is not only higher above the Earth but has also acquired some speed v. We're considering a time when the barbell is still headed upward, before the woman has brought the barbell to a stop above her.

For convenience in the following discussion, let E_w represent the following energy terms:

E_w = chemical energy of woman +
kinetic energy of woman (moving arm) +
gravitational energy of woman and Earth +
thermal energy of woman

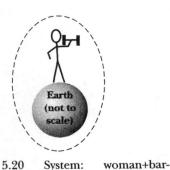

Figure 5.20 System: woman+barbell+Earth.

Woman + barbell + Earth:

If we choose the system to consist of the woman, the barbell, and the Earth (Figure 5.20), there are no energy transfers into or out of our system, so energy does not change, and the energy equation looks like this as the woman lifts the barbell a distance h, starting from rest (we neglect the tiny kinetic energy of the recoiling Earth):

$$\Delta E_{\text{sys}} = \Delta E_w + \Delta U_{\text{g, barbell and Earth}} + \Delta K_{\text{barbell}} = 0$$

? Write the energy equation for the system of woman + barbell + Earth, in terms of the known quantities m, g, h, the constant force F, p, and E_w.

Here is the energy equation for the combined system:

$$\Delta E_w + (+mgh) + \left(\frac{p^2}{2m}\right) = 0$$

The energy term ΔE_w represents changes in energy associated with the woman, mainly a decrease in her stored chemical energy, because she must perform chemical reactions with food (or with chemicals stored in body tissues after eating food) in order to be able to move her arm. This term also includes an increase in the kinetic energy of her arm, an increase in gravitational potential energy of woman and Earth associated with her arm being higher, and an increase in her thermal energy (because her temperature rises a bit when she exerts herself).

? The energy term ΔE_w includes one decrease and some increases. What is the net sign of ΔE_w, positive or negative?

The net effect must be a decrease in her energy, because the barbell and the Earth are farther apart, and the barbell goes faster, both of which represent energy increases. The woman has to burn more energy than the rest of the system gets, since she has to supply energy both to the rest of the system and to supply her own energy increases (increased kinetic energy and gravitational potential energy associated with her arm, and increased thermal energy).

Barbell only:

If we instead choose the system to consist just of the barbell (Figure 5.21), the woman does (positive) work $+Fh$ on the system, and the Earth does (negative) work $-mgh$ on the system (because mg is down and the displacement h is up)

? Write the energy equation for the system consisting solely of the barbell.

Here is what we get for this choice of the barbell only as the system:

$$\Delta E_{sys} = \Delta K_{barbell} = W_{done\ by\ woman} + W_{done\ by\ Earth}$$

$$\left(\frac{p^2}{2m}\right) = (+Fh) + (-mgh)$$

Figure 5.21 System: barbell only.

Compare this equation to the equation we obtained when considering the system of woman+barbell+Earth. We see that energy terms have moved to the other side of the equation as a result of our different choice of system.

? Which is bigger, F or mg? Why?

Since the barbell's upward momentum increases, F must be larger than mg, since the net force in the upward direction is $F-mg$.

? Compare the equations for system 1 (woman + barbell + Earth) and system 2 (barbell only). In terms of F, calculate ΔE_w.

By comparing the two equations, we see that the change in the energy associated with the woman must be $-Fh$. The sign seems right: we had concluded earlier that her energy change must be negative.

Warning! It is tempting to include a term for the change in the gravitational energy of the barbell and to write the energy equation for the barbell alone like this:

$$\left(\frac{p^2}{2m}\right) + (mgh) = (+Fh) + (-mgh) \quad (Wrong!)$$

? Why is this wrong (in addition to giving the wrong result)?

The error here is in assigning some gravitational energy to the barbell alone. Gravitational potential energy is associated with a change in the shape of a multi object system; in particular, to changes in the separations of pairs of interacting objects. When we choose the barbell alone as the system, the system does not change shape. Instead of a change of gravitational potential energy we have an external force exerted by the Earth on our barbell system. Instead of a $+mgh$ on the left of the equation we have a $-mgh$ on the right. We shouldn't double-count and include both terms!

Barbell + Earth:

If we choose the system to consist of the barbell plus the Earth (Figure 5.22), this system changes shape as the woman pushes the two pieces of the system apart. It is almost as though she is stretching an invisible spring connecting the two objects.

? Write the energy equation for this choice of system (barbell + Earth).

Here is the energy equation for the barbell + Earth:

$$\Delta E_{sys} = \Delta U_{g,\ barbell\ and\ Earth} + \Delta K_{barbell} = W_{done\ by\ woman}$$

$$(+mgh) + \left(\frac{p^2}{2m}\right) = (+Fh)$$

Figure 5.22 System: barbell+Earth.

? Does the woman do any work *on the Earth*? Why or why not?

The woman's feet push down on the Earth but do negligible work, because there is essentially no displacement there (the Earth hardly moves). Work done by a force is calculated by taking into account the displacement of the point where the force is applied. In this case, the point of application hardly moves, so the force does hardly any work.

Reflection

Looking back over the equations obtained for different choices of system, you can see how energy terms that represent transfers across the system boundary for one choice of system become changes of energy *inside* the system for a different choice of system. Comparing equations for different choices of system can be useful in determining an unknown quantity (such as the energy change in the woman, $-Fh$, in the example you just worked through). Also, analyzing a process for more than one choice of system is a good check on your calculations and your understanding.

Ex. 5.8 Consider a harmonic oscillator (mass on a spring without friction). Taking the mass alone to be the system, how much work is done on the system as the spring contracts from its maximum stretch A to its relaxed length? What is the change in kinetic energy of the system during this motion? For what choice of system does energy remain constant during this motion?

5.7 Energy dissipation

The total energy of the Universe does not change, but useful energy is often "dissipated" into forms less useful to us. Push a chair across the floor, and some of the work that you do goes into heating the floor and the chair, rather than into kinetic energy of the chair. Throw a ball up into the air, and some of the initial energy is dissipated into increased microscopic energy of the air. Sliding friction, air resistance, viscous friction—all of these phenomena are examples of energy dissipation discussed in the rest of the chapter.

5.8 Air resistance

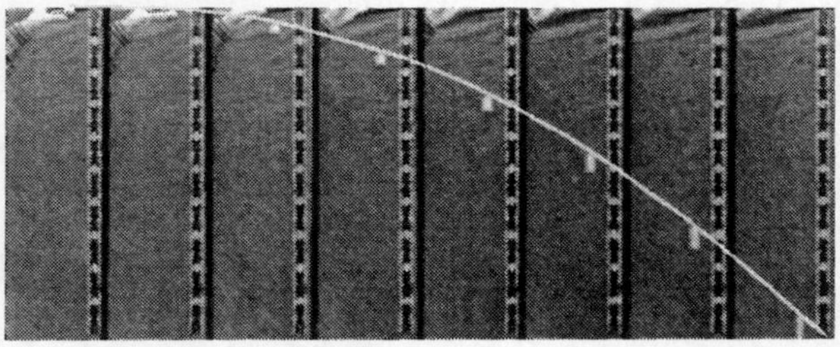

We begin our study of energy dissipation by considering air resistance. Air resistance isn't a major factor when you drop a metal ball a short distance. A video sequence of a falling metal ball shows that the ball moves faster and faster as it falls, an effect that is hard to observe by eye alone. The time between adjacent frames in Figure 5.23 is 1/15 second, and the increasing distances between heights of the ball in adjacent frames show that the speed is getting faster and faster. (Also note the increasing blur due to faster motion of the ball.) The gravitational attraction of the Earth acting on the ball makes the momentum of the ball continually increase. A curve is drawn along the tops of the ball images. The major visible marks on the vertical meter sticks are 10 cm apart. In 7/15 s the ball falls about 1 meter.

Figure 5.23 A falling metal ball goes faster and faster.

Terminal speed

In contrast, it is observed that although a sky diver initially speeds up due to the gravitational force acting downward, the sky diver's speed does *not* keep getting bigger and bigger. Rather the sky diver eventually reaches a "terminal speed" and falls thereafter at constant speed despite the gravitational force. This terminal speed for falling humans is so high (about 60 m/s, or about 135 miles per hour) that hitting the ground without a parachute is normally fatal, although there have been cases of people hitting deep snow at terminal speed and surviving.

Similarly, if you drop a bowl-shaped paper coffee filter, a video sequence of the falling filter shows that the filter's speed does increase at first, but instead of continuing to gain speed it quickly reaches a constant terminal speed despite the gravitational force acting on it. In the video sequence below the time between adjacent frames is again 1/15 second. In 16/15 s the filter falls about 1 meter. A curve is drawn through the centers of the coffee filter images. The nearly straight line in the later part of the motion indicates motion at constant speed.

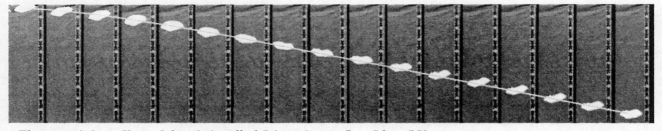

The restraining effect of the air is called "air resistance" or "drag." You yourself have probably observed that it is harder to move something quickly through a fluid such as water or air than to move it slowly. This suggests that air resistance acting on a falling object might depend on the speed of the object.

? Consider a falling coffee filter at two different times: 1) while the filter's speed is still increasing, and 2) after terminal speed has been reached. How do you know the air resistance is larger at the second instant than at the first instant?

While the filter's speed is increasing there must be a nonzero net force, which means that the air resistance is smaller than the gravitational force (Figure 5.25). But later the net force is zero, because the speed is no longer increasing, which means that the force of the air has grown to be as large as the gravitational force (which is nearly constant at mg and in fact increases very slightly as the coffee filter gets closer to the Earth). Since the change in the density of the air is small over the distance the filter falls, and the change in the gravitational force is very small, it is reasonable to conclude that the change in the air resistance is due to a speed dependence of that force.

Dependence on cross-sectional area

We observe that the effect of air resistance increases as the cross sectional area of an object increases—a sky diver falls much more slowly with an open parachute than with a closed parachute. The low-density paper coffee filter has a large cross-sectional area, and a small gravitational force acts on it, so air resistance plays a major role. In contrast, a high-density metal ball has a small cross-sectional area, but a large gravitational force acts on it, and air resistance may be negligible in comparison with gravity.

Applications

Drop a ball. At the instant when the ball has nearly reached the ground, the gravitational potential energy of the Universe has decreased, with a corresponding increase in the kinetic energy of a portion of the Universe (the ball). In the presence of air, kinetic plus gravitational potential energy is only approximately constant. As the ball falls, it has to push aside the air underneath it, and the air molecules acquire additional kinetic energy K_{air}.

Figure 5.24 A falling coffee filter speeds up briefly, then falls at constant speed.

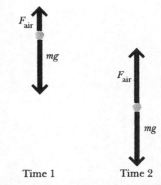

Figure 5.25 Forces acting on a falling coffee filter at Time 1 when the filter's speed is increasing and at Time 2 when terminal speed has been reached.

Ex. 5.9 Therefore, in air will the *actual* final speed of the ball be greater or less than you would predict from the constancy of kinetic plus gravitational energy? Why?

Ex. 5.10 A coffee filter of mass 1.4 grams dropped from a height of 2 m reaches the ground with a speed of 0.8 m/s. How much kinetic energy K_{air} did the air molecules gain from the falling coffee filter?

5.8.1 Mechanism of air resistance

We would like to understand the details of the interaction of the air with a falling object. In particular, we'd like to see if we can find a reason that the air resistance would depend on the speed of the object.

Here is a possible microscopic model to explain air resistance. When you hold a coffee filter stationary, air molecules hit it nearly equally from above and below, as seen in Figure 5.26. (Actually there are slightly more collisions per second on the lower surface than on the upper surface, leading to a small upward buoyant force which is very small compared to air resistance.) But when the filter is moving downward due to the gravitational attraction of the Earth, the bottom side of the filter is running into the air molecules, while the top side is moving away from the air molecules.

On average, the bottom side will have a significantly increased number of collisions per second with air molecules, and with greater impact. On the top side there will be a reduced number of collisions per second, and with less impact. There is a net upward push on the filter. This air resistance force increases with increasing downward speed of the filter, because higher speed increases the rate and impact of collisions on the bottom, and decreases the rate and impact of collisions on the top.

Eventually, when the downward speed of the coffee filter has become big enough, the net upward push by the air molecules becomes as large as the downward gravitational pull of the Earth. From then on the filter falls at a constant speed (terminal speed).

While this model captures the qualitative aspects of the situation, a simple quantitative calculation along these lines (using techniques discussed in a later chapter on gases) predicts a terminal speed that is *a thousand times smaller* than is observed! This is a spectacular failure of a simple model. The problem is that the falling coffee filter establishes macroscopic "wind"-like motions in the air, and the random motions of the air molecules at any particular location are relative to the bulk wind motion at that location. The techniques of "fluid dynamics" are capable of analyzing such phenomena, but these techniques are mathematically quite difficult and well beyond the scope of this introductory course. (Our later study of gases will address simpler situations, where there is no large-scale wind motion.)

5.8.2 Dependence of air resistance on speed

We established that the force of air resistance is dependent on the speed of a falling object.

? Did we establish that the force F_{air} of the air is proportional to the speed v of the falling object?

All we know is that the air resistance increases with increasing speed, but it need not be simply proportional to v. For all we know the force could be proportional to $\sqrt{v}$ *or* v^2 *or* v^3. There is a simple experiment you can do to

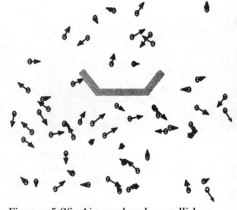

Figure 5.26 Air molecules collide with a falling coffee filter.

determine the form of the speed dependence, at least for low speeds. This is the topic of the second part of the following experimental problem.

Problem 5.2 Experiment—falling coffee filters
A paper coffee filter in the shape of a truncated cone falls in a stable way and quickly reaches terminal speed because it has a large area (big air resistance) but low mass (small gravitational force). Use a stopwatch to time the drop of coffee filters from as high a starting position as you can conveniently manage. If a stairwell is available, drop the coffee filters there to time a longer fall. This experiment is easier to do with a partner.

(a) By stacking coffee filters you can change the mass of a falling object without changing the shape. With this scheme you can explore how the terminal speed depends on the mass. Time the fall for different numbers of stacked filters, taking some care that the shape of the bottom filter is always the same (the filters tend to flatten out when removed from a stack). Start the measurement after the filters have fallen somewhat, to allow them to reach terminal speed. Average the results of several repeated measurements of time and height. Can you think of a simple experiment you can do to verify that the coffee filters do in fact reach terminal speed before you start timing?

(b) We want to know quantitatively how air resistance depends on speed. Plot your data for the air resistance force *vs.* the terminal speed. (The air resistance force is equal to the gravitational force when terminal speed has been reached, so it is proportional to the number of filters in a stack). How does the air resistance force depend on v?

There is an important constraint on the graph of speed dependence: Should the curve of the air resistance force vs. terminal speed go through the origin (force and terminal speed both very small), or not?

Analysis hint: If you suspect that the force is proportional to v^3, you might plot force vs. v^3 and see whether a straight line fits the data.

Other aspects of air resistance

? If you could double the cross-sectional area without changing the mass or the shape of the filter (by making the filter out of thinner paper), what change would you expect in the magnitude of the air resistance force at a particular speed? If you double the density of the air through which the filter falls, what change would you expect?

Measurements of a variety of ordinary-sized objects moving through air or other fluids show that the friction force is in fact proportional to the cross-sectional area A (hitting more molecules per second.) Air resistance is also proportional to the density ρ of the air (again, hitting more molecules per second), so there is less air resistance at higher altitudes. For ordinary-sized blunt objects such as a coffee filter moving at ordinary speeds, the force is found to be approximately proportional to the square of the speed, which is probably what you found in your own experiment. Combining these effects, the air resistance is approximately proportional to $\rho A v^2$.

? There is another important effect. From your own experience, would you expect larger air resistance (at a particular speed) for a pointed object or a blunt object?

As you might expect, the air resistance force depends not only on density, cross-sectional area, and speed, but also on the shape of the object. There is a larger air resistance force on a blunt object such as a coffee filter than on a ball or a streamlined object such as an arrow. This effect, and the dependence on v^2, are not easily explainable in quantitative detail in terms of a molecular model without performing a large molecular dynamics calculation.

The shape effect is usually taken into account by measuring the air resistance force (for example, by measuring the terminal speed) and determining an experimental "drag coefficient" C such that the following formula expresses the experimental results for the friction force:

APPROXIMATE AIR RESISTANCE FORCE (EMPIRICAL)

$F_{\text{air}} \approx \frac{1}{2}C\rho A v^2$, where $0.3 \leq C \leq 1.0$ (blunt objects, ordinary speeds)

The direction is opposite to the velocity: $\vec{F}_{\text{air}} \approx -\frac{1}{2}C\rho A v^2 \hat{v}$

Blunter objects have a higher value of C.

It is important to keep in mind that this is *not* a fundamental force law like the law of gravitation. Air resistance is the average result of a huge number of momentary contact electric interactions of air molecules hitting atoms on the surfaces of the falling object, and the average net air resistance is very difficult to calculate from fundamental principles at the molecular level.

5.9 Motion through Earth's atmosphere

How important is the effect of air resistance in the everyday world? Let's see to what extent we can predict the motion of a ball that is thrown or kicked with an initial speed v_0 at an angle θ_0 to the horizontal (Figure 5.27). Air resistance and the associated energy dissipation affect the motion. In order to see how large these effects are, we will first analyze the motion of projectiles without including the effects of air resistance. Then we will revise our analysis to take air resistance into account, and compare.

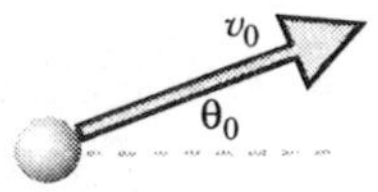

A major league pitcher can throw a fast ball with an initial speed of nearly 100 miles per hour, and we might ask how far a pitcher can throw a baseball, and at what angle to the horizontal the ball should be thrown to get the maximum range. Or we can turn the question around: if you see a football or soccer ball kicked a distance of 50 yards, what was its initial speed? If you throw a basketball from one end of the court toward the other end, how long does the ball take to get there?

? Neglecting air resistance for the moment, what forces act on a baseball at a time shortly after it has left the pitcher's hand? Draw a system diagram for the baseball.

Your diagram should show only a single force, a downward gravitational force on the baseball exerted by the Earth. If we temporarily neglect air resistance, there are no other significant forces acting on the ball. The analysis of the motion of baseballs and footballs and basketballs, neglecting air resistance, is a concrete though approximate example of an important class of motion—motion under the influence of a single constant force. Other examples include the motion of an electron moving through a region of approximately uniform "electric field" such as is found between two large oppositely charged metal plates.

? Is the gravitational force on a baseball in fact constant in magnitude and direction?

In its flight the baseball's distance from the center of the Earth changes, so the magnitude of the gravitational attraction of the Earth for the ball changes. However, the variation in height (some tens of meters) is very small compared to the radius of the Earth (6.4×10^6 m), so it is a good approximation

Figure 5.27 A ball is given an initial speed v_0 at an angle θ_0.

to consider the magnitude of the gravitational force to be nearly constant, with magnitude *mg*.

Also, with respect to fixed *x* and *y* axes, the direction of the gravitational force changes slightly since the force always points toward the center of the Earth (Figure 5.28). However, for distances that are short compared to the radius of the Earth, the direction of the gravitational force is nearly in the $-y$ direction, and the *x* component of the force is nearly zero. Approximately, $\vec{F} \approx 0\hat{i} + (-mg)\hat{j}$. Note that the gravitational field strength *g* (force per unit mass) is a positive number: $g = +9.8$ N/kg.

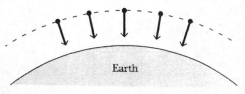

Figure 5.28 The gravitational force is only approximately constant in direction.

Air resistance may not be negligible

The force due to air resistance is not necessarily small compared to the gravitational force, especially at high speeds. Nevertheless we gain considerable insight into projectile motion by first considering a simplified model in which we ignore air resistance, and then later refining our model by including air resistance and seeing what difference that makes in the predicted path of the ball. Our simplified model would be a good model for projectile motion on the airless Moon.

5.9.1 Modeling projectile motion without air resistance

Consider a ball in flight. We need to identify interactions and draw forces on a physics diagram.

Non-contact interactions

The Earth exerts a gravitational force on the ball. Since the ball is near the Earth's surface we can approximate the gravitational force by a downward force *mg* (Figure 5.29). We neglect variation in the magnitude of the gravitational force with height, and we neglect variation in the direction of the gravitational force. We will neglect gravitational interactions with objects other than the Earth.

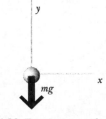

Figure 5.29 Forces on a ball, neglecting air resistance (and buoyancy).

Contact interactions

There is a small buoyant force upward on the ball due to the variation in the pressure of the air with height, but this buoyant force is negligibly small for a baseball. Since the ball is moving, there is an air resistance force on the ball. For the moment we will neglect air resistance, though later we will add air resistance to our model, because it can have a significant effect on the motion of a high-speed ball.

Write the momentum principle

If we neglect air resistance we can obtain an analytical solution for this problem, in the approximation that we ignore the small variation in the magnitude and direction of the gravitational force. Here is the momentum principle:

$$\frac{d\vec{p}}{dt} = <0, -mg, 0>$$

In our approximate model, all of the forces are known, so we can predict the motion into the future rather than have to use motion information to determine unknown forces.

Predict the motion

Let the initial location of the projectile at time $t = 0$ be (x_0, y_0), the initial speed be v_0, and the initial angle to the horizontal be θ_0. Figure 5.30 shows how to obtain the *x* and *y* components of the initial velocity. Because there

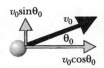

Figure 5.30 Resolving the initial velocity into *x* and *y* components.

is no x component of force, we deduce that the x component of velocity must be constant, and therefore equal to the initial value:

$$v_x(t) = \text{constant} = v_0 \cos\theta_0$$

The y component of the motion is a bit more complicated. From our equation for the momentum principle we can see that the y component of velocity is continually decreasing, at a constant rate $-g$:

$$\frac{dp_y}{dt} = -mg \quad \Rightarrow \quad \frac{dv_y}{dt} = -g$$

The initial value of v_y is $v_0\sin\theta_0$ at $t = 0$. At some later time t, v_y has decreased to the following value:

$$v_y(t) = v_0 \sin\theta_0 - gt$$

By $v_y(t)$ we mean "the value of v_y at time t." So for any time t we know both the x and y components of the velocity. What remains is to find the x and y positions at this same time t. The value of x as a function of time is easy, since v_x is constant and equal to $v_0\cos\theta_0$. At the initial time $t = 0$ we called the initial position x_0, so we have

$$x(t) = x_0 + (v_0 \cos\theta_0)t$$

Since $v_y = dy/dt$, the easiest way to determine the value of y at time t is to search for an expression $y(t)$ whose derivative dy/dt is equal to the known expression for v_y, $v_y(t) = v_0\sin\theta_0 - gt$.

? Verify that the derivative dy/dt of the following expression for y does give the correct value for v_y, and that it also gives the correct value for y at time $t = 0$, if y_0 is the initial height:

$$y(t) = y_0 + (v_0 \sin\theta_0)t - \frac{g}{2}t^2$$

Note that in order to predict the future position y and velocity v_y we need the two initial values of these variables (y_0 and $v_0\sin\theta_0$); similarly for x and v_x (x_0 and $v_0\cos\theta_0$). This is related to the fact that the derivative form of the momentum principle is actually a "second-order" differential equation, involving a second derivative, since

$$v_y = \frac{dy}{dt}, \text{ so } \frac{dv_y}{dt} = \frac{d}{dt}\left(\frac{dy}{dt}\right) = \frac{d^2y}{dt^2}$$

What have we neglected?

The results we just obtained are only approximate. We have neglected air resistance, which can have a sizable effect, and we have neglected the variation in the gravitational force associated with its inverse-square-distance property and its radial direction. We have neglected buoyancy. We have also neglected the (presumably very small) effects of a non-spherical Earth (for example, the sideways pull of a nearby building or mountain), and the tiny gravitational pulls exerted by other planets and the Sun.

We have also failed to take into account the fact that the Earth is rotating on its axis, and moving in a mutual orbit with the Moon, and curving around the Sun, all of which mean that the surface of the Earth is only approximately an "inertial frame of reference" in which the momentum principle is valid. Despite all these approximations, the formulas we have derived do give a useful picture of the main features of projectile motion. Just don't think that we have obtained an "exact" result!

Example: Neglecting air resistance, calculate the range R (horizontal distance along a level field) when θ_0 is the initial angle to the horizontal.

Solution: First we find the time $t_{1/2}$ it takes to reach maximum height; this is also the time it takes to come down, so the total time the projectile is in the air is $2t_{1/2}$. Once we know the total time, the range is just the (constant) horizontal component of velocity times the total time.

$$v_y = v\sin\theta_0 - gt_{1/2} = 0 \quad (v_y = 0 \text{ at maximum } y)$$

$$t_{1/2} = \frac{v\sin\theta_0}{g}$$

$$R = (v\cos\theta_0)\left(\frac{2v\sin\theta_0}{g}\right) = \frac{v^2}{g}(2\sin\theta_0\cos\theta_0) = \frac{v^2}{g}(\sin 2\theta_0)$$

Ex. 5.11 Considering these results, what initial angle will give the maximum range? What is this maximum range?

Ex. 5.12 A soccer ball is kicked at an angle of 60° to the horizontal with an initial speed of 20 m/s. Assume for the moment that we can neglect air resistance. For how much time is the ball in the air? How far does it go (horizontal distance along the field)? How high does it go?

5.9.2 Adding air resistance to a model of projectile motion

In the following problem, we will explore how significant the effect of air resistance is on a projectile such as a baseball.

Problem 5.3 Numerical integration of projectile motion with friction
Use a computer to carry out the following step-by-step quantitative calculations for a baseball. Remember that the direction of the air resistance force on an object is opposite to the direction of the object's velocity.

A baseball has a mass of 155 grams and a diameter of 7 cm. The drag coefficient C for a baseball is about 0.35, and the density of air at sea level is about 1.3×10^{-3} grams/cm^3, or 1.3 kg/m^3.

(a) Including the effect of air resistance, compute and plot the trajectory of a baseball thrown or hit at an initial speed of 100 miles per hour, at an angle of 45° to the horizontal. How far does the ball go? Is this a reasonable distance?

(A baseball field is about 400 feet from home plate to the fence in center field. An outfielder cannot throw a baseball in the air from the fence to home plate.)

(b) On the same display as (a), plot a graph of $(K + U_g)$ vs. time.

(c) Temporarily neglect air resistance (set $C = 0$ in your computations) and determine the range of the baseball. Check your computations by using the analytical solution for motion of a projectile without air resistance. Compare your result with the range found in part (a), where the effect of air resistance was taken into account. Is the effect of air resistance significant? (The effect is surprisingly large—about a factor of 2!)

(d) On the same display as (c), neglecting air resistance, plot a graph of $(K + U_g)$ vs. time.

(e) In Denver, a mile above sea level, the air is about 83% as dense as the air at sea level. Including the effect of air resistance, use your computer model to predict the trajectory and range of a baseball thrown in Denver. How does the range compare with the range at sea level?

5.9.3 A spinning ball

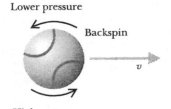

Figure 5.31 A spinning ball experiences additional forces.

Even when we include air resistance in our calculations, we still have ignored a force that can have a major effect on the trajectory. If a ball has spin, there is an effect of fluid flow around the ball that raises the air pressure on the side where the rotational motion is in the same direction as the ball's velocity, and lowers the air pressure on the other side, where the rotational motion is in the opposite direction to the velocity. If the force points to the side, the ball curves; this is an important effect in baseball pitching. If the force points upward (due to "backspin," as in Figure 5.31), it lifts the ball and extends the range. If the force points downward (due to "topspin") it decreases the range. The calculations involved in modeling this effect are quite complex, and involve "fluid dynamics" (the study of fluid flow).

5.9.4 Viscous friction

Very small particles such as dust or droplets of fog falling slowly through air, or small objects moving through a thick liquid such as honey, experience a friction force that is proportional to v rather than v^2. This is called "viscous" friction. The approximate dependence on v or v^2 for air resistance is roughly valid only for specific situations. The formulas for friction do not have the wide applicability of such major physical laws as the law of gravitation. However, even approximate formulas for friction make it possible to make much better predictions than we could make if we ignore friction completely.

5.10 Motion across a surface: Sliding friction

Figure 5.32 Stretch the spring without moving the block.

Sliding friction between two solid objects is a familiar interaction that involves energy dissipation. We'll consider it in some detail. Connect a spring to a block and stretch the spring an amount s to the right without moving the block (Figure 5.32). We need to identify interactions and draw forces on the block on a physics diagram.

Non-contact forces

? What objects exert non-contact forces on the block?

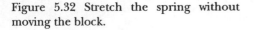

Figure 5.33 Physics diagram (incomplete) showing all non-contact forces.

The Earth exerts a force on the block straight down, in the $-y$ direction (Figure 5.33). Your hand doesn't touch the block, so if it exerts a force on the block it must exert a non-contact force. How could it do this? The hand can exert a very small gravitational force on the block, but this force is so small that we can neglect it. The hand and block are not electrically charged. There is no mechanism for a significant interaction between the hand and the block, so we can't list the hand as an object that interacts with the block.

Contact forces

? What objects exert contact forces on the block?

The spring touches the block, and exerts a force to the right (in the $+x$ direction) of magnitude $k_s s$. The table touches the block. What is the direction of the force exerted on the block by the table? Clearly the table exerts a force on the block upward, in the $+y$ direction (Figure 5.34). In Chapter 3 we examined the interactions between a brick and a table, and found that the table exerted a force upward on the brick as the atoms in the table were compressed by the brick. A similar interaction will happen between the block and the table, and the upward force by the table on the block can balance the downward gravitational force exerted on the block by the Earth.

Write the momentum principle and motion information

? Is the momentum of the block changing? Is your answer consistent with the system diagram in Figure 5.34?

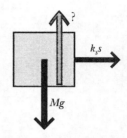

Figure 5.34 Physics diagram (incomplete) showing some contact forces as well as non-contact forces.

We are left with a problem. Although the force exerted on the block by the table is upward, we can understand why the y component of the block's momentum is not changing, because there is a downward gravitational force. However, there appears to be a component of the net force on the block to the right, yet the block is at rest—the block's momentum is not changing. This is not consistent with the momentum principle. We must have failed to identify a force.

What could be exerting a force to the left on the block? The only object that can possibly be exerting a force to the left to balance the spring force is the table. How does it do that? Atoms of the block "run into" atoms in the table, and the resulting interactions stretch and compress interatomic bonds in the objects. Figure 5.35 provides an atomic-level picture of such an interaction.

This image is a single frame from a molecular dynamics computer simulation of sliding friction between two objects, performed by Judith A. Harrison and coworkers at the United States Naval Academy in Annapolis, Maryland. The computations are similar to the numerical integrations you have done, but involving a large number of objects (atoms). The forces between the atoms are modeled by spring-like forces.

In this image, a diamond tip is dragged to the right across a diamond surface. The large spheres represent carbon atoms, and the small ones are hydrogen atoms on the carbon surfaces. Note that the projecting tip has "caught" on the lower surface, causing interatomic bonds in the two objects to deform. The stretching and compression of interatomic bonds produces the horizontal component of the force exerted on the moving object by the lower object. For more information about Harrison's work, see

http://chem.mathsci.usna.edu/jah/ResearchStuff/Friction_page.html

Figure 5.35 A computer simulation of sliding friction between a diamond tip and diamond surface. Image courtesy of Judith A. Harrison, U.S. Naval Academy.

Figure 5.37 shows a complete physics diagram for the block. The force exerted by the table acts at an angle. It is common practice to replace the force vector exerted by the table with two equivalent vectors, a compression force F_N normal to the surface and a friction force f parallel to the surface, as shown in Figure 5.37. This turns out to be very useful for many reasons, but it is important to remember that an atom in the table exerts one interatomic force on a neighboring atom in the block, not two. It is for our own convenience that we split this force into two equivalent forces.

We can construct the component equations for the momentum principle by reading the forces off of our physics diagram. We choose axes with x to the right and y up (z is outward). We know that the block remains at rest

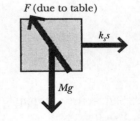

Figure 5.36 A complete physics diagram for the block.

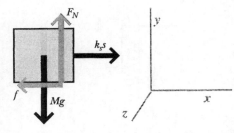

Figure 5.37 Complete physics diagram for the block; the force on the block by the table has been resolved into components normal to and parallel to the surface.

even though we are pulling on it, so we can write motion information in addition to the momentum principle:

$$\frac{d\vec{p}}{dt} = <(k_s s - f), (F_N - Mg), 0>$$

$$p_x = 0 \text{ at all times, so } \frac{dp_x}{dt} = 0$$

$$p_y = 0 \text{ at all times, so } \frac{dp_y}{dt} = 0$$

Solve equations

? What is the magnitude of the compression force F_N? What is the magnitude of the friction force f?

From $(F_N - Mg) = 0$ we deduce that the compression force is just Mg; the table supports the block. From $(k_s s - f) = 0$ we deduce that the friction force must be equal to the spring force. If you reduce the stretch of the spring, the friction force f gets smaller, too, because the sideways interatomic forces are smaller. As long as the block remains at rest, the friction force has a range of values equal to the range of spring forces applied to the block. However, this is not the case when the block is in motion.

Note that this is another example of a situation where we don't have a force law for one of the forces, the force exerted by the table (which we split into two components). We are able to deduce that $F_N = Mg$ and $f = k_s s$ thanks to the motion information that supplements our incomplete force information.

5.10.1 Block in motion with sliding friction

Now slowly stretch the spring more and more, and eventually the block moves to the right, which means that during the instant when it began moving, its momentum changed. By the momentum principle, that implies that the spring force $k_s s$ to the right became greater than the friction force f to the left. Evidently there must be a maximum possible friction force f_{max} for this object on this table (it is different for different surfaces).

If we apply a larger spring force than this maximum friction force, we break temporary interatomic bonds between the block and the table, and we can speed up the block. Measurements on sliding objects show that the friction force remains approximately equal to f_{max}, independent of the speed of sliding. This implies that the friction force is related to the strength of the interatomic bonds, which are continually made and continually broken as the block slides along.

In contrast, remember that air resistance depends on the speed of the moving object. The mechanism for air resistance is very different from that of sliding friction and leads to a dependence on the speed.

Coefficient of friction

It is approximately true for many objects on many surfaces that the maximum friction force f_{max} is proportional to the amount of compression F_N, with a proportionality constant μ that is a property of the two surfaces that are in contact (Figure 5.38):

$$f_{max} \approx \mu F_N$$

Typical values for the "coefficient of friction" μ for wooden blocks sliding on wooden tables are in the range of 0.25 to 0.5 (μ is Greek lowercase mu).

These approximate properties of sliding friction do not have the status of fundamental physical laws such as the law of gravitation. For example, the

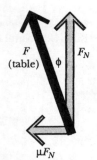

Figure 5.38 Sliding friction approximately independent of the speed.

$$\tan\phi \approx \frac{\mu F_N}{F_N} = \mu$$

details of motion with friction depend critically on how clean the surface is. Very flat, very clean metal or glass plates may bind to each other so strongly that it is almost impossible to slide them.

When two very clean and flat metal surfaces are brought in contact, the interface region looks much like the interior of the metals, and the two metal objects may behave pretty much like a single block. Also, when surfaces are very flat, it is possible that air will be squeezed out from between the objects, in which case air pressure on the outer surfaces raises the compression force F_N to a very large value, and the maximum friction force μF_N is then very large. This is an example of a situation where F_N is much greater than the weight Mg of the block. What matters in the formula $f_{max} = \mu F_N$ is how much the two surfaces are squeezed together, not the weight of the block.

Dragging with constant speed

You can experience some of the vagaries of sliding friction by trying to drag a block at a constant speed. You will need to keep the stretch of the spring constant, so $k_s s = f_{max}$, and there should be no acceleration—no change in the constant velocity. You may find that it is very difficult to maintain a constant stretch of the spring (and a constant velocity of the block), in part because the block skips and loses some contact with the table. It can be easier to maintain constant velocity if the object is quite massive and therefore less subject to sudden large changes in velocity.

You may observe that you have to pull harder to break the block loose from the table and start it moving than you do to keep it moving at a constant speed. For some surfaces the maximum friction force for a block at rest is larger than the maximum friction force for a moving block. We will look at sliding friction in more detail later in the course, but a full understanding of this complex phenomenon is the subject of ongoing contemporary research.

An intriguing recent experimental result is that with extremely clean surfaces ("atomically" clean), it matters whether the interatomic spacings of the two objects are commensurate or not; that is, whether the grid of atoms of one surface fit into the grid of atoms of the other surface. If they do fit, like nested egg cartons, it takes a sizable force to pull the block up out of the atomic grid, and a smaller force to keep it going. But if there isn't a good fit, the slightest force will move the block.

Dragging with continually increasing speed

To drag with continually increasing speed, it is necessary for the pulling force to be greater than the friction force (Figure 5.39). Try dragging a block across a table by stretching the spring enough that $k_s s$ is greater than f_{max}, and try to maintain a constant stretch s in order to obtain a constant nonzero dv_x/dt. You may find that it is quite difficult to maintain a constant acceleration for very long. A practical difficulty is that as the speed increases, you need to move your hand faster and faster to maintain a constant large stretch in the spring, and you also may run out of room on the table. In Figure 5.39 we have assumed, based on experiments, that during sliding the frictional force remains approximately equal to f_{max}.

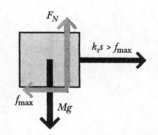

Figure 5.39 Pulling with a force greater than the maximum friction force.

Ex. 5.13 If the mass of the block is M, what is the acceleration $d\vec{v}/dt$ of the block? (Apply the momentum principle along both the x and y axes.)

5.10.2 Energy dissipation

If you pull the block with a constant force F at constant speed through a distance d, you do an amount of work Fd, but the block's speed doesn't increase. The making and breaking of interatomic bonds at the sliding interface increases the oscillation of atoms around their equilibrium positions. This increased energy is then shared with many other atoms in the block and table, with the result that the average energy of atoms in both objects increases. This is the same thing as saying that the temperature rises in the block and in the table. You can observe such a rise in temperature by rubbing your hands together.

The energy is dissipated throughout the block and table, and cannot easily be recovered. There is a kind of irreversibility in a process where friction plays a role, because the dissipated or dispersed energy can't be brought back to the interface region and used to move the block.

5.10.3 A paradox involving friction

If we are careful in our energy bookkeeping for a system involving motion with friction, we encounter a paradox that cannot easily be resolved. Consider the situation shown in Figure 5.40.

You exert a constant force F, and pull a block a distance d across a table. Because there is friction between the block and the table, the block travels at constant speed. Since the momentum of the block does not change, we conclude correctly that the net force on the block is zero, and that therefore the magnitude of the friction force must also be F.

In terms of energy inputs, you have done an amount of work Fd on the block. The friction force, acting in the opposite direction, has apparently done an amount of work $-Fd$. The net work done on the block therefore would seem to be zero. This is consistent with the fact that the kinetic energy of the block did not change.

It is not, however, consistent with the fact that both the block and the table get hotter as they rub together! If the net work done on the block was zero, where did that increased thermal energy come from?

This is not a "trick question," but a genuine puzzle. We will return to this issue in Chapter 7, when we will deal with it in enough detail to resolve the paradox.

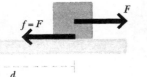

Figure 5.40 Forces in the x direction on a block pulled with constant speed across a table. (Forces in the y direction are not shown).

5.11 Including friction in modeling a spring-mass system

One of the most obvious things about the motion of a real macroscopic spring-mass system is that the amplitude gets smaller and smaller with time. Our model of a spring-mass system that we developed in Chapter 3 is too simple: we should include the effects of friction.

Problem 5.4 Numerical integration of spring-mass system with friction
Modify your numerical integration of the spring-mass system to include the effects of friction. Such a system is called a "damped oscillator."

Simulate the effect of all three kinds of friction: sliding friction (independent of v), viscous friction (proportional to v), and air resistance for large blunt objects (proportional to v^2). Choose friction to be small enough that you get at least two full cycles of oscillation, but large enough that you see an effect. If the friction is so large that the graph never crosses the axis, the system is said to be "over-damped." For each kind of friction, display the following:

      (1) Display an animation of the motion of the system
      (2) Plot a graph of position *vs.* time

(3) Plot graphs of *K*, *U*, and (*K+U*) *vs.* time

On the energy graphs, when is the energy loss rate large and small, and why?

In the case of "viscous" friction, where the friction force is $\vec{f} = -c\vec{v}$, it is possible with elaborate math (or by guessing the solution and plugging it into the momentum principle for the system) to find an analytical solution, in which a cosine function is multiplied by an exponential factor:

$$Ae^{-\frac{c}{2m}t}\cos(\omega t)$$

The amplitude dies away exponentially with time. The angular frequency is

$$\omega = \sqrt{\frac{k_s}{m} - \left(\frac{c}{2m}\right)^2}$$

which is approximately $\sqrt{k_s/m}$ for low friction.

There is no such simple analytical solution if the friction force is independent of the speed (sliding friction) or proportional to v^2 (air resistance for large blunt objects).

5.12 Friction and irreversibility

When there is friction, processes are not reversible, in contrast with situations describable in terms of potential energy. If energy dissipation is negligible, a bouncing ball continually returns to nearly its original position and momentum, and the process would run just as well backwards as forwards. A movie of the bouncing ball could be played backwards and it would look the same as when the movie is played forwards. But in the real world, energy is dissipated in the collision with the floor, and due to air resistance, so that the bouncing ball does *not* return to its initial position and momentum, and the process is not reversible. A movie of the process played backward would look quite odd, because the ball would bounce higher and higher.

Energy dissipation

In section 5.2 we saw that potential energy is independent of path. In the absence of air resistance, a ball thrown up into the air will return to your hand with exactly the same kinetic energy it had when it left your hand. The situation is entirely different in the presence of friction. You throw the ball up in the air, and as it goes up it loses some energy to the air (air resistance), and on the way down it loses some more energy to the air. When it returns to your hand the net change in the kinetic energy is negative despite the fact that the net change in the gravitational energy is zero (same position). The round trip difference in the kinetic energy is *not* zero.

Similarly, in the presence of friction, when a mass on a spring returns to the same *s*, it has a smaller *v*. Again, the change in the kinetic energy of the mass during the round trip is not zero, even though the spring potential energy has not changed.

A particularly compelling case involving friction is that of pulling a block at constant speed to the right across a table with friction and then to the left back to its starting point. On the way to the right the block and table get hot, and on the way back to the left the block and table get even hotter (Figure 5.41 and Figure 5.42). In order to have no change in the round trip, the block and table would have had to cool down as you pulled the block to the left; this would be a very surprising effect indeed.

Note that the friction force exerted by the table points to the left on the way out, and then to the right on the way back. The direction of the friction force depends on the direction of the motion, not merely on position. An

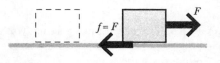

Figure 5.41 Pull to the right; the block and table get hot.

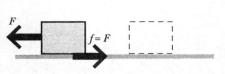

Figure 5.42 Pull back to the left; the block and table get even hotter.

interaction that we can associate with potential energy would not behave this way. The force would have the same direction on the way out and on the way back ($F_x = -dU/dx$); one energy change would be positive and the other negative, making a net change of zero. Friction is a good example of an interaction for which we cannot define a potential energy, because the friction force depends on the direction of motion.

Forces which depend only on position, and for which it is possible to define an associated potential energy (with the force equal to the negative gradient of the potential energy), are called "conservative forces." Friction is not a conservative force.

A key issue is that we define potential energy in terms of the interactions of point particles, or particle-like objects whose internal state doesn't change. But a block being dragged across a table is a complicated multiparticle system. Individual atoms interact through electric forces associated with the interatomic potential energy, and at the atomic level these interactions are reversible. But energy spreads throughout the block and floor, and macroscopically this looks to us like energy dissipation.

Irreversibility has far-reaching consequences for the statistical or probabilistic behavior of macroscopic objects consisting of large numbers of atoms. We will consider issues of reversibility and irreversibility in more detail later in the course.

Ex. 5.14 You move a block slowly to the right a distance of 0.2 m on a table. You apply a constant force of 20 N. How much work do you do? You then move slowly back to the starting point; how much work do you do in this second move? What change has there been in the thermal energy of the block and table?

Ex. 5.15 A block is attached to a spring whose stiffness is 40 N/m. You move the block slowly from $s = 0$ to $s = 0.2$ m, with negligible friction present. How much work do you do? You then move slowly back to the starting point; how much work do you do in this second move? What change has there been in the thermal energy of the block?

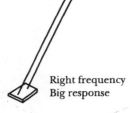

Right frequency
Big response

Wrong frequency
Small response

Figure 5.43 When you push a swing, you get a big response if you match the natural frequency of the swing.

5.13 *Resonance

Imagine pushing somebody in a swing (Figure 5.43). If you time your pushes to match the natural frequency of the swing, you can build up an increasingly large amplitude of the swing. But larger amplitude means higher speed and a higher rate of energy dissipation.

? You keep pushing with constant (small) amplitude, and the swing eventually reaches a "steady state" with constant (large) amplitude. What determines the size of this steady-state amplitude?

The steady state is established when the energy dissipated through air resistance on each cycle has grown to be equal to the energy input you make on each cycle. The large steady-state oscillation needs only little pushes at the right times from you to make up for energy dissipation, and the amplitude of the oscillation can be much larger than the amplitude of the pushes.

On the other hand, if you deliberately push at the wrong times, not matching the natural frequency of the swing, the amplitude of the swing won't continually build up. To get a large response from an oscillating system, inputs should be made at the natural, free-oscillation frequency of that system. This important feature of driven oscillating systems is called "resonance."

Problem 5.5 Experiment—observing resonance

You can observe the main effects of resonance with very simple experiments. Hold a spring vertically with a mass suspended at the other end, and observe the frequency of "free" oscillations with your hand kept still. Then stop the oscillations, and move your hand *extremely* slowly up and down in a kind of slow sinusoidal motion. You will see that the mass moves up and down with the same very low frequency.

(a) How does the amplitude (plus or minus displacement from the center location) of the mass compare with the amplitude of your hand? (Notice that the phase shift of the oscillation is $0°$; the mass moves up when your hand moves up.)

(b) Next move your hand up and down at a significantly higher frequency than the free-oscillation frequency. How does the amplitude of the mass compare to the amplitude of your hand? (Notice that the phase shift of the oscillation is $180°$; the mass moves down when your hand moves up.)

(c) Finally, move your hand up and down at the free-oscillation frequency. How does the amplitude of the mass compare with the amplitude of your hand? (It is hard to observe, but the phase shift of the oscillation is $90°$; the mass is at the midpoint of its travel when your hand is at its maximum height.)

(d) Change the system in some way so as to increase the air resistance significantly. For example, attach a piece of paper to increase drag. At the free-oscillation frequency, how does this affect the size of the response?

A strong dependence of the amplitude and phase shift of the system to the driving frequency is called resonance.

5.13.1 *Analytical treatment of resonance

Problem 5.27 (page 209) is a numerical integration of a sinusoidally driven spring-mass oscillator (Figure 5.44). This problem lets you study the buildup of the steady state as well as the steady state itself. For the steady-state portion of the motion, it is possible to obtain an analytical solution, if the energy dissipation is due to viscous friction. Here is the momentum principle for the driven oscillator in Figure 5.44, with viscous friction force $-cv$:

$$\frac{dp}{dt} = -k_s[x - D\sin(\omega_D t)] - cv$$

where x is the position of the mass; D is the amplitude of the sinusoidal motion and ω_D is the variable angular frequency of the driven end of the spring. For more details on how this equation was obtained, see the discussion in Problem 5.27.

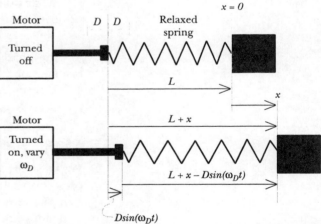

Figure 5.44 Sinusoidal driving of a spring-mass system; ω_D can be varied.

In the steady state x must be a sinusoidal function, with amplitude A and phase shift ϕ depending strongly on the driving angular frequency ω_D:

$$x = A\sin(\omega_D t + \phi)$$

By taking derivatives of this expression to obtain $v = dx/dt$ and dv/dt as a function of time t, and plugging these derivatives into the momentum principle, it can be shown that this expression for x as a function of time is a solution in the steady state if the amplitude A and phase shift ϕ have values determined by the following expressions:

$$A = \frac{\omega_F^{\,2}}{\sqrt{\left(\omega_F^{\,2} - \omega_D^{\,2}\right)^2 + \left(\dfrac{c}{m}\omega_D\right)^2}} D$$

$$\cos\phi = \frac{\left(\omega_F^{\,2} - \omega_D^{\,2}\right)}{\sqrt{\left(\omega_F^{\,2} - \omega_D^{\,2}\right)^2 + \left(\dfrac{c}{m}\omega_D\right)^2}}$$

The detailed proof is rather complicated, so we omit it. Instead, in the exercises below we give you the opportunity to see that these expressions for amplitude and phase shift are in agreement with your experimental observations and the computer modeling of Problem 5.27. Figure 5.45 is a graph of the amplitude A as a function of the driving angular frequency ω_D, for two different values of the viscous friction parameter c.

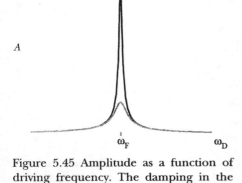

A

ω_F ω_D

Figure 5.45 Amplitude as a function of driving frequency. The damping in the lower curve is five times greater than in the upper curve.

Ex. 5.16 Using the formula for the amplitude A, show that if the viscous friction is small, the amplitude is large when ω_D is approximately equal to ω_F.

Ex. 5.17 Show that with small viscous friction, the amplitude A drops to $1/\sqrt{2}$ of the peak amplitude when the driving angular frequency differs from resonance by this amount:

$$\left|\omega_F - \omega_D\right| \approx \frac{c}{2\,m}\omega_F.$$

(Hint: note that near resonance $\omega_D \approx \omega_F$, so $\omega_F + \omega_D \approx 2\omega_F$.)

Ex. 5.18 Given the results of the previous exercise, how does the width of the resonance peak depend on the amount of friction? What would the resonance curve look like if there were very little friction?

Ex. 5.19 Using the formula involving the phase shift ϕ, show that the phase shift is approximately 0° for very low driving frequency ω_D, approximately 180° for very high driving frequency ω_D, and 90° at resonance, consistent with your experiment.

5.13.2 *Resonance in other systems

We have discussed resonance in the context of a particular system—a mass on a spring driven by a motor. We study resonance because it is an important phenomenon in a great variety of situations.

When you choose a radio or television station you adjust the parameters of an electrically oscillating circuit so that the circuit has a narrow resonance at the chosen station's main frequency, and other stations with significantly different frequencies have little effect on the circuit.

Magnetic resonance imaging (MRI) is based on the phenomenon of nuclear magnetic resonance (NMR) in which nuclei acting like toy tops precess in a strong magnetic field. The precessing nuclei are significantly affected by radio waves only at the precession frequency. Because the precession frequency depends slightly on what kind of atoms are nearby, resonance occurs for different radio frequencies, which makes it possible to identify different kinds of tissue and produce remarkably detailed images.

5.14 Summary

Fundamental physical principles

No new fundamental principles were introduced in this chapter.

New concepts

Spring (elastic) energy $U_s = \frac{1}{2} k_s s^2 + U_0$

Thermal energy = average energy of atoms in an object, proportional to temperature of the object

Power = $\dfrac{\text{energy}}{\text{time}}$ (J/s = watts); also calculable as $\vec{F} \bullet \vec{v}$

Thermal energy transfer across the system boundary, Q

Heat capacity = energy per gram required to raise temperature 1K

$$\Delta E_{\text{thermal}} = mC\Delta T; \quad C = \frac{\Delta E_{\text{thermal}}}{m\Delta T} \quad (C \text{ in J/K per gram, } m \text{ in grams})$$

Open system: energy flows into and/or out of system.

Closed system: no energy flows into or out of system.

The choice of system (open or closed) affects energy accounting.

Dissipation of energy

Terminal speed

Air resistance: collisions between air molecules and moving object

Sliding friction force due to forming and breaking of bonds between two solid objects

Path independence of potential energy
    Impossibility of describing friction in terms of potential energy

Resonance: sensitivity of driven oscillator to driving frequency

Results

Air resistance

Air resistance is approximately proportional to speed squared, and in a direction opposite to the velocity:

$$\vec{F}_{\text{air}} \approx -\frac{1}{2} C\rho A v^2 \hat{v}$$

Viscous friction

For small particles ("viscous" friction), fluid friction is approximately proportional to v.

Sliding friction

Sliding friction force $f \le \mu F_N$, where the coefficient of friction μ depends on the surfaces, and F_N is the normal force compressing the two surfaces together.

Sliding friction force does not depend on speed (approximate).

Motion formulas

Analytical solution for projectile motion without air resistance:

$$x(t) = x_0 + (v_0 \cos \theta_0) t$$

$$y(t) = y_0 + (v_0 \sin \theta_0) t - \frac{g}{2} t^2$$

Numerical solution for projectile motion with air resistance.

Numerical solution for motion of free and driven spring-mass system with friction.

Analytical solution for driven spring-mass system: $x = A \sin(\omega_D t + \phi)$, where $\omega_F = \sqrt{k_s / m}$; viscous friction cv, driven at ω_D:

$$A = \frac{\omega_F^2}{\sqrt{(\omega_F^2 - \omega_D^2)^2 + \left(\frac{c}{m} \omega_D\right)^2}} D, \quad \cos \phi = \frac{(\omega_F^2 - \omega_D^2)}{\sqrt{(\omega_F^2 - \omega_D^2)^2 + \left(\frac{c}{m} \omega_D\right)^2}}$$

5.15 Example problems

5.15.1 Example 1: A rebounding block

A metal block of mass 3 kg is moving downward with speed 2 m/s when the bottom of the block is 0.8 m above the floor (Figure 5.46). When the bottom of the block is 0.4 m above the floor, it strikes the top of a relaxed vertical spring 0.4 m in length. The stiffness of the spring is 2000 N/m.

(a) The block continues downward. When the bottom of the block is 0.3 m above the floor, what is its speed?

(b) The block eventually heads back upwards, loses contact with the spring, and continues upward. What is the maximum height reached by the bottom of the block above the floor?

(c) What approximations did you make?

Solution

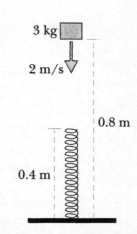

Figure 5.46 A block rebounds from a vertical spring.

(a) Take the Earth, block, and spring as the system to which we'll apply the energy principle. There are no energy transfers to this system from the surroundings, so the energy of the system is a constant. We will calculate E_i and E_f and set them equal to each other. In the initial state the energy is gravitational potential energy plus kinetic energy; in the final state the energy is gravitational plus kinetic plus spring potential energy (we omit the rest energy, which does not change).

$$E_i = E_f$$

$$mgy_i + \frac{1}{2}mv_i^2 = mgy_f + \frac{1}{2}mv_f^2 + \frac{1}{2}k_s s_f^2$$

Solve for the final speed:

$$v_f^2 = v_i^2 + 2g(y_i - y_f) - \frac{k_s}{m}s_f^2$$

$$v_f = \sqrt{(2\ \text{m/s})^2 + 2(9.8\ \text{N/kg})(0.5\ \text{m}) - \frac{(2000\ \text{N/m})}{(3\ \text{kg})}(0.1\ \text{m})^2}$$

$$v_f = 2.7\frac{\text{m}}{\text{s}}$$

(b) Here we need to consider a different final state. Take as the initial state the original initial state, and as final state the instant when the block is at the top of its trajectory, where $v = 0$ and spring stretch $s = 0$:

$$mgy_i + \frac{1}{2}mv_i^2 = mgy_f$$

Divide by mg:

$$y_f = y_i + \frac{v_i^2}{2g} = (0.8\ \text{m}) + \frac{(2\ \text{m/s})^2}{2(9.8\ \text{N/kg})} = 1.0\ \text{m}$$

(c) We assumed negligible air resistance and dissipation in the spring. We used the approximate expression $U_g = mgy$ near Earth's surface. We neglected the extremely tiny kinetic energy change of the Earth. The gravitational and spring forces on the Earth and on the block are equal and opposite, so the momentum changes are equal and opposite. But since $K = p^2/(2m)$, for the same momentum p the kinetic energy of the Earth is extremely small compared to the kinetic energy of the block.

Another way to look at this is to say that while the forces on the Earth and the block have the same magnitudes, the Earth hardly moves, so the force does extremely little work on the Earth and gives it little kinetic energy.

5.15.2 Example 2: Heat capacity

A 300 gram block of aluminum at temperature 500 K is placed on a 650 gram block of iron at temperature 350 K in an insulated enclosure. At these temperatures the heat capacity of aluminum is approximately 1.0 J/K/gram, and the heat capacity of iron is approximately 0.42 J/K/gram. Within a few minutes the two metal blocks reach the same common temperature T. Calculate T.

Solution

System: the two blocks
Initial state: different temperatures, not in contact
Final state: blocks have come to thermal equilibrium, same T.
Energy principle: $\Delta E_{Al} + \Delta E_{Fe} = 0$
(The total energy of the two blocks does not change, because there is no energy transfer from or to the surroundings.)

$$(300 \text{ gram})(1.0 \text{ J/K/gram})(T - 500) +$$
$$(650 \text{ gram})(0.42 \text{ J/K/gram})(T - 350) = 0$$

Solving for the final temperature, we find $T = 429 \text{ K}$.

In words, what happens in this process is that the aluminum temperature falls from 500 K to 429 K, and the iron temperature rises from 350 K to 429 K. The thermal energy decrease in the aluminum is equal to the thermal energy increase in the iron.

5.16 Review questions

Definitions

RQ 5.1 An oil company included in its advertising the following phrase: "Energy—not just a force, it's power!" In technical usage, what are the differences among the terms energy, force, and power?

Energy conservation

RQ 5.2 50 grams of water whose temperature is 25° Celsius are added to a thin glass containing 800 grams of water at 20° Celsius (about room temperature). What is the final temperature of the water? What simplifying assumptions did you have to make in order to determine your approximate result?

Spring potential energy

RQ 5.3 A spring has a relaxed length of 6 cm and a stiffness of 100 N/m. How much work must you do to change its length from 5 cm to 9 cm?

Power

RQ 5.4 Electricity is billed in kilowatt-hours. Is this energy or power? How much is one kilowatt-hour in standard physics units? (The typical cost of one kilowatt-hour is 5 to 10 cents.)

Open and closed systems

RQ 5.5 Which of the following are open systems with respect to energy, and which are closed? Explain why.

a car; a person; an insulated picnic chest; the Universe; the Earth

RQ 5.6 A horse whose mass is M gallops at constant speed v up a long hill whose vertical height is h, taking an amount of time t to reach the top. The horse's hooves do not slip on the rocky ground, so the work done by the force of the ground on the hooves is zero (no displacement of the force). When the horse started running, its temperature rose quickly to a point at which from then on, thermal energy transfer from the horse to the air keeps the horse's temperature constant.

(a) First consider the horse as the system of interest: What objects exert forces on the horse? How much work is done on the horse by each of these forces? What non-work energy transfers are there? What is the change in the energy of the horse? In what forms?

(b) Next consider the Universe as the system of interest: What objects exert forces on this system? How much work is done on the system by each of these forces? What non-work energy transfers are there? What is the change in the energy of the system? In what forms?

Heat capacity

RQ 5.7 Substance A has a large heat capacity (per gram), while substance B has a smaller heat capacity. If the same amount of energy is put into a 100 gram block of each substance, and if both blocks were initially at the same temperature, which one will now have the higher temperature?

Energy dissipation

RQ 5.8 When a falling object reaches terminal speed, its kinetic energy reaches a constant value. However, the gravitational energy of the system consisting of object plus Earth continues to decrease. Does this violate the principle of conservation of energy? Explain why or why not.

Air resistance

RQ 5.9 Describe a situation in which it would be appropriate to neglect the effects of air resistance.

RQ 5.10 Describe a situation in which neglecting the effects of air resistance would lead to significantly wrong predictions.

RQ 5.11 List all the approximations that are made in order to find a simple analytical solution for the motion of a projectile near the Earth's surface.

RQ 5.12 You are standing at the top of a 50 m cliff. You throw a rock as hard as you can, in the horizontal direction. Approximately where does it hit on the flat plain below? Is your answer too large or too small?

RQ 5.13 Write the computational statements required at the heart of a numerical integration of projectile motion that includes air resistance.

RQ 5.14 If you let a mass at the end of a string start swinging, at first the maximum swing decreases rather quickly, but once the swing has become small it takes a long time for further significant decrease to occur. Try it! Explain this simple observation.

Sliding friction

RQ 5.15 You drag a block with constant speed v across a table with friction. Explain in detail what you have to do in order to change to a constant speed of $2v$ on the same surface. (That is, the puzzle is to explain how it is possible to drag a block with sliding friction at different constant speeds.)

Damped oscillators

RQ 5.16 Sketch a graph of position vs. time for a spring-mass oscillator subject to friction (not over-damped).

Potential energy

RQ 5.17 A particle moves inside a circular glass tube under the influence of a tangential force of constant magnitude F (Figure 5.47). Explain why we cannot associate a potential energy with this force (which is therefore a "nonconservative" force). How is this situation different from the case of a block on the end of a string, which is swung in a circle?

RQ 5.18 In Figure 5.47, calculate the change in electric energy along two different paths in moving charge q away from charge Q from A to B along a radial path, then to C along a circle centered on Q, then to D along a radial path. Also calculate the change in energy in going directly from A to D along a circle centered on Q. Specifically, what are $U_B - U_A$, $U_C - U_B$, $U_D - U_C$, and there sum? What is $U_D - U_A$? Also, calculate the round-trip difference in the electric energy when moving charge q along the path from A to B to C to D to A.

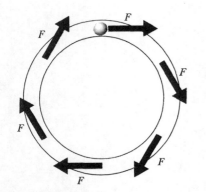

Figure 5.47 A particle moves in a circular glass tube (RQ 5.17).

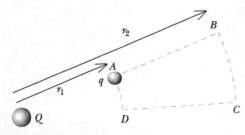

Figure 5.48 Change in electric energy along two different paths (RQ 5.18).

5.17 Problems

5.17.1 In-line problems from this chapter

Problem 5.1 (page 169) Energy in a spring-mass system
Energy calculations provide a powerful check on the accuracy of a numerical integration. Modify your numerical integration of the motion of a mass on a spring that you did in Chapter 3, Problem 3.4 (page 87). In addition to plotting graphs of position versus time, also plot graphs of kinetic energy, of spring energy, and of the sum of the kinetic energy and the spring energy, versus time during the motion. It is up to you to choose some value for the potential energy U_0 at the bottom of the well. Label the energy graphs with numerical values and units. Make sure that the energy values are reasonable for your system.

(a) Does the sum of kinetic and potential energy remain constant as a function of time? What if you vary the step size (which varies the accuracy of the numerical integration)?

(b) Increase the energy by increasing the initial speed. How does the amplitude A change? How does the angular frequency ω change?

Problem 5.2 (page 183) Experiment—falling coffee filters
A paper coffee filter in the shape of a truncated cone falls in a stable way and quickly reaches terminal speed because it has a large area (big air resistance) but low mass (small gravitational force). Use a stopwatch to time the drop of coffee filters from as high a starting position as you can conveniently manage. If a stairwell is available, drop the coffee filters there to time a longer fall. This experiment is easier to do with a partner.

(a) By stacking coffee filters you can change the mass of a falling object without changing the shape. With this scheme you can explore how the terminal speed depends on the mass. Time the fall for different numbers of stacked filters, taking some care that the shape of the bottom filter is always the same (the filters tend to flatten out when removed from a stack). Start the measurement after the filters have fallen somewhat, to allow them to reach terminal speed. Average the results of several repeated measurements of time and height. Can you think of a simple experiment you can do to verify that the coffee filters do in fact reach terminal speed before you start timing?

(b) We want to know quantitatively how air resistance depends on speed. Plot your data for the air resistance force *vs.* the terminal speed. (The air resistance force is equal to the gravitational force when terminal speed has been reached, so it is proportional to the number of filters in a stack). How does the air resistance force depend on v?

There is an important constraint on the graph of speed dependence: Should the curve of the air resistance force vs. terminal speed go through the origin (force and terminal speed both very small), or not?

Analysis hint: If you suspect that the force is proportional to v^3, you might plot force vs. v^3 and see whether a straight line fits the data.

Problem 5.3 (page 187) Numerical integration of projectile motion with friction
Use a computer to carry out the following step-by-step quantitative calculations for a baseball. Remember that the direction of the air resistance force on an object is opposite to the direction of the object's velocity.

A baseball has a mass of 155 grams and a diameter of 7 cm. The drag coefficient C for a baseball is about 0.35, and the density of air at sea level is about 1.3×10^{-3} grams/cm^3, or 1.3 kg/m^3.

(a) Including the effect of air resistance, compute and plot the trajectory of a baseball thrown or hit at an initial speed of 100 miles per hour, at an angle of 45° to the horizontal. How far does the ball go? Is this a reasonable distance?

(A baseball field is about 400 feet from home plate to the fence in center field. An outfielder cannot throw a baseball in the air from the fence to home plate.)

(b) On the same display as (a), plot a graph of $(K + U_g)$ vs. time.

(c) Temporarily neglect air resistance (set $C = 0$ in your computations) and determine the range of the baseball. Check your computations by using the analytical solution for motion of a projectile without air resistance. Compare your result with the range found in part (a), where the effect of air resistance was taken into account. Is the effect of air resistance significant? (The effect is surprisingly large—about a factor of 2!)

(d) On the same display as (c), neglecting air resistance, plot a graph of $(K + U_g)$ vs. time.

(e) In Denver, a mile above sea level, the air is about 83% as dense as the air at sea level. Including the effect of air resistance, use your computer model to predict the trajectory and range of a baseball thrown in Denver. How does the range compare with the range at sea level?

Problem 5.4 (page 192) Numerical integration of spring-mass system with friction

Modify your numerical integration of the spring-mass system to include the effects of friction. Such a system is called a "damped oscillator."

Simulate the effect of all three kinds of friction: sliding friction (independent of v), viscous friction (proportional to v), and air resistance for large blunt objects (proportional to v^2). Choose friction to be small enough that you get at least two full cycles of oscillation, but large enough that you see an effect. If the friction is so large that the graph never crosses the axis, the system is said to be "over-damped." For each kind of friction, display the following:

(1) Display an animation of the motion of the system
(2) Plot a graph of position vs. time
(3) Plot graphs of K, U, and $(K+U)$ vs. time

On the energy graphs, when is the energy loss rate large and small, and why?

Problem 5.5 (page 195) Experiment—observing resonance

You can observe the main effects of resonance with very simple experiments. Hold a spring vertically with a mass suspended at the other end, and observe the frequency of "free" oscillations with your hand kept still. Then stop the oscillations, and move your hand *extremely* slowly up and down in a kind of slow sinusoidal motion. You will see that the mass moves up and down with the same very low frequency.

(a) How does the amplitude (plus or minus displacement from the center location) of the mass compare with the amplitude of your hand? (Notice that the phase shift of the oscillation is 0°; the mass moves up when your hand moves up.)

(b) Next move your hand up and down at a significantly higher frequency than the free-oscillation frequency. How does the amplitude of the mass compare to the amplitude of your hand? (Notice that the phase shift of the oscillation is 180°; the mass moves down when your hand moves up.)

(c) Finally, move your hand up and down at the free-oscillation frequency. How does the amplitude of the mass compare with the amplitude of your hand? (It is hard to observe, but the phase shift of the oscillation is 90°; the mass is at the midpoint of its travel when your hand is at its maximum height.)

(d) Change the system in some way so as to increase the air resistance significantly. For example, attach a piece of paper to increase drag. At the free-oscillation frequency, how does this affect the size of the response?

A strong dependence of the amplitude and phase shift of the system to the driving frequency is called resonance.

5.17.2 Additional problems

Problem 5.6 Energy in a spring-mass system

A horizontal spring-mass system has low friction, spring stiffness 200 N/m, and mass 0.4 kg. The system is released with an initial compression of the spring of 10 cm and an initial speed of the mass of 3 m/s.

(a) What is the maximum stretch during the motion?

(b) What is the maximum speed during the motion?

(c) Now suppose that there is energy dissipation of 0.01 J per cycle of the spring-mass system. What is the average power input in watts required to maintain a steady oscillation?

Problem 5.7 Vertical spring-mass system

Write an equation for the total energy of a system consisting of a mass suspended vertically from a spring, and include the Earth in the system. Place the origin for gravitational energy at the equilibrium position of the mass, and show that the changes in energy of a vertical spring-mass system are the same as the changes in energy of a horizontal spring-mass system.

Problem 5.8 Hot water

400 grams of boiling water (temperature 100° C, heat capacity 4.2 J/gram/K) are poured into an aluminum pan whose mass is 600 grams and initial temperature 20° C (the heat capacity of aluminum is 0.9 J/gram/K). After a short time, what is the temperature of the water? Explain your calculation. What simplifying assumptions did you have to make?

Problem 5.9 Place a mass on a vertical spring

A spring with stiffness k_s and relaxed length L stands vertically on a table. You hold a mass M just barely touching the top of the spring.

(a) You *very slowly* let the mass down onto the spring a certain distance, and when you let go, the mass doesn't move. How much did the spring compress? How much work did *you* do?

(b) Now you again hold the mass just barely touching the top of the spring, and then let go. What is the maximum compression of the spring? State what approximations and simplifying assumptions you made.

(c) Next you push the mass down on the spring so that the spring is compressed an amount s, then let go, and the mass starts moving upward and goes quite high. When the mass is a height of $2L$ above the table, what is its speed?

Problem 5.10 Bungee jumping

Design a "bungee jump" apparatus. A bungee jumper falls from a high platform with two elastic cords tied to the ankles. The jumper falls freely for a while, with the cords slack. Then the jumper falls an additional distance with the cords increasingly tense. Assume that you have cords that are 10 m long, and that the cords stretch in the jump an additional 20 m for an average jumper.

(a) Make a series of simple diagrams, like a comic strip, showing the platform, the jumper, and the two cords at various times in the fall and the rebound. On each diagram, draw and label vectors representing the forces acting on the jumper, and the jumper's velocity. Make the relative lengths of the vectors reflect their relative magnitudes.

(b) At what instant is there the greatest tension in the cords? How do you know?

(c) What is the jumper's speed at this instant?

(d) Is the jumper's momentum changing at this instant or not? (That is, is dp/dt nonzero or zero?) Explain briefly.

(e) Focus on this instant, and use the principles of this chapter to determine the spring stiffness k_s for each cord. Explain your analysis.

(f) What is the maximum tension that each cord must support without breaking?

(g) What is the maximum acceleration (in "g's") that the jumper experiences? What is the direction of this maximum acceleration?

(h) State clearly what approximations and estimates you have made in your design.

Problem 5.11 Heating a house

During three hours one winter afternoon, when the outside temperature was 0° C (32° F), a house heated by electricity was kept at 20° C (68° F) with the expenditure of 45 kwh (kilowatt·hours) of electric energy. What was the average energy leakage in joules per second through the walls of the house to the environment (the outside air and ground)?

The rate of thermal energy transfer between two systems is often proportional to their temperature difference. Assuming this to hold in this case, if the house temperature had been kept at 25° C (77° F), how many kwh of electricity would have been consumed?

Problem 5.12 Experiment—your power capability

(a) If you follow a diet of 2000 food calories per day (2000 kilocalories), what is your average power consumption in watts? (A food or "large" calorie is a unit of energy equal to 4.2×10^3 J; a regular or "small" calorie is equal to 4.2 J.) Compare with the power consumption in a table lamp.

(b) You can produce much higher power for short periods. Make appropriate measurements as you run up some stairs, and report your measurements. Use these measurements to estimate your power output (this is in addition to your basal metabolism—the power needed when resting). Compare with a horsepower (which is about 750 watts) or a toaster (which is about 1000 watts).

(c) How many days of a diet of 2000 large calories are equivalent to the gravitational energy difference for you between sea level and the top of Mount Everest, 8848 m above sea level? (However, the body is not anywhere near 100% efficient in converting chemical energy into change in altitude. Also note that this is in addition to your basal metabolism.)

Problem 5.13 Human energy consumption

(a) A typical candy bar provides 280 calories (one "food" or "large" calorie is equal to 4.2×10^3 joules). How many candy bars would you have to eat to replace the chemical energy you expend doing 100 sit-ups? Explain your work, including any approximations or assumptions you make. (In a sit-up, you go from lying on your back to sitting up.)

(b) Humans have about 60 milliliters (60 cm^3) of blood per kilogram of body mass, and blood makes a complete circuit in about 20 seconds, to keep tissues supplied with oxygen. Make a crude estimate of the *additional* power output of your heart (in watts) when you are standing compared with when you are lying down. Note that we're not asking you to estimate the power output of your heart when you are lying down, just the change in power when you stand up. You will have to estimate the values of some of the relevant parameters. Because we're only looking for a crude estimate, try to construct a model that is as simple as possible. Describe what approximations and estimates you made.

Problem 5.14 Graph of energy with dissipation

Figure 5.49 is a portion of a graph of energy terms vs. time for a mass on a spring, subject to air resistance. Identify and label the three curves as to what kind of energy each represents. Explain briefly how you determined which curve represented which kind of energy.

Problem 5.15 Drop coffee filters from a tall building

You drop a single coffee filter from a very tall building, and it takes 52 seconds to reach the ground. Next you drop a stack of 3 of these coffee filters. About how long will they take to hit the ground? Explain briefly, including any approximations or simplifying assumptions you had to make.

Problem 5.16 Launch a package from an asteroid

A package of mass m sits at the equator of an airless asteroid of mass M, radius R, and spin angular speed ω (Figure 5.50). We want to launch the package in such a way that it will never come back, and when it is very far from the asteroid it will be traveling with speed v. We have a powerful spring whose stiffness is k_s. How much must we compress the spring?

Problem 5.17 A mass hangs from a spring

A mass of 0.3 kg hangs motionless from a vertical spring whose length is 0.8 m and whose unstretched length is 0.65 m. Next the mass is pulled down to where the spring has a length of 0.9 m and given an initial speed upwards of 1.2 m/s. What is the maximum length of the spring during the motion that follows? What approximations or simplifying assumptions did you make?

Problem 5.18 A sky diver

Terminal speed for a falling sky diver has been measured to be about 60 m/s. Using your computer model, determine how far (in meters) and how long (in seconds) a sky diver falls before reaching terminal speed. (You will need to think about the meaning of "terminal speed" and how to test for it.)

Plot graphs of the sky diver's speed *vs.* time and position *vs.* time.

Problem 5.19 Experiment—coefficient of friction

Design and carry out an experiment to determine the coefficient of sliding friction for a block on a flat plate.

Problem 5.20 Experiment—speed dependence of sliding friction

For this problem you will need measurements of the position vs. time of a block sliding on a table, starting with some initial velocity, slowing down, and coming to rest. If you do not have an appropriate laboratory setup for making these measurements, your instructor will provide you with such data. Analyze these data to see how well they support the assertion that the force of sliding friction is essentially independent of the speed of sliding. (Recall that a reason was given for this assertion—the friction force is determined by the strength of temporary interatomic bonds that must be broken.)

Problem 5.21 Block dragged at an angle

A block of mass M is dragged along a table by a force applied at an angle θ to the horizontal by a spring of stiffness k_s, stretched a constant amount s

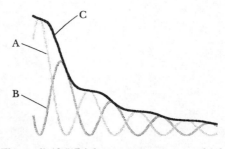

Figure 5.49 Which curve represents which kind of energy (Problem 5.14)?

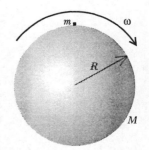

Figure 5.50 Launch a package from an asteroid (Problem 5.16).

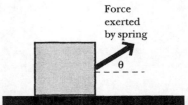

Figure 5.51 A block is dragged by a force at an angle to the horizontal (Problem 5.21).

(Figure 5.51). The coefficient of friction between block and table is μ. At $t = 0$, the speed of the block was v_0. What is its speed v at time t?

Problem 5.22 A basketball pass

You have probably seen a basketball player throw the ball to a teammate at the other end of the court, 30 m away. Estimate a reasonable initial angle for such a throw, and then determine the corresponding initial speed. For your chosen angle, how long does it take for the basketball to go the length of the court? What is the highest point along the trajectory, relative to the thrower's hand?

Problem 5.23 A free throw in basketball

Determine two different possible ways for a player to make a free throw in basketball. In both cases give the initial speed, initial angle, and initial height of the basketball. The rim of the basket is 10 feet (3.0 m) above the floor. It is 14 feet (4.3 m) along the floor from the free-throw line to a point directly below the center of the basket.

Problem 5.24 The case of the falling flower pot

You are a detective investigating why someone was hit on the head by a falling flowerpot. One piece of evidence is a home video taken in a 4th-floor apartment, which happens to show the flowerpot falling past a tall window. Inspection of individual frames of the video shows that in a span of 6 frames the flowerpot falls a distance that corresponds to 0.85 of the window height seen in the video (note: standard video runs at a rate of 30 frames per second). You visit the apartment and measure the window to be 2.2 m high. What can you conclude? Under what assumptions? Give as much detail as you can.

Problem 5.25 A car crash

A car traveling 25 m/s (about 55 miles per hour) crashes into a concrete wall. The front of the car crumples and collapses, so that the car is now 0.5 meter shorter than before.

(a) Approximately how long was the duration of the impact?
(b) What was the approximate magnitude of the acceleration during the crash?
(c) Explain briefly your approximations or simplifying assumptions.

Problem 5.26 Electron motion in a CRT

In a cathode ray tube (CRT) used in oscilloscopes and television sets, a beam of electrons is steered to different places on a phosphor screen, which glows at locations hit by electrons (Figure 5.52). The CRT is evacuated, so there are few gas molecules present for the electrons to run into. Electric forces are used to accelerate electrons of mass m to a speed v_0, after which they pass between positively and negatively charged metal plates which deflect the electron in the vertical direction (upward in Figure 5.52, or downward if the sign of the charges on the plates is reversed).

While an electron is between the plates, it experiences a uniform vertical force F, but when the electron is outside the plates there is negligible force on it. The gravitational force on the electron is negligibly small compared to the electric force in this situation. The length of the metal plates is d, and the phosphor screen is a distance L from the metal plates. Where does the electron hit the screen? (That is, what is y_f?)

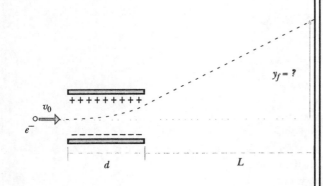

Figure 5.52 A cathode ray tube (Problem 5.26).

Problem 5.27 Computer modeling of resonance

You can study resonance in a driven oscillator in detail by modifying your computation for the spring-mass system with friction. Let one end of the spring be moved back and forth sinusoidally by a motor, with a motion given by $D\sin(\omega_D t)$ (Figure 5.53). Here D is the amplitude of the motor motion and ω_D is the angular frequency of the motor, which can be varied. (The free-oscillation angular frequency $\omega_F = \sqrt{k_s/m}$ of the spring-mass system has a fixed value, determined by the spring stiffness k_s and the mass m.)

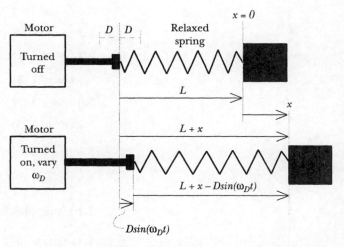

Figure 5.53 A driven spring-mass system (Problem 5.27).

We need an expression for the stretch s of the spring in order to be able to calculate the force $-k_s s$ of the spring on the mass. We have to do a bit of geometry to figure out the length of the spring when the mass is displaced a distance x from the equilibrium position, and the motor has moved the other end of the spring by an amount $D\sin(\omega_D t)$.

On the diagram, the spring gets longer when the mass moves to the right $(+x)$, but the spring gets shorter when the motor moves to the right $(-D\sin(\omega_D t))$. The new length of the spring is $L + x - D\sin(\omega_D t)$, and the net stretch of the spring is this quantity minus the unstretched length L, yielding $s = x - D\sin(\omega_D t)$. A check that this is the correct expression for the net stretch of the spring is that if the motor moves to the right the same distance as the mass moves to the right, the spring will have zero stretch. Replace x in your computer computation with the quantity $[x - D\sin(\omega_D t)]$.

(a) Use viscous damping (friction proportional to v). The damping should be small. That is, without the motor driving the system ($D = 0$), the mass should oscillate for many cycles. Set ω_D to $0.9\omega_F$ (that is, 0.9 times the free-oscillation angular frequency, $\omega_F = \sqrt{k_s/m}$), and let x_0 be 0. Graph position x vs. time for enough cycles to show that there is a transient buildup to a "steady state." In the steady state the energy dissipation per cycle has grown to exactly equal the energy input per cycle.

Vary ω_D in the range from 0 to $2.0\omega_F$, with closely spaced values in the neighborhood of $1.0\omega_F$. Have the computer plot the position of the motor end as well as the position of the mass as a function of time. For each of these driving frequencies, record for later use the steady-state amplitude and the phase shift (the difference in angle between the sinusoids for the motor and for the mass). If a harmonic oscillator is lightly damped it has a large response only for driving frequencies near its own free-oscillation frequency.

(b) Is the steady-state angular frequency equal to ω_F or ω_D for these various values of ω_D? (Note that during the transient buildup to the steady state the frequency is not well-defined, because the motion of the mass isn't a simple sinusoid.)

(c) Sketch graphs of the steady-state amplitude and phase shift vs. ω_D. Note that when $\omega_D = \omega_F$ the amplitude of the mass can be much larger than D, just as you observed with your hand-driven spring-mass system. Also note the interesting variation of the phase shift as you go from low to high driving frequencies—does this agree with the phase shifts you observed with your hand-driven spring-mass system?

(d) Repeat part (c) with viscous friction twice as large. What happens to the resonance curve (the graph of steady-state amplitude vs. angular frequency ω_D)?

5.18 Answers to exercises

5.1 (page 169)   If $k_s = 0.6$ N/m and $s = 0.2$ m, $U_s = 0.012$ J

5.2 (page 169)   0.5 m/s

5.4 (page 173)   0.12 K (that is, 0.12 Kelvin)

5.5 (page 174) 1.3×10^5 J; 1.3×10^5 J; -1.3×10^5 J

5.6 (page 175)   About 500 J; about 2000 kg per second

5.7 (page 175)   2.5 m/s^2

5.8 (page 180) $\frac{1}{2}kA^2$; $\frac{1}{2}kA^2$; spring + mass

5.9 (page 182)   less; some of the initial energy is now in the air

5.10 (page 182) $[2.7 \times 10^{-2} - 4.5 \times 10^{-4}]$ J $\approx 2.7 \times 10^{-2}$ J

5.11 (page 187) $45°$, $R = v^2/g$

5.12 (page 187)   time in air = 3.5 seconds; horizontal distance = 35 m (about 38 yards); maximum height = 15.3 m

5.13 (page 191) $\dfrac{dv_x}{dt} = \dfrac{(k_s s - f_{max})}{M}$ and $\dfrac{dv_y}{dt} = 0$.

5.3 (page 170)   No change; same stretch

5.14 (page 194)   +4 J; +4J; 8 J

5.15 (page 194)   +0.8 J; –0.8 J; 0 J

5.16 (page 196)   Denominator gets very small when $\omega_D = \omega_F$.

5.17 (page 196)   When $\omega_D = \omega_F$, denominator $= \dfrac{c}{m}\omega_D$.

Denominator $= \sqrt{2}\dfrac{c}{m}\omega_D$ when $\left|(\omega_F^{\,2} - \omega_D^{\,2})\right| = \dfrac{c}{m}\omega_D$.

But since near resonance $\omega_D \approx \omega_F$, we have

$$(\omega_F^{\,2} - \omega_D^{\,2}) = (\omega_F + \omega_D)(\omega_F - \omega_D) \approx 2\omega_D(\omega_F - \omega_D)$$

Hence when $|\omega_F - \omega_D| \approx \dfrac{c}{2m}\omega_D \approx \dfrac{c}{2m}\omega_F$, amplitude down by a factor of $1/\sqrt{2}$.

5.18 (page 196)   Larger friction, wider resonance peak. If friction very small, resonance peak is a very narrow spike.

5.19 (page 196)   Low frequency, $\cos\phi \approx \dfrac{(\omega_F^{\,2})}{\sqrt{(\omega_F^{\,2})^2 + \left(\dfrac{c}{m}\omega_D\right)^2}} \approx 1$, so

$\phi \approx 0°$.

High frequency, $\cos\phi \approx \dfrac{(-\omega_D^{\,2})}{\sqrt{(-\omega_D^{\,2})^2 + \left(\dfrac{c}{m}\omega_D\right)^2}} \approx -1$, so

$\phi \approx 180°$.

Resonance frequency, $\cos\phi = 0$, so $\phi \approx 90°$.

Chapter 6

Energy Quantization

Chapter 6

Energy Quantization

In the early 20th century, as scientists learned more about the detailed behavior of atoms they were excited to discover that small systems such as atoms can have only certain energies, called energy "levels," and cannot have energies in between these discrete levels. The energy changes associated with interactions therefore come in finite packets, or "quanta."

6.1 Energy quantization

Discrete energy levels were not noticed for a long time because the energy differences between these levels are quite small. A typical separation between energy levels in an atom is about 10^{-19} joules. The energies of macroscopic objects are on the order of a joule; an energy increase of 10^{-19} joules is completely undetectable even by the most sensitive measurements of motion or temperature. However, it is possible to detect these tiny differences by observing the interaction of light and matter, as we will see.

The discovery of discrete energy levels is analogous to the discovery of the atomic nature of matter. We now know that matter is not continuous but lumpy, being made up of atoms. For most of human history this lumpiness was not noticed because the atoms are so tiny. Similarly, the noncontinuous, lumpy nature of energy was not noticed because the quantized energies are so very close together that energy seemed continuous.

6.1.1 Electronic energy levels

An example of discrete energy levels is found in the "electronic" energy associated with the motion of electrons around the nucleus of an atom. You probably know from your study of chemistry that electronic energy levels are discrete—an atom can have only certain energies. The electronic energy levels in atomic hydrogen are shown in Figure 6.1 (atomic hydrogen is a single H atom, composed of one proton and one electron, not the usual diatomic molecule H_2).

The diagram shows the familiar electric potential energy of the proton-electron system. The horizontal lines represent the possible bound energy states for the hydrogen atom; bound states with other values of $(K+U)$ are not observed. Unbound states however $(K+U \geq 0)$ are not quantized.

The electric potential energy of an electron and a proton bound together in a hydrogen atom is negative for all separations:

$$U_e = \frac{1}{4\pi\varepsilon_0}\frac{q_1 q_2}{r} = \frac{1}{4\pi\varepsilon_0}\frac{(+e)(-e)}{r} = -\frac{1}{4\pi\varepsilon_0}\frac{e^2}{r}$$

A quantum mechanical calculation of the energy levels of a system involves solving a differential equation called the Schrödinger equation, which includes a potential energy term. Solving this equation for the hydrogen atom predicts that the discrete energy levels for bound states of hydrogen (ignoring the rest energies of the proton and the electron) will be these:

ENERGY LEVELS FOR THE HYDROGEN ATOM

$$E_N = K + U_e = -\frac{13.6\text{ eV}}{N^2}, \ N = 1, 2, 3, \text{ etc.}$$

An electron-volt (eV) is an energy unit described below.

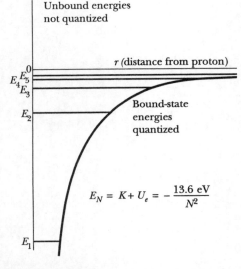

Figure 6.1 Electronic energy levels for a hydrogen atom. The potential energy curve describes the potential energy of the electron-proton system as a function of distance.

This arrangement of hydrogen energy levels has been verified by observations of the interaction of light and hydrogen, as we will see.

These energy levels denote different bound states of the electron+proton system. Recall that a bound state has a negative value of $K+U$, since a potential energy of zero is associated with a very large separation between the electron and the proton. We have seen that bound gravitational orbits are also associated with negative values of $K+U$. As you can see from the energy diagram, in higher energy states the average location of the electron is farther from the proton. Note that there is a lowest energy state; a hydrogen atom cannot have $K+U < -13.6$ eV!

Before the development of the fully quantum mechanical Schrödinger equation, Niels Bohr made a simple model for the hydrogen atom which also yielded the energy formula given above. We will study the Bohr model later in Chapter 9.

6.1.2 Electron volts (eV)

In our analysis of quantum systems we will frequently be considering very small changes in energy, on the order of 10^{-19} joules. A common unit of energy used in descriptions of quantum phenomena is the "electron volt," abbreviated "eV." In terms of joules:

$$1 \text{ eV} = 1.6 \times 10^{-19} \text{ J}$$

Here are some typical energies measured in electron-volts:
Typical chemical reaction: about 1 eV.
Typical nuclear reaction: about 1 MeV (mega-electron-volt, or 10^6 eV).
Rest mass of the electron: about 0.5 MeV.
Rest mass of the proton: about 1 GeV (giga-electron-volt, or 10^9 eV).

Example: How much energy in electron-volts is required to raise the energy of a hydrogen atom from its lowest level ($N = 1$) to the next lowest level ($N = 2$)?

Solution: Calculate the required increase in energy:

$$\Delta E = E_2 - E_1 = \left(-\frac{13.6 \text{ eV}}{2^2} \right) - \left(-\frac{13.6 \text{ eV}}{1^2} \right) = 10.2 \text{ eV}$$

Note that you cannot add less than 10.2 eV to a hydrogen atom in its lowest energy level; any smaller amount fails to be absorbed.

Ex. 6.1 How much energy in electron-volts is required to ionize a hydrogen atom (that is, remove the electron from the proton), if initially the atom is in its lowest energy level?

6.2 The quantum model of interaction of light and matter

What is the evidence for energy quantization? A key piece of evidence is that atomic systems emit and absorb light only with quantized packets of energy, implying a quantized loss or gain of energy in the system. The study of the interaction of light and matter is called spectroscopy.

6.2.1 A particle model of light

Early in the 20th century it was discovered that light can be considered to be lumpy, being made up of packets of energy called photons. As with atoms and energy, the noncontinuous, lumpy nature of light was not noticed for a long time because the amount of energy carried by one photon in an ordinary beam of light is extremely small compared to the total energy of the beam. It has now become a routine matter to detect individual photons in a very weak beam of light, and to measure the energy of an individual photon.

You may already be familiar with the classical model of light, in which light is described as a travelling wave. The classical wave model of light makes it possible to account for important aspects of the behavior of light, including interference and diffraction phenomena. To account for other properties of light, including the observation that packets of light corresponding to certain fixed energies can be absorbed or emitted by quantum systems, we need a quantum model of light that describes light as particles (photons). In this chapter we will emphasize the quantum model of light, and we will be concerned with the energy of photons emitted or absorbed as quantum systems go from one energy state to another.

The electromagnetic spectrum

Visible light is an example of electromagnetic radiation, a phenomenon associated with the production and detection of electric and magnetic fields of a particular kind. The instruments needed to detect and measure electromagnetic radiation depend on the photon energies involved. The spread of observable energies is called the electromagnetic spectrum (Figure 6.2).

That part of the electromagnetic spectrum for which your eye is a good detector is called "visible" light (Figure 6.3). Its energy ranges from around 1.8 eV per photon (red light) to 3.1 eV per photon (violet light). Photon energies greater than 3 eV include ultraviolet, x-rays, and gamma rays. Electromagnetic radiation with photon energies less than 1.8 eV includes infrared, radio waves, and microwaves. Sometimes the term "light" is used to refer just to visible electromagnetic radiation, that part of the spectrum detectable by human eyes.

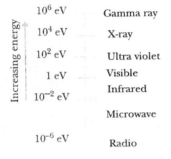

Figure 6.2 The electromagnetic spectrum.

Figure 6.3 The visible portion of the electromagnetic spectrum.

Ex. 6.2 The photon energy for green light lies between that for red and violet light. What is the approximate energy of the photons in green light?

Ex. 6.3 The intensity of sunlight is about 1400 watts (J/s) per square meter. That is, when sunlight hits perpendicular to a square meter of area, about 1400 watts of energy can be absorbed. Using the average photon energy found in the previous exercise, about how many photons per second strike an area of one square meter? (This is why the lumpiness of light was not noticed for so long.)

6.2.2 A spectrum

A beam of light can be composed of photons of many different energies. When the photons in the beam are separated according to their energies, the resulting distribution is called the spectrum of that beam of light. One way to separate a beam of visible light into its component energies is to pass a narrow beam of light through a prism. You have probably seen the continuous rainbow spectrum seen when ordinary white light goes through a prism, as in Figure 6.4. Your eye reacts differently to photons of different energies, which is why you see a colorful spectrum.

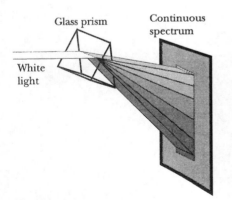

Figure 6.4 A prism separates a narrow beam of white light into a continuous rainbow of colors.

6.2.3 Emission spectra

In contrast to white light, when a narrow beam of light emitted by electrically excited atomic hydrogen is passed through a prism it does not look like a continuous spectrum but rather appears as a few widely spaced bright lines of particular colors (Figure 6.5). Such discontinuous spectra are strong evidence for energy quantization in the systems that are emitting the light.

When a hydrogen atom drops from a higher energy level E_4 to a lower energy level E_1 it emits a photon with the associated energy $E_4 - E_1$ (Figure 6.6). A collection of hydrogen atoms can emit photons whose energies correspond to the differences in energy between any two electronic energy levels of the hydrogen atom. The light emitted by excited atomic hydrogen is called the "hydrogen emission spectrum," and the bright lines of light of particular energies in the spectrum are called "spectral lines." The observed energies of light emitted by excited atomic hydrogen are consistent with the energy level differences predicted by quantum mechanics. Other atoms have different quantized energy levels, and a line spectrum can be used to identify the emitting atoms by the pattern of emitted photon energies.

If an isolated hydrogen atom is initially in a high energy level, over time it will spontaneously emit photons and drop to lower and lower states, eventually ending up in the lowest energy level, which is called the "ground state." Once an atom has dropped to the ground state, it cannot lose any more energy, nor emit photons, because there is no lower energy level than the ground state. This is quite different from the classical (prequantum) picture, in which a system can have an arbitrarily low energy.

Example: Consider a collection of hydrogen atoms. Initially each individual atom is in one of the lowest three energy states ($N = 1, 2, 3$). What will be the energies of photons emitted by this collection of hydrogen atoms?

Solution: There are three possibilities: transitions from $N = 3$ to $N = 2$, from $N = 2$ to $N = 1$, or from $N = 3$ to $N = 1$. The photon energies for the three transitions are these:

$$N = 3 \text{ to } N = 2: \quad \left(-\frac{13.6 \text{ eV}}{3^2}\right) - \left(-\frac{13.6 \text{ eV}}{2^2}\right) = 1.9 \text{ eV}$$

$$N = 2 \text{ to } N = 1: \quad \left(-\frac{13.6 \text{ eV}}{2^2}\right) - \left(-\frac{13.6 \text{ eV}}{1^2}\right) = 10.2 \text{ eV}$$

$$N = 3 \text{ to } N = 1: \quad \left(-\frac{13.6 \text{ eV}}{3^2}\right) - \left(-\frac{13.6 \text{ eV}}{1^2}\right) = 12.1 \text{ eV}$$

Ex. 6.4 How many different photon energies would emerge from a collection of hydrogen atoms which occupy the lowest four energy states ($N = 1, 2, 3, 4$)? (You need not calculate the energies of these transitions.)

6.2.4 Absorption spectra

An atom can drop from a higher energy level to a lower energy level and emit a photon. An atom can also absorb a photon of an appropriate energy and jump from a lower energy level to a higher energy level (Figure 6.7). If an atom is in its ground state, only photon absorption is possible; photon

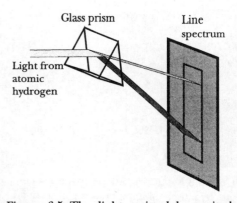

Figure 6.5 The light emitted by excited atomic hydrogen contains only a few separate colors. Light in a spectrum of a particular color and energy is called a "spectral line."

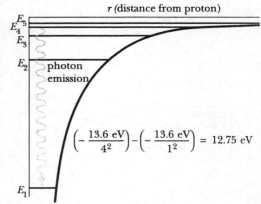

Figure 6.6 An atom can drop to a lower energy level by emitting a photon whose energy corresponds to the difference between the energies of the levels. By convention a squiggly line with an arrowhead indicates a photon.

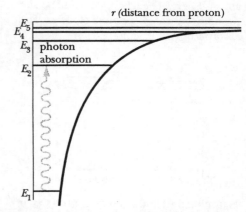

Figure 6.7 An atom can gain energy by absorbing a photon whose energy corresponds to the difference between two energy levels.

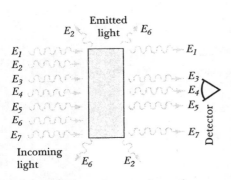

Figure 6.8 Measuring an absorption spectrum. A sample of a material is placed in a transparent container. Incoming light is absorbed, leading to emission. The outgoing light is depleted of the absorption energies.

emission cannot occur. Energy absorption can occur if the photon energy corresponds to the difference between two energy levels of the atom. If we shine white light (containing a mixture of different photon energies) through a collection of atoms (such as a container of hydrogen gas), only photons whose energy is the difference of atomic energy levels will be strongly absorbed, while other photons can pass right through the material. In this case, the spectral lines are dark lines (missing energies) on a light background, instead of the bright lines on a dark background observed in an emission spectrum. This is called an "absorption spectrum."

The atoms that have jumped to higher energy levels through photon absorption will eventually drop to lower energy levels with the emission of photons. However, these photons are emitted in all directions, not just in the direction of the original beam of light, so the light that has passed through the material will have little intensity at photon energies corresponding to the strongly absorbed energies (Figure 6.8).

> *Example:* At room temperature in a gas of atomic hydrogen almost all of the atoms are in the ground state. Why might you expect the gas to be completely transparent to visible light?
>
> *Solution:* As we saw in the example on page 213, the energy required to go from the ground state of atomic hydrogen to the first excited state is 10.2 eV, but visible light consists of photons whose energies lie between 1.8 eV and 3.1 eV. So visible light cannot be absorbed by ground-state hydrogen atoms. (In a collision between a photon of visible light and an atom, the atom can acquire some kinetic energy, but no internal energy change is possible.)
>
> *Ex. 6.5* Suppose you had a collection of hypothetical quantum objects whose individual energy levels were –4.0 eV, –2.3 eV, and –1.6 eV. If nearly all of the individual objects were in the ground state, what would be the energies of dark spectral lines in an absorption spectrum if visible white light (1.8 to 3.1 eV) passes through the material?

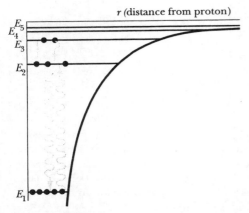

Figure 6.9 A dot represents the energy state of one hydrogen atom in a group of atoms. Energy is exchanged with the surroundings through photon emission and absorption.

6.2.5 The effects of temperature

If an atom absorbs radiation coming from elsewhere, or absorbs energy from being jostled by a neighboring atom, it may jump to a higher energy level, and eventually emit photons and drop again to a lower level (Figure 6.9). In thermal equilibrium with its surroundings, an atom's energy continually goes up and down, but with an average energy determined by the temperature of the collection of atoms.

In Figure 6.9 each dot represents one hydrogen atom in a container of atomic hydrogen. If three dots are on a line, this means three atoms are in this particular state. Each hydrogen atom has the same allowed energy levels, and at any particular instant some of the atoms will be in the ground state, some in the first excited state, etc.

For a particular temperature, at any instant there is a certain probability that one particular atom will be in state 1, 2, 3, etc. High temperature is associated with a significant probability that the various atoms are in various states, from the ground state up to a high energy level. Very low temperature is associated with small probability that atoms are in any state other than the ground state (Figure 6.10). If the temperature is near absolute zero, the atoms will nearly always be found in the ground state. As we will see in Chapter

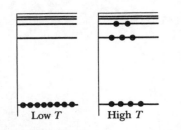

Figure 6.10 At a low temperature (system on the left) all atoms are in low energy states. At a high temperature (system on the right) some atoms are in higher energy states.

10 on "statistical mechanics," the population of a level whose energy is E above the ground state is proportional to $\exp(-E/kT)$, where k is the Boltzmann constant $(1.4\times10^{-23}\,\text{J/K})$ and T is the absolute temperature in kelvins.

? Can you observe an emission spectrum if the material is very cold? Can you observe an absorption spectrum if the material is very cold?

Very cold matter will normally be in the ground state and will not emit, but an absorption spectrum will show gaps (dark lines) corresponding to photons being absorbed and lifting an atom above its ground state. (In order to observe an absorption spectrum you would of course need to use a beam of photons whose energies are sufficiently high to raise the system energy at least to the first excited state above the ground state.)

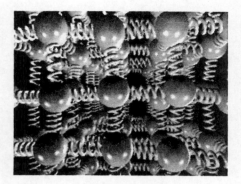

Figure 6.11 Model of a solid as a network of balls and springs.

6.3 Quantized vibrational energy levels

For concreteness we have described energy quantization for the electronic states of atomic hydrogen. Many other kinds of microscopic systems have quantized energy. We will next discuss vibrational energy quantization in atomic oscillators.

6.3.1 Vibrational energy in a classical oscillator

We have modeled a solid as a network of classical harmonic oscillators, balls and springs (Figure 6.11). Although this classical model explains many aspects of the behavior of solids, the predictions it makes regarding the thermal properties of solids do not correspond to experimental measurements at low temperatures. The classical model also fails to explain some aspects of the interaction of light with solid matter. To explain these phenomena we need to make our model of a solid more sophisticated, by incorporating the quantum nature of these atomic oscillators into our model.

The vibrational energy of a classical (non-quantum) spring-mass system can have *any* value (the gray region in Figure 6.12). Equivalently, the amplitude A (maximum stretch) can have *any* value. We can express the energy as $E = \frac{1}{2}k_sA^2$, because at the maximum stretch A the speed is zero, so the kinetic energy is zero, and the energy $K+U$ is just the potential energy at that instant. Classically, the amplitude A can have any value whatsoever, and the energy is continuous, not quantized.

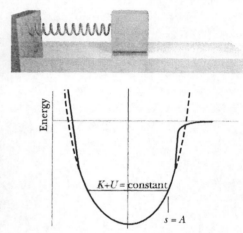

Figure 6.12 A classical harmonic oscillator (spring-mass system) can vibrate with any amplitude, and hence can have any energy. $K+U$ of the system could have any value in the gray region.

6.3.2 Vibrational energy in a quantum mechanical oscillator

Historically, the first hint that energy might be quantized came from an attempt by Max Planck in 1900 to explain some puzzling features of the spectrum of light emitted through a small hole in a hot furnace (so-called "blackbody radiation"). Planck found that he could correctly predict the properties of blackbody radiation if he assumed that transfers of energy in the form of electromagnetic radiation were quantized—that is, an energy transfer could take place only in fixed amounts, rather than having a continuous range of possible values.

Later, with the full development of quantum mechanics, it was recognized that one could model an atom in the wall of the furnace as a tiny spring-mass system whose energy levels are discrete (Figure 6.13). Classically, the energy of a mass on a spring is related to the amplitude of its oscillation ($E = \frac{1}{2}k_sA^2 + U_0$), so saying that a spring-mass oscillator can only have certain discrete amounts of energy is equivalent to saying that only certain amplitudes are allowed. (To see this, note that we have $K + U = \frac{1}{2}p^2/m + \frac{1}{2}k_ss^2 + U_0 = \frac{1}{2}k_sA^2 + U_0$, since when the stretch s is equal to the amplitude A, the momentum is zero.)

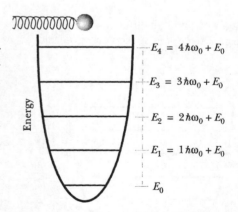

Figure 6.13 A quantum harmonic oscillator can have only certain energies, indicated by the horizontal lines on the graph. This idealized potential energy well represents the situation near the bottom of a real well (which is at a negative energy U_0).

By solving the quantum mechanical Schrödinger equation, one can show that the spacing ΔE between the allowed, quantized energies has the following value:

$$\Delta E = \hbar \sqrt{\frac{k_s}{m}}, \text{ or } \Delta E = \hbar \omega_0, \text{ where } \omega_0 = \sqrt{\frac{k_s}{m}}$$

The quantity k_s is the interatomic spring constant, which we estimated in Chapter 3, m is the mass, and $\omega_0 = \sqrt{k_s/m}$ is the angular frequency of the oscillator (in radians per second), as we learned in our studies in Chapter 3. The quantity $\hbar$, pronounced "h-bar," is equal to h, "Planck's constant," divided by 2π. Planck's constant h has an astoundingly small value:

PLANCK'S CONSTANT

$$h = 6.6 \times 10^{-34} \text{ joule} \cdot \text{second}$$

$$\hbar = \frac{h}{2\pi} = 1.05 \times 10^{-34} \text{ joule} \cdot \text{second}$$

The quantized energies for an atomic harmonic oscillator are these:

QUANTIZED VIBRATIONAL ENERGY LEVELS

$$E_N = N\hbar\omega_0 + E_0, \ N = 0, 1, 2, \text{ etc., where } \omega_0 = \sqrt{\frac{k_s}{m}}$$

In other words, the energy can have only the following values:
$$E = 1\hbar\omega_0 + E_0, \ 2\hbar\omega_0 + E_0, \ 3\hbar\omega_0 + E_0, \text{ etc.}$$

It is an unusual and special feature of the quantized harmonic oscillator that the spacing between adjacent energy levels is the same for all levels ("uniform spacing"). This is quite different from the uneven spacing of energy levels in the hydrogen atom.

It is important to understand that in both the quantum and the classical picture, the angular frequency $\omega_0 = \sqrt{k_s/m}$ of a particular oscillator is fixed and is the same whether the energy of the oscillator is large or small. A useful way to think about the situation is to say that the frequency is determined by the parameters k_s and m, while the energy is determined by the initial conditions (or equivalently by the amplitude A). In the quantum world, it is as though the amplitude can have only certain specific values.

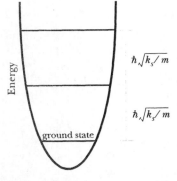

Figure 6.14 A quantum oscillator with a large k_s has widely spaced energy levels.

Ex. 6.6 Show that $(\hbar\sqrt{k_s/m})$ has units of joules.

Ex. 6.7 Suppose a collection of quantum harmonic oscillators occupies the lowest 4 energy levels, and the spacing between levels is 0.4 eV. What is the complete emission spectrum for this system? That is, what photon energies will appear in the emissions? Include all energies, whether or not they fall in the visible region of the electromagnetic spectrum.

6.3.3 Spring stiffness k_s

The energy quantum depends on the stiffness k_s of the "spring." A stiffer spring (larger value of k_s, Figure 6.14) means a steeper, narrower potential curve and larger spacing between the allowed energies. A smaller value of k_s (Figure 6.15) means a shallower, wider potential curve and smaller spacing between the allowed energies.

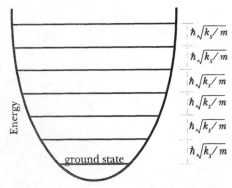

Figure 6.15 A quantum oscillator with a small k_s has closely spaced energy levels.

Ex. 6.8 You may have measured the properties of a simple spring-mass system in the lab. Suppose you found $k_s = 0.7$ N/m and $m = 0.02$ kg, and you observed an oscillation with an amplitude of 0.2 m. What is the approximate value of N, the "quantum number" for this oscillator? (That is, how many levels above the ground state is this oscillator?)

Ex. 6.9 Summarize the differences and similarities between different energy levels in a quantum oscillator. Specifically, for the first two levels in Figure 6.14, compare the angular frequency $\sqrt{k_s/m}$, the amplitude A, and the kinetic energy K at the same value of s. (In a full quantum-mechanical analysis the concepts of angular frequency and amplitude require reinterpretation. Nevertheless there remain elements of the classical picture. For example, larger amplitude corresponds to a higher probability of observing a large stretch.)

Ex. 6.10 What is the energy of the photon emitted when a harmonic oscillator with stiffness k_s and mass m loses 3 of its quanta of energy?

Ex. 6.11 Use typical values for the interatomic spring stiffness k_s and mass m to calculate the energy of the photon in Exercise 6.10. To what region of the spectrum (x-ray, visible, microwave, etc.) does the photon belong? (See Figure 6.2 on page 214.)

6.3.4 The ground state of a quantum oscillator

The lowest possible quantized energy E_0 is not zero relative to the bottom of the potential curve, but is $\frac{1}{2}(\hbar\sqrt{k_s/m})$ above the bottom of the well. This is closely related to the "Heisenberg uncertainty principle." When applied to the harmonic oscillator, this principle implies that if you knew that the mass on the spring were exactly at the center of the potential, its speed would be completely undetermined, whereas if you knew that the speed of the mass were exactly zero, the position of the mass could be literally anywhere! The minimum energy $\frac{1}{2}(\hbar\sqrt{k_s/m})$ above the bottom of the well can be viewed as a kind of a compromise, with neither the speed nor the position fixed at zero. The minimum energy is called the ground state of the system.

In most of our work with the quantized oscillator we will measure energy from the ground state rather than from the bottom of the potential well, so the offset from the bottom of the well won't come into our calculations.

6.3.5 Energy levels for the interatomic potential energy

If we model the atoms in a solid as simple harmonic oscillators, whose potential energy curve is a parabola, then the energy levels of these systems are uniformly spaced. However, if we use a more realistic interatomic potential energy such as is shown in Figure 6.16, the energy levels of the system follow a more complex pattern. The lowest energy levels are nearly evenly spaced, corresponding to the fact that for small amplitude the bottom of the potential energy curve can be fit rather well by a curve of the form $\frac{1}{2}k_s s^2 + U_0$. However, for large energies the energy spacing is quite small. And of course there are no bound states for energies larger than the binding energy. The spacing of the vibrational energy levels of a diatomic molecule (Figure 6.17)

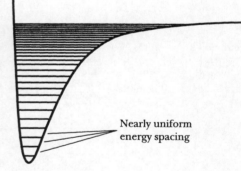

Nearly uniform energy spacing

Figure 6.16 Energy levels for the interatomic potential energy, which describes the interaction of two neighboring atoms. This diagram is not to scale; the number of vibrational energy levels in the uniform region may be very large.

Figure 6.17 The vibrational energy of a diatomic molecule is quantized.

typically corresponds to photons in the infrared region of the spectrum, as you found in Exercise 6.11.

Energy spacing in various systems

Most quantum objects have energy level spacings that decrease as you go to higher levels. The harmonic oscillator potential energy is unusual in leading to uniform energy level spacing. Even less common is the situation where the energy level spacing actually increases with higher energy. One such exception is the so-called "square well" with very steep sides and a flat bottom. The energy levels in a square well actually get farther and farther apart for higher energies.

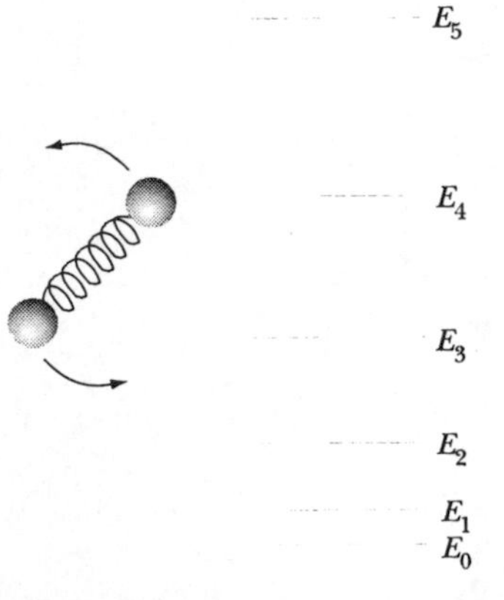

6.3.6 Rotational energy levels

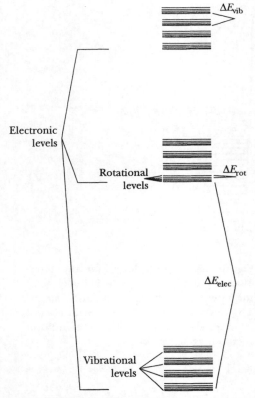

Figure 6.18 Rotational kinetic energy levels of a diatomic molecule.

A molecule in the gas phase has not only electronic energy and vibrational energy, but rotational energy as well. Like electronic and vibrational energy levels, rotational energy levels are also discrete—only certain rotational energies are permitted. As we will see in Chapter 9, the angular momentum of a molecule is quantized, and this quantization gives rise to discrete rotational energy levels. In this case all the energy of rotation is kinetic energy—there is no potential energy involved. Rotational energy levels for a diatomic molecule such as N_2 are shown in Figure 6.18. Note that the spacing between levels actually increases with increasing energy, which is the opposite of the trend seen in many atomic systems. Rotational energy levels are typically close together; transitions between these levels usually involve photons in the microwave region of the electromagnetic spectrum.

6.4 Case study: The diatomic molecule

Analyzing a diatomic molecule such as N_2 provides a good review of three important types of quantized energy levels: electronic, vibrational, and rotational.

Only certain configurations of the electron clouds in a diatomic molecule are possible, corresponding to discrete electronic energy levels. The spacings of the low-lying molecular electronic energy levels are typically in the range of 1 eV or more, like the spacings of the low-lying energy levels in the hydrogen atom.

For each major configuration of the electron clouds, many different vibrational energy states are possible, with energy spacings of the order of 10^{-2} eV (see Exercise 6.11).

For each major configuration of the electron clouds, and each vibrational energy state, many different rotational kinetic energy states are possible, with energy spacings of the order of 10^{-4} eV, as we will see in Chapter 9.

These three different families of energy levels, with their very different energy level spacings, give rise to energy "bands" and a distinctive "band" spectrum of emitted photons. See Figure 6.19, which is not to scale due to the large differences in level spacing for the three kinds of quantized energy levels. An emitted photon can have the energy of an electronic transition plus a smaller-energy vibrational transition plus an even smaller-energy rotational transition. The various possible energy differences give rise to "bands" in the observed emission spectrum.

6.5 Nuclear energy levels

Just as atoms have discrete states corresponding to allowed configurations of the electron cloud, so nuclei have discrete states corresponding to allowed configurations of the nucleons (protons and neutrons). An excited nucleus can drop to a lower state with the emission of a photon. The energy

Figure 6.19 The energy bands of a diatomic molecule (not to scale). This energy level structure gives rise to a band spectrum of emitted photons.

spacing of nuclear levels is very large, of the order of 10^6 eV (1 MeV), so the emitted photons have energies in the range of millions of electron-volts. Such energetic photons are called "gamma rays."

Ex. 6.12 A gamma ray with energy greater than or equal to 2.2 MeV can dissociate a deuteron (the nucleus of deuterium, heavy hydrogen) into its constituents, a proton and a neutron. What is the nuclear binding energy of the deuteron?

6.6 Hadronic energy levels

Atoms have energy levels associated with the configurations of electrons in the atom, and nuclei have energy levels associated with the configurations of nucleons in the nucleus. Hadrons are particles made of quarks. In a simple model we represent baryons (particles like protons) as a combination of three quarks, and mesons (particles like pions) as a combination of a quark and an antiquark. Different configurations of quarks correspond to different energy levels of the multiquark system. These different configurations are the different hadrons we observe, and the corresponding energy levels are the observed rest energies of these particles. The energy spacing is of the order of hundreds of MeV, or about 10^8 eV. (Leptons, which include electrons, muons, and neutrinos, are not made of quarks.)

An example of a hadron is the Δ^+ particle, which has a rest energy of 1232 MeV. It can be considered an excited state of the three-quark system whose ground state is the proton, whose rest energy is 938 MeV. The Δ^+ particle can decay into a proton with the emission of a very high energy photon. However, this "electromagnetic" decay occurs in only about 0.5% of the decays. In most cases the Δ^+ decays into a proton plus a pion, a strong-interaction decay which has a much higher rate, corresponding to the much stronger interaction.

An excited electronic state of hydrogen is still called hydrogen, but for historical reasons we continue to give excited quark states distinctive names, such as Δ^+.

Ex. 6.13 What is the approximate energy of the photon emitted when the Δ^+ decays electromagnetically?

6.7 Comparison of energy level spacings

Here is a summary of typical energy scales for various quantum objects:

Type of state	Typical energy level spacing
hadronic	10^8 eV
nuclear	10^6 eV
electronic (atoms, molecules)	1 eV
vibrational (molecules)	10^{-2} eV
rotational (molecules)	10^{-4} eV

It is important to distinguish among hadronic, nuclear, electronic, vibrational, rotational, and other kinds of quantized energy situations. We can generalize the main issues to any kind of generic quantum object. The key features of all these different kinds of objects are these:
- discrete energy levels
- discrete emissions whose energies equal differences in energy levels

Some of the homework problems at the end of the chapter ask you to analyze generic objects, independent of the details of what gave rise to the particular discrete energy levels.

6.8 *Case study: How a laser works

A laser offers an interesting case study of the interaction of light and atomic matter. We will give a basic introduction to lasers, but we can only scratch the surface of a vast topic which continues to be an active field of research. Laser action depends on a process called "stimulated emission" and on creating an "inverted population" of quantum states, which can lead to a chain reaction and produce a special kind of light, called "coherent" light.

Stimulated emission

If an atom can be placed into an excited state, that is, a state of higher energy than the ground state, normally it drops quickly to a lower energy state and emits a photon in the process. For example, in a neon sign high-energy electrons collide with ground-state neon atoms and raise them to an excited state. An excited neon atom quickly drops back to the ground state, emitting a photon whose energy is the difference in energy between the two quantized energy levels, and you see the characteristic orange color corresponding to these photons.

We call this emission process "spontaneous emission." Spontaneous emission happens at an unpredictable time. All that quantum mechanics can predict is the *probability* that the excited atom will drop to the ground state in the next small time interval. It is truly not possible to predict the exact time of the emission, just as it is impossible to predict exactly when a free neutron will decay into a proton, electron, and antineutrino.

A collection of excited atoms such as those in a neon sign emit spontaneously, at random times. In the full quantum-mechanical description of light, light has both the properties of a particle (the photon picture) and the properties of a wave; see Section 6.9. Emissions at random times have random phases in a wave description: sine, cosine, and all phases between a sine and a cosine. We say that such light is "incoherent."

"Stimulated emission" of light is a different process which produces coherent light. Consider an atom in an energy level E above the ground state. A photon of this same energy E can interact with the atom in a remarkable way, immediately causing the atom to drop to the ground state with the emission of a second, "clone" photon. Where there had been one photon of energy E, there are now two photons of exactly the same energy E, and in a wave description they have exactly the same phase (Figure 6.20). That is, the two waves are both sines, or both cosines, or both with exactly the same phase somewhere between a sine and a cosine. Such light is called coherent light, and it has valuable properties for such applications as making holograms.

Chain reaction

The phenomenon of stimulated emission can be exploited to create a chain reaction. Start with one photon of the right energy E to match the energy difference between the ground state and an excited state of the atoms in the

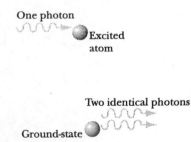

One photon

Excited atom

Two identical photons

Ground-state atom

Figure 6.20 Stimulated emission produces two identical photons from one.

system. If that photon causes stimulated emission from an excited atom, there are now two identical photons, both capable of triggering stimulated emission of other excited atoms, which yields four identical photons, then 8, 16, 32, 64, 128, 256, 512, 1024, etc.

However, the chain reaction won't proceed if all of the photons escape from the system, so some kind of containment mechanism is required. In a common type of gas laser the gas is in a long tube with mirrors at both ends (Figure 6.21). Photons that happen to be going in the direction of the tube are continually reflected back and forth through the tube, which increases the probability that they will interact with excited atoms and cause stimulated emission. Since the stimulated-emission photons are clones, they too are headed in the direction of the tube and they too are likely to interact. One of the mirrors is deliberately made to be an imperfect reflector, so a fraction of the photons leaks out and provides a coherent beam of light.

Unfortunately, the stimulated-emission photons also have just the right energy E to be absorbed by an atom in the ground state, in which case they can no longer contribute to the chain reaction. For this reason we need an "inverted population," with few atoms in the ground state to absorb the photons and stop the chain reaction. We discuss this issue next.

Inverted population

As we have discussed earlier in this chapter, in a normal collection of atoms more atoms are in the ground state than in any other state, and the number of atoms decreases in each successively higher state. In fact, for a collection of atoms in thermal equilibrium the fraction of atoms in a particular energy state decreases exponentially with the energy above the ground state, as we will see in Chapter 10.

By clever means it is possible to invert this population scheme, so that there are more atoms in an excited state than in the ground state (this is not a state of thermal equilibrium). An inverted population is crucial to laser action, so that stimulated emission with cloning of photons will dominate over simple absorption of photons by atoms in the ground state.

There are many different schemes for creating an inverted population. For example, the scheme used in the common helium-neon laser is to "pump" atoms to a high energy level (Figure 6.22). There are diverse energy input schemes, including electric discharges and powerful flashes of ordinary light. Atoms in this high energy level rapidly drop to an intermediate energy level, higher than the ground state. This intermediate state happens to have a rather long lifetime for spontaneous emission, so that it is possible to sustain an inverted population, with more atoms in the intermediate state than in the ground state. Stimulated emission drives these excited atoms down to the ground state, from which they are again pumped up to the high-energy state. Spontaneous emission from the high-energy state to the intermediate state maintains the inverted population.

6.9 *Wavelength of light

In this chapter we have emphasized the particle model of light, in terms of photons and their energy. For completeness we will mention a connection between the particle model and the wave model of light. Light consisting of photons whose energy is E can also be treated as a wave, and the energy E and wavelength λ are related as follows:

ENERGY AND WAVELENGTH OF LIGHT

$$E_{\text{photon}} = \frac{hc}{\lambda_{\text{light}}}$$

Figure 6.21 A gas laser with mirrors at the ends. Photons leak through one of the mirrors.

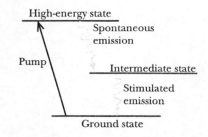

Figure 6.22 A three-state scheme for sustaining an inverted population.

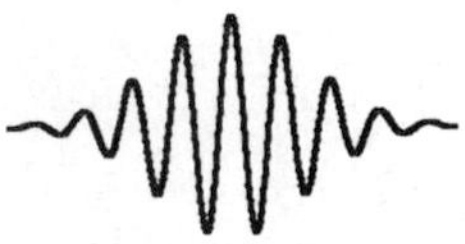

Figure 6.23 A wave packet is a travelling wave of finite spatial extent.

where h is Planck's constant. A model for light that incorporates both the features of a wave and a localized particle is a "wave packet," a wave whose magnitude is nonzero only in a small region of space (Figure 6.23). We will return to this topic in Chapter 24.

6.10 Summary

Fundamental Physical Principles

At the microscopic level the energy of bound states is discrete, not continuous. (Unbound states have a continuous range of energies.)

New concepts

Quantized systems can have only certain energies. They can emit or absorb photons when they change energy levels.

The temperature is a measure of the average energy of a collection of objects, so high temperature is associated with significant probability of finding particles in any of many states, from the ground state up to a high energy level. Very low temperature is associated with little probability of the particles being in any state other than the ground state.

Results

Discrete hydrogen atom energy levels:

$E_N = -(13.6 \text{ eV})/N^2$, where N is a nonzero positive integer (1, 2, 3...)

Discrete harmonic oscillator energy levels:

$$E_N = N\hbar\omega_0 + E_0, \quad N = 0, 1, 2, \text{ etc., where } \omega_0 = \sqrt{\frac{k_s}{m}}$$

Planck's constant $h = 6.6 \times 10^{-34}$ joule $\cdot$ second

$$\hbar = \frac{h}{2\pi} = 1.05 \times 10^{-34} \text{ joule} \cdot \text{second}$$

$1 \text{ eV} = 1.6 \times 10^{-19}$ J

Photon energy and wavelength: $E_{\text{photon}} = \dfrac{hc}{\lambda_{\text{light}}}$

6.11 Example problem: Observed photon emission

You observe photon emissions from a collection of quantum objects, each of which is known to have just four quantized energy levels. The collection is kept at a high temperature, and you detect emitted photons using a detector sensitive to photons in the energy range from 2.5 eV to 30 eV. With this detector, you observe photons emitted with energies of 3 eV, 6 eV, 8 eV, and 9 eV, but no other energies.

(a) It is known that the ground state energy of each of these objects is −10 eV. Propose two possible arrangement of energy levels that are consistent with the experimental observations. Explain in detail, using diagrams.

(b) You obtain a second detector that is sensitive to photon energies in the energy range from 0.1 eV to 2.5 eV. What additional photon energies do you observe to be emitted? Explain briefly.

(c) You cool the collection of objects to a very low temperature and send a beam of photons with a wide range of energies through the material. Using both detectors to determine the absorption spectrum, what are the photon energies of the dark lines? How can this information be used to choose between your two proposed energy level schemes? Explain briefly.

Solution

Fundamental principle: the energy principle.
System: some unknown object in an excited energy state
Initial state: object in excited state
Final state: object in lower energy state, plus photon
Energy principle: $E_{\text{obj,i}} = E_{\text{obj,f}} + E_{\text{photon}}$
So the energy of any emitted photon must be equal to the difference in energy between an excited state and a lower energy state.

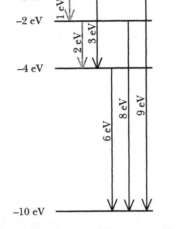

(a) Figure 6.24 shows a possible energy-level scheme that fits the emission data. We found this scheme essentially by trial and error. We knew the energy differences between levels, so we tried various arrangements. The arrows are labeled with the energies of photons emitted in transitions between the two states indicated. The 1 eV and 2 eV emissions would not be detected by a detector that is sensitive to the photon energy range 2.5 eV to 30 eV.

Figure 6.25 shows another possible scheme that fits the emission data. This is somewhat less likely, because it is more common to find the energy level spacing decreasing at higher energies, as shown in Figure 6.24.

Figure 6.24 A possible energy-level scheme that fits the emission data. Lighter arrows indicate photons not detectable by this detector.

(b) With this detector we observe the 1 eV and 2 eV photon emissions.

(c) At very low temperature almost all of the objects are in the ground state. If the energy levels are as shown in Figure 6.24, the dark lines will be at 6 eV, 8 eV, and 9 eV, corresponding to absorption in the ground state. If the energy levels are as shown in Figure 6.25, the dark lines will be at 1 eV, 3 eV, and 9 eV. Therefore the low temperature absorption spectrum allows us to distinguish between the two possible energy level schemes.

Comment: Since this was an energy analysis, we didn't need to know what kind of object this was, or whether the energy levels were electronic, vibrational, or rotational energy levels.

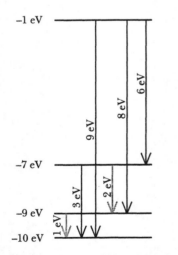

Figure 6.25 Another possible energy-level scheme that fits the emission data.

6.12 Review questions

Energy quantization

RQ 6.1 If you double the amplitude, what happens to the frequency in a classical (non-quantum) harmonic oscillator? In a quantum harmonic oscillator?

RQ 6.2 If you try to increase the amplitude of a quantum harmonic oscillator by adding an amount of energy

$$\frac{1}{2}\hbar\sqrt{\frac{k_s}{m}}$$

the amplitude doesn't increase. Why not?

RQ 6.3 When starlight passes through a cold cloud of hydrogen gas, some hydrogen atoms absorb energy, then reradiate it in all directions. As a result, a spectrum of the star shows dark absorption lines at the energies for which less energy from the star reaches us. How does the spectrum of dark absorption lines for very cold hydrogen differ from the spectrum of bright emission lines from very hot hydrogen?

6.13 Problems

Problem 6.1 Determining energy levels
Suppose we have reason to suspect that a certain quantum object has only three quantum states. When we excite such an object we observe that it emits electromagnetic radiation of three different energies: 2.48 eV (green), 1.91 eV (orange), and 0.57 eV (infrared). Propose two possible energy-level schemes for this system.

Problem 6.2 Emission and absorption
A certain material is kept at very low temperature. It is observed that when photons with energies between 0.2 and 0.9 eV strike the material, only photons of 0.4 eV and 0.7 eV are absorbed. Next the material is warmed up so that it starts to emit photons. When it has been warmed up enough that 0.7 eV photons begin to be emitted, what other photon energies are also observed to be emitted by the material? Explain briefly.

Problem 6.3 Predicting a spectrum
Suppose a hypothetical object has just four quantum states, with the energies shown on the potential energy diagram in Figure 6.26.

(a) Suppose that the temperature is high enough that in a material containing many such objects, at any instant some objects are found in all of these states. What are all the energies of photons that could be strongly emitted by the material? (In actual quantum objects there are often "selection rules" that forbid certain emissions even though there is enough energy; assume that there are no such restrictions here.)

(b) If the temperature is very low and electromagnetic radiation with a wide range of energies is passed through the material, what will be the energies of photons corresponding to missing ("dark") lines in the spectrum? (Assume that the detector is sensitive to a wide range of photon energies, not just energies in the visible region.)

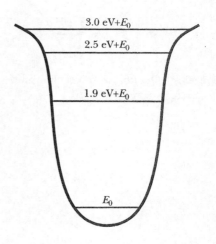

Figure 6.26 Potential energy diagram for Problem 6.3.

Problem 6.4 Emission from hot metal

A hot bar of iron glows a dull red. Using our simple model of a solid, answer the following questions, explaining in detail the processes involved. The mass of one mole of iron is 56 grams.

(a) What is the energy of the lowest-energy spectral emission line? Give a numerical value.

(b) What is the approximate energy of the highest-energy spectral emission line? Give a numerical value.

(c) What is the quantum number of the highest-energy occupied state?

(d) Predict the energies of two other lines in the emission spectrum of the glowing iron bar.

(Note: our simple model is too simple—the actual spectrum is more complicated. However, this simple analysis gets at some important aspects of the phenomenon.)

Problem 6.5 Potential energy in a molecule

(a) Figure 6.27 is a graph of potential energy vs. interatomic distance for a particular molecule. What is the direction of the associated force at location *A*? At location *B*? At location *C*? Rank the magnitude of the force at locations *A*, *B*, and *C* (that is, which is greatest, which is smallest, are any of these equal to each other). For the energy level shown on the graph, draw a line whose height is the kinetic energy when the system is at location *D*.

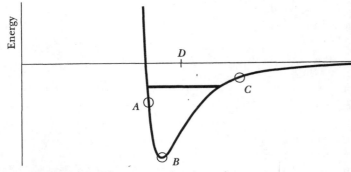

Figure 6.27 Potential energy vs. interatomic distance for a molecule (Problem 6.5a).

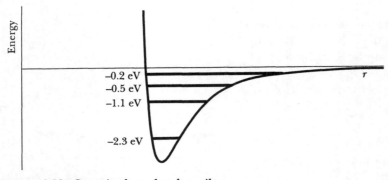

Figure 6.28 Quantized molecular vibrational energies (Problem 6.5b).

(b) Figure 6.28 shows all of the quantized energies (bound states) for one of these molecules. The energy for each state is given on the graph, in electron-volts (1 eV = 1.6×10^{-19} J). How much energy is required to break a molecule apart, if it is initially in the ground state? (Note that the final state must be an unbound state; the unbound states are not quantized.)

(c) If the temperature is high enough, in a collection of these molecules there will be at all times some molecules in each of these states, and light will be emitted. What are the energies in electron-volts of the emitted light?

(d) The "inertial" mass of the molecule is the mass that appears in Newton's second law, and which determines how much acceleration will result from applying a given force. Compare the inertial mass of a molecule in the top energy state and the inertial mass of a molecule in the ground state. If there is a difference, briefly explain why and calculate the difference. If there isn't a difference, briefly explain why not.

Problem 6.6 A microscopic oscillator

Consider a microscopic spring-mass system whose spring stiffness is 50 N/m, and the mass is 4×10^{-26} kg.

(a) What is the smallest amount of vibrational energy that can be added to this system?

(b) What is the difference in mass (if any) of the microscopic oscillator between being in the ground state and being in the first excited state?

(c) In a collection of these microscopic oscillators, the temperature is high enough that the ground state and the first three excited states are occupied. What are the possible energies of photons emitted by these oscillators?

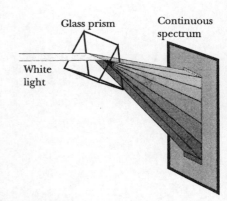

Figure 6.29 Spectrum of white light.

Problem 6.7 The spectrum of atomic hydrogen

The eye is sensitive to photons with energies in the range from about 1.8 eV, corresponding to red light, to about 3.1 eV, corresponding to violet light. White light is a mixture of all the energies in the visible region. If you shine white light through a slit onto a glass prism, you can produce a rainbow spectrum on a screen, because the prism bends different colors of light by different amounts (Figure 6.29).

If you replace the source of white light with an electric-discharge lamp containing excited atomic hydrogen, you will see only a few lines in the spectrum, rather than a continuous rainbow (Figure 6.30).

Predict how many lines will be seen in the visible spectrum of atomic hydrogen, and specify the atomic transitions that are responsible for these lines, given that the energies of the quantized states in atomic hydrogen are given by $E_N = -(13.6 \text{ eV})/N^2$, where N is a nonzero positive integer (1, 2, 3...).

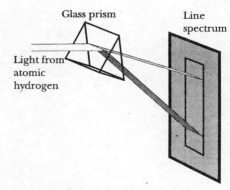

Figure 6.30 Atomic emission spectrum (Problem 6.7).

Problem 6.8 An experiment with quantum objects

Some material consisting of a collection of microscopic objects is kept at a high temperature. A photon detector capable of detecting photon energies from infrared through ultraviolet observes photons emitted with energies of 0.3 eV, 0.5 eV, 0.8 eV, 2.0 eV, 2.5 eV, and 2.8 eV. These are the only photon energies observed.

(a) Draw and label a possible energy-level diagram for one of the microscopic objects, which has 4 bound states. On the diagram, indicate the transitions corresponding to the emitted photons. Explain briefly.

(b Would a spring-mass model be a good model for these microscopic objects? Why or why not?

(c) The material is now cooled down to a very low temperature, and the photon detector stops detecting photon emissions. Next a beam of light with a continuous range of energies from infrared through ultraviolet shines on the material, and the photon detector observes the beam of light after it passes through the material. What photon energies in this beam of light are observed to be significantly reduced in intensity ("dark absorption lines")? Explain briefly.

6.14 Answers to exercises

6.1 (page 213) 13.6 eV

6.2 (page 214) About 2.5 eV

6.3 (page 214) About 3.5×10^{21} photons per square meter per second.

6.4 (page 215) Six different photon energies, corresponding to transitions from 4 to 3, 4 to 2, 4 to 1, 3 to 2, 3 to 1, and 2 to 1.

6.5 (page 216) Just one dark line at 2.4 eV

6.7 (page 218) 0.4 eV (three transitions), 0.8 eV (two transitions), and 1.2 eV (one transition)

6.8 (page 219) $N \approx 10^{31}$. A system in a very high quantum state behaves like a classical system.

6.9 (page 219) Upper level has same ω, larger A, larger K at same s.

6.10 (page 219) $3\hbar \sqrt{\dfrac{k_s}{m}}$

6.11 (page 219) About 0.05 eV

(assuming $k_s = 20$ N/m and $m = 20 \cdot 2 \times 10^{-27}$ kg); this is in the infrared region of the spectrum (see Figure 6.2 on page 214)

6.12 (page 221) 2.2 MeV

6.13 (page 221) 294 MeV

Chapter 7

Multiparticle Systems

Chapter 7

Multiparticle Systems

At times we have treated a block of metal as though it were a single particle. At other times we have modeled a block as being made up of a huge number of atoms connected by "springs" and in continual motion around their equilibrium positions.

When is it valid to treat a complex, multiparticle system as though it were one big particle? Why is this *ever* valid? In this chapter we will look in detail at multiparticle systems.

7.1 The motion of a multiparticle system

There are some major complications in applying the momentum principle to systems consisting of many particles. Suppose you pull on a thin wire that is attached to a metal block, and you accelerate the block across a low-friction surface (Figure 7.1).

? Actually, you are not touching the block and are not exerting a force on the atoms in the block. How do these atoms know that they should accelerate?

When you first start pulling on your end of the wire, you very slightly lengthen the interatomic bonds between neighboring atoms in the nearby section of the wire (Figure 7.2). As a result of your pulling, these atoms stretch the bonds of their neighbors, and very quickly this new interatomic bond-stretching propagates all the way to the other end of the wire, and the whole wire is then in tension.

This process is very fast, but it is not instantaneous. These changes in the average lengths of the interatomic bonds propagate at the speed of sound in the material. (See page 94 in Chapter 3.)

Eventually the interatomic bonds at the left end of the wire will be stretched. The left-most atoms, which are attached to the metal block, will pull on atoms in the block. Now the block itself begins to stretch. As with the wire, initially, only those atoms of the block that are in direct contact with the wire are pulled to the right, but quickly (at the speed of sound in the material that the block is made of) these effects propagate throughout the block. Eventually it must be the case that all of the interatomic bonds in the block are stretched, but not equally.

Consider a slice of the block, indicated by dashed lines in Figure 7.3. Since this slice is accelerating to the right, the interatomic forces acting on the right face of this slice (F_1) must be larger than those on the left face (F_2). This requires that the interatomic bonds on the right be stretched more than those on the left. While the block accelerates to the right, the pattern of strain in the block must look something like Figure 7.4 (strain is $\Delta L/L$, or fractional stretch). Actually, even this description is oversimplified, because there will be maximum strain near the point where the wire is attached, and the lines of strain will be somewhat curved.

For a metal block, the strain is normally too small to see with the unaided eye. Nevertheless, the situation is pretty complicated! And there is a similar gradient of strain in the wire. There is also random thermal motion of the

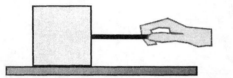

Figure 7.1 You pull on a wire attached to a block.

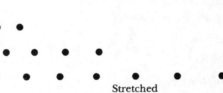

Figure 7.2 Strain in the wire increases with time.

t_0

t_1

t_2

t_3

Unstretched   Stretched

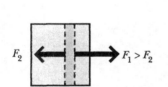

F_2 ⟵   ⟶ $F_1 > F_2$

Figure 7.3 Forces acting on a slice of the block.

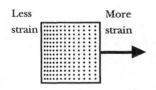

Less   More
strain   strain

Figure 7.4 Pattern of strain in a block accelerating to the right.

atoms within the block. Given all this complexity, what hope is there of treating the block as though it were just one big, rigid particle?

Ex. 7.1 If you let go and allow the block to coast along at nearly constant speed, what must happen to the strain throughout the block? What is the mechanism for that happening at the atomic level?

Energy in multiparticle systems

Just as there are great complexities in dealing with forces in multiparticle systems, so also there are puzzling aspects of work and energy. For example, when you jump upwards the floor pushes up on your feet, yet this force does no work, because there is no displacement of the point of contact (Figure 7.5). However, it is clear that your kinetic energy has increased. Evidently there are some subtle issues here that we need to address.

Figure 7.5 The floor does no work on the jumper, yet K increases.

7.2 The momentum principle for multiparticle systems

In Chapter 2 (page 43) we derived an equation that predicts the overall motion of a complex multiparticle system, in terms of its total momentum $\vec{P}_{tot}$:

THE MOMENTUM PRINCIPLE FOR MULTIPARTICLE SYSTEMS

$$\frac{d\vec{P}_{tot}}{dt} = \vec{F}_{net,ext}$$

The basis for this result is that internal forces between pairs of particles cancel for gravitational and electric forces (reciprocity; Newton's third law). This result is what makes it possible for many purposes to treat a complex system as though it were a simple point particle. A concept that simplifies analyses of this kind is the "center of mass," which we discuss next.

7.2.1 Center of mass

For a system of particles moving slowly compared to the speed of light, with total mass M, we can show that $\vec{P}_{tot} = M\vec{v}_{cm}$, where $\vec{v}_{cm}$ is the velocity of the "center of mass" of the system. Since $d\vec{P}_{tot}/dt = \vec{F}_{net,ext}$, we can treat a complicated system as though it were a single particle of mass M, located at the center of mass of the system.

The center of mass can be thought of as a weighted average of the locations of all the particles in the system. In many cases it is simply the geometrical center of a system. For example, the center of mass of a disk or sphere or cube is the geometrical center of the object, if the object has uniform density. Let $\vec{r}_1$ be the location (relative to some arbitrary origin) of mass m_1, $\vec{r}_2$ be the location of mass m_2, etc., as shown in Figure 7.6. The center of mass is located at position $\vec{r}_{cm}$, defined like this:

CENTER OF MASS

$$\vec{r}_{cm} \equiv \frac{m_1\vec{r}_1 + m_2\vec{r}_2 + m_3\vec{r}_3 + \ldots}{m_1 + m_2 + m_3 + \ldots}$$

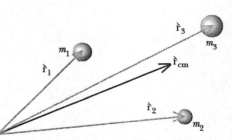

Figure 7.6 Location of the center of mass of the three-particle system.

Let's see whether this makes sense as a definition for the location of the center of mass. We'll consider several simple concrete situations and check that this definition yields results that are consistent with what we would intuitively think of as the center of mass.

? Consider the case where there are three masses very close to each other. Does the definition

$$\vec{r}_{cm} \equiv \frac{m_1\vec{r}_1 + m_2\vec{r}_2 + m_3\vec{r}_3 + \ldots}{m_1 + m_2 + m_3 + \ldots}$$

make sense in this case?

In this case, $\vec{r}_1 = \vec{r}_2 = \vec{r}_3$, and the definition of the location of the center of mass leads to the result that $\vec{r}_{cm} = \vec{r}_1$, which makes sense. If all three masses are located at the origin, then according to the definition the center of mass is also at the origin, which makes sense.

? Consider the case where there are two particles, with equal masses, located at <–5, 0, 0> and <+5, 0, 0>. Does the definition make sense in that case?

In this case we find that $\vec{r}_{cm} = 0$, which makes sense; the center of mass is equidistant between the two equal masses.

? What if there is a very small mass located at <–5, 0, 0> and a very large mass located at <+5, 0, 0>. Does the definition make sense in that case?

In this case the formula yields $\vec{r}_{cm} \approx$ <+5, 0, 0>, corresponding to the fact that almost all of the total mass is concentrated at that location.

If M is the total mass of the system, we can write these two equivalent equations:

$$\vec{r}_{cm} = \frac{m_1\vec{r}_1 + m_2\vec{r}_2 + m_3\vec{r}_3 + \ldots}{M}$$

$$M\vec{r}_{cm} = m_1\vec{r}_1 + m_2\vec{r}_2 + m_3\vec{r}_3 + \ldots$$

Center of mass of several large objects

Often we need to know the location of the center of mass of a system consisting of several large objects, each of whose center of mass is known. In Figure 7.7, $\vec{R}_1$ is the location (relative to the origin) of the center of mass of the object whose total mass is M_1, etc.

For each of the large objects we have an equation for its center of mass of the form $M_1\vec{R}_1 = m_{11}\vec{r}_{11} + m_{12}\vec{r}_{12} + \ldots$, $M_2\vec{R}_2 = m_{21}\vec{r}_{21} + m_{22}\vec{r}_{22} + \ldots$, etc.

Group terms in the calculation of the location of the center of mass of the total system:

$$\vec{r}_{cm} \equiv \frac{(m_{11}\vec{r}_{11} + m_{12}\vec{r}_{12}) + \ldots + (m_{21}\vec{r}_{21} + m_{22}\vec{r}_{22}) + \ldots}{(m_{11} + m_{12} + \ldots) + (m_{21} + m_{22} + \ldots) + \ldots}$$

$$\vec{r}_{cm} = \frac{M_1\vec{R}_1 + M_2\vec{R}_2 + M_3\vec{R}_3 + \ldots}{M_1 + M_2 + M_3 + \ldots}$$

This looks just like the equation for finding the center of mass of a bunch of atoms, except that the $\vec{R}$ vectors refer to the centers of mass of large objects.

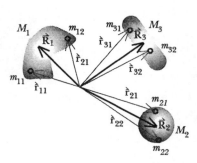

Figure 7.7 Finding the center of mass of a system consisting of several large objects.

Ex. 7.2 Determine the location of the center of mass of a "T" whose thin vertical and horizontal members have the same length L and the same mass M. Use the formal definition to find the x and y

coordinates, and check your result by doing the calculation with respect to two different origins, one at the bottom of the vertical member and one at the left end of the horizontal member.

Ex. 7.3 Relative to an origin at the center of the Earth, where is the center of mass of the Earth-Moon system? The mass of the Earth is 6×10^{24} kg, the mass of the Moon is 7×10^{22} kg, and the distance from the center of the Earth to the center of the Moon is 4×10^{8} m. The radius of the Earth is 6400 km. One can show that the Earth and Moon orbit each other around their center of mass (for example, see Problem 7.4).

7.2.2 Total momentum and center of mass motion

Now that we have introduced the concept of center of mass, we can make a useful connection between the total momentum and the motion of the center of mass. The velocity of the center of mass of a system is obtained by taking the time derivative of the definition of the position of the center of mass, where $\vec{v}_1 = d\vec{r}_1/dt$, etc. Take the important case where all the individual masses are constants. In that case, since

$$M\vec{r}_{cm} = m_1\vec{r}_1 + m_2\vec{r}_2 + m_3\vec{r}_3 + ...$$

we have this:

$$M\vec{v}_{cm} = M\frac{d\vec{r}_{cm}}{dt} = m_1\vec{v}_1 + m_2\vec{v}_2 + m_3\vec{v}_3 + ... \text{ (constant masses)}$$

But $(m_1\vec{v}_1 + m_2\vec{v}_2 + m_3\vec{v}_3 + ...)$ is the total (nonrelativistic) momentum of the system, so we have shown that

$$\vec{P}_{tot} = M\vec{v}_{cm} \text{ (if } v \ll c, \text{ and constant mass)}$$

In summary, we restate the important results:

THE MOMENTUM PRINCIPLE FOR MULTIPARTICLE SYSTEMS

$$\frac{d\vec{P}_{tot}}{dt} = \vec{F}_{net,ext} \text{ (in general)}$$

$$\vec{P}_{tot} = M\vec{v}_{cm} \text{ (if } v \ll c, \text{ and constant mass)}$$

For example, if you apply an external force F to one side of a block (through the tension in a wire, for example), the center of mass of the block moves as though that force were applied at the center of mass, producing an acceleration of the center of mass $dv_{cm}/dt = F/M$, even though the detailed motions of the many atoms in the block aren't simple (different amounts of strain in different parts of the block, thermal agitation of the atoms, etc.).

? A high diver jumps outward, then tucks into a tight ball and spins rapidly, then straightens out before hitting the water. What is the path of the diver's center of mass?

Although the diver's overall motion is very complex, the center of mass of the diver must move like that of a simple projectile! We do not yet have enough tools to analyze the diver's complicated rotation and stretching relative to the center of mass, yet we can easily analyze the motion of the center of mass.

7.2.3 Application: An ice skater pushes off from a wall

All of this may sound quite plausible, but it can lead to some rather odd consequences. Consider a woman on ice skates who pushes off from a wall with a nearly constant force F_N (normal to the wall) and moves backwards with increasing speed. Her center of mass is marked on each frame of the following sequence:

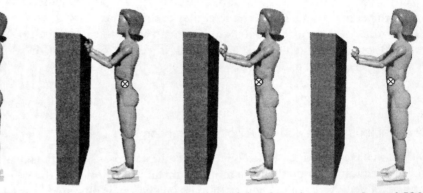

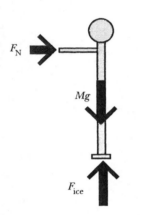

Figure 7.8 Force diagram for the skater.

What would we predict for the motion of the skater's center of mass? If M is her total mass, and if we can assume that the frictional force of the ice on the skates is negligibly small, we predict this from the force diagram (Figure 7.8):

$$\frac{d\mathbf{P}_{tot}}{dt} = F_N$$

What's odd about this simple and intuitively appealing result is that during the time that the wall exerts a force F_N on the skater's hand, her hand does not move! (Note that the work done by the force exerted by the wall is zero; more about this later.) The place where the force F_N is applied is very far away from the mathematical point (the center of mass) whose rate of change of speed we predict. It is only because the electric interactions between neighboring atoms in the skater's body obey the principle of reciprocity (Newton's third law) that these internal forces cancel and lead to a simple prediction for the motion of the center of mass:

$$\frac{d\vec{\mathbf{P}}}{dt} = \vec{\mathbf{F}}_{net,ext}$$

Note that as the skater straightens her arms, the location in her body of the center of mass shifts toward her hands a little; it is not a point fixed at some place in her body. The motion that we predict is the motion of this mathematical center of mass point.

7.2.4 Application: Pull on two hockey pucks

We'll look at another odd consequence of this multiparticle version of the momentum principle. Tie a string to the center of a hockey puck, and wrap another string around the outside of a second hockey puck, as shown in Figure 7.9. Then two people pull on the two strings so that both strings have the same tension F_T. As shown in the figure, in a time Δt hand 2 pulling the bottom puck has traveled farther, because the string has unrolled from the puck a bit.

The pucks have the same mass M. Assume that friction with the ice rink is negligible. It would be reasonable to suppose that it would make a difference where you attach the string to the puck.

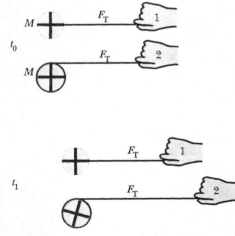

Figure 7.9 One puck is pulled by a string attached to its center. The other puck is pulled by a string wrapped around its edge, which unrolls as the puck is pulled.

? What will be the velocity (direction and magnitude) of the center of the first puck after a short time Δt? Of the center of the second puck? How do you justify your predictions?

Both pucks are subjected to exactly the same external force, so their centers of mass must move in exactly the same way. After a short time Δt, the momentum of the center of mass must be $F_T \Delta t$, in the direction of the string, for both pucks. Despite the fact that one puck is pulled from the side, the change in momentum of its center of mass must be in the direction of the net external force, so it cannot acquire a momentum component at right angles to that force. Doesn't this seem a bit odd? For example, one might have expected the second puck to have some kind of sideways motion.

Ex. 7.4 Is there any difference at all in the motion of the two pucks? Think about the motion of your hands at the other ends of the strings. Which hand does more work? Which puck has more energy? In what forms? Does the momentum principle for the motion of the center of mass predict all the details of the motion of both pucks?

Ex. 7.5 A meter stick whose mass is 300 grams lies on ice (Figure 7.10). You pull at one end of the meter stick, at right angles to the stick, with a force of 6 newtons. The ensuing motion of the meter stick is quite complicated, but what are the initial magnitude and direction of the rate of change of the momentum of the stick, $d\mathrm{P}_{\mathrm{tot}}/dt$, when you first apply the force? What is the magnitude of the initial acceleration of the center of the stick?

Figure 7.10 Pull one end of a meter stick that is lying on ice.

7.3 Energy in multiparticle systems

We have seen that we can predict the motion of a multiparticle object by treating it as if all its mass were concentrated at the center of mass, acted upon by a force that is the (vector) sum of all the external forces acting on the object. In a number of important situations we find that the energy of a multiparticle system can also be analyzed very simply.

7.3.1 Gravitational energy of a multiparticle system

Our first example is the gravitational energy of a block of mass M and the Earth, near the Earth's surface. The gravitational energy associated with the i-th atom in the block is $m_i g y_i$ (Figure 7.11).

We calculate the total gravitational energy by adding up the gravitational energy associated with the interaction of each of the N atoms with the Earth:

$$U_g = m_1 g y_1 + m_2 g y_2 + m_3 g y_3 + \ldots = g(m_1 y_1 + m_2 y_2 + m_3 y_3 + \ldots)$$

But the last expression in parentheses appears in the calculation of the y component of the location of the center of mass:

$$M y_{\mathrm{cm}} = m_1 y_1 + m_2 y_2 + m_3 y_3 + \ldots$$

Therefore for a block made up of many atoms, near the Earth's surface, we have this:

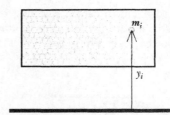

Figure 7.11 The i-th atom in the block is a distance y_i from the Earth's surface.

GRAVITATIONAL ENERGY U_g

$$U_g = M g y_{\mathrm{cm}} \text{ near the Earth's surface}$$

In other words, the gravitational energy of a block of material and the Earth, near the Earth's surface, can be evaluated as though the block were a tiny particle of mass M located at the center of mass of the block. This simple formula is not valid if the object is so large that the strength of the gravitational field is significantly different at different locations in the object, but in that case mg would not be a good approximation for the gravitational force, nor mgy a good approximation for the gravitational energy associated with each atom.

? In a calculation of gravitational energy in the voyage of a spacecraft to the Moon is it a valid approximation to treat the spacecraft as if all its mass were concentrated at one point?

This is a good approximation, because the spacecraft is very small compared to the distances to the Earth and Moon, so the gravitational force on each atom is nearly the same.

7.3.2 Kinetic energy of a multiparticle system

We can show that the total kinetic energy K_{tot} of a multiparticle system (that is, the sum of the kinetic energies of all the particles in the system) can usefully be split into two parts: the "translational" kinetic energy $K_{trans} = P_{tot}^2/(2M)$ of a fictitious particle of mass M located at the center of mass and moving with total momentum P_{tot} (which is equal to Mv_{cm}), plus the kinetic energy K_{rel} of the various atoms relative to the center of mass.

MULTIPARTICLE KINETIC ENERGY

$$K_{tot} = K_{trans} + K_{rel}, \text{ where } K_{trans} = \frac{P_{tot}^2}{2M} = \tfrac{1}{2}Mv_{cm}^2$$

For example, the kinetic energy of a rotating, vibrating diatomic molecule such as oxygen (O_2) is the sum of the kinetic energy that the molecule would have if it were not rotating or vibrating (this simple motion is called "translational" motion), plus additional kinetic energy of rotation around the center of mass and kinetic energy of vibration relative to the center of mass (Figure 7.12).

This split of the total kinetic energy into different parts sounds reasonable. A formal derivation of this result is given in section 7.6 at the end of this chapter.

Rotation and translation

As an example, suppose you spin a bicycle wheel on its axis, and hold the axle stationary, as in Figure 7.13. The spinning wheel has some kinetic energy K_{rel}, which in this case we will call K_{rot}, because the atoms are moving in rotational motion relative to the center of mass. Almost all of the mass M of the wheel is in the rim, so if the rim is traveling at a (tangential) speed v_{rel}, the kinetic energy of the wheel is approximately $K_{rot} \approx \tfrac{1}{2}Mv_{rel}^2$. (This isn't exact, because there is some mass in the spokes, and the atoms in the spokes are moving at speeds smaller than v.)

With the wheel still spinning, now move the axle with some speed v_{cm}, without touching the wheel to the ground, as in Figure 7.14. When the center of mass moves, we say that there is *translational* motion. At the top of the wheel the speed of an atom is $v_{cm} + v_{rel}$, because it shares the overall translational motion of the wheel but is also moving forward relative to the axle. At the bottom of the wheel the speed of an atom is $v_{cm} - v_{rel}$, because it shares the overall translational motion of the wheel but is also moving backward relative to the axle. The motions of atoms at other locations around the rim are somewhat complicated. Nevertheless, when we add up all the ki-

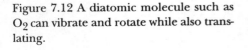

Figure 7.12 A diatomic molecule such as O_2 can vibrate and rotate while also translating.

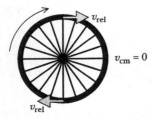

Figure 7.13 A bicycle wheel spinning on its axle, which is at rest. Atoms in the rim have a speed v_{rel} relative to the center of mass.

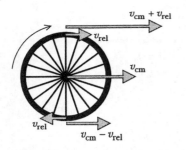

Figure 7.14 The velocity of an atom in the rim is the vector sum of the center-of-mass velocity and the velocity relative to the center of mass.

netic energies of all the atoms, we find that $K_{tot} = \frac{1}{2}Mv_{cm}^2 + K_{rot}$. The wheel has kinetic energy corresponding to its overall motion (this is kinetic energy that it would have even if it weren't rotating) plus kinetic energy corresponding to its rotation (this is kinetic energy that it would have even if the center of mass weren't moving). It is appropriate to call the rotational kinetic energy a kind of "internal energy" of the system.

If you drop the spinning bicycle wheel, the energy principle can be expressed like this:

$$\left(\frac{P_{tot}^2}{2M} + K_{rot} + Mgy_{cm}\right)_f = \left(\frac{P_{tot}^2}{2M} + K_{rot} + Mgy_{cm}\right)_i$$

If you simply drop the wheel, nothing will make the wheel spin faster or slower around the axle, so K_{rot} doesn't change, and the energy equation for this very complicated motion, involving both translation and rotation, reduces to the energy principle for a simple point particle:

$$\left(\frac{P_{tot}^2}{2M} + Mgy_{cm}\right)_f = \left(\frac{P_{tot}^2}{2M} + Mgy_{cm}\right)_i$$

Vibration and translation

Another kind of energy that is internal to a system is vibrational energy, both elastic and kinetic. Consider an oxygen molecule (O_2) that has no translational motion (that is, the center of mass is stationary), but is vibrating (Figure 7.15), with elastic energy and kinetic energy continually interchanging, but with the sum of the two energies remaining constant, like the mass and spring you modeled in Chapter 3.

If in addition the oxygen molecule is translating (Figure 7.16), the total energy of the molecule is the sum of the translational kinetic energy, the vibrational kinetic energy (in terms of velocities of the two atoms relative to the center of mass), and the elastic energy of the "spring" holding them together, and the rest energies of the constituent atoms:

$$E_{tot} = \frac{P_{tot}^2}{2M} + K_{vib} + \frac{1}{2}k_s s^2 + U_0 + 2mc^2$$

Note that the elastic energy $\frac{1}{2}k_s s^2$ is clearly unaffected by the translational motion of the center of mass, because it depends only on the stretch of the distance between the two atoms.

A related example is the motion of a hot block of metal. In addition to its translational kinetic energy $P_{tot}^2/(2M)$ there is vibrational kinetic and elastic energy of the atoms around their equilibrium positions, which we call thermal energy (or internal energy E_{int}).

Rotation, vibration, and translation

The most general motion of an oxygen molecule involves rotation, vibration, and translation. It is easy to show that the internal (non-translational) kinetic energy separates cleanly into rotational and vibrational contributions. Consider one atom with momentum p relative to the center of mass, and resolve its momentum into "tangential" and "radial" components (Figure 7.17). By "radial" component we mean a component in the direction of the line from the center of mass to the particle. By "tangential" component we mean a component perpendicular to that line. If the particle is in circular motion around the center of mass, the radial component is zero, so tangential momentum is associated with rotation. Radial momentum is associated with vibration.

We work out what the kinetic energy is in terms of these two momentum components:

Figure 7.15 A vibrating oxygen molecule whose center of mass is at rest.

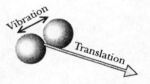

Figure 7.16 A translating, vibrating oxygen molecule.

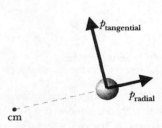

Figure 7.17 Radial and tangential components of momentum of a single atom, relative to the center of mass of an object.

$$\frac{\langle p^2 \rangle}{2m} = \frac{(p^2{}_{\text{tangential}} + p^2{}_{\text{radial}})}{2m} = K_{\text{rot}} + K_{\text{vib}}$$

Therefore the kinetic energy relative to the center of mass splits into two terms, rotational energy and vibrational energy:

ROTATIONAL + VIBRATIONAL ENERGY

$$K_{\text{rel}} = K_{\text{rot}} + K_{\text{vib}}$$

The total energy of a translating, rotating, vibrating oxygen molecule can be written like this:

$$E_{\text{tot}} = K_{\text{trans}} + K_{\text{rot}} + K_{\text{vib}} + \tfrac{1}{2}k_s s^2 + U_0 + 2mc^2 \text{ , where } K_{\text{trans}} = \frac{P^2_{\text{tot}}}{2M}$$

(If there are changes in height near the Earth's surface, there is also a term of the form Mgy_{cm} for the energy of the system of Earth plus molecule.)

If the object can fly apart into separate pieces, we might call the kinetic energy relative to the center of mass $K_{\text{explosion}}$! For example, a fireworks rocket has translational kinetic that doesn't change from just before to just after its explosion, and the pieces move at high speeds relative to the center of mass of the rocket. These large kinetic energies relative to the center of mass come from the chemical energy used up in the explosion.

7.4 The "point-particle system"

In analyzing the motion and energy of deformable multiparticle systems (systems whose shape can change), we encounter situations in which it would be extremely convenient to be able to calculate only the change in the translational kinetic energy K_{trans} of the system. Perhaps this might be all we are interested in, or perhaps we might wish to subtract this energy change from the change of the total energy of the system to find out how much energy has gone into vibration or rotation. Fortunately, it is quite simple to calculate ΔK_{trans} by itself.

We have established that the center of mass of a multiparticle system moves exactly like a simple "point" particle whose mass is the mass of the entire system, under the influence of the net external force applied to the entire system. Pretend you could crush the real system down to a very tiny ball and apply to this "point particle" the same forces that acted on the real system. What would the motion of this point-particle system look like? It would look exactly like the motion of the center of mass of the real system (Figure 7.18).

These two different systems have the same total mass M, and both are acted on by the same net force, so the two paths are exactly the same. The difference is that the real system may rotate and stretch and vibrate due to the effects of the forces acting at different locations on the extended object. In contrast, the point-particle system has no rotational motion, no vibrational motion, no internal energy of any kind. All of the forces act at the location of the point particle, and these forces don't stretch or rotate it. The only energy the point-particle system can have is translational kinetic energy, and this is exactly the same as the translational kinetic energy of the real multiparticle system:

Real system
Forces act at different locations

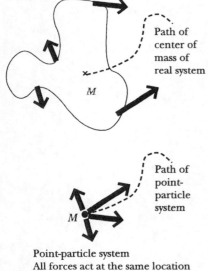

Path of center of mass of real system

M

Path of point-particle system

M

Point-particle system
All forces act at the same location

Figure 7.18 The path of the point-particle system is exactly the same as the path of the center of mass of the real system.

$$K_{\text{trans}} = \frac{P^2_{\text{tot}}}{2M} = \frac{1}{2}M_{\text{tot}}v^2_{\text{cm}} \text{ for the real system or the point-particle system}$$

An analysis of the point-particle system gives the translational kinetic energy of the real system. This is why the point-particle system is a useful concept. The force and energy equations for the fictitious point-particle are these:

POINT-PARTICLE SYSTEM (MASS *M* AT CM LOCATION)

$$\frac{d\vec{P}_{tot}}{dt} = \vec{F}_{net,ext}$$

$$\Delta K_{trans} = \Delta\left(\frac{P^2_{tot}}{2M}\right) = \int_i^f \vec{F}_{net,ext} \bullet d\vec{r}_{cm}$$

The integral is the work done on the point-particle system by the net force, which acts at the location of the point itself. In Chapter 4 we showed that starting from the momentum principle for a single point particle we could derive the result that the change in the kinetic energy of the particle is equal to the work done on the particle by the net force. Here, we consider the net force to be applied to the fictitious point particle located at the center-of-mass point, doing work to increase the kinetic energy of the point particle. A more formal derivation of the energy equation for the point-particle system may be found at the end of this chapter.

These equations tell us about the motion of the center of mass of a multiparticle system, but they tell us nothing about the vibration or rotation or thermal changes of the system. For a full treatment we need to combine analyses of the point-particle system with analyses of the real system.

7.4.1 Application: Jumping up

As an example of how to exploit the point-particle system, we'll analyze jumping (Figure 7.19). You crouch down and then jump straight upward. We want to calculate how fast your center of mass v_{cm} is moving at the instant that your feet leave the floor. Let *h* be the distance through which your center of mass rises in this process. Note that there is a change of shape of the system.

Let F_N be the electric contact force that the atoms in the floor exert on the bottom of your foot, and let *M* be your total mass. We don't know exactly how the floor force F_N varies with time, but to get an idea of what happens let's make the crude approximation that it is constant as long as you are in contact with the floor (of course it falls suddenly to zero as your foot leaves the floor). Consider the point-particle system shown in Figure 7.20, which does not change shape.

Figure 7.19 As the jumper leaves the floor, the center of mass has risen *h*. The system changes shape.

Point-particle system

? What is the net force on you? What is the energy equation for the point-particle system?

The net force in the upward direction is $(F_N - Mg)$ (Figure 7.20), and the energy equation for the point-particle system is

$$\Delta K_{trans} = \frac{P^2_{tot}}{2M} - 0 = \int_i^f \vec{F}_{net,ext} \bullet d\vec{r}_{cm} = (F_N - Mg)h$$

As we will see, this is not the same as the energy equation for the real system consisting of you, the jumper! The reason F_N shows up in the point-particle energy equation is that it contributes to the net force, which affects the center of mass motion. The work done on the point-particle system involves the net force multiplied by the displacement of the center-of-mass point.

Real system

? What are some of the energy changes that occur during the jump that are not represented in this point-particle energy equation?

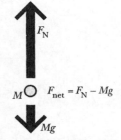

Figure 7.20 Force diagram for your body, considered as a point particle. This system does not change shape.

There is change of kinetic energy of your moving arms and legs relative to the center of mass, decrease of chemical energy inside you, and increase of thermal energy inside you (your temperature rises somewhat).

? In the real system, how much work is done by the floor force F_N?

The floor force F_N is applied to the bottom of your foot, and the contact point does not move during the entire process leading to lift-off, so there is no work. The definition of the work done by a force involves (the parallel component of) the displacement of the point of contact, at the place where the force is applied to the object. No displacement, no work done on you.

The energy equation for the real system (you, the jumper) looks something like the following, ignoring thermal energy transfer Q between you and the surrounding air:

$$\Delta E_{\text{sys}} = W = \sum_i \int_i^f \vec{\mathbf{F}}_i \bullet d\vec{\mathbf{r}}_i \quad (W \text{ is sum of work done by all external forces})$$

$$\Delta K_{\text{trans}} + \Delta K_{\text{other}} + \Delta E_{\text{thermal}} + \Delta E_{\text{chemical}} = -Mgh$$

The floor force F_N does not appear in the energy equation for the real system, because this force does no work (no displacement of the point of contact; $d\vec{\mathbf{r}}_{\text{floor}} = 0$). (The floor force does however appear in the energy equation for the point-particle system.) As before, the gravitational force does negative work on you.

ΔK_{other} includes the rotation of your upper and lower legs and the swinging of your arms. $\Delta E_{\text{thermal}}$ corresponds to the higher temperature that your body runs due to the exertion (which ultimately will lead to thermal energy transfer Q out of your body into the surrounding, cooler air). $\Delta E_{\text{chemical}}$ represents the burning of chemical energy stored in your body, in order to support this activity. None of these terms appear in the energy equation for the point-particle system, because the point-particle equation deals just with the motion of the mathematical center of mass point, which doesn't rotate or stretch or get hot. A crucial difference is that you change shape (legs unbend, etc.), but the point-particle system doesn't.

The point-particle equation gives you $\Delta K_{\text{trans}} = (F_N - Mg)h$. Substitute this into the energy equation for the real system and solve for the other terms, $\Delta K_{\text{other}} + \Delta E_{\text{thermal}} + \Delta E_{\text{chemical}}$.

Comparison of the point-particle and real energy equations

The important thing to notice is that the energy equation for the real system and for the point-particle system are quite different. Most energy terms do not appear at all in the point-particle equation, and the work that appears in the energy equation for the real system is not the same as the work that appears in the energy equation for the point-particle system.

? Which of the energy changes in the energy equation shown above for the real system are positive? Which are negative?

All of these energy changes of your body are positive except for the change in chemical energy, which decreases to pay for all the other increases.

7.4.2 Application: Stretching a spring

Let's check to make certain this kind of analysis gives us a sensible answer in a simple situation. You pull on one end of a spring (Figure 7.21) with a force F_L to the left, making a short displacement Δr_L to the left, and you pull on the other end with an equal but opposite force F_R and equal but opposite short displacement Δr_L. The system (the spring) changes shape. Our analysis should yield the expected result: the spring should not gain translational kinetic energy!

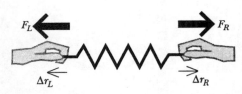

Figure 7.21 You pull on both ends of a spring with equal and opposite forces. The system (the spring) changes shape.

The real system

Consider the spring itself as the system.

? Does your right hand do any work on the spring? Of what sign? Does your left hand do any work on the spring? Of what sign? Is there any change in the energy of the spring?

Your left hand exerts a force through a short displacement Δr_L and does some (positive) work $W_L = F_L \Delta r_L \cos(0°) = +F_L \Delta r_L$.

Your right hand also exerts a force through a short displacement Δr_R and does some (positive) work $W_R = F_R \Delta r_R \cos(0°) = +F_R \Delta r_R$.

The total work increases the energy of the spring, in the form of spring (elastic) energy corresponding to increased stretch of the spring. Note carefully that both hands do positive work. The energy equation for the real system undergoing this process (the spring) is this:

$$\Delta[\tfrac{1}{2}k_s s^2] = W_L + W_R$$

The point-particle system

Next consider the "point-particle system" consisting of a fictitious particle located at the center of mass of the spring (Figure 7.22). This point-particle system does not change shape.

? What is the net force $\vec{F}_{net,ext}$? How far does the center of mass move? How much work $\sum \vec{F}_{net,ext} \bullet \Delta \vec{r}_{cm}$ is done? What is the change in $K_{trans} = P_{tot}^2/(2M)$?

The net force is zero ($F_R - F_L = 0$). Reassuringly, the center of mass does not move. The work done on the fictitious point particle is zero, so there is no change in the speed of the center of mass, so no change in the translational kinetic energy $K_{trans} = P_{tot}^2/(2M)$. The energy equation reduces to $0 = 0$, which is certainly correct but provides no detailed information.

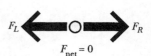

Figure 7.22 Force diagram for the "point-particle system," which does not change shape.

Comparison of the point-particle and real energy equations

The key point here is that in the real system, work is calculated by adding up the amount of work done by *each* force: the work done by your right hand *plus* the work done by your left hand. The work done by each force is the force times the actual displacement of the point of application of that force, $F_L \Delta r_L$ and $F_R \Delta r_R$, both of which are positive.

But in the point-particle system, work is calculated by first adding up all the forces to get the *net* force before calculating the work, and the net force is multiplied by the displacement of the center of mass. In the present example the net force is zero, and the displacement of the center of mass is zero, so the work done on the point-particle system is zero.

Note also that the actual work done by an individual force involves the displacement of the point where the force is applied. In contrast, work done on the point-particle system involves the displacement of the center of mass point, a point which may be very far away from any of the actual locations where the forces are applied, with a motion that is quite different from the motion of the point of application of an individual force.

Another way of looking at the situation is that the real multiparticle system (the spring) changes shape, and there is potential energy associated with that change of shape. The point-particle system does not change shape, and there is no change in potential energy.

The energy equation for the real system and the energy equation for the point-particle system are both valid and give correct results for any process, but they provide different kinds of information:

Real system: Info on change of total energy of system

Point-particle system: Info on change of translational kinetic energy

The differences are what make it useful to analyze certain kinds of phenomena both ways. Note that the two energy equations are different for this spring example because the system is deformable and changes shape, which leads to very different calculations of the work in the two systems.

Ex. 7.6 If a system such as a large block does not change shape (or rotate faster and faster due to tangential forces, and there are no changes of identity or temperature), the energy equations for the real system and for the point-particle system are identical. Explain why.

Ex. 7.7 A runner whose mass is 50 kg accelerates from a stop to a speed of 10 m/s in 3 seconds. (A good sprinter can run 100 meters in about 10 seconds, with an average speed of 10 m/s.) What is the average horizontal component of the force that the ground exerts on the runner's shoes?

Ex. 7.8 How much displacement is there of the force that acts on the sole of the runner's shoes, assuming that there is no slipping? Therefore, how much work is done on the real system (the runner) by the force you calculated in the previous exercise? How much work is done on the point-particle system by this force?

Ex. 7.9 The kinetic energy of the runner in the previous exercises increases—what kind of energy decreases? By how much?

Ex. 7.10 A very similar situation is the acceleration of a car on dry pavement, if there is no slipping. The axle moves at speed v, and the outside of the tire moves at speed v relative to the axle. The instantaneous velocity of the bottom of the tire is zero. How much work is done by the force exerted on the tire by the road? What is the source of the energy that increases the car's translational kinetic energy?

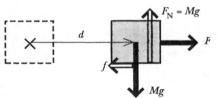

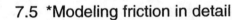

Figure 7.23 The center of mass of the block moves a distance d under the influence of a net force $F-f$.

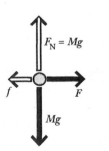

Figure 7.24 The block considered as a point particle.

7.5 *Modeling friction in detail

We can use what we've learned about complex systems to model the nature of sliding friction, which is a complicated phenomenon. In Chapter 5 we described a seeming paradox involving friction, that when you drag a block at constant speed across a table it seems that no work is done on the block, yet its temperature increases, indicating an increase in its thermal energy. With the new tools developed in this chapter we are in a position to analyze this phenomenon in more detail and resolve the paradox.

Point-particle system

A key issue that we will address is the question of how much work is done on the block by the friction force f exerted by the table. One of the new tools is the energy equation for the point-particle system (Figure 7.23 and Figure 7.24). You apply a force F to the right and the table exerts a force f to the left. (Pretend for the moment that you don't know the magnitude of the friction force f.) The change in the kinetic energy of the fictitious center-of-mass point particle is given by the product of the net force $(F-f)$ times the displacement d of the center-of-mass point:

$$\Delta K_{\text{trans}} = (F-f)d$$

Since the speed of the block is constant, $\Delta K_{trans} = 0$, so $F - f = 0$, which shows that the friction force is equal and opposite to the force you apply: $f = F$. The thermal energy does not appear in this energy equation for the point-particle system, because this equation deals merely with the motion of the (mathematical) center of mass, not with the many kinds of energy in the real system.

We could also have obtained the result $f = F$ directly from the momentum principle: $dP_{tot}/dt = F - f = 0$. The momentum principle is closely related to the energy equation for the point-particle system, because both involve the net force acting on a system.

Real system

The energy equation for the real system of the block can be written as follows, where Fd is the work done by you, and W_{fric} is the work done by the table:

$$\Delta K_{trans} + \Delta E_{th} = Fd + W_{fric}$$

where ΔE_{th} is the rise in the thermal energy of the block. We're assuming that the process takes a short enough time that there is negligible thermal energy transfer Q between the system of the block and the surroundings (the table), because energy transfer due to a temperature difference between table and block is a relatively slow process. (To remove such energy transfer as a possible complication, we could consider a block sliding not on a table but on an identical block. Then the symmetry of the situation is such that there cannot be any net thermal energy transfer into or out of the upper block. This tactic—reducing the complexity of a model—is a useful approach to complex problems.)

Since the speed of the center of mass v_{cm} does not change, $\Delta K_{trans} = 0$, and the energy equation reduces to

$$\Delta E_{th} = Fd + W_{fric}$$

If we conclude that the friction force does an amount of work $-Fd$, then we would have to conclude that the thermal energy of the block doesn't change, which is absurd. The block definitely gets hotter, indicating an increase in its thermal energy. We need to find a way around the entirely plausible but apparently incorrect conclusion that the friction force does an amount of work $-Fd$.

? Considering the sign of the change of thermal energy, must the magnitude of the work done by the friction force be greater or less than Fd?

Since the thermal energy change ΔE_{th} is surely positive (the block gets hotter, not colder), the magnitude of the work done by the friction force must be less than Fd. Since the friction force is definitely equal to F, the friction force must act through some effective distance d_{eff} *that must be less than d!*

Evidently the energy equation for the real system has the following form, with d_{eff} less than d:

$$\Delta E_{th} = Fd - Fd_{eff}$$

7.5.1 *A physical model for dry friction

How can the effective distance through which the friction force does work be less than the distance through which the block moves? We can understand this by looking at a microscopic picture of what happens on the surfaces in contact.

When a metal block slides on a metal surface, the block is supported by as few as three protruding "teeth," called "asperities" in the literature on

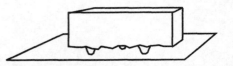

Figure 7.25 A block slides on a small number of protruding teeth (vertical scale exaggerated).

friction (Figure 7.25). The very high load per unit area on these teeth makes the material partially melt and flow, and high local temperatures produced during sliding lead to adhesion (welding) in the contact regions. The frictional force divided by the tiny contact area corresponds to the large "shear" (sideways) stress required to break these welds. This shearing of contact welds is the dominant friction mechanism for a dry metal sliding on the same metal.

Because the tooth tips can become stronger than the bulk metal due to a process called "work-hardening" (which introduces dislocations in the otherwise regular geometrical arrangements of atoms), shearing often occurs in the weaker regions of the teeth, away from the tip. This is a major effect when the two objects are made of the same material, and chunks of metal can break off and embed in the other surface. Nevertheless, this wear will be ignored in the further discussion. It is in any case a symmetrical effect for identical blocks. If the metal surfaces have oxide coatings, this can reduce the shear stress required to break the temporary weld (which reduces the friction force) and can prevent the breaking off of chunks of metal, if the oxide contact area is the weakest section.

This is the model of dry friction developed in a classic treatise on friction, *The Friction and Lubrication of Solids*, Part 1 and Part II, F. P. Bowden and D. Tabor (Oxford University Press, 1950 and 1964). The physics and chemistry of friction continues to be an active field of research, because the effects of friction can be quite complex, and there is high practical interest in controlling friction. A recent textbook is *Friction, Wear, Lubrication*, K. C Ludema, (CRC Press, 1996).

Having briefly reviewed a basic model of dry friction, we proceed to use this model to calculate the work done by frictional forces exerted at the contact points. The key issue is that the surface is deformable, which leads to differences in the energy equation for the real system compared with the energy equation for the point-particle system.

Actual work done by friction forces

We'll consider a microscopic picture of the contact region between two identical blocks. Figure 7.26 shows in a schematic way two teeth that have temporarily adhered to each other. The vertical scale has been greatly exaggerated for clarity—machined surfaces have much gentler slopes.

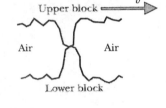

As the top block is dragged to the right, the teeth continue to stick together for a while. Both teeth must deform as a result; the top tooth is stretched backward, and the bottom tooth is stretched forward. In Figure 7.27 we see that when the top block has moved a distance d to the right, the point where the teeth touch each other has moved a distance of only d_{eff}, because the bottom block hasn't moved. For two identical blocks made of the same material, on average the two teeth will bend the same amounts, and d_{eff} will be equal to $d/2$.

Figure 7.26 Two projecting teeth of identical blocks, which have temporarily adhered to each other. (Exaggerated vertical scale).

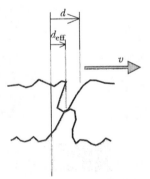

Figure 7.27 The top block has moved a distance d to the right, but the point of contact has moved only d_{eff}.

This is the key issue: The point of contact where the friction force is applied moves a shorter distance (d_{eff}) than the block itself moves (d). This means that the work done by the friction force is less than one might expect.

Eventually the weld breaks, and the top tooth snaps forward, so all the atoms in the tooth now catch up with the rest of the top block, but during this part of their journey, there is no external force acting on them (Figure 7.28).

The time average of the contact forces is indeed F (as determined by the fact that the net force must be zero, $F - F$), but the effective displacement d_{eff} at the point of contact of the frictional force is less than the displacement of the center of mass of the upper block. For two identical blocks we expect to find that $d_{eff} = d/2$.

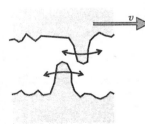

Figure 7.28 After the temporary weld breaks, the top tooth catches up to the rest of the block. Now no force acts on the teeth. Both teeth vibrate.

Internal energy of the blocks

Once the weld has broken, the teeth can vibrate, as we saw in Figure 7.28. The vibration of the tooth belonging to the top block and thermal conduction upward from the hot tip into the main body of the block contribute to the increase in the thermal energy of the upper block. Similarly, vibration and thermal conduction in the tooth belonging to the bottom block end up as increase in the thermal energy of the bottom block.

A more physical picture

Having two long teeth in contact is rather unphysical. A better picture is the one shown in Figure 7.29 (again with a greatly exaggerated vertical scale), where we show one of the top block's longest teeth in contact with the average surface level of the bottom block, and one of the bottom block's longest teeth similarly in contact with the average surface level of the top block. The two teeth shown here are to be taken as representative of time and space averages of the sliding friction. The frictional force F exerted on the top block is divided on the average into two forces each of magnitude $F/2$ at the ends of the two sets of long teeth.

In Figure 7.30 we see that when the top block moves a distance d to the right the upper contact also moves a distance d, whereas the lower contact does not move at all. The frictional work is therefore $(-F/2)(0) + (-F/2)(d)$, which is $-Fd/2 = -Fd_{eff}$, so d_{eff} is equal to $d/2$.

Summary of dry friction

The fundamental reason why the friction force can act through a distance less than d is that the block is deformable. All atoms in the top block eventually move the same distance d, but because of the stick/slip contact between the blocks the friction force F acts only for a portion of the displacement ($d/2$ for identical blocks). The point of contact for the friction force does not move in the same way as the center of mass. Note again that the energy equation for the point-particle system is not the same as the energy equation for the real system if the system changes shape.

7.5.2 *A model-independent calculation of the effective distance

In the special case of a block sliding on an identical block we can calculate d_{eff}, independent of the particular model of the surfaces. Consider a system consisting of both blocks together (Figure 7.31). The bottom block is held stationary by applying a force to the left (to prevent it being dragged to the right). This constraining force acts through no distance (the bottom block stands still), so this force does no work. The only work done on the two-block system is done by you, of magnitude Fd. So the energy equation for the real two-block system is this:

$$\Delta E_{th,1} + \Delta E_{th,2} = Fd$$

Since the two blocks are identical, half of this increased thermal energy shows up in the upper block ($Fd/2$), and half in the lower block ($Fd/2$). We can plug this into our earlier energy equation for the upper block:

$$\Delta E_{th,1} = \frac{Fd}{2} = Fd - Fd_{eff}$$

? Calculate d_{eff}.

We find that the effective distance through which the friction force acts is half the distance d that the block moves: $d_{eff} = d/2$. This is in agreement with the model of dry friction with bending teeth that we examined earlier, but our result is a general one that applies to any kind of model of friction surfaces for two symmetrical blocks.

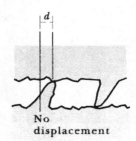

Figure 7.29 A more representative picture of contact between two surfaces, with projecting teeth in contact with flat areas.

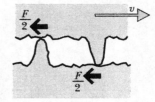

Figure 7.30 During this time interval the upper contact moves a distance d, while the lower contact does not move at all.

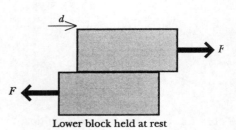

Lower block held at rest

Figure 7.31 The lower block is held at rest, while the upper block slides a distance d. Forces in the vertical direction are not shown.

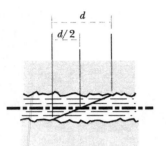

Figure 7.32 Two objects separated by a film of oil, which is modeled as a stack of fluid layers that can move relative to each other.

7.5.3 *Lubricated friction

It is interesting to see that the model-independent result $d_{\text{eff}} = d/2$ is also consistent with the case of lubricated friction. We separate the two blocks with a film of viscous lubricating oil, so that the two blocks do not make direct contact, as shown in Figure 7.32.

It is a property of simple fluid flow that fluid layers immediately adjacent to the blocks are constrained to share the motion of the blocks. Also, for a common type of flow called "laminar" flow, the displacement profile in the oil is linear. In particular, at the midplane the fluid moves half as far as the top block moves. As a result, as the top block is pulled a distance d to the right, the top layer of the oil is dragged along and moves a distance d to the right, the bottom layer of oil doesn't move, and the layer in the midplane (halfway between the blocks) moves a distance $d/2$.

If we take as the symmetrical systems of interest the top block with the upper half of the oil, and the bottom block with the lower half of the oil, we see that the shear force between the two systems (at the midplane in the oil) acts through a distance which is again half the displacement of the top block: $d_{\text{eff}} = d/2$. (Of course the magnitude of the friction force is much reduced by the lubrication, and the applied force F must be much smaller if the velocity is to be constant.)

This discussion of friction is based on an article "Work and heat transfer in the presence of sliding friction," by B. Sherwood and W. Bernard, *American Journal of Physics* volume **52**, number 11, Nov. 1984, pages 1001-1007, which in turn draws on an earlier article "Real work and pseudowork," by B. Sherwood, *American Journal of Physics* volume **51**, number 7, July 1983, pages 597-602.

7.6 *Derivation: Kinetic energy of a multiparticle system

In this section we derive the important result that $K_{\text{tot}} = K_{\text{trans}} + K_{\text{rel}}$, where $K_{\text{trans}} = P_{\text{tot}}^2/(2M)$ and K_{rel} is the kinetic energy relative to the center of mass. This result sounds entirely plausible, but the formal proof is rather difficult.

As in the case of calculating the gravitational energy of a multiparticle object, the derivation of the kinetic energy of a multiparticle system hinges on the definition of the center of mass point of a collection of atoms. The (vector) location of the i-th atom of the object can be expressed as the sum of two vectors, one from the origin to the center of mass ($\vec{r}_{\text{cm}}$), plus another from the center of mass to the i-th atom ($\vec{r}_i$).

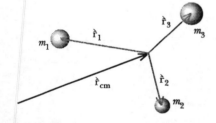

Figure 7.33 Center of mass of a multiparticle system.

The kinetic energy $K_i = \frac{1}{2}mv_i^2$ of the i-th atom is this, where we write $\vec{v}_i = d\vec{r}_i/dt$, the velocity of the i-th particle relative to the center of mass:

$$K_i = \frac{1}{2}m_i\left|\frac{d}{dt}(\vec{r}_{\text{cm}} + \vec{r}_i)\right|^2$$

$$= \frac{1}{2}m_i\left|\vec{v}_{\text{cm}} + \vec{v}_i\right|^2$$

This can be expanded, using the fact that the magnitude squared of a vector can be written as a vector dot product, $\left|\vec{C}\right|^2 = \vec{C} \bullet \vec{C} = C^2\cos(0) = C^2$:

$$K_i = \frac{1}{2}m_i[\vec{v}_{\text{cm}} + \vec{v}_i] \bullet [\vec{v}_{\text{cm}} + \vec{v}_i]$$

$$= \frac{1}{2}m_iv_{\text{cm}}^2 + 2\frac{1}{2}m_i\vec{v}_{\text{cm}} \bullet \vec{v}_i + \frac{1}{2}m_iv_i^2$$

Now that we have the kinetic energy of the i-th atom in terms of its position $\vec{r}_i$ relative to the center of mass, we need to add up the total kinetic energy of all the atoms. The total kinetic energy K_{tot} can be written in the following compact way, where Σ (Greek capital sigma) means "sum," and the sum goes from $i = 1$ through $i = N$:

$$K_{\text{tot}} = \sum_{i=1}^{N} \frac{1}{2} m_i v_{\text{cm}}^2 + \sum_{i=1}^{N} 2(\frac{1}{2} m_i) \vec{v}_{\text{cm}} \bullet \vec{v}_i + \sum_{i=1}^{N} \frac{1}{2} m_i v_i^2$$

First term   **Second term**   **Third term**

First term

The first term in this summation turns out to be the kinetic energy of a single particle of mass M, moving at the speed of the center of mass:

$$\sum_{i=1}^{N} \frac{1}{2} m_i v_{\text{cm}}^2 = \frac{1}{2} v_{\text{cm}}^2 \left(\sum_{i=1}^{N} m_i \right) = \frac{1}{2} v_{\text{cm}}^2 (m_1 + m_2 + m_3 + \ldots)$$

$$= \frac{1}{2} M v_{\text{cm}}^2 = \frac{P_{\text{tot}}^2}{2M}$$

Second term

The second term can be shown to be zero:

$$\sum_{i=1}^{N} 2(\frac{1}{2} m_i) \vec{v}_{\text{cm}} \bullet \vec{v}_i = \vec{v}_{\text{cm}} \bullet \sum_{i=1}^{N} m_i \vec{v}_i = \vec{v}_{\text{cm}} \bullet \frac{d}{dt} \left(\sum_{i=1}^{N} m_i \vec{r}_i \right) = 0$$

This result follows from the way that we calculate the location of the center of mass:

$$\vec{r}_{\text{cm}} = \frac{m_1 \vec{r}_1 + m_2 \vec{r}_2 + m_3 \vec{r}_3 + \ldots}{m_1 + m_2 + m_3 + \ldots} = \frac{\displaystyle\sum_{i=1}^{N} m_i \vec{r}_i}{M}$$

We are measuring the location $\vec{r}_i$ of the i-th atom relative to the center of mass, and since the distance from the center of mass to the center of mass is of course zero, we have

$$\vec{r}_{\text{cm}} = 0 = \frac{\displaystyle\sum_{i=1}^{N} m_i \vec{r}_i}{M} \text{ and therefore } \sum_{i=1}^{N} m_i \vec{r}_i = 0.$$

Third term

The third term by definition is the kinetic energy of the atoms relative to the (possibly moving) center of mass:

$$\sum_{i=1}^{N} \frac{1}{2} m_i v_i^2 = \frac{1}{2} m_1 v_1^2 + \frac{1}{2} m_2 v_2^2 + \frac{1}{2} m_3 v_3^2 + \ldots = K_{\text{rel}}$$

Putting the three pieces together, we find that the total kinetic energy splits into two parts: a term associated with the overall motion of the center of mass, plus the kinetic energy relative to the center of mass:

$$K_{\text{tot}} = K_{\text{trans}} + K_{\text{rel}}, \text{ where } K_{\text{trans}} = \frac{P_{\text{tot}}^2}{2M} = \frac{1}{2} M v_{\text{cm}}^2$$

7.7 *Derivation: The point-particle energy equation

In this chapter we showed that the energy equation for the point-particle version of the system follows from the fact that the motion of the center of mass is just like that of a point particle with the total mass of the real system and subjected to the *net* force acting on the real system. Here we give a more formal derivation of this important result.

Start from the x component of the momentum principle for a multiparticle system whose center of mass is moving at nonrelativistic speed:

$$\frac{dP_{tot,x}}{dt} = M_{tot}\frac{dv_{cm,x}}{dt} = F_{net,ext,x}$$

Integrate through the x displacement of the center of mass:

$$M_{tot}\int_i^f \frac{dv_{cm,x}}{dt}dx_{cm} = \int_i^f F_{net,ext,x}\,dx_{cm}$$

Switch dv and dx:

$$M_{tot}\int_i^f \frac{dx_{cm}}{dt}dv_{cm,x} = \int_i^f F_{net,ext,x}\,dx_{cm}$$

But dx_{cm}/dt is the x component of the center of mass velocity:

$$M_{tot}\int_i^f v_{cm,x}\,dv_{cm,x} = \int_i^f F_{net,ext,x}\,dx_{cm}$$

The integral on the left can be carried out:

$$M_{tot}\left[\frac{1}{2}v_{cm,x}^2\right]_i^f = \int_i^f F_{net,ext,x}\,dx_{cm}$$

$$\Delta\left[\frac{1}{2}M_{tot}v_{cm,x}^2\right] = \int_i^f F_{net,ext,x}\,dx_{cm}$$

We could repeat exactly the same argument for the y and z motions:

$$\Delta\left[\frac{1}{2}M_{tot}v_{cm,y}^2\right] = \int_i^f F_{net,ext,y}\,dy_{cm}$$

$$\Delta\left[\frac{1}{2}M_{tot}v_{cm,z}^2\right] = \int_i^f F_{net,ext,z}\,dz_{cm}$$

Note that $\frac{1}{2}M_{tot}v_{cm,x}^2 + \frac{1}{2}M_{tot}v_{cm,y}^2 + \frac{1}{2}M_{tot}v_{cm,z}^2 = \frac{1}{2}M_{tot}v_{cm}^2$.

Moreover, $F_{net,ext,x}\,dx_{cm} + F_{net,ext,y}\,dy_{cm} + F_{net,ext,z}\,dz_{cm} = \vec{F}_{net,ext}\bullet d\vec{r}_{cm}$.

Adding the three equations together (the sum of the left sides is equal to the sum of the right sides), we have the following equation for the translational kinetic energy of the system:

$$\Delta\left[\frac{1}{2}M_{tot}v_{cm}^2\right] = \int_i^f \vec{F}_{net,ext}\bullet d\vec{r}_{cm}$$

In words, the change in the translational kinetic energy of a system is equal to the integral of the *net* force acting through the displacement of the center of mass point.

The derivation that we have just carried out shows that although this equation looks like an energy equation, it is actually closely related to the momentum principle from which it was derived. The common element is the *net* force.

In contrast, the actual energy equation for the real system involves the work done by each individual force through the displacement of the point of application of that force. If the system deforms or rotates, these displacements of the individual forces need not be the same as the displacement of the center of mass.

7.8 Summary

Fundamental principles

No new fundamental principles.

New concepts

Center of Mass

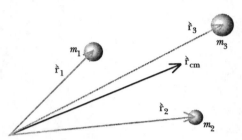

$$\vec{r}_{cm} \equiv \frac{m_1 \vec{r}_1 + m_2 \vec{r}_2 + m_3 \vec{r}_3 + \ldots}{m_1 + m_2 + m_3 + \ldots}$$

Center of mass

Problem Solving Techniques

Combined use of the energy equations for the real system and for the fictitious "point-particle system" in analyzing complex phenomena involving multiparticle systems.

   Point-particle system: Info on change of translational kinetic energy
   Real system: Info on change of total energy

Results

The momentum principle for multiparticle systems, derived from Newton's second and third laws of motion, was extended by defining the center of mass:

THE MOMENTUM PRINCIPLE FOR MULTIPARTICLE SYSTEMS

$$\frac{d\vec{P}_{tot}}{dt} = \vec{F}_{net,ext} \ \text{(in general)}$$

$$\vec{P}_{tot} = M\vec{v}_{cm} \ \text{(if } v \ll c \text{, and constant mass)}$$

Gravitational energy of multiparticle systems plus Earth, near the Earth's surface:

$$U_g = Mgy_{cm}$$

Kinetic energy of multiparticle systems:

$$K_{tot} = K_{trans} + K_{rel} \, , \text{where } K_{trans} = \frac{P^2_{tot}}{2M} = \tfrac{1}{2} M v^2_{cm}$$

Kinetic energy relative to center of mass can be split into two terms:

$$K_{rel} = K_{rot} + K_{vib}$$

POINT-PARTICLE SYSTEM

$$\Delta K_{trans} = \Delta \left(\frac{P^2_{tot}}{2M} \right) = \int_i^f \vec{F}_{net,ext} \bullet d\vec{r}_{cm}$$

Sliding friction can deform the contact surfaces, with the result that the frictional force may act through a distance that is different from the distance through which the center-of-mass point moves. As with any deformable system, the energy equations for the real system and for the point-particle system differ (though both are correct).

7.9 Example problem: A box containing a spring

A thin box in outer space contains a large ball of clay of mass M, connected to an initially relaxed spring of stiffness k_s (Figure 7.34). The mass of the box and spring are negligible compared to M. The apparatus is initially at rest. Then a force of constant magnitude F is applied to the box. When the box has moved a distance b, the clay makes contact with the left side of the box and sticks there, with the spring stretched an amount s. See the diagram for distances.

(a) Immediately after the clay sticks to the box, how fast is the box moving?

(b) What is the increase in thermal energy of the clay?

Solution

The fundamental principle here is the energy principle. We will apply it both to the real system and to the point-particle system (a system where we imagine crushing the real system done to a small dense point, and apply all of the forces at that point).

(a) Since we're just interested in the translational speed, consider the point-particle version of the system. Imagine you could crush the box and its contents down to a tiny ball with the total mass M of the real system. Apply the same forces to the point-particle system that were applied to the real system. The center of mass of the real system is at the center of the ball of clay, since essentially all of the mass is in the clay. The center of mass moves a distance $(b - s)$. The net force is F.

$$\Delta K_{\text{trans}} = F\Delta x_{\text{cm}} \quad \text{(energy principle applied to point-particle system)}$$

$$(\tfrac{1}{2}Mv^2 - 0) = F(b - s)$$

$$v = \sqrt{\frac{2F(b - s)}{M}}$$

There is nothing about the spring or thermal energy in these equations, because they pertain to the point-particle system, which doesn't stretch or get hot.

(b) Since we're interested in internal energy, we should apply the energy principle to the real system (the point-particle system has no internal energy). The work done on the real system is Fb, because the point of application of the force moves a distance b.

$$\Delta E_{\text{system}} = W + Q \quad \text{(energy principle applied to real system)}$$

Assume that the process takes place so quickly that there isn't time for any significant thermal energy transfer Q to or from the surroundings:

$$\Delta(K_{\text{trans}} + U_{\text{spring}} + E_{\text{thermal}}) = Fb$$

Substitute the value of ΔK_{trans} obtained from the point-particle analysis:

$$F(b - s) + \tfrac{1}{2}k_s s^2 + \Delta E_{\text{thermal}} = Fb$$

$$\Delta E_{\text{thermal}} = Fs - \tfrac{1}{2}k_s s^2$$

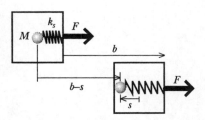

Figure 7.34 A thin box containing a spring and a ball of clay.

7.10 Review questions

The motion of a multiparticle system

RQ 7.1 Discuss qualitatively the motion of the atoms in a block of steel that falls onto another steel block. Why and how do large-scale vibrations damp out?

Center of mass

RQ 7.2 Can you give an example of a system that has no atoms located at its center of mass?

The momentum principle for multiparticle systems

RQ 7.3 Two people with different masses but equal speeds slide toward each other with little friction on ice with their arms extended straight out to the side (so each has the shape of a "t"). Her right hand meets his right hand, they hold hands and spin 90°, then release their holds and slide away. Make a rough sketch of the path of the center of mass of the system consisting of the two people, and explain briefly. (It helps to mark equal time intervals along the paths of the two people, and of their center of mass.)

RQ 7.4 The momentum principle for multiparticle systems would seem to say that the center of mass of a system moves just as though the system were a point particle. This led Chris to ask, "Wouldn't that mean that a piece of paper ought to fall in the same way as a small metal ball, if they have the same mass?" Explain carefully to Chris the resolution of this puzzle.

Gravitational energy of a multiparticle system

RQ 7.5 Consider the voyage to the Moon that you studied in Chapter 2. Would it make any difference, even a very tiny difference, whether the spacecraft is long or short, if the mass is the same? Explain briefly.

Kinetic energy of a multiparticle system

RQ 7.6 Outline how you would calculate the kinetic energy of the Earth as it moves through space.

The point-particle system

RQ 7.7 Under what conditions does the energy equation for the point-particle system differ from the energy equation for the real system? Give two examples of such a situation. Give one example of a situation where the two equations look exactly alike.

7.11 Problems

Problem 7.1 Jumping straight up—experiment

(a) Crouch down and jump straight up, as high as you can. Measure important heights in this process, and report your measurements. Include a rough sketch on which you clearly label what heights you measured. You might want to have a friend help you make the measurements.

(b) Analyze this process as fully as possible, using all the theoretical tools now available to you, especially the concepts in this chapter. Include a calculation of the force of the floor on your feet, and the change in your chemical energy. Be sure to explain clearly what approximations and simplifying assumptions you made in modeling the process.

Problem 7.2 Ice skater pushes away from a wall

On page 236 we discussed a woman ice skater who pushes away from a wall.

(a) Estimate the speed she can achieve just after pushing away from the wall. Then estimate the average acceleration during this process. How many "g's" is this? (That is, what fraction or multiple of 9.8 m/s² is your estimate?) Be sure to explain clearly what approximations and simplifying assumptions you made in modeling the process.

(b) For this process, choose the woman as the system of interest and discuss the energy transfers, and the changes in the various forms of energy. Estimate the amount of each of these, including the correct signs.

Problem 7.3 Translation and rotation

A hoop of mass M and radius R rolls without slipping down a hill (Figure 7.35). The lack of slipping means that when the center of mass of the hoop has speed v, the tangential speed of the hoop relative to the center of mass is also equal to v, since in that case the instantaneous speed is zero for the part of the hoop that is in contact with the ground ($v - v = 0$).

(a) The initial speed of the hoop is v_i, and the hill has a height h. What is the speed v_f at the bottom of the hill?

(b) Replace the hoop with a bicycle wheel whose rim has mass M and whose hub has mass m (Figure 7.36). The spokes have negligible mass. What would the bicycle wheel's speed be at the bottom of the hill?

Problem 7.4 Moving in a canoe

A man whose mass is 80 kg and a woman whose mass is 50 kg sit at opposite ends of a canoe 5 m long, whose mass is 30 kg.

(a) Relative to the man, where is the center of mass of the system consisting of man, woman, and canoe? (Hint: Choose a specific coordinate system with a specific origin.)

(b) Suppose the man moves quickly to the center of the canoe and sits down there. How far does the canoe move in the water? Explain your work and your assumptions.

Problem 7.5 Pulling a disk from the side

A string is wrapped around a disk of mass M and radius R. Starting from rest, you pull the string with a constant force F along a nearly frictionless surface (Figure 7.37). At the instant when the center of the disk has moved a distance d, a length L of string has unwound off the disk.

(a) At this instant, what is the speed of the center of mass of the disk?

(b) At this instant, how much rotational kinetic energy does the disk have relative to its center of mass?

If you are completely stuck, check the hint at the end of this chapter.

Problem 7.6 Two disks colliding

Two disks are initially at rest, each of mass M, connected by a string between their centers (Figure 7.38). The disks slide on low-friction ice as the center of the string is pulled by a string with a constant force F through a distance d. The disks collide and stick together, having moved a distance b horizontally.

(a) What is the final speed of the stuck-together disks?

(b) When the disks collide and stick together, their temperature rises. Calculate the increase in thermal energy of the disks, assuming that the process is so fast that there is insufficient time for the disks to lose much thermal energy to the ice. (Also, ignore the small amount of energy radiated away as sound produced in the collisions between the disks.)

If you are completely stuck, check the hint at the end of this chapter.

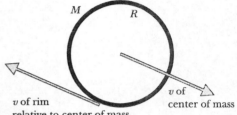

Figure 7.35 A hoop of mass M and radius R rolls down a hill (Problem 7.3a).

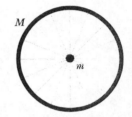

Figure 7.36 A bicycle wheel rolls down a hill (Problem 7.3b). The rim has mass M and the hub has mass m.

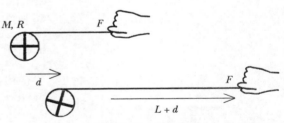

Figure 7.37 Pull a disk by a string wrapped around the disk (Problem 7.5).

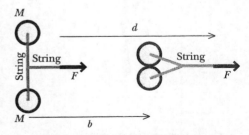

Figure 7.38 Pull a string that is tied to a string connecting two disks (Problem 7.6).

Problem 7.7 An uncoiling chain

A chain (mass M) of metal links is coiled up in a tight ball on a frictionless table. You pull on a link at one end of the chain with a constant force F. Eventually the chain straightens out to its full length L, and you keep pulling until you have pulled your end of the chain a total distance d.

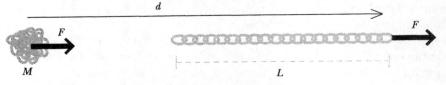

(a) What is the speed of the chain at this instant?

(b) In straightening out, the links of the chain bang against each other, and their temperature rises. Calculate the increase in thermal energy of the chain, assuming that the process is so fast that there is insufficient time for the chain to lose much thermal energy to the table. (Also, ignore the small amount of energy radiated away as sound produced in the collisions among the links.)

If you are completely stuck, check the hint at the end of this chapter.

Problem 7.8 Friction

It is sometimes claimed that friction forces always slow an object down, but this is not true. If you place a box of mass M on a moving horizontal conveyor belt, the friction force of the belt acting on the bottom of the box speeds up the box. At first there is some slipping, until the speed of the box catches up to the speed v of the belt. The coefficient of friction between box and belt is μ.

(a) What is the distance d (relative to the floor) that the box moves before reaching the final speed v? Use energy arguments, and explain your reasoning carefully.

(b) How much time does it take for the box to reach its final speed?

(c) The belt and box of course get hot. Is the effective distance through which the friction force acts greater than or less than d? Give as quantitative an argument as possible. You can assume that the process is quick enough that you can neglect thermal energy transfer Q between the belt and the box. Do not attempt to use the *results* of the friction analysis in this chapter; rather, apply the *methods* of that analysis to this different situation.

(d) Explain the result of part (c) qualitatively from a microscopic point of view, including physics diagrams.

Problem 7.9 Falling person

You hang by your hands from a tree limb that is a height L above the ground, with your center of mass a height h above the ground, and your feet a height d above the ground (Figure 7.39). You then let yourself fall. You absorb the shock by bending your knees, ending up momentarily at rest in a crouched position with your center of mass a height b above the ground. Your mass is M. You will need to draw labeled physics diagrams for the various stages in the process.

(a) What is the net internal energy change ΔE_{int} in your body (chemical plus thermal)?

(b) What is your speed v at the instant when your feet first touch the ground?

(c) What is the approximate average force F exerted by the ground on your feet during the time when your knees are bending?

(d) How much work is done by this force F?

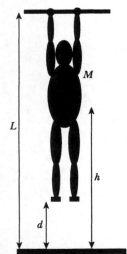

Figure 7.39 You hang from a tree limb and then drop to the ground (Problem 7.9).

Problem 7.10 Pulling a box containing an apparatus

A box and its contents have a total mass M. A string passes through a hole in the box, and you pull on the string with a constant force F (this is in outer space—there are no other forces acting).

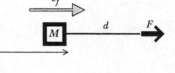

(a) Initially the speed of the box was v_i. After the box had moved a long distance w, your hand had moved an additional distance d (a total distance of $w + d$), because additional string of length d came out of the box. What is now the speed v_f of the box?

(b) If we could have looked inside the box, we would have seen that the string was wound around a hub that turns on an axle with negligible friction (Figure 7.40). Three masses, each of mass m, are attached to the hub at a distance r from the axle. Initially the angular speed of this apparatus was ω_i. In terms of the given quantities, what is the final angular speed ω_f of the apparatus?

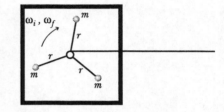

Figure 7.40 The mechanism inside the box (Problem 7.10).

Problem 7.11 Moving blocks connected by a spring

Two identical 0.1 kg blocks (labeled 1 and 2) were at rest on a nearly frictionless surface, connected by an unstretched spring whose stiffness is 100 N/m. Then a constant force of 5 N to the right was applied to block 2, and at a later time the blocks are in the new positions shown in the lower diagram.

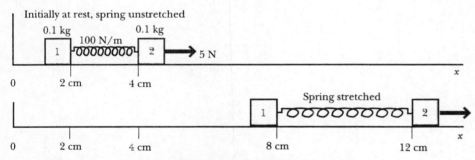

(a) At this later time, what is $K+U$ of the two-block system?

(b) What is the translational kinetic energy of the two-block system?

(c) What is the speed of the center of mass of the two-block system?

(d) What is the vibrational kinetic energy of the two-block system? Note that the spring is now stretched.

Problem 7.12 Spring and two masses

You exert an upward force of $2Mg$ to hold an object at rest that consists of two masses, each of mass M, connected by a low-mass spring whose stiffness is k_s (Figure 7.41). The initial stretch of the spring is s_i (equal to Mg/k_s). Then you suddenly start applying a larger force, of constant magnitude $F > 2Mg$. The diagram shows the situation some time later, when the blocks have moved upward, and the spring stretch has increased to s_f.

(a) What is now the speed of the center of mass of the two blocks? As part of your explanation, on a diagram show all of the forces that are acting. Also, be clear and explicit about what system you are analyzing.

(b) What is the vibrational kinetic energy of the two blocks (their kinetic energy relative to the center of mass)? Be clear and explicit about what system you are analyzing.

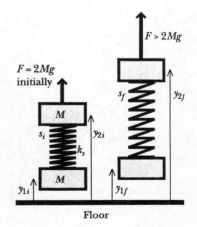

Figure 7.41 Pull up on a system consisting of two masses connected by a spring (Problem 7.12).

Problem 7.13 A binary star system

(a) About half of the visible "stars" are actually binary star systems, two stars that orbit each other with no other objects nearby. Describe the motion of the center of mass of a binary star system. Briefly explain your reasoning.

(b) For a particular binary star system, telescopic observations repeated over many years show that one of the stars (whose unknown mass we'll call M_1) has a circular orbit with radius $R_1 = 6 \times 10^{11}$ m, while the other star (whose unknown mass we'll call M_2) has a circular orbit of radius $R_2 = 9 \times 10^{11}$ m about the same point. Make a sketch of the orbits, and show the positions of the two stars on these orbits at some instant. Label the two stars as to which is which, and label their orbital radii. Indicate on your sketch the location of the center of mass of the system, and explain how you know its location, using the concepts and results of this chapter.

(c) This double star system is observed to complete one revolution in 40 years. What are the masses of the two stars? (For comparison, the distance from Sun to Earth is about 1.5×10^{11} m, and the mass of the Sun is about 2×10^{30} kg.) This method is often used to determine the masses of stars. The mass of a star largely determines many of the other properties of a star, which is why astrophysicists need a method for measuring the mass.

7.12 Answers to exercises

7.1 (page 233)   No acceleration, no force, no strain. Once you are no longer exerting a force on the right-most atoms in the block, atoms just to the left catch up and the interatomic force relaxes. This relaxation propagates to the left through the block, until finally there is no strain, and the block coasts along at a constant speed.

7.2 (page 234)   In the vertical member, $3L/4$ up from its bottom.

7.3 (page 235)   4600 km (so inside the Earth)

7.4 (page 237)   The puck with the string wound around it is rotating and so has additional kinetic energy. You have to pull farther, by the amount of string that unwinds, so you do more work, which corresponds to this puck having more kinetic energy. The momentum principle for the motion of the center of mass tells us nothing about the rotational aspects of the motion.

7.5 (page 237)   6 N to the right; 20 m/s^2

7.6 (page 244)   All of the applied forces act through the same displacement, which is also the displacement of the center of mass. So it makes no difference whether you calculate the work done by each individual force, and then add them up, or first add up all the forces, and then calculate the work. Also, the only energy change in such a system is the overall kinetic energy.

7.7 (page 244)   Average force = 167 newtons.

7.8 (page 244)   Work done on real system = 0 joules. Work done on point-particle system = 2500 joules.

7.9 (page 244)   Change in chemical energy of runner = −2500 joules

7.10 (page 244)   The instantaneous speed of the bottom of the tire is zero. The force by the road on the tire does no work. The increased translational kinetic energy of the car comes from a decrease in the chemical energy of the gasoline.

Hint for Problems 7.5, 7.6, and 7.7: Carefully label distances on a diagram of the process, and write the energy equations for the process both for the real system and for the point-particle system.

Chapter 8

Collisions:
Exploring the Nucleus

Chapter 8

Collisions:
Exploring the Nucleus

In high-energy particle accelerators, subatomic particles are hurled at each other. Changes of momentum and changes in identity of the particles involved in the collisions can reveal a great deal about the properties of the particles, and about their mutual interactions. We may not be able to observe what happens during a brief collision, but by applying conservation of momentum we can analyze the situation before and after the collision, and can obtain much useful information.

By "collision" we mean an interaction that takes place in a very short time, with little interaction before or after the collision. Neutral molecules in a gas experience collisions—the interatomic electric force is effectively short range, so there is little interaction until the molecules come nearly into contact. Similarly, collisions between macroscopic objects are quite common: before a ball hits a wall it has little interaction with the wall, and the contact between ball and wall lasts a very short time.

Although we have not yet talked very much about conservation of momentum, it is already familiar to us because it is simply a consequence of the momentum principle and the reciprocity of electric and gravitational forces. In this chapter we will learn how to apply conservation of momentum to the analysis of collisions. Our study of collisions will lead naturally into studying how processes look from different reference frames. We will see what we can learn about interactions by observing the probability distribution of deflection angles. We will investigate "elastic" collisions (no change of kinetic energy) and "inelastic" collisions (some energy is changed from kinetic to other forms of energy).

8.1 Momentum and impulse

In a collision an object receives a very short-duration impulse $\vec{F}\Delta t$ and its momentum changes by an amount $\Delta\vec{p} = \vec{F}\Delta t$. The following exercises review the relation between impulse and momentum change.

Example: A ball whose mass is 0.1 kg hits the floor with a speed of 6 m/s and rebounds upward with a speed of 5 m/s. What was the magnitude and direction of the impulse delivered to the ball by the floor? (Remember that momentum and impulse are vectors.)

Solution: With the y axis pointing upward,

$$\vec{p}_i = (0.1 \text{ kg})(<0, -6, 0> \text{ m/s})$$

$$\vec{p}_f = (0.1 \text{ kg})(<0, 5, 0> \text{ m/s})$$

$$\vec{p}_f - \vec{p}_i = <0, +1.1, 0> \text{ kg·m/s} = \text{the impulse}$$

Ex. 8.1 In the preceding exercise, if the ball was in contact with the floor for 1 ms (10^{-3} s), what was the average magnitude of the force exerted on the ball by the floor? Compare with the weight of the ball *mg*.

This illustrates the class of interactions we're talking about. We can't neglect the gravitational *mg* force before or after the collision, but *during* the brief collision the force exerted by the floor is so large that the other forces are negligible.

8.2 Conservation of momentum

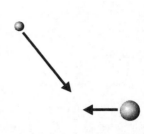

Figure 8.1 Two particles moving toward each other, free from external forces.

Momentum conservation is particularly useful in analyzing collisions between two particles. As an example, consider two particles in outer space (Figure 8.1), free of external forces, which move toward each other and eventually collide. During the short time of the collision, the particles exert large forces on each other (Figure 8.2), which according to the momentum principle will cause changes in each particle's momentum (Figure 8.3).

Consider the system consisting of both particles. The net external force on this combined system is zero. Therefore there is no change in the total momentum of the system:

$$\Delta \vec{P}_{tot} = \Delta(\vec{p}_1 + \vec{p}_2) = 0$$

Figure 8.2 During the collision the two particles exert large forces on each other.

Another way to predict the same result is to note that by the principle of reciprocity (Newton's third law), the force (and impulse) exerted on particle 1 by particle 2 is equal and opposite to the force (and impulse) exerted on particle 2 by particle 1, so the change in the momentum of particle 1 is equal and opposite to the change of the momentum of particle 2. Therefore the net change in the total momentum is zero. (Note that it was the principle of reciprocity that allowed us to derive the multiparticle version of the momentum principle.)

Here is a statement of this very important principle:

CONSERVATION OF MOMENTUM

$$\Delta \vec{P}_{system} + \Delta \vec{P}_{surroundings} = 0$$

Figure 8.3 After the collision each particle's momentum has changed.

A system that is free of external forces exerted by the surroundings will have a total momentum that does not change.

? Are there any systems of particles (smaller than the Universe itself) that are completely free of external forces?

No object is completely free of gravitational interactions, so the momentum for systems of interest may change slightly, yet a constant momentum can be an excellent approximation to the real situation, especially in collisions. Why? A collision takes a very short time Δt, from just before to just after the initial contact. The nonzero external forces are small compared to the large forces the colliding objects exert on each other, and therefore have only a small effect during the short duration of the collision, since $\Delta \vec{P}_{tot} = \vec{F}_{net,ext}\Delta t$.

As long as Δt is very short, and the net external force is of ordinary magnitude, the external impulse will be small and the change it causes in the total momentum of the two-object system will usually be negligible. Moreover, the change in the total momentum is likely to be negligible compared to the change in the momentum of one of the colliding particles, which receives a large impulse due to a very large force. An example is a collision of two baseballs in midair (Figure 8.4).

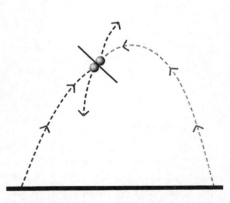

Figure 8.4 Midair collision between two baseballs.

? Look at the curving trajectories before the collision. Does the total momentum of the two baseballs change during the long period before the collision? What is the cause?

The total momentum of the two-ball system changes, because gravitational and air-resistance forces act on the system.

The collisional contact lasts a very short time, and involves very large contact forces, as is evident from the abrupt changes of directions of both balls.

? How does the external impulse $\vec{F}_{net,ext}\Delta t$ (gravitational force, air resistance) acting on just one of the baseballs compare with the impulse due to the other baseball, during contact? Is it a good approximation to say that from just before contact to just after contact, the total momentum of the two balls is (nearly) constant?

The external impulse involves an ordinary force for a very short time, while the impulse due to the other ball involves a huge force acting for that same short time. Therefore the momentum change of one of the balls is very nearly the opposite of the momentum change of the other ball. Approximately, during the short contact time, the total momentum of the two-ball system is constant.

These considerations justify an approach to the analysis of collisions that proves fruitful. Up until the collision, and after the collision, we do need to take into account external forces, which affect the total momentum. But during the brief time interval of the collision itself, we can use the approximation that the external impulse is small, and the change in total momentum of the interacting objects is negligible.

"Collisions" without contact

Consider the electric interaction of a proton and electron, or the gravitational interaction of two asteroids, as shown in Figure 8.5. There is no contact—no touching—yet we call such an interaction a "collision." A collision is any process in which there is little interaction before and after a short time interval, and large interactions during that short time interval.

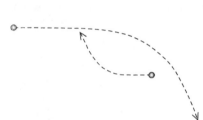

Figure 8.5 Electric or gravitational interactions without contact; these too are considered to be "collisions."

8.2.1 A simple example

In the following sections we'll apply the principle of conservation of momentum to several different kinds of processes. In this section we consider a simple but instructive example.

A ball of mass m is thrown with speed v against a wall and rebounds with nearly the same speed v (Figure 8.6).

? Is the momentum of the ball constant?

No. Even though the magnitude of the momentum doesn't change, the x component of the momentum vector does change, from positive (going to the right) to negative (going to the left).

Figure 8.6 A ball rebounds off a wall with nearly the same speed v.

? What caused the momentum of the ball to change?

The wall exerted a large force to the left on the ball for a short time, and this impulse changed the momentum of the ball.

? What was the magnitude and direction of the momentum change, $\Delta\vec{p}$?

The initial momentum in the x direction was $+p$ and the final momentum in the x direction was $-p$ (Figure 8.7), so there was a change of x component of momentum of the following amount:

$$\Delta p_x = (p_f - p_i) = -2p$$

? What was the magnitude and direction of the impulse exerted by the wall on the ball?

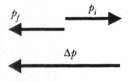

Figure 8.7 Change in momentum of the ball.

Since change of momentum is equal to the impulse, evidently the impulse must have had a magnitude of $2p$, and a direction to the left; the x component of the impulse was $-2p$.

? But isn't something constant when the ball bounced off the wall? After all, the speed hardly changed.

Yes, the energy of the ball hardly changed—specifically, the kinetic energy before and after was nearly the same. When the ball compressed against the wall, momentarily the kinetic energy went to zero, but elastic energy in the ball restored the kinetic energy on the rebound. Note that energy is a scalar represented by a single number, whereas momentum is a vector represented by three numbers: $< p_x, p_y, p_z >$.

? But isn't momentum always unchanged in a process?

No, the momentum of a system changes if an impulse (a force acting for a time) is applied to the system by the surroundings, in which case the momentum of the surroundings has an equal and opposite change (momentum conservation). Here the wall exerted a force on the ball, so the momentum of the ball was not constant. However, if you consider a larger system consisting of the Earth (including the wall) plus the ball, the momentum of the combined system *is* constant in the collision. When the ball bounces to the left, the wall (and the Earth to which it is connected) recoil to the right! In the following sections we will develop a method of analysis that can be used to determine the recoil speed of the Earth.

8.2.2 A head-on collision

Later in this chapter we will study a famous experiment that led to the discovery of the nucleus inside the atom. The experiment was carried out in 1911 by a group led by Ernest Rutherford. In that experiment high-speed alpha particles (now known to be helium nuclei, consisting of two protons and two neutrons) were shot at a thin gold foil.

In the original Rutherford experiment, the velocity of the alpha particles was not high enough to allow them to make actual contact with the nucleus. It is nonetheless appropriate to think of the process as a collision, because there are sizable electric forces acting between the alpha particle and the gold nucleus for a very short time. We call any process of this kind a collision, even if there is no direct contact but only gravitational, electric, or magnetic interactions at a distance.

As an example of the use of the momentum concept, we will analyze the head-on collision of an alpha particle and a gold nucleus (consisting of 79 protons and 118 neutrons). We treat a head-on collision first because it is a simple but important case, with only x components of momentum.

We make "before" and "after" diagrams of the relevant momenta in a head-on collision (Figure 8.8 and Figure 8.9). Let m be the mass of the alpha particle and M be the mass of the gold nucleus. The incoming alpha particle has a known momentum $\vec{p}_1$, the gold nucleus initially has a known momentum $\vec{p}_2 = 0$, the outgoing alpha particle has a momentum $\vec{p}_3$, and the gold nucleus picks up a momentum $\vec{p}_4$. Our task is to calculate p_3 and p_4.

In the Rutherford experiment the speeds of all of the particles were small compared to the speed of light, so we expect to be able to use the low-speed approximations for momentum and energy to analyze the collision. The outgoing speed of the gold nucleus is quite small despite the large momentum vector, because it has a large mass: $v_4 = p_4 / M$.

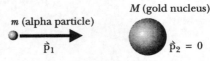

Figure 8.8 Before a head-on collision between an alpha particle and a gold nucleus.

Figure 8.9 After the collision.

? What can you say about the relation of the total momentum of the alpha particle plus gold nucleus *before* the collision to their total momentum *after* the collision? Why?

During the collision external forces are negligible, so the total momentum of the two-particle system will not change. Therefore we can write the following momentum conservation equation relating the total momentum before the collision to the total momentum after the collision:

$$\vec{p}_1 + \vec{p}_2 = \vec{p}_3 + \vec{p}_4$$

$$p_{1x} + 0 = p_{3x} + p_{4x}$$

We expect to find that p_{3x} is negative, corresponding to the alpha particle bouncing straight back when it hits the massive gold nucleus head-on. Note that if the alpha particle's momentum essentially turns around, $p_{3x} \approx -p_{1x}$. the gold nucleus picks up a lot of momentum:

$$p_{4x} = p_{1x} - p_{3x} \approx p_{1x} - (-p_{1x}) = 2p_{1x}$$

The outgoing gold nucleus has about twice the momentum of the incoming alpha particle, as suggested in Figure 8.8 and Figure 8.9. However, the speed of the gold nucleus, about $2p_{1x}/M$, is small because its mass is very large.

The conservation of momentum has given us one equation in two unknowns, p_{3x} and p_{4x} (the incoming momentum p_{1x} is known for these alpha particles). We don't have enough information to solve for the unknowns and check our qualitative conclusions. Can we use the energy principle to get a second equation involving the two unknown momenta?

8.2.3 Elastic collisions

Since conservation of energy is an important principle in its own right, and separate from conservation of momentum, it may be fruitful to apply energy conservation to the analysis of a head-on collision between an alpha particle and a gold nucleus in the Rutherford experiment. The conditions of this experiment were such that there was no change in the internal energy of either the alpha particle or the gold nucleus.

There do exist "excited" states of nuclei, in which there is a change in the configuration of the nucleons corresponding to a higher energy than the ground state, the normal configuration of the nucleus. The internal energy of the nucleus is quantized, and there are no states with energy intermediate between these quantized energies (Figure 8.10). In the Rutherford experiment, the interaction did not raise the gold nucleus nor the alpha particle to an excited state; both nuclei remained in their ground states. This happy accident made it much easier for Rutherford to analyze the experiment. Such a collision, in which there is no change in the internal energy of the colliding particles, is called an "elastic" collision.

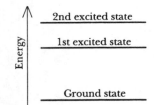

Figure 8.10 Energy levels in a nucleus. The potential energy curve is not shown.

ELASTIC COLLISION

In an "elastic" collision, the internal energy of the objects in the system does not change: $\Delta E_{\text{int}} = 0$

We want to write an energy equation for the collision of the alpha particle and the gold nucleus. This is an elastic collision; the internal energy of the system does not change.

What about the electric potential energy associated with the electric forces? Long before the collision the two nuclei were far from each other, and long after the collision they are again far from each other. Like gravitational energy, electric energy is proportional to $1/r$, where r is the distance between the two charged objects. Therefore from long before the collision to long after the collision the change in electric energy is very nearly zero; that is, $\Delta U_{el} \approx 0$.

Can we in fact claim that in the initial and final states the particles are far enough apart to justify the approximation $\Delta U_{el} \approx 0$? Because the electrons of the atom are much less massive than the alpha particle they do not affect its path very much, so we can start our analysis when the alpha particle is deep inside the atom, with the electrons of the gold atom mostly far outside the region of interest. An atomic diameter is about 10^{-10} m, whereas the diameter of a gold nucleus is much smaller, less than 10^{-14} m, so we might start our analysis of the collision when the particles are about 10^{-12} m apart, a distance that is very small compared to an atom and very large compared to a nucleus. At this distance the electric potential energy is

$$U_{el} = \frac{1}{4\pi\varepsilon_0} \frac{(2e)(79e)}{r} = \left(9 \times 10^9 \, \frac{N \cdot m^2}{C^2}\right) \frac{(2)(79)(1.6 \times 10^{-19} \, C)^2}{(10^{-12} \, m)}$$

$$U_{el} = (3.6 \times 10^{-14} \, J) \frac{(1 \, eV)}{(1.6 \times 10^{-19} \, J)} = 0.2 \times 10^6 \, eV = 0.2 \, MeV$$

The radioactive source used by the Rutherford group provided alpha particles whose kinetic energy was 10 MeV, so if we take as the initial state a separation of 10^{-12} m the electric energy is only 2% of the kinetic energy. And if we also take as the final state a separation of 10^{-12} m we have $\Delta U_{el} = 0$. Evidently the only significant energy changes in this experiment are changes in the kinetic energies of the alpha particle and the gold nucleus.

Here then is the additional equation we get from energy conservation, involving only kinetic energies:

$$K_1 + 0 = K_3 + K_4$$

Rewrite in terms of the momentum components:

$$\frac{p_{1x}^2}{2m} = \frac{p_{3x}^2}{2m} + \frac{p_{4x}^2}{2M}$$

Remember that momentum conservation gave us this equation:

$$p_{1x} = p_{3x} + p_{4x}$$

Now we have two equations in two unknowns, p_{3x} and p_{4x} .

Ex. 8.2 Try to solve the two equations to determine the final momenta, in terms of the known initial momentum p_{1x}. A good way to start is to solve the momentum equation for p_{4x} (the final momentum of the gold nucleus) and plug this into the energy equation, then solve for the single unknown p_{3x} (the final momentum of the alpha particle).

8.2.4 Light projectile, heavy target

In the preceding exercise, you probably used the quadratic formula to find the following result for the final momentum p_{3x} of the alpha particle:

$$p_{3x} = \left[\frac{m \pm M}{m + M}\right] p_{1x}$$

This equation has two solutions. If you take the "+" sign, you get $p_{3x} = p_{1x}$, implying that instead of bouncing back the alpha particle kept going in the forward direction, with its original momentum p_{1x}. This solution simply corresponds to passing right through the gold nucleus without interacting! It is interesting that this is a possible solution, since it would be valid if a neutrino instead of an alpha particle were the projectile, since typically a neutrino would pass right through a gold nucleus without interacting. It also corresponds to the case where the alpha particle passes very far from the gold nucleus and so is hardly deflected by the tiny electric force.

The other solution, with the "−" sign, is the interesting one for the alpha particle final momentum:

$$p_{3x} = \left[\frac{m-M}{m+M}\right] p_{1x}$$

? Should the magnitude of the final momentum p_3 of the alpha particle be greater or smaller than the initial momentum p_1? Why? Is this solution consistent with that prediction?

Since the gold nucleus acquires some nonzero kinetic energy, the kinetic energy of the alpha particle must decrease, and that means that the magnitude of the alpha particle momentum must decrease. Since the magnitude of $(m-M)/(m+M)$ is less than one, the result is consistent with the loss of kinetic energy to the gold nucleus.

? We assumed that the low-mass alpha particle would bounce backward upon colliding with the high-mass gold nucleus, and that p_{3x} should be negative. Is it?

Yes, $(m-M)$ is negative, since $m < M$.

Approximate final momentum of the light projectile

? What approximation can we make to get a simpler expression for the final momentum of the alpha particle?

The alpha particle contains 4 nucleons (2 protons and 2 neutrons), while the gold nucleus contains 197 nucleons (79 protons and 118 neutrons). Since protons and neutrons have nearly the same mass, the mass of a nucleus containing N nucleons is approximately equal to the mass of N nucleons (neglecting the small mass difference associated with the binding energy of the nucleus).

Since the mass of the alpha particle is very small compared to the mass of the gold nucleus ($m \ll M$), $(m-M)/(M+m)$ is nearly equal to −1, and

$$p_{3x} \approx -p_{1x}$$

? Is this consistent with what happens when you throw a (low-mass) ping pong ball at a (high-mass) bowling ball?

Our approximate formula predicts that a fast lightweight object hitting a heavy object at rest will bounce back with nearly the initial momentum (and speed), which is what we observe when a ping pong ball bounces off a bowling ball (which is a nearly elastic collision).

Rutherford, who was familiar with the result we have just derived, was extremely surprised when his group observed some alpha particles bounce straight back from a gold foil at high speed.

? Why might Rutherford have been surprised? Given what you know (and Rutherford didn't) about the structure of atoms, is his observation surprising to you?

Rutherford didn't know about the nucleus: he was about to discover it! He didn't know that there was a very small, massive core at the heart of the atom. Knowing that a gold nucleus is much more massive than an alpha particle, it isn't surprising to us that the alpha particle could bounce straight back.

It is important to note that we didn't need to know what kind of force was involved in the collision. Our results would be valid for nonelectric forces.

Approximate final momentum of the heavy target

If you solve the momentum and energy conservation equations for the momentum of the recoiling gold nucleus you get the following:

$$p_{4x} = \frac{2M}{M+m}p_{1x}$$

Since the mass M of the gold nucleus is much larger than the mass m of the alpha particle, $2M/(M+m)$ is approximately equal to 2, and the gold nucleus recoils with about twice the momentum of the alpha particle:

$$p_{4x} \approx 2p_{1x}$$

This justifies our original qualitative analysis. If the alpha particle simply turns around, its momentum change is about $-2p_{1x}$, and the gold nucleus acquires a momentum of about $2p_{1x}$.

Ex. 8.3 We can use our results for head-on elastic collisions to analyze the recoil of the Earth when a ball bounces off a wall embedded in the Earth. Suppose a professional baseball pitcher hurls a baseball ($m = 155$ grams) with a speed of 100 miles per hour ($v_1 = 44$ m/s) at a wall, and the ball bounces back with little loss of kinetic energy. What is the recoil speed of the Earth ($M = 6\times10^{24}$ kg)?

Ex. 8.4 Calculate the recoil kinetic energy of the Earth and compare to the kinetic energy of the baseball. The Earth gets lots of momentum (twice the momentum of the baseball) but very little kinetic energy.

8.2.5 Heavy projectile, light target

We have seen what happens when a light object collides with a stationary massive one. For completeness, it is interesting to consider the case of a gold nucleus hitting head-on an alpha particle that is initially at rest (Figure 8.11). We can determine the final speeds and directions without completely repeating our analysis. Merely switch M and m in the final results! In that case we predict that after the collision the momentum of the gold nucleus is

$$p_{3x} = \left[\frac{M \pm m}{M+m}\right]p_{1x} \approx p_{1x} \text{ (since } M \text{ is much larger than } m)$$

Taking the "+" sign, we find exactly $p_{3x} = p_{1x}$, which would be the case if there were no interaction (or the gold nucleus passes far from the alpha particle). Taking the "−" sign, $(M-m)/(M+m)$ is only very slightly less than +1, which means that the gold nucleus keeps going with nearly its initial momentum. This makes sense: if you throw a heavy bowling ball at a light ping pong ball, you expect the bowling ball to keep going, and not to lose much speed.

What about the low-mass target—the alpha particle (or ping pong ball)? We use our previous result for the final target momentum, but interchange M and m:

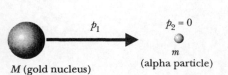

Figure 8.11 A gold nucleus collides head on with an alpha particle that is initially at rest.

$$p_{4x} = \frac{2\,m}{m + M}p_{1x} \approx \frac{2\,m}{M}p_{1x}$$

In terms of the final speed v_4 of the alpha particle (ping pong ball) and the initial speed v_1 of the gold nucleus (bowling ball), this yields the following:

$$mv_4 \approx \frac{2\,m}{M}Mv_1 \ \text{ or } \ v_4 \approx 2\,v_1$$

This is a surprisingly simple result. The low-mass target moves at *twice* the speed of the incoming high-mass projectile, independent of the precise ratio of the masses. If you throw a bowling ball with speed v_1 at a ping-pong ball, the bowling ball keeps going with a speed almost equal to v_1, and the ping pong ball moves out ahead of the bowling ball with a speed that is approximately $2v_1$.

Viewed from a different reference frame

It is possible to understand this simple result, $v_4 \approx 2\,v_1$, by considering the original case of an alpha particle hitting a stationary gold nucleus, in which case we know that the alpha particle rebounds with nearly unchanged speed v_1. Run along with the velocity of the incoming alpha particle, and you will see it at rest, with an oncoming gold nucleus. The outgoing alpha particle heads to the left with a speed of approximately $2v_1$ from your point of view, because it is headed to the left with a speed of nearly v_1 while you are headed to the right with a speed v_1.

8.3 Scattering

We considered the simplest case of collisions between an alpha particle and a gold nucleus, a head-on collision. What happens if the particles don't collide head-on? Figure 8.12 shows what the trajectories look like in this case.

In atomic or nuclear collisions, we can't observe in detail the curving trajectories inside the tiny interaction region. We only observe the trajectories before and after the collision, when the particles are far apart and their mutual interaction is very weak, so they are traveling in nearly straight lines.

The alpha particle is deflected through some final angle θ, and the gold nucleus recoils at some other final angle ϕ. The final angles are measured relative to the initial direction of the alpha particle (Figure 8.13 and Figure 8.14). The process is called "scattering," and we say that the alpha particle "scattered" through an angle θ.

We'll again assume that there is no internal energy change inside either the alpha particle or the gold nucleus, so that the initial kinetic energy (before the collision) should be equal to the final kinetic energy (after the collision). We'll take the "before" time early enough and the "after" time late enough that the two particles are sufficiently far away from each other that we can neglect their electric potential energy. As for conservation of momentum, we must write two equations, for the x and y components of momentum conservation.

? Try to write the two momentum-conservation equations, and the energy-conservation equation. Choose x in the direction of the incoming alpha particle, and y at right angles to it. Consider ϕ to be a positive angle.

We write the three equations in terms of the magnitudes of the momenta:

$p_1 = p_3\cos\theta + p_4\cos\phi$ momentum conservation; x components

$0 = p_3\sin\theta - p_4\sin\phi$ momentum conservation; y components

$\dfrac{p_1^2}{2\,m} = \dfrac{p_3^2}{2\,m} + \dfrac{p_4^2}{2\,M}$ energy conservation; scalar

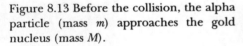

Figure 8.12 A collision that is not head-on.

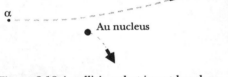

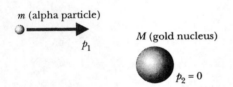

Figure 8.13 Before the collision, the alpha particle (mass m) approaches the gold nucleus (mass M).

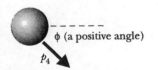

Figure 8.14 After the collision, when the particles are far apart and moving with nearly uniform velocities. Final angles are measured relative to the initial direction of the alpha particle.

Make absolutely sure that you understand every detail of these three equations! In particular, pay attention to the fact that since momentum is a vector, the momentum equations are written in terms of x and y components of the momentum.

? It is fruitful to reflect on which of the quantities in these three equations are known and which are unknown. We know the masses of the alpha particle and the gold nucleus, and we know the initial speed of the alpha particle. Which quantities in these three equations are unknown?

You should have identified the two final momenta and the two final angles as unknown quantities (p_3, p_4, θ, and ϕ, all of which represent positive numbers). But if there are four unknowns and only three equations, how can we solve for the unknowns? The simple answer is, we can't!

Nevertheless, there are many situations where these three equations can be exploited. For example, if we measure the final direction of motion of the alpha particle, the angle θ, we can solve for the other three unknown quantities and thereby predict the final momenta p_3 and p_4 of the two particles and the direction of motion ϕ of the gold nucleus, even without measuring all of these quantities.

It is worth mentioning that in experimental particle physics the momentum is often measured in a very direct way. Charged particles curve in a magnetic field due to magnetic forces that act on moving charges, and the radius of curvature is a direct measure of the momentum of the particle, if you know how much charge it has, which you do know if you know what kind of particle it is. Usually the momentum of a particle can be measured more accurately than the velocity, though once you have measured the momentum you can deduce the velocity if you know what kind of particle it is, since in that case you know its mass.

Example: Elastic collision between identical particles, one initially at rest

An interesting example of an elastic collision is one involving particles of equal mass, such as an elastic collision between two alpha particles, or two gold nuclei, or two billiard balls (whose collision is nearly elastic), in which one of the two objects (the "target") is initially at rest (Figure 8.15 and Figure 8.16).

Because the vector momentum is conserved, we can write $\vec{p}_1 = \vec{p}_3 + \vec{p}_4$, then calculate p_1^2 by expanding the dot product:

$$p_1^2 = \vec{p}_1 \bullet \vec{p}_1 = (\vec{p}_3 + \vec{p}_4) \bullet (\vec{p}_3 + \vec{p}_4) = p_3^2 + p_4^2 + 2p_3 p_4 \cos A$$

Divide by $2m$:

$$\frac{p_1^2}{2m} = \frac{p_3^2}{2m} + \frac{p_4^2}{2m} + \frac{2p_3 p_4 \cos A}{2m}$$

Since the kinetic energy K can be written as $p^2/2m$, we have

$$K_1 = K_3 + K_4 + \frac{2p_3 p_4 \cos A}{2m}$$

But energy conservation for this elastic collision says that $K_1 = K_3 + K_4$.

? What can you conclude about $\cos A$? What is the angle A?

In this special situation, an elastic collision between equal particles one of which is initially at rest, the angle between the final velocities must be $90°$, as long as neither p_3 nor p_4 is zero (in that case the equation reduces to $0\cos A = 0$, in which case $\cos A$ is indeterminate). If one of the final momenta

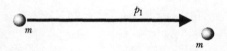

Figure 8.15 Before a collision of two identical particles, one initially at rest.

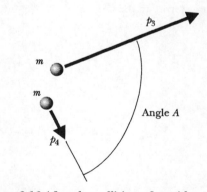

Figure 8.16 After the collision of two identical particles, one initially at rest.

is indeed zero, the other particle must have all the momentum of the initial particle. This can happen either because the incoming particle missed the target entirely and kept going in the original direction, or because the incoming particle hit the target dead center, in which case the only consistent solution of the equations is for the incoming particle to come to rest and the target particle to move forward with the initial momentum.

8.3.1 Impact parameter

The distance between centers perpendicular to the incoming velocity is called the "impact parameter" and is usually denoted by b. A head-on collision has an impact parameter of zero. Here are possible elastic collisions between two billiard balls, for various impact parameters:

If the impact parameter, b, is too large, there is no interaction at all between two billiard balls.

A grazing impact results in small angle scattering.

A medium impact parameter results in symmetric scattering.

In a head-on collision the impact parameter $b = 0$. In this case all the momentum is transferred to the target.

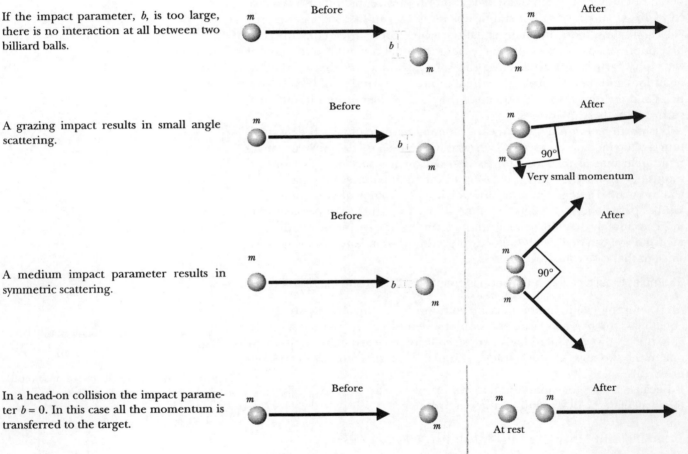

As you can see, the smaller the impact parameter, the more severe is the collision, and the larger the deflection angle of the incoming particle (larger "scattering"), except for a head-on collision, where the incoming particle stops dead and the target particle gets the entire momentum.

8.4 Discovering the nucleus inside atoms

In 1911 Ernest Rutherford and his coworkers decided to perform a scattering experiment in order to probe the internal structure of matter. They were familiar with the relationships of momenta in collisions that we have just studied, and they reasoned that by shooting microscopic particles at a thin layer of metal, they might be able to learn something about the microscopic structure of the metal by looking at patterns of scattering.

8.4.1 The plum pudding model of an atom

For about a decade, the prevailing model of an atom had been what was called a "plum pudding" model proposed by J. J. Thomson. Electrons were discovered in 1897 by Thomson, using a cathode ray tube, a device similar to a television picture tube. Thomson was able to show that the "cathode rays" that he and others had observed in electric gas discharges were in fact particles that were negatively charged, by the already existing convention for labeling electrical charges. This was the first conclusive demonstration of the existence of what we now call "electrons," one of the major constituents of all atoms.

Since these "electrons" were extracted from ordinary matter, which is normally not electrically charged, Thomson's discovery implied that matter also contained something having a positive charge. Because the electrons apparently had little mass, it seemed likely that most of the mass in ordinary matter was charged positively. Thomson proposed what was called a "plum pudding" model of atoms: a positively charged substance of uniform, rather low density, with tiny electrons distributed like raisins throughout it. Rutherford's group expected to gather evidence supporting this model from their collision experiments. Rutherford was extremely surprised by the results of these experiments, which led to a radical revision of the model.

8.4.2 The Rutherford experiment

A new source of fast-moving particles was provided through the discoveries of natural radioactivity by Henri Becquerel and by Marie and Pierre Curie. These newly discovered particles included high-speed, massive, positively charged "alpha particles," which we now know to be the nuclei of helium atoms, consisting of two protons and two neutrons bound together. Alpha particles are emitted in the radioactive decay of heavy elements. Alpha particles were convenient microscopic projectiles, and Rutherford's group decided to shoot alpha particles at a very thin piece of gold foil and observe where they reappeared.

? Based on the plum pudding model, what would Rutherford and his colleagues have expected to see in their collision experiments?

The researchers expected the alpha particles to be deflected only very slightly by interactions with the low-mass electrons and with the low-density "pudding" of positively charged matter in the gold atoms. They could detect a deflected ("scattered") alpha particle by a pulse of light that was emitted when the particle struck a fluorescent material (zinc sulfide). They moved their detector around at various angles to the incoming beam of alpha particles and measured what fraction of the alpha particles were scattered through various angles (Figure 8.17).

Rutherford and his coworkers were astounded to observe that an alpha particle, which is much more massive than an electron, sometimes bounced straight backwards, as though it had encountered an even more massive yet very tiny positive particle inside the gold foil, rather than being slightly deflected in its passage through the "pudding" of positive charge.

"It was quite the most incredible event that has ever happened to me in my life," Rutherford said. "It was almost as incredible as if you fired a 15-inch [artillery] shell at a piece of tissue paper and it came back and hit you."

He continued, "On consideration, I realized that this scattering backward must be the result of a single collision, and when I made the calculations I saw that it was impossible to get anything of that order of magnitude unless you took a system in which the greater part of the mass of the atom was concentrated in a minute nucleus. It was then that I had the idea of an atom with a minute massive center, carrying a charge." (See Figure 8.18.)

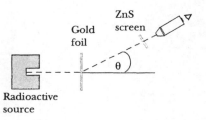

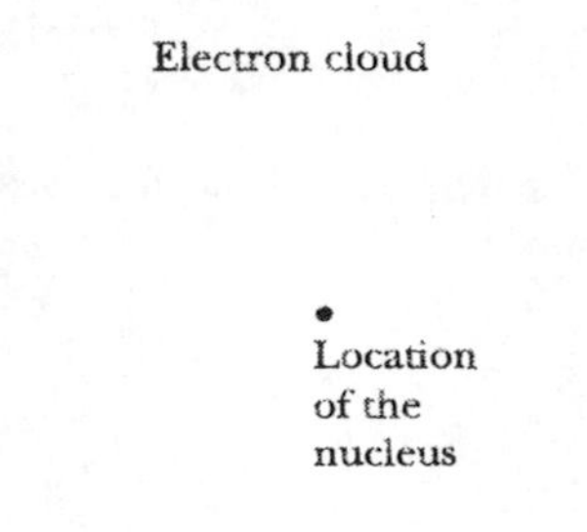

Figure 8.17 Schematic diagram of the Rutherford experiment. A radioactive source emits alpha particles, which are directed at a thin piece of gold foil. Particles scattered by gold nuclei in the gold foil hit a screen coated with ZnS, which fluoresces, giving off light, when hit by an energetic particle. The light was observed by the experimenter.

Electron cloud

•
Location
of the
nucleus

Figure 8.18 This diagram of an atom is not to scale; the radius of the nucleus is only about 1/100000 the radius of the electron cloud, so the nucleus would not be visible at all on this scale.

8.4.3 Computer modeling of the Rutherford experiment

Since the details of the electric interaction are now known, we can model the Rutherford experiment in detail. Remember that Coulomb's law for electric forces is very similar to Newton's law of gravitation. The electric force F_e between two small charged particles carrying amounts of charge q_1 and q_2, and located a distance r apart is this:

$$F_e = \frac{1}{4\pi\varepsilon_0} \frac{q_1 q_2}{r^2}$$

The charges q_1 and q_2 are measured in units of coulombs, and the proportionality factor is measured to be $1/(4\pi\varepsilon_0) = 9\times10^9 \mathrm{N \cdot m^2/coulomb^2}$. The electric charge of the proton is $e = +1.6\times10^{-19}$ coulomb, and the electric charge of the electron is $-e = -1.6\times10^{-19}$ coulomb.

━━

Problem 8.1 The Rutherford experiment

Use numerical integration techniques to explore possible trajectories of an alpha particle when it collides with a gold nucleus at different impact parameters. Your program should display the paths of the alpha particle and of the gold nucleus, and also plot the x and y components of momentum of both particles as a function of time. We will consider motion only in two dimensions (a slice through a 3-D beam of alpha particles).

An alpha particle is a helium nucleus (two protons and two neutrons, charge +2e); in the Rutherford experiment its initial kinetic energy was about 10 MeV (remember that 1 eV = 1.6×10^{-19} J). The gold nucleus is inside a piece of gold foil and is initially stationary; its charge is +79e and it contains 197 nucleons. Initially we will ignore the interactions of the alpha particle with the cloud of electrons surrounding the nucleus, which are in fact small.

(a) In this model it is important to pick reasonable initial conditions and display scales—otherwise you may not see anything meaningful. Before beginning your program, calculate or estimate reasonable values for the following and report your results:

• *Closest approach.* For a head-on collision, calculate approximately the distance of closest approach, given that the alpha particle and the gold nucleus repel each other. Assume that the gold atom's electrons are scattered away by the incoming alpha particle.

• *Initial position of the alpha particle.* Given the distance of closest approach, what would be a reasonable initial x for the alpha particle? What would be a good range to explore for the impact parameter b (initial y coordinate of the alpha particle)? Use these estimates to decide on a scale for the coordinates used to display the trajectories.

• *Value of Δt.* Trial and error alone will not be very helpful in estimating Δt. You can calculate the initial speed of the alpha particle, and you know the approximate size scale for the collision. Choose a value of Δt such that the alpha particle doesn't move very far in one step.

• *Time scale.* To plot p_x and p_y vs. t during the interaction, you must select a scale for the time axis that allows you to observe the entire interaction. About how long should the time axis be?

• *Momentum scale.* What should the maximum value of the p_x axis be? You should scale the p_y axis to the same value, to make comparisons easy.

• *Exit criterion.* You will need to exit from your computational loop when the particles have moved far apart. How will you test for this?

(b) Write a program that displays:

• The paths of the alpha particle and the gold nucleus.

• The value of the impact parameter b, and the value of the corresponding scattering angle (angle between the initial path of the alpha par-

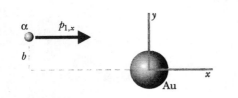

Coordinate system for Rutherford scattering calculations.

ticle and its final path). The angle can be calculated by taking the arctangent of p_y/p_x, using the final values.

- A plot of p_x for each particle and total p_x, and a similar plot for p_y. (Consider what these plots should look like if the program is working properly.)

Report what value of Δt gives adequate accuracy. What is your criterion? Report values of the impact parameter b that produce the following:

- A very small scattering angle
- An angle around 90 degrees
- A large deflection angle (back-scattering)

(c) What other physical quantity could you plot that would provide a check on the accuracy of your calculation? Carry out this check and report what value of Δt provides good accuracy according to this check.

(d) To see what a different model of matter would have predicted, replace the gold nucleus in your model by a proton. Describe briefly what you observe. (Note that if the protons were uniformly spread out throughout the volume of the atom, they would be far enough apart that on average an alpha particle would interact with only one of them.)

8.4.4 Distribution of scattering angles

In general, we have three equations describing an elastic collision—two momentum equations (in the x and y directions) and one energy equation, but there are four unknown quantities (the outgoing speeds and directions for each particle). The concept of impact parameter provides an additional quantity, and you have seen on page 272 how the impact parameter determines the scattering angle θ of the incoming particle, from which you can predict the other final values (p_3, p_4, and ϕ).

In a macroscopic situation, such as shooting billiard balls, you can choose an impact parameter simply by carefully aiming at a point a distance b from the center of the target. Unfortunately, in experiments where you fire alpha particles or other subatomic projectiles at nuclei, you can't aim at a specific impact parameter because the target is so incredibly small. All you can do is fire lots of projectiles at the subatomic target and expect a probabilistic distribution of impact parameters

When we fire a projectile at this microscopic target, the probability of hitting one particular ring (whose radius corresponds to the impact parameter) is proportional to the area of that ring (Figure 8.19). If you calculate the range of scattering angles that corresponds to hitting somewhere within a particular ring, you can predict the angular distribution of scattering angles that will be observed.

Cross section

If a target particle such as a gold nucleus has a radius R, we say that its "geometrical cross section" is πR^2. A high-speed alpha particle that hits within this cross-sectional area will interact strongly with the gold nucleus. Quantum-mechanical effects can make the effective target size bigger or smaller than the geometrical cross section, and the effective target size is called the "cross section" for interactions of this kind. Cross sections are measured experimentally by observing how often projectiles are deflected by an interaction, instead of passing straight through the material without interacting. Cross sections have units of square meters.

A "differential" cross section for scattering within a particular range of angles is the area in square meters of a ring bounded by impact parameters corresponding to the given angular range.

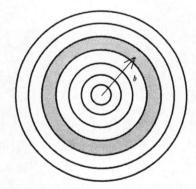

Figure 8.19 The probability of hitting a particular ring is proportional to the area of the ring.

8.4.5 Conservation laws *vs.* details of the interaction

We used general principles to determine the final momenta of two colliding particles (conservation of momentum, conservation of energy). We didn't have to say what kind of force was involved. But doesn't it make a difference whether the interaction is electric or nuclear? Yes, the type of interaction determines the *distribution* of scattering angles. For example, in what fraction of the collisions is the scattering angle between 0 and 10 degrees? Between 10 and 20 degrees? Between 20 and 30 degrees? These fractions are different for different forces.

To put it another way, if the scattering angle is 20 degrees, the momenta are determined. The question is, how often is the scattering angle in fact near 20 degrees? Measurement of the distribution of scattering angles in collisions at modern particle accelerators is one of the ways used to study the nature of the strong and weak interactions.

Rutherford used Coulomb's law of electric force to make a detailed prediction of the distribution of scattering angles, assuming that the positive nucleus was point-like and very massive. That is, he tried to predict what fraction of the alpha particles would be scattered by less than 10 degrees, by an angle between 10 and 20 degrees, by an angle between 20 and 30 degrees, etc. The observed distribution of scattering angles agreed with the predicted distribution (Figure 8.20 shows a portion of this distribution).

The calculation that Rutherford did involves a lot of complicated geometry, algebra, and probability (with a random distribution of impact parameters), so we won't go through it. But the main features of Figure 8.20 are easily understood. Because the gold nucleus is very small, most of the time the alpha particle will miss the nucleus by a large distance. With a large impact parameter the electric force is very small, and the deflection (scattering angle) is very small. For that reason small scattering angles are common. Large scattering angles occur only for nearly head-on collisions (small impact parameter), and these events are rare.

Since Rutherford's prediction assumed a point-like nucleus, the observations were consistent with the nucleus being very small. Also, the good fit between prediction and observation strengthened the belief that the interaction was purely electric in nature (the strong interaction played no role because it is a very short-range interaction, and the alpha particle was not going fast enough to overcome the electric forces and contact the gold nucleus).

If the alpha particle has high enough energy, it can overcome the electric repulsion and make contact with the gold nucleus. In that case the strong interaction comes into play, and the angular distribution changes. The radius of the gold nucleus can be determined by finding out how much energy an alpha particle has to have in order to obtain an angular distribution that deviates from what is expected due just to the electric interaction.

? For a particular scattering angle, can we use momentum and energy conservation to determine the unknown angle and momenta, even if the alpha particle makes contact with the gold nucleus?

Yes. These conservation principles are completely general and apply to all types of interactions. But they cannot by themselves predict the angular distribution—that is, how often a particular range of scattering angles will occur. To do that you have to have a mathematical description for the specific interaction, such as the $1/r^2$ force law for electric interactions.

8.5 Relativistic momentum and energy

Rutherford's original experiments were carried out with alpha particles whose kinetic energy was about 10 MeV (10 million electron-volts), with a

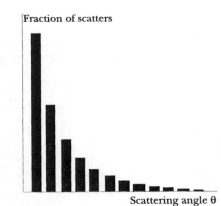

Fraction of scatters

Scattering angle θ

Figure 8.20 For an electric interaction, the fraction of scatters at different scattering angles falls rapidly with angle.

corresponding speed that was small compared to the speed of light. In order to probe the nucleus more deeply, physicists later built special machines, called "particle accelerators," which use electric forces to accelerate charged particles to very much higher energies, with speeds that may be as much as 99.9999% of the speed of light. For such high speeds we cannot use the low-speed approximate formulas for momentum and energy.

Physicists analyzing collisions produced by high-energy accelerators routinely do calculations based on the relativistic energy and momentum formulas. Consider a collision in which a particle of mass m_1 and momentum p_1 strikes a stationary particle of mass m_2 (Figure 8.21). There is a change of identity, with the production of two new particles of mass m_3 and m_4, with momenta p_3 and p_4, at angles θ and ϕ to the horizontal (Figure 8.22).

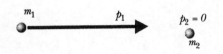

Figure 8.21 Before a high-energy collision.

We will write down relativistically correct equations for momentum conservation and energy conservation, in terms of momentum p and energy E, knowing that we can determine E from p, since $E^2 - (pc)^2 = (mc^2)^2$:

$p_1 = p_3\cos\theta + p_4\cos\phi$ momentum conservation; x components

$0 = p_3\sin\theta - p_4\sin\phi$ momentum conservation; y components

$E_1 + m_2 c^2 = E_3 + E_4$ energy conservation; scalar

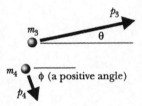

Figure 8.22 After the collision.

The energy equation is equivalent to

$$\sqrt{(p_1 c)^2 + (m_1 c^2)^2} + m_2 c^2 = \sqrt{(p_3 c)^2 + (m_3 c^2)^2} + \sqrt{(p_4 c)^2 + (m_4 c^2)^2}$$

? Suppose the initial momentum p_1 of the projectile is known, and all four masses are known (because we were able to identify the particles). How many unknown quantities are there, considering that we can write the energy equation in terms of p?

There are four unknown quantities, the magnitudes of the momenta and the directions of the two outgoing particles (p_3, p_4, θ, and ϕ), but only three equations. A typical situation in a high-energy particle experiment is that one of these unknown quantities is measured, and momentum and energy conservation are used to calculate the other three quantities.

As in the Rutherford experiment, measurements of the outgoing particle momenta do not tell us anything about the detailed nature of the interaction involved in the collision, because they are determined by general momentum and energy conservation laws. The identity changes of the particles tell us a lot, and the distribution of scattering angles also tells us a lot.

Photons and neutrinos

If a particle has zero mass (the photon and possibly the neutrino), the relationship $E^2 - (pc)^2 = (mc^2)^2$ reduces simply to $E = pc$. For the photon we can't write its energy as $mc^2 / \sqrt{1 - v^2/c^2}$, because both the numerator and denominator are zero, and $0/0$ is indeterminate. Nevertheless, $E = pc$ is valid: a photon of energy E carries momentum $p = E/c$.

Identifying a particle

A variation on the relativistic collision analysis can be used to identify one of the outgoing particles. Suppose we measure p_3 and θ, and we know m_3 but not m_4. We can use our three equations in the three unknowns (m_4, p_4, and ϕ) to solve for the unknown mass m_4. This is a very powerful technique when we are unable to observe the particle directly. For example, neutrinos don't have electromagnetic or strong interactions and therefore typically fail to register in ordinary detectors. But if we measure the other outgoing particle we can determine the properties of the unseen particle.

The discovery of the neutrino

A specific example of such analyses was the original motivation for proposing the existence of the neutrino. The neutron (either free or when bound into some nuclei) was observed to decay with the creation of a proton and an electron, both charged particles that are easy to detect. But the amount of energy and momentum carried by the proton and electron varied from one decay to the next, which made no sense:

$$n \rightarrow p^+ + e^- \text{, violating momentum and energy conservation??}$$

It was proposed that there was an unseen third particle, called the neutrino (or antineutrino in neutron decay), which carried energy and momentum that would salvage the conservation laws:

$$n \rightarrow p^+ + e^- + \bar{\nu} \text{, conserving momentum and energy}$$

It was possible to solve for the properties of the unseen particle by the techniques outlined above, and it was found that the neutrino must have zero mass and must travel at the speed of light. (Recent experimental evidence suggests that the neutrino may have a very small nonzero mass. If so, it must travel at slightly less than the speed of light.)

For many years the neutrino was just a theoretical construct which balanced the accounts, but eventually the neutrino was detected directly. The neutrino is very hard to detect because it does not have electromagnetic or strong interactions, only weak interactions, so the most probable thing for it to do is to pass right through any detector you place in its way. In fact, neutrinos emitted in nuclear reactions in the Sun mostly pass right through the entire Earth! However, with extremely intense neutrino beams produced in nuclear reactors or accelerators, or with gigantic underground detectors used to identify neutrinos coming from the Sun, it is possible to have a measurable rate of neutrino reactions. (A current puzzle is that many fewer neutrinos are detected coming from the Sun than expected.)

8.6 Inelastic collisions

Collisions in which the initial and final internal energies of the particles are the same are called "elastic" collisions. The Rutherford experiment involved elastic collisions, because the internal energies of the gold nucleus and the alpha particle didn't change during the interaction. When there is an internal energy change, or the production of additional particles or the emission of light, we call the collision "inelastic."

If we shoot alpha particles at a gold nucleus with higher energy than the energy of alpha particles in the Rutherford experiment, we may be able to excite the gold nucleus to a higher quantum energy level, an amount ΔE above the ground state of the gold nucleus (Figure 8.23). Such a state corresponds to a different configuration of the nucleons inside the gold nucleus.

Suppose that you know the incoming momentum p_1 of the alpha particle, and you are able to measure its outgoing momentum p_3 and scattering angle θ (Figure 8.24).

? What is the form of the energy-conservation and momentum-conservation equations in this case? Don't assume that the speeds are small compared to the speed of light. Note that the rest energy of the recoiling gold nucleus is $M_4 c^2 + \Delta E$ because it is in an excited state.

Here are the relevant equations:

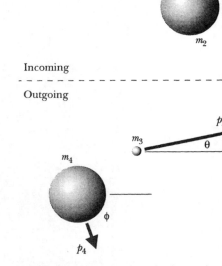

Incoming

Outgoing

Figure 8.24 You measure the incoming and outgoing momenta of the alpha particle and deduce something about the energy levels in the gold nucleus.

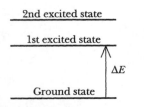

2nd excited state

1st excited state

ΔE

Ground state

Figure 8.23 Energy levels for a gold nucleus. A sufficiently energetic alpha particle may have enough energy to raise the nucleus above its ground state.

$$p_1 = p_3 \cos\theta + p_4 \cos\phi \,,$$

where p_4 and ϕ refer to the gold nucleus, which is not observed in the experiment.

$$0 = p_3 \sin\theta - p_4 \sin\phi$$

$$E_1 + Mc^2 = E_3 + E_4 \,, \text{ where } E_4 = \sqrt{(p_4 c)^2 + (M_4 c^2 + \Delta E)^2}$$

? How many unknown quantities are there in the energy and momentum equations? Is there enough information to be able to determine the internal energy difference ΔE? (This is one of the methods used to study the energy levels of nuclei. This energy level determination can be verified by measuring the energy of a gamma-ray photon emitted when the nucleus drops back to the ground state.)

There are three unknowns (p_4, ϕ, and ΔE), and there are three equations (energy, x and y momentum), so we could determine ΔE.

8.6.1 Change of identity

Instead of merely exciting the target particle to a higher internal energy state, suppose that there is a change of identity in which the target particle turns into some other particle. An example is the following particle reaction, which was used in the 1950's to discover the Δ^+ particle of mass m_Δ, which is a particle more massive than a proton:

$$\pi^- + p^+ \rightarrow \pi^- + \Delta^+$$

A beam of negative pions (π^-) of known mass m_π and known kinetic energy, produced by a particle accelerator, is directed at a container of hydrogen gas, and the targets for the collisions are the protons of mass m_p that are the nuclei of the hydrogen atoms. Sometimes an incoming pion interacts through the strong interaction with a proton to produce a Δ^+ particle. The momentum of the scattered pion can be measured by measuring the radius of curvature of its curving trajectory in a magnetic field.

? What is the form of the energy-conservation and momentum-conservation equations in this case? (Don't assume that the speeds are small compared to the speed of light.)

Here are the relativistically correct equations:

$$p_1 = p_3 \cos\theta + p_4 \cos\phi$$

$$0 = p_3 \sin\theta - p_4 \sin\phi$$

$$E_1 + m_p c^2 = E_\pi + E_\Delta \,, \text{ where } E^2 - (pc)^2 = (mc^2)^2$$

? How many unknown quantities are there in the energy and momentum equations? Is there enough information to be able to determine the mass of the Δ^+ particle?

There are three unknowns (p_4, ϕ, and E_Δ) and three equations, so it is possible to determine the mass of the Δ^+ particle.

8.6.2 What we learn from inelastic collisions

The analysis for finding the mass change in the preceding exercises is essentially the same as the analysis for determining the internal energy change of an excited gold nucleus. In fact, we can say that the Δ^+ particle is an excited state of three quarks, with the proton representing the ground state of three quarks (Figure 8.25). The excited state (the Δ^+ particle) can decay to the ground state (the proton) with the emission of a neutral pion (π^0), which is a quark-antiquark pair. The extra kinetic energy is equal to $(m_\Delta c^2 - m_p c^2)$. Scattering experiments have been an important tool for discovering new particles (or new excited states of quark systems).

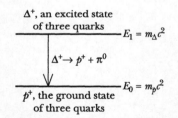

Figure 8.25 The Δ^+ particle can be considered to be an excited state of three quarks, with the proton representing the ground state. A decay to the ground state can occur with the emission of a neutral pion (π^0).

8.6.3 A sticking collision

To illustrate the wide applicability of the momentum principle, we'll consider an inelastic collision in which two objects stick together when they make contact. A macroscopic example is the collision of a truck and a car in an icy intersection (Figure 8.26). In the figure we have shown velocity vectors rather than momentum vectors; a typical truck would have much more momentum than the car, given the truck's much greater mass.

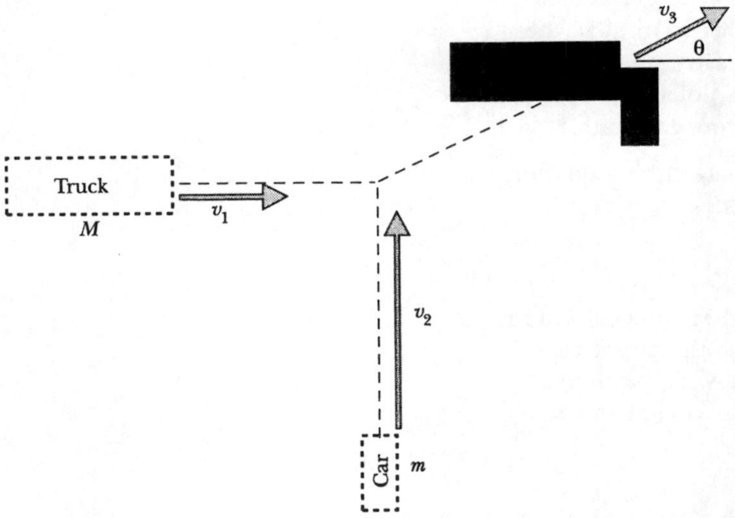

Let M be the mass of the truck, m the mass of the car, and θ the angle of their mutual final velocity, as shown in the diagram. Also let v_1 be the initial speed of the truck, v_2 the initial speed of the car, and v_3 the final speed when they're stuck together and sliding on the ice.

? How does the fact that the intersection is icy affect our ability to apply conservation of momentum to the truck-car collision?

Friction with the road is small if the intersection is icy, so we can presumably ignore road friction and air resistance during the short time of the actual collision. Therefore the total momentum of the truck+car system should be (approximately) unchanged. Momentum conservation is really three equations, one each for the x, y, and z directions. In the z direction (vertically upward) the ice simply supports the weights of the vehicles, so the net force is zero, and the zero z component of momentum remains zero throughout the collision.

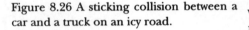

Figure 8.26 A sticking collision between a car and a truck on an icy road.

? In terms of the named quantities, write two equations for the x and y components of the momentum conservation equation, equating momentum components before the collision to momentum components after the vehicles stick together.

Here are the x and y components of the momentum conservation:

$$Mv_1 = (M+m)v_3\cos\theta$$
$$mv_2 = (M+m)v_3\sin\theta$$

Assume that we know the initial speeds of the truck and car (v_1 and v_2). Then we have two equations in two unknowns (v_3 and θ), and we can solve for the final speed and angle. First, divide the y equation by the x equation:

$$\frac{(M+m)v_3\sin\theta}{(M+m)v_3\cos\theta} = \frac{mv_2}{Mv_1} = \tan\theta, \text{ so } \theta = \arctan\left(\frac{mv_2}{Mv_1}\right)$$

Next square the equations and add them:

$$(Mv_1)^2 + (mv_2)^2 = [(M+m)v_3]^2[(\sin\theta)^2 + (\cos\theta)^2]$$

$$\sqrt{(Mv_1)^2 + (mv_2)^2} = (M+m)v_3$$

$$v_3 = \frac{\sqrt{M^2v_1^2 + m^2v_2^2}}{M+m}$$

We used the trig identity $\sin^2\theta + \cos^2\theta = 1$, which corresponds to the Pythagorean theorem applied to a triangle whose hypotenuse has length 1.

Energy in a sticking collision

A sticking collision is unusual in that momentum conservation alone gives us enough equations (two) to be able to solve for the unknown final speed and unknown direction. If the car and truck bounce off each other instead of sticking, there are four unknowns instead of two, because there are two final velocities, each with a speed and a direction to be calculated. In that case momentum conservation alone doesn't provide enough equations to be able to analyze the situation fully.

There is an important issue associated with energy in a sticking collision such as this one. Using our result for the final speed, we can write the kinetic energy after the collision:

$$\tfrac{1}{2}(M+m)v_3^2 = \tfrac{1}{2}(M+m)\left[\frac{(Mv_1)^2+(mv_2)^2}{(M+m)^2}\right]$$

After minor rearrangements we get the following:

$$\tfrac{1}{2}(M+m)v_3^2 = \left(\frac{M}{M+m}\right)(\tfrac{1}{2}Mv_1^2)+\left(\frac{m}{M+m}\right)(\tfrac{1}{2}mv_2^2)$$

Since $M/(M+m)$ is less than 1, and $m/(M+m)$ is also less than 1, the final kinetic energy is less than the initial kinetic energy.

? Where is the "missing" kinetic energy?

Part of the energy was radiated away as sound waves in the air, produced when the vehicles collided with a bang. Some of the energy went into raising the temperature of the vehicles, so that their thermal energy is now higher than before (larger random motions of the atoms).

Some of the energy went into deforming the metal bodies of the truck and car. A bent fender has a higher internal energy, associated with a change in the configuration of the atoms. Sometimes you can get such energy back (as when you compress a spring, changing the configuration of the atoms in the spring, and later let it push back, returning to its original configuration), and we speak of "elastic" energy. But sometimes you can't get the energy back, because the configuration change gets "locked in."

A sticking collision is an example of an "inelastic" collision, referring to the effective "loss" of kinetic energy into forms that are inaccessible.

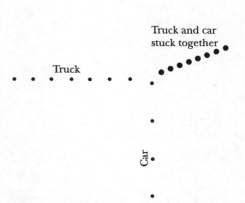

Figure 8.27 View of collision from hovering (stationary) helicopter.

Ex. 8.5 Two cars collide, one headed north and one headed west. They stick together and leave skid marks on the pavement, which show that car 1 was deflected 30° (so car 2 was deflected 60°. What can you conclude about the cars before the collision?

8.7 Viewing collisions from other reference frames

We can show that the truck-car collision process looks a lot simpler from the vantage point of a helicopter flying over the scene with the final velocity of the truck+car combined system. This will lead to new insights. Suppose the accident takes place at night, and from the helicopter all you can see are lights mounted on top of the two vehicles. Figure 8.27 shows a multiple exposure of the scene that you would see from a helicopter that is hovering over the intersection.

Suppose instead that your helicopter flies over the intersection with the same velocity that the stuck-together truck and car share after their collision, as shown in Figure 8.28. The view thus obtained is in an important sense "simpler," as shown in Figure 8.29. From the moving observation platform you see the truck and car run into each other and then stand still (since you and they now have the same velocity). In particular, in this (moving) reference frame it is obvious that there is a loss of kinetic energy into

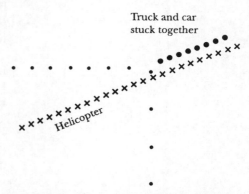

Figure 8.28 Path of moving helicopter that produces the simple view seen in Figure 8.29.

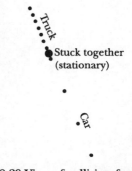

Figure 8.29 View of collision from helicopter moving with velocity of stuck-together wreck.

other forms of energy, because after the collision the kinetic energy is clearly zero. In contrast, we had to go through a fair amount of algebra in the non-moving reference frame in order to establish the loss of kinetic energy.

The reason that we imagine making these observations at night is so that you won't be distracted by the moving background. If you actually view this collision from a moving helicopter in daylight, there is a strong tendency to compensate mentally for your motion and to put yourself (mentally) into the reference frame of the road, in which case you could miss the important fact that *relative to you*, the car and truck are stationary after the collision.

You may have had an experience related to this situation. You're sitting in a bus or train, waiting to leave. You think you've started forward, but then you realize that you were fooled by a neighboring bus or train moving backwards. For a moment you were mentally in the reference frame of the moving vehicle.

8.7.1 Total momentum is zero in the center-of-mass frame

Viewing a collision from a moving reference frame can provide insights into all kinds of collisions, not just sticking collisions. The key point that leads to a useful generalization is that the total momentum of the truck+car system after their collision is zero when the vehicles are motionless relative to your helicopter.

? In this moving reference frame, if the total momentum is zero after the collision, what was the total momentum before the collision? Why?

The helicopter is moving at constant velocity (constant speed and direction), and we know that the momentum principle is valid in this uniformly moving reference frame. So the principle of momentum conservation still holds, and the initial total momentum of the car+truck system must be zero if the final total momentum is zero. That's what makes the process look so simple: the total momentum is zero before and after the collision in the moving reference frame.

In viewing the scene from the moving helicopter, we have subtracted from each of the velocities an unchanging, constant vector velocity, the velocity $\vec{v}_h$ of the helicopter, and this subtraction makes no change in the momentum principle for the car (or truck):

$$\frac{d[\,m(\vec{v} - \vec{v}_h)\,]}{dt} = \frac{d[\,m\vec{v}\,]}{dt} = \vec{F}_{\text{net}}$$

The net force is the same in both reference frames because the contact electric interaction forces depend on interatomic distances, which are the same in both reference frames.

Evidently we gain insight into a collision by viewing the collision in a moving reference frame where the total momentum is zero. How do we determine the appropriate velocity for such a reference frame? Let this velocity be $\vec{v}_{\text{fr}}$, which is to be subtracted from all the original velocities to obtain the velocities as viewed in the moving reference frame. We want the total momentum to add up to zero in the moving frame:

$$\vec{P}_{\text{tot,new frame}} = m_1(\vec{v}_1 - \vec{v}_{\text{fr}}) + m_2(\vec{v}_2 - \vec{v}_{\text{fr}}) = 0$$

$$(m_1 + m_2)\vec{v}_{\text{fr}} = m_1\vec{v}_1 + m_2\vec{v}_2$$

$$\vec{v}_{\text{fr}} = \frac{m_1\vec{v}_1 + m_2\vec{v}_2}{m_1 + m_2}$$

We have solved for a particular velocity for the moving reference frame that makes the total momentum in that frame be zero. But look at the form of this equation.

? Does this equation look familiar?

Evidently $\vec{v}_{fr} = \vec{v}_{cm}$, the velocity of the center of mass of the system, since

$$\vec{r}_{cm} = \frac{m_1\vec{r}_1 + m_2\vec{r}_2}{m_1 + m_2}$$

We have obtained an important result. In a reference frame moving with the center of mass of a system, the total momentum of the system is zero. If no external forces act, the total momentum is unchanged, and the total momentum is zero before and after a collision.

We can also express the frame velocity directly in terms of the total momentum:

$$\vec{v}_{cm} = \frac{m_1\vec{v}_1 + m_2\vec{v}_2}{m_1 + m_2} = \frac{\vec{P}_{tot}}{M_{tot}}$$

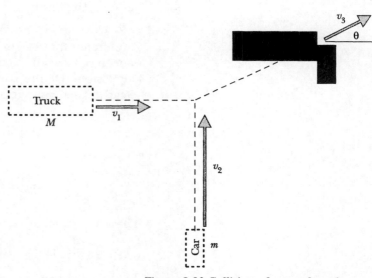

Figure 8.30 Collision of car and truck.

This provides another way to calculate the frame velocity and shows why this is often called the "center-of-momentum" velocity.

Let's reanalyze the car-truck collision from this new point of view. Again let M be the mass of the truck, m the mass of the car, and θ the angle of their mutual final velocity, as shown in Figure 8.30. Also let v_1 be the initial speed of the truck, v_2 the initial speed of the car, and v_3 the final speed when they're stuck together and sliding on the ice.

? What are the x and y components of $\vec{v}_{cm}$ in terms of v_1 and v_2?

$$\vec{v}_{cm} = <\frac{Mv_1}{M+m}, \frac{mv_2}{M+m}, 0>$$

? In the moving center-of-momentum reference frame, what are the x and y components of the car's velocity before the collision? Remember that you find the car's velocity (or its components) in the center-of-momentum frame by subtracting the center-of-momentum velocity (or its components) from that of the car. (That is, the view from the helicopter is obtained by subtracting the helicopter's velocity from the velocities of the car and truck.)

$$<-\frac{Mv_1}{M+m}, v_2 - \frac{mv_2}{M+m}, 0> = <-\frac{Mv_1}{M+m}, \frac{Mv_2}{M+m}, 0>$$

? In the moving center-of-momentum reference frame, what are the x and y components of the truck's velocity before the collision? Show that this velocity is in a direction opposite to the velocity of the car, and that the total momentum is zero.

$$<v_1 - \frac{Mv_1}{M+m}, -\frac{mv_2}{M+m}, 0> = <\frac{mv_1}{M+m}, -\frac{mv_2}{M+m}, 0>$$

The x and y components of velocity are in the opposite direction to those of the car (multiply the car's velocity by $-m/M$ and you have the truck's velocity). Also, the total momentum is zero in the center-of-momentum frame:

$$<\left[m\left(-\frac{Mv_1}{M+m}\right) + M\left(\frac{mv_1}{M+m}\right)\right], \left[m\left(\frac{Mv_2}{M+m}\right) + M\left(-\frac{mv_2}{M+m}\right)\right], 0> = 0$$

? In the moving center-of-momentum reference frame, what is the speed of the car after the collision? Of the truck? Why?

Both are zero; in the center-of-mass frame the total momentum is zero, and with the car and truck stuck together the only way for this to be true is for them to be at rest. Because of this, the kinetic energy loss to other kinds of energy is equal to the total kinetic energy in the center-of-momentum frame before the collision.

? Now, after the collision, transform back to the (non-moving) reference frame of the intersection, by adding back in the $\vec{v}_{cm}$ that had been subtracted. What are the speed and direction of the car after the collision? Of the truck? Compare with the results obtained in our earlier analysis of this collision.

$$\vec{v}_3 = <\frac{Mv_1}{M+m}, \frac{mv_2}{M+m}, 0>, \text{ which yields}$$

$$v_3 = \frac{\sqrt{M^2 v_1^2 + m^2 v_2^2}}{M+m}; \ \theta = \arctan\left(\frac{mv_2}{Mv_1}\right)$$

These are the same results we obtained in our earlier analysis.

8.7.2 Energy in the two reference frames

As far as energy is concerned, note that the kinetic energy of the car+truck system is as usual

$$K_{tot} = K_{trans} + K_{rel}, \text{ where } K_{trans} = \frac{P_{tot}^2}{2M_{tot}}$$

In the original reference frame, $K_{trans} > 0$ but is the same before and after the collision, since P_{tot} doesn't change. In the center-of-momentum reference frame, $K_{trans} = 0$ both before and after the collision, because $P_{tot} = 0$ in the center-of-momentum reference frame.

In either reference frame, the kinetic energy relative to the center of mass, K_{rel}, decreases to zero in the collision, because the car and truck after the collision are traveling with the center of mass. It is ΔK_{rel} that represents the loss of kinetic energy in this inelastic collision, and this quantity is the same in both reference frames.

Also note that distances between atoms are the same in both reference frames, so potential energy associated with bending of metal parts is the same in both frames.

Ex. 8.6 A 1000 kg car moving east at 30 m/s runs head-on into a 3000 kg truck moving west at 20 m/s. The vehicles stick together. Use the concept of the center-of-momentum frame to determine how much kinetic energy is lost.

8.8 Summary

Fundamental principles

CONSERVATION OF MOMENTUM

$$\Delta \vec{P}_{system} + \Delta \vec{P}_{surroundings} = 0$$

New concepts

The notion of a "collision"
Scattering, and scattering distributions
The center-of-mass reference frame

$$\vec{v}_{cm} = \frac{m_1 \vec{v}_1 + m_2 \vec{v}_2}{m_1 + m_2} = \frac{\vec{P}_{tot}}{M_{tot}}$$

Results

Evidence for the nuclear model of an atom

Problem solving techniques

Conservation of momentum applied to collisions
Simplification of calculations in the center-of-mass reference frame

8.9 Example problem: Photodissociation of the deuteron

A photon of energy E is absorbed by a stationary deuteron. As a result, the deuteron breaks up into a proton and a neutron. The photon energy is sufficiently low that the proton and neutron have speeds small compared to the speed of light. The proton is observed to move in a direction at an angle θ to the direction of the incoming photon.

Write equations that could be solved to determine the unknown magnitudes of the proton momentum p_p and neutron momentum p_n, and the positive angle α of the neutron to the direction of the incoming photon. Let the precise masses of the particles be M_d, M_p, and M_n for the deuteron, proton, and neutron. Explain clearly on what principles your equations are based. Provide a physics diagram of the reaction and label the angles. Do not attempt to solve the equations.

Solution

Fundamental principles: momentum and energy conservation.

Momentum conservation (Figure 8.31):

$$\vec{\mathbf{P}}_{\text{photon}} = \vec{\mathbf{P}}_p + \vec{\mathbf{P}}_n$$

A photon with energy E has momentum $p = E/c$:

$$\left\langle \left(\frac{E}{c}\right), 0, 0\right\rangle = \left\langle (p_p\cos\theta + p_n\cos\alpha), (p_p\sin\theta - p_n\sin\alpha), 0\right\rangle$$

This is equivalent to two component equations:

$$\frac{E}{c} = (p_p\cos\theta + p_n\cos\alpha)$$

$$0 = (p_p\sin\theta - p_n\sin\alpha)$$

Energy conservation:

$$E + M_d c^2 = \left(M_p c^2 + \frac{p_p^2}{2M_p}\right) + \left(M_n c^2 + \frac{p_n^2}{2M_n}\right)$$

We have three equations (two momentum component equations and one energy equation) in three unknowns: p_p, p_n, and α. So in principle we can solve for the unknown quantities, though the algebra may be messy.

We could also have written the energy equation relativistically, using the relationship $E^2 - (pc)^2 = (mc^2)^2$ which gives $E = \sqrt{(pc)^2 + (mc^2)^2}$:

$$E + M_d c^2 = \sqrt{(p_p c)^2 + (M_p c^2)^2} + \sqrt{(p_n c)^2 + (M_n c^2)^2}$$

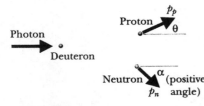

Figure 8.31 Physics diagram for photodissociation of deuteron.

8.10 Review questions

Momentum and impulse

RQ 8.1 A force <30, 0, 0> N acts for 0.2 seconds on an object of mass 1.2 kg whose initial velocity was <0, 20, 0> m/s. What is the new velocity?

Conservation of momentum

RQ 8.2 Under what conditions is the momentum of a system constant? Can the x component of momentum be constant even if the y component is changing? In what circumstances? Give an example of such behavior.

Collisions

RQ 8.3 What happens when a high-mass object hits a low-mass object head-on? What happens when a low-mass object hits a high-mass object head-on?

RQ 8.4 What properties of the alpha particle and the gold nucleus in the original Rutherford experiment were responsible for the collisions being elastic collisions?

RQ 8.5 Give an example of what we can learn about matter through the use of momentum and energy conservation applied to scattering experiments. Explain what it is that we cannot learn this way, for which we need to measure the distribution of scattering angles.

Sticking collisions

RQ 8.6 In an elastic collision involving known masses and initial momenta, how many unknown quantities are there after the collision? How many equations are there? In a sticking collision involving known masses and initial momenta, how many unknown quantities are there after the collision? Explain how you can determine the amount of kinetic energy change.

RQ 8.7 In order to close a door, you throw an object at the door. Which would be more effective in closing the door, a 50 gram tennis ball or a 50 gram lump of sticky clay? Explain clearly what physics principles you used to draw your conclusion.

RQ 8.8 A bullet of mass m traveling horizontally at a very high speed v embeds itself in a block of mass M that is sitting at rest on a nearly frictionless surface.

(a) What is the speed of the block after the bullet embeds itself in the block?

(b) Calculate the total translational kinetic energy before and after the collision. Compare the two results and explain why there is a difference.

Changing reference frame

RQ 8.9 What is it about analyzing collisions in the center-of-mass frame that simplifies the calculations?

8.11 Homework problems

8.11.1 In-line problems from this chapter

Problem 8.1 (page 274) The Rutherford experiment

Use numerical integration techniques to explore possible trajectories of an alpha particle when it collides with a gold nucleus at different impact parameters. Your program should display the paths of the alpha particle and of the gold nucleus, and also plot the x and y components of momentum of both particles as a function of time. We will consider motion only in two dimensions (a slice through a 3-D beam of alpha particles).

An alpha particle is a helium nucleus (two protons and two neutrons, charge +2e); in the Rutherford experiment its initial kinetic energy was about 10 MeV (remember that 1 eV = 1.6×10^{-19} J). The gold nucleus is inside a piece of gold foil and is initially stationary; its charge is +79e and it contains 197 nucleons. Initially we will ignore the interactions of the alpha particle with the cloud of electrons surrounding the nucleus, which are in fact small.

(a) In this model it is important to pick reasonable initial conditions and display scales—otherwise you may not see anything meaningful. Before beginning your program, calculate or estimate reasonable values for the following and report your results:

• *Closest approach.* For a head-on collision, calculate approximately the distance of closest approach, given that the alpha particle and the gold nucleus repel each other. Assume that the gold atom's electrons are scattered away by the incoming alpha particle.

• *Initial position of the alpha particle.* Given the distance of closest approach, what would be a reasonable initial x for the alpha particle? What would be a good range to explore for the impact parameter b (initial y coordinate of the alpha particle)? Use these estimates to decide on a scale for the coordinates used to display the trajectories.

• *Value of Δt.* Trial and error alone will not be very helpful in estimating Δt. You can calculate the initial speed of the alpha particle, and you know the approximate size scale for the collision. Choose a value of Δt such that the alpha particle doesn't move very far in one step.

• *Time scale.* To plot p_x and p_y vs. t during the interaction, you must select a scale for the time axis that allows you to observe the entire interaction. About how long should the time axis be?

• *Momentum scale.* What should the maximum value of the p_x axis be? You should scale the p_y axis to the same value, to make comparisons easy.

• *Exit criterion.* You will need to exit from your computational loop when the particles have moved far apart. How will you test for this?

(b) Write a program that displays:

• The paths of the alpha particle and the gold nucleus.

• The value of the impact parameter b, and the value of the corresponding scattering angle (angle between the initial path of the alpha particle and its final path). The angle can be calculated by taking the arctangent of p_y/p_x, using the final values.

• A plot of p_x for each particle and total p_x, and a similar plot for p_y. (Consider what these plots should look like if the program is working properly.)

Report what value of Δt gives adequate accuracy. What is your criterion? Report values of the impact parameter b that produce the following:

• A very small scattering angle

• An angle around 90 degrees

• A large deflection angle (back-scattering)

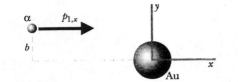

Coordinate system for Rutherford scattering calculations.

Continued on next page.

(c) What other physical quantity could you plot that would provide a check on the accuracy of your calculation? Carry out this check and report what value of Δt provides good accuracy according to this check.

(d) To see what a different model of matter would have predicted, replace the gold nucleus in your model by a proton. Describe briefly what you observe. (Note that if the protons were uniformly spread out throughout the volume of the atom, they would be far enough apart that on average an alpha particle would interact with only one of them.)

8.11.2 Additional problems

Problem 8.2 Deviations from Rutherford scattering

(a) A gold nucleus contains 197 nucleons packed tightly against each other. A single nucleon (proton or neutron) has a radius of about 10^{-15} m. Calculate the approximate radius of the gold nucleus.

(b) Rutherford correctly predicted the angular distribution for 10-MeV (kinetic energy) alpha particles colliding with gold nuclei. What kinetic energy must alpha particles have in order to make contact with a gold nucleus? (In this case the angular distribution will deviate from that predicted by Rutherford, which was based solely on electric interactions.)

Problem 8.3 The Rutherford experiment in the center-of-mass frame

Redo the analysis of the Rutherford experiment, this time using the concept of the center-of-momentum reference frame. Let m = the mass of the alpha particle and M = the mass of the gold nucleus. Consider the specific case of the alpha particle rebounding straight back. The incoming alpha particle has a momentum p_1, the outgoing alpha particle has a momentum p_3, and the gold nucleus picks up a momentum p_4.

(a) Determine the velocity of the center of momentum of the system.

(b) Transform the initial momenta to that frame (by subtracting the center-of-momentum velocity from the original velocities).

(c) Show that if the momenta in the center-of-momentum frame simply turn around (180°), with no change in their magnitudes, both momentum and energy conservation are satisfied, whereas no other possibility satisfies both conservation principles. (Try drawing some other momentum diagrams.)

(d) After the collision, transform back to the original reference frame (by adding the center-of-momentum velocity to the velocities of the particles in the center-of-mass frame). Show that the momenta of the alpha particle and gold nucleus are consistent with the results obtained earlier, on page 267 and page 269. Note that although using the center-of-momentum frame may be conceptually more difficult, the algebra for solving for the final speeds is much simpler.

Problem 8.4 The bouncing ball

A steel ball of mass m falls from a height h onto a scale calibrated in newtons. The ball rebounds repeatedly to nearly the same height h. The scale is sluggish in its response to the intermittent hits and displays an *average* force F_{avg}, such that

$$F_{avg} T = F\Delta t$$

where $F\Delta t$ is the brief impulse that the ball imparts to the scale on every hit, and T is the time between hits.

Calculate this average force in terms of m, h, and physical constants. Compare your result with what the scale reads if the ball merely rests on the scale. Explain your analysis carefully (but briefly).

Problem 8.5 Moving the Earth

Suppose all the people of the Earth go to the North Pole and, on a signal, all jump straight up. Estimate the recoil speed of the Earth. The mass of the Earth is 6×10^{24} kg, and there are about 6 billion people (6×10^9).

Problem 8.6 The car and the mosquito

A car of mass M moving in the x direction at high speed v strikes a hovering mosquito of mass m, and the mosquito is smashed against the windshield.

(a) What is the approximate momentum change of the mosquito? Give magnitude and direction. Explain any approximations you make.

(b) At a particular instant during the impact, when the force exerted on the mosquito by the car is F, what is the magnitude of the force exerted on the car by the mosquito?

(c) What is the approximate momentum change of the car? Give magnitude and direction. Explain any approximations you make.

(d) Qualitatively, why is the collision so much more damaging to the mosquito than to the car?

Problem 8.7 Collisions and springs

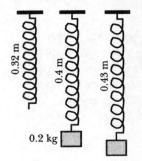

(a) A spring has an unstretched length of 0.32 m. A block with mass 0.2 kg is hung at rest from the spring, and the spring becomes 0.4 m long. Next the spring is stretched to a length of 0.43 m and the block is released from rest. Air resistance is negligible. How long does it take for the block to return to where it was released?

(b) Next the block is again positioned at rest, hanging from the spring (0.4 m long). A bullet of mass 0.003 kg traveling at a speed of 200 m/s straight upward buries itself in the block, which then reaches a maximum height h above its original position. What is the speed of the block immediately after the bullet hits?

(c) Now write an equation that could be used to determine how high the block goes after being hit by the bullet (a height h), but you need not actually solve for h.

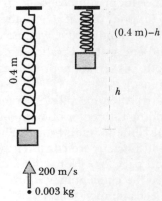

Problem 8.8 Solar sailing

It has been proposed to propel spacecraft through the Solar System with a large sail that is struck by photons from the Sun.

(a) Which would be more effective, a black sail that absorbs photons or a shiny sail that reflects photons back toward the Sun? Explain briefly.

(b) Suppose N photons hit a shiny sail per second, perpendicular to the sail. Each photon has energy E. What is the force on the sail? Explain briefly.

Problem 8.9 Gamma-ray emission

A ^{57}Fe nucleus is at rest and in its first excited state, 14.4 keV above the ground state (14.4×10^3 eV, where 1 eV = 1.6×10^{-19} J). The nucleus then decays to the ground state with the emission of a gamma ray (a high-energy photon).

(a) What is the recoil speed of the nucleus?

(b) Calculate the slight difference in eV between the gamma-ray energy and the 14.4 keV difference between the initial and final nuclear states.

(c) The "Mössbauer effect" is the name given to a related phenomenon discovered by Rudolf Mössbauer in 1957, for which he received the 1961 Nobel prize for physics. If the ^{57}Fe nucleus is in a solid block of iron, occasionally when the nucleus emits a gamma ray the entire solid recoils as one object. This can happen due to the fact that neighboring atoms and nuclei are connected by the electric interatomic force. In this case, repeat the calculation of part (b) and compare with your previous result. Explain briefly.

Problem 8.10 Slowing down neutrons

In a nuclear fission reactor, each fission of a uranium nucleus is accompanied by the emission of one or more high-speed neutrons which travel through the surrounding material. If one of these neutrons is captured in another uranium nucleus, it can trigger fission, which produces more fast neutrons, which could make possible a chain reaction.

However, fast neutrons have low probability of capture and usually scatter off uranium nuclei without triggering fission. In order to sustain a chain reaction, the fast neutrons must be slowed down in some material, called a "moderator" (Figure 8.32). For reasons having to do with the details of nuclear physics, slow neutrons have a high probability of being captured by uranium nuclei.

In the following analyses, remember that neutrons have almost no interaction with electrons. Neutrons do however interact strongly with nuclei, either by scattering or by being captured and made part of the nucleus. Therefore you should think about neutrons interacting with nuclei, not atoms.

(a) Based on what you now know about collisions, explain why fast neutrons moving through a block of uranium experience little change in speed.

(b) Explain why carbon should be a much better moderator of fast neutrons than uranium.

(c) Should water be a better or worse moderator of fast neutrons than carbon? Explain briefly.

Background: The first fission reactor was constructed in 1941 in a squash court under the stands of Stagg Field at the University of Chicago by a team led by the physicist Enrico Fermi. The moderator consisted of blocks of graphite, a form of carbon. The graphite had to be exceptionally pure because certain kinds of impurities have nuclei that capture neutrons with high probability, removing them from contributing to the chain reaction. Many reactors use ordinary "light" water as a moderator, though sometimes a proton captures a neutron and forms a stable deuterium nucleus, in which case the neutron is lost to contributing to the chain reaction. Heavy water, D_2O, in which the hydrogen atoms are replaced by deuterium atoms, actually works better as a moderator than light water, because the probability of a deuterium nucleus capturing a neutron to form tritium is quite small.

Problem 8.11 Fusion reaction revisited

Here is a modified version of a problem from Chapter 4, before we had discussed collisions. One of the thermonuclear or fusion reactions that takes place inside a star such as our Sun is the production of helium-3 (^{3}He, with two protons and one neutron) and a gamma ray (high-energy photon) in a collision between a proton (^{1}H) and a deuteron (^{2}H, the nucleus of "heavy" hydrogen, consisting of a proton and a neutron):

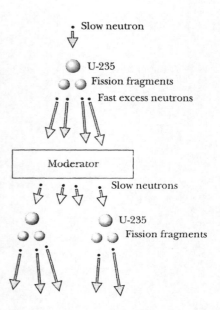

Figure 8.32 A slow neutron induces fission of U-235, with the emission of additional (fast) neutrons. The moderator is some material that slows down the fast neutrons, enabling a chain reaction.

$$^1\text{H} + {}^2\text{H} \rightarrow {}^3\text{He} + \gamma$$

The rest mass of the proton is 1.0073 u (unified atomic mass unit, $1.66{\times}10^{-27}$ kg), the rest mass of the deuteron is 2.0136 u, the rest mass of the helium-3 nucleus is 3.015 u, and the gamma ray is a massless photon.

The strong interaction has a very short range and is essentially a contact interaction. For this fusion reaction to take place, the proton and deuteron have to come close enough together to touch. The approximate radius of a proton or neutron is about 10^{-15} m.

(a) In the center-of-momentum frame (that is, the frame in which the total momentum is zero), draw a diagram showing the momentum vectors before the reaction and after the reaction. Clearly label the momentum vectors.

(b) In the center-of-momentum frame, what minimum kinetic energy must the proton have, and what minimum kinetic energy must the deuteron have, in order for the reaction to take place? Express your results in eV. (In the lab frame where the deuteron was at rest, the proton must have more kinetic energy, because the deuteron moves away due to electric repulsion and the proton has to catch up to it.)

(c) In this case, still in the center-of-momentum frame, what will be the kinetic energy of the helium-3 nucleus and the energy of the gamma ray in eV? Hint: You will find that the speed of the helium-3 nucleus is very small compared to the speed of light. You may find it useful to use the relationship $E^2 - (pc)^2 = (mc^2)^2$. Remember that the fundamental definition of the center-of-momentum frame is that it is the frame in which the total momentum is zero. This is the definition that must be used when dealing with zero-mass particles such as photons.

Problem 8.12 Particle decay

There is an unstable particle called the "sigma-minus" (Σ^-), which can decay into a neutron and a negative pion (π^-): $\Sigma^- \rightarrow n + \pi^-$. The mass of the Σ^- is 1196 MeV/c^2, the mass of the neutron is 939 MeV/c^2, and the mass of the π^- is 140 MeV/c^2. Write equations that could be used to calculate the momentum and energy of the neutron and the pion. You do not need to solve the equations, which would involve some messy algebra. But be clear in showing that you have enough equations that you could in principle solve for the unknown quantities in your equations.

It is advantageous to write the equations not in terms of v but rather in terms of E and p; remember that $E^2 - (pc)^2 = (mc^2)^2$.

Problem 8.13 Pion production of a particle

A beam of high energy π^- (negative pions) is shot at a flask of liquid hydrogen, and sometimes a pion interacts through the strong interaction with a proton in the hydrogen, in the reaction $\pi^- + p^+ \rightarrow \pi^- + X^+$, where X^+ is a positively-charged particle of unknown mass.

The incoming pion momentum is 3 GeV/c (1 GeV = 1000 MeV = 10^9 electron-volts). The pion is scattered through 40°, and its momentum is measured to be 1510 MeV/c (this is done by observing the radius of curvature of its circular trajectory in a magnetic field). A pion has a rest energy of 140 Mev, and a proton has a rest energy of 938 Mev.

What is the rest mass of the unknown X^+ particle, in MeV/c^2? Explain your work carefully.

It is advantageous to write the equations not in terms of v but rather in terms of E and p; remember that $E^2 - (pc)^2 = (mc^2)^2$.

8.12 Answers to exercises

8.1 (page 262) $\quad$ 1.1×10^3 N; $mg = 0.98$ N

8.3 (page 269) $\quad$ 2.3×10^{-24} m/s (!) So it is an excellent approximation to say that the Earth doesn't recoil.

8.4 (page 269) $\quad$ $K_{\text{Earth}} = 1.6 \times 10^{-23}$ J; $K_{\text{baseball}} = 150$ J

8.5 (page 281) $\quad$ Car 1 must have had more momentum than car 2

8.6 (page 284) $\quad$ We outline the entire solution. First find $v_{\text{cm}, x}$:

$$v_{\text{cm}, x} = \frac{(1000 \text{ kg})(30 \text{ m/s}) + (3000 \text{ kg})(-20 \text{ m/s})}{(4000 \text{ kg})} = -7.5 \text{ m/s}$$

Transform to center-of-momentum frame:

$$v_{\text{car}, x} = (30 \text{ m/s}) - (-7.5 \text{ m/s}) = 37.5 \text{ m/s}$$

$$v_{\text{truck}, x} = (-20 \text{ m/s}) - (-7.5 \text{ m/s}) = -12.5 \text{ m/s}$$

Check to make sure that the momenta are equal and opposite in this frame:

$$p_{\text{car}, x} = (1000 \text{ kg})(37.5 \text{ m/s}) = 37500 \text{ kg·m/s}$$

$$p_{\text{truck}, x} = (3000 \text{ kg})(-12.5 \text{ m/s}) = -37500 \text{ kg·m/s}$$

In this frame the kinetic energy before the collision is this;

$$K = \frac{(37500 \text{ kg·m/s})^2}{2(1000 \text{ kg})} + \frac{(37500 \text{ kg·m/s})^2}{2(3000 \text{ kg})} = 9.4 \times 10^5 \text{ J}$$

After the collision, the velocities are zero in this frame, so kinetic energy in this frame goes to zero, and there is an increase in the internal energy of the mangled car and truck of amount 9.4×10^5 J. There must be the same internal energy increase in the original reference frame, so the kinetic energy lost in the original reference frame was 9.4×10^5 J.

Chapter 9

Angular Momentum

Chapter 9

Angular Momentum

Rotational motion is such an important type of motion that special techniques have been developed to describe and predict it. A particularly important concept that arises from the study of rotational motion is "angular momentum," the rotational analogue of the ordinary "linear" momentum with which we are already familiar. Angular momentum is conserved, just as energy and linear momentum are. In the atomic world angular momentum is quantized, which has far-reaching consequences for the structure of atoms and nuclei.

9.1 Angular momentum

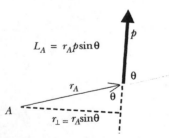

Figure 9.1 A child jumps onto a playground ride, and the disk spins. What is the rate of spin?

A playground ride consists of a disk of mass M and radius R mounted on a low-friction axle (Figure 9.1). A child of mass m runs at speed v on a line tangential to the disk and jumps onto the outer edge of the disk. The disk spins, but at what rate?

The momentum principle can be used to determine the impulse exerted by the axle on the disk, which prevents the disk from translating sideways, but it tells us nothing about the spin rate. The energy principle can be used to determine how much thermal energy is produced in the inelastic sticking collision of the child and the disk, but it tells us nothing about the spin rate. We need a new principle: the angular momentum principle.

In Figure 9.2 we see a top view of the child heading toward the disk along several different paths, with different effects. Evidently the child's effectiveness in spinning the disk depends on the impact parameter $r_\perp$, which measures how far off-center is the child's initial path. Presumably if the child has more mass m and/or more speed v, these too will affect the spin rate.

Putting these influences together, the spin rate of the disk should be proportional to $r_\perp mv$, and we call this the magnitude of the "angular momentum" of the child relative to the center of the disk. The term "angular momentum" is used to reflect the fact that it is often associated with something rotating through an angle (in this case, the disk), and it is proportional to the familiar momentum mv. When necessary to emphasize the difference between angular momentum and ordinary momentum, the latter is called "linear momentum."

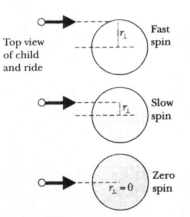

Figure 9.2 How much the disk spins depends on the impact parameter $r_\perp$.

More generally, we define L_A, the magnitude of the angular momentum of a particle relative to a location A like this, as shown in Figure 9.3:

MAGNITUDE OF ANGULAR MOMENTUM OF A PARTICLE RELATIVE TO LOCATION *A*

$$L_A = r_\perp p = r_A p \sin\theta$$

Here $\vec{r}_A$ is a position vector of the particle measured from location A to the particle, $\vec{p}$ is the ordinary "linear" momentum of the particle, and θ is the angle between the position vector and the momentum vector. $r_\perp$ is the perpendicular distance from location A to the line of motion.

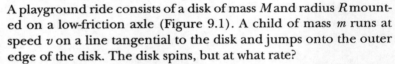

Figure 9.3 Magnitude of the angular momentum of a particle relative to location A.

We will find that the sum of the angular momentum of the child plus that of the disk is constant, which is what makes angular momentum useful.

9.1.1 The cross product

The magnitude of angular momentum isn't everything: the disk may rotate clockwise or counterclockwise, depending on the child's path (Figure 9.4). How do we take this into account in a definition of the child's angular momentum? The angular momentum of a particle is defined as a vector, using the "cross product," to be explained in a moment:

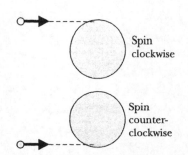

Figure 9.4 The disk will rotate clockwise or counterclockwise, depending on the child's path.

ANGULAR MOMENTUM OF A PARTICLE RELATIVE TO LOCATION *A*

$$\vec{L}_A = \vec{r}_A \times \vec{p}$$

The cross product $\vec{A} \times \vec{B}$ is a vector in a direction perpendicular to the plane defined by $\vec{A}$ and $\vec{B}$. In Figure 9.5, both $\vec{A}$ and $\vec{B}$ lie in the plane, and the vector $\vec{A} \times \vec{B}$ is perpendicular to the plane.

CROSS PRODUCT

$\vec{A} \times \vec{B}$ is a vector with

magnitude $|\vec{A} \times \vec{B}| = AB\sin\theta$ and

direction given by a "right-hand rule"

$$\vec{A} \times \vec{B} = <(A_y B_z - A_z B_y), (A_z B_x - A_x B_z), (A_x B_y - A_y B_x)> \text{ (see Section 9.1.3)}$$

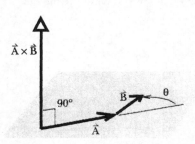

Figure 9.5 The cross product produces a vector that is perpendicular to the two original vectors.

In the particular use of the cross product to describe angular momentum, angular momentum is a vector $\vec{L}_A = \vec{r}_A \times \vec{p}$ that is perpendicular to $\vec{r}_A$ and perpendicular to $\vec{p}$. The angular momentum vector lies along the axis about which there would be rotation if the particle struck a disk whose axle is at location *A*, and which way it points along this rotation axis is related to whether the disk would rotate clockwise or counterclockwise (more on this in a moment). The magnitude of the angular momentum vector is proportional to how much spin the particle would produce in the disk.

9.1.2 Right-hand rule for cross products

There are several different schemes for evaluating cross products. We will describe a scheme that has significant advantages for our purposes. We will illustrate a general "right-hand rule" for the particular case of angular momentum, $\vec{L}_A = \vec{r}_A \times \vec{p}$.

1) With your right hand open and flat (fingers and thumb in the same plane), with your thumb perpendicular to your fingers, point your fingers in the direction of the first vector ($\vec{r}_A$). See Figure 9.6.

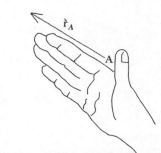

Figure 9.6 Open right hand, fingers point in direction of $\vec{r}_A$.

2) Bend your fingers toward the second vector ($\vec{p}$), noting the angle of bend θ, which is always less than 180° (Figure 9.7). You may have to rotate your wrist to get your hand into a position that lets you bend your fingers this way; this wrist rotation is an essential part of the right-hand rule. You must get your right hand into a position that permits bending your fingers from pointing along $\vec{r}_A$ to pointing along $\vec{p}$.

3) Your thumb now points in the direction of the cross-product vector $\vec{r}_A \times \vec{p}$. This direction is perpendicular to the plane defined by $\vec{r}_A$ and $\vec{p}$. The fingers of your right hand curl in the direction that a disk with center at *A* would rotate if struck by the particle. It may seem odd to have the angular momentum vector perpendicular to the motion, but the point is that it lies along the (possible) axis of rotation.

4) The magnitude of the cross-product vector is the magnitude of the first vector, times the magnitude of the second vector, times the sine of the angle of bend of your fingers: $L_A = r_A p \sin\theta$. One of the advantages of this particular way of dealing with cross products is that you have a visual representation of the angle θ whose sine goes into the magnitude of the cross product.

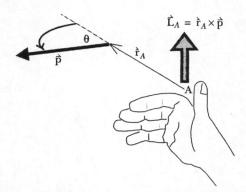

Figure 9.7 Bend fingers toward lining up with $\vec{p}$. Thumb points in direction of $\vec{r}_A \times \vec{p}$.

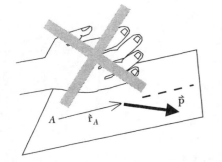

Figure 9.8 You can't bend your fingers backward. You must rotate the wrist into a position that lets you bend the fingers.

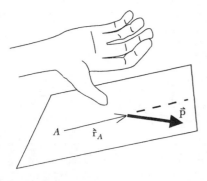

Figure 9.9 After rotating the wrist, it is possible to bend the fingers.

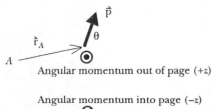

Angular momentum out of page (+z)

Angular momentum into page (−z)

Figure 9.10 2-D projections: symbols for angular momentum out of and into page.

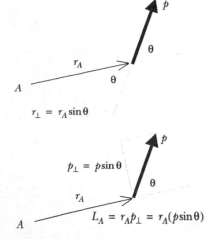

$$r_\perp = r_A \sin\theta$$

$$p_\perp = p\sin\theta$$

$$L_A = r_A p_\perp = r_A(p\sin\theta)$$

Figure 9.11 Two different ways to calculate the magnitude of the cross product.

Things to watch out for

The rotation of the wrist is an important part of the right-hand rule. Consider the situation in Figure 9.8. With your right hand in this position, you can't bend your fingers backwards from the first vector $\hat{r}_A$ toward the second vector $\vec{p}$—it is a physical impossibility.

You need to rotate your wrist into a position from which it is possible to bend the fingers, as shown in Figure 9.9. The direction of your thumb is down, which is the correct direction for the cross product in this situation. The right-hand rule forces you to rotate your wrist into an orientation that permits bending your fingers from the first vector toward the second vector, in which case your thumb will point in the correct direction of the cross product.

Important: Pay attention to the size of the angle through which you bend your fingers. This is the angle whose sine is part of the definition of the magnitude of the cross product. Also note that this angle should never be more than 180°. If it is, you have made a mistake in orienting your hand; probably you need to rotate your wrist.

Two-dimensional projections

Because it is difficult to sketch a situation in three dimensions, whenever possible we will work with two-dimensional projections onto the *x-y* plane. If $\hat{r}_A$ and $\vec{p}$ lie in the *x-y* plane, the angular momentum $\hat{r}_A \times \vec{p}$ points in the +z direction (out of the page) or in the −z direction (into the page). A special notation is used to indicate "out of the page" (⊙) and "into the page" (⊗); see Figure 9.10. The symbol ⊙ can be thought of as showing the tip of an arrow (a vector) pointing out at you, while the symbol ⊗ represents the feathers of an arrow (a vector) pointing away from you.

Calculating the magnitude

It is often easiest to calculate the magnitude $r_A p \sin\theta$ of the angular momentum by associating the sine of the angle with r_A or with p. In Figure 9.11, note that the perpendicular component $r_\perp$ of the "lever arm" is $r_A \sin\theta$, while the perpendicular component $p_\perp$ of the momentum is $p \sin\theta$. Study Figure 9.11 to learn two efficient methods for calculating the magnitude of the angular momentum of a particle.

Practice

Be sure that you do and check the following exercises before going on, because you need to be able to use the concept of cross product quickly and easily in later discussions.

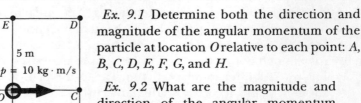

Ex. 9.1 Determine both the direction and magnitude of the angular momentum of the particle at location *O* relative to each point: *A, B, C, D, E, F, G,* and *H.*

Ex. 9.2 What are the magnitude and direction of the angular momentum about location *K* below?

9.1.3 The cross product and unit vectors

It is also possible to evaluate the cross product in terms of unit vectors. First, note that $\hat{\imath}\times\hat{\imath} = 0$, $\hat{\jmath}\times\hat{\jmath} = 0$, and $\hat{k}\times\hat{k} = 0$, since when we cross a vector with itself the angle between the two vectors is zero, and $\sin 0° = 0$.

Second, $\hat{\imath}\times\hat{\jmath} = \hat{k}$, since the angle is 90° and the right-hand rule gives a result in the +z direction (out of the page; Figure 9.12). On the other hand, $\hat{\jmath}\times\hat{\imath} = -\hat{k}$, because the right-hand rule gives a result in the −z direction (into the page). Similarly, $\hat{\jmath}\times\hat{k} = \hat{\imath}$, $\hat{k}\times\hat{\jmath} = -\hat{\imath}$, $\hat{k}\times\hat{\imath} = \hat{\jmath}$, and $\hat{\imath}\times\hat{k} = -\hat{\jmath}$. Putting this all together, we obtain the following general result:

$$\vec{A}\times\vec{B} = (A_yB_z - A_zB_y)\hat{\imath} + (A_zB_x - A_xB_z)\hat{\jmath} + (A_xB_y - A_yB_x)\hat{k} \text{ or}$$

$$\vec{A}\times\vec{B} = <(A_yB_z - A_zB_y), (A_zB_x - A_xB_z), (A_xB_y - A_yB_x)>$$

This approach to calculating a cross product is particularly useful in computer calculations. Note the cyclic nature of the subscripts: *xyz, yzx, zxy*.

Ex. 9.3 Given that $\vec{r}_A = <x, y, z>$ and $\vec{p} = <p_x, p_y, p_z>$, show that angular momentum can be calculated in the following way:

COMPONENT CALCULATION OF ANGULAR MOMENTUM

$$\vec{L}_A = <(yp_z - zp_y), (zp_x - xp_z), (xp_y - yp_x)>$$

9.2 Angular momentum of multiparticle systems

Later we will show that when the child in Figure 9.1 jumps on the disk-shaped playground ride, the initial angular momentum of the child is equal to the sum of the final angular momenta of the child and disk. In order to calculate the angular momentum of a multiparticle system such as the disk, we need additional tools.

Recall that the kinetic energy of a multiparticle system can be split into overall translational kinetic energy plus rotational and vibrational kinetic energy relative to the center of mass:

$$K_{tot} = K_{trans} + K_{rel} = K_{trans} + K_{rot} + K_{vib}$$

In an analogous fashion, angular momentum can be split into a term associated with the motion of the center of mass (which we will call translational angular momentum, sometimes called "orbital" angular momentum) plus a term associated with rotation relative to the center of mass (which we will call rotational angular momentum, sometimes called "spin" angular momentum):

$$\vec{L}_A = \vec{L}_{trans,A} + \vec{L}_{rot}$$

Splitting the angular momentum into these two terms is very useful in analyzing multiparticle systems such as the Earth, which orbits the Sun but also rotates on its axis. You may also know that electrons orbiting the nucleus of an atom have both translational and rotational angular momentum.

We will calculate the angular momentum relative to location A of a 3-mass system (Figure 9.13). We specify the positions of each mass relative to location A by going from A to the center of mass (vector $\vec{r}_{cm}$) and then to the individual masses. (The location A need not be in the same plane as the masses, nor are the momentum vectors necessarily in the plane of the masses.)

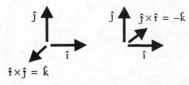

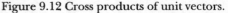

Figure 9.12 Cross products of unit vectors.

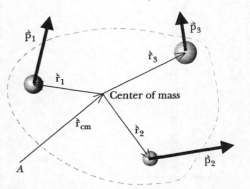

Figure 9.13 Position vectors relative to the center of mass.

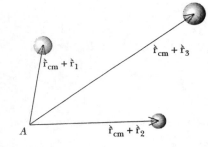

Figure 9.14 Position vectors relative to location *A*.

The position of mass m_1 relative to location A is $\overset{\leftarrow}{\vec{r}}_{cm} + \overset{\leftarrow}{\vec{r}}_1$, the position of mass m_2 is $\overset{\leftarrow}{\vec{r}}_{cm} + \overset{\leftarrow}{\vec{r}}_2$, and the position of mass m_3 is $\overset{\leftarrow}{\vec{r}}_{cm} + \overset{\leftarrow}{\vec{r}}_3$ (Figure 9.14). Using these vectors we calculate the total angular momentum relative to location A:

$$\vec{L}_A = (\overset{\leftarrow}{\vec{r}}_{cm} + \overset{\leftarrow}{\vec{r}}_1) \times \vec{p}_1 + (\overset{\leftarrow}{\vec{r}}_{cm} + \overset{\leftarrow}{\vec{r}}_2) \times \vec{p}_2 + (\overset{\leftarrow}{\vec{r}}_{cm} + \overset{\leftarrow}{\vec{r}}_3) \times \vec{p}_3$$

$$\vec{L}_A = [\overset{\leftarrow}{\vec{r}}_{cm} \times (\vec{p}_1 + \vec{p}_2 + \vec{p}_3)] + [\overset{\leftarrow}{\vec{r}}_1 \times \vec{p}_1 + \overset{\leftarrow}{\vec{r}}_2 \times \vec{p}_2 + \overset{\leftarrow}{\vec{r}}_3 \times \vec{p}_3]$$

Since $(\vec{p}_1 + \vec{p}_2 + \vec{p}_3)$ is the total momentum $\vec{P}_{tot}$, we have the following:

$$\vec{L}_A = \underbrace{[\overset{\leftarrow}{\vec{r}}_{cm} \times \vec{P}_{tot}]}_{\text{"translational"}} + \underbrace{[\overset{\leftarrow}{\vec{r}}_1 \times \vec{p}_1 + \overset{\leftarrow}{\vec{r}}_2 \times \vec{p}_2 + \overset{\leftarrow}{\vec{r}}_3 \times \vec{p}_3]}_{\text{"rotational"}}$$

The first term in the total angular momentum,

$$\overset{\leftarrow}{\vec{r}}_{cm} \times \vec{P}_{tot}$$

is the angular momentum the system would have if all the mass were concentrated at the center of mass. We call this term the "translational" angular momentum $\vec{L}_{trans,A}$, and it is what we used in analyzing the speed of a comet around the Sun. This term is also called the "orbital" angular momentum.

The second term in the total angular momentum,

$$\overset{\leftarrow}{\vec{r}}_1 \times \vec{p}_1 + \overset{\leftarrow}{\vec{r}}_2 \times \vec{p}_2 + \overset{\leftarrow}{\vec{r}}_3 \times \vec{p}_3$$

is the angular momentum of the system relative to the center of mass of the system. We call this term the "rotational" angular momentum $\vec{L}_{rot}$. An example of rotational angular momentum is the angular momentum due to the rotation of the Earth on its axis. This term is also called the "spin" angular momentum. In Figure 9.15 we show both the translational and rotational angular momentum vectors for the Earth relative to the Sun.

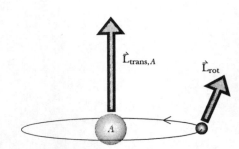

Figure 9.15 Translational angular momentum of the Earth relative to the Sun, with rotational angular momentum relative to the Earth's center of mass. At this instant the linear momentum of the Earth is into the page.

TRANSLATIONAL PLUS ROTATIONAL ANGULAR MOMENTUM

$$\vec{L}_A = \vec{L}_{trans,A} + \vec{L}_{rot}$$

Translational: $\vec{L}_{trans,A} = \overset{\leftarrow}{\vec{r}}_{cm,A} \times \vec{P}_{tot}$

Rotational: $\vec{L}_{rot} = \overset{\leftarrow}{\vec{r}}_1 \times \vec{p}_1 + \overset{\leftarrow}{\vec{r}}_2 \times \vec{p}_2 + \overset{\leftarrow}{\vec{r}}_3 \times \vec{p}_3$

The translational angular momentum differs for different choices of the location A, so we need the "A" subscript on the translational angular momentum and the total angular momentum to remind us of this dependence. The rotational angular momentum doesn't need a subscript because it is calculated relative to the center of mass, and this calculation is unaffected by our choice of the point A, and by the motion of the center of mass, which may even be accelerating.

For a point particle, or the "collapsed" point-particle version of a multi-particle system, the translational angular momentum is the only kind of angular momentum there is, just as translational kinetic energy is the only kind of kinetic energy a point particle can have.

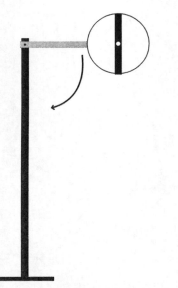

Figure 9.16 A rod rotating in the vertical plane with a wheel attached.

Ex. 9.4 A rod rotates in the vertical plane around a horizontal axle. A wheel is free to rotate on the rod, as shown in Figure 9.16. A vertical stripe is painted on the wheel. As the rod rotates clockwise, the vertical stripe on the wheel remains vertical. Is the translational angular momentum of the wheel zero or nonzero? If nonzero, what is its direction? Is the rotational angular momentum of the wheel zero or nonzero? If nonzero, what is its direction?

Ex. 9.5 Consider a system similar to that in the previous exercise, but with the wheel welded to the rod (not free to turn). As the rod rotates clockwise, does the stripe on the wheel remain vertical? Is the translational angular momentum of the wheel zero or nonzero? If nonzero, what is its direction? Is the rotational angular momentum of the wheel zero or nonzero? If nonzero, what is its direction?

Ex. 9.6 Pinocchio rides a horse on a merry-go-round turning counterclockwise as viewed from above, with his long nose always pointing forward, in the direction of his velocity. Is Pinocchio's translational angular momentum zero or nonzero? If nonzero, what is its direction? Is his rotational angular momentum zero or nonzero? If nonzero, what is its direction?

9.2.1 Rotational angular momentum: moment of inertia

An important special case of rotational motion is one where a system is rotating on an axis, with all the atoms in the system sharing the same "angular speed" in radians per second but with different linear speeds in meters per second, depending on their distances from the axis (Figure 9.17). This is the situation for a rigid object.

As we saw in Chapter 2, angular speed, normally denoted by ω (lowercase Greek omega), is a measure of how fast something is rotating. If an object makes one complete turn of 360 degrees (2π radians) in a time T, we say that its angular speed is $\omega = 2\pi/T$ radians per second. The time T is called the period.

We showed in Chapter 2 that the magnitude of the time rate of change of a rotating vector $\vec{X}$ is $|d\vec{X}/dt| = \omega X$. One application of this general result is that the speed v of an atom rotating around an axis at a distance r from the axis is $|d\vec{r}/dt| = v = \omega r$. We measure angular speed ω in the "natural" units of radians per second rather than degrees per second, because the fundamental geometrical relationship between angle and arc length (arc length = $r\theta$) is valid only if the angle θ is measured in radians.

An advantage of using angular speed can be seen in calculating the rotational angular momentum of a rotating rigid object, all of whose parts share a common angular speed ω, but have different linear speeds. We define an angular velocity vector:

ANGULAR VELOCITY VECTOR $\vec{\omega}$

magnitude: radians/second

direction: see Figure 9.18

Start with your fingers headed radially outward, bend them in the direction of the rotation, and your thumb points in the direction of the angular velocity, along the axis of rotation (angular velocity is also shown in Figure 9.17).

The magnitude of the angular momentum of each piece is this, where $r_\perp$ is the perpendicular distance of a piece from the axis:

$$|\vec{r} \times \vec{p}| = r_\perp mv = r_\perp m(\omega r_\perp) = mr_\perp^2 \omega$$

So the magnitude of the rotational angular momentum is this:

$$|\vec{L}_{\text{rot}}| = [m_1 r_{\perp 1}^2 + m_2 r_{\perp 2}^2 + m_3 r_{\perp 3}^2 + m_4 r_{\perp 4}^2]\omega$$

The quantity in brackets is called the "moment of inertia" and is usually denoted by the letter I:

MOMENT OF INERTIA

$$I = m_1 r_{\perp 1}^2 + m_2 r_{\perp 2}^2 + m_3 r_{\perp 3}^2 + m_4 r_{\perp 4}^2 + ...$$

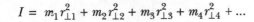

Figure 9.17 A case of rigid rotation about an axis with angular speed ω. Note the different speeds, with $v_i = \omega r_i$.

Figure 9.18 Finding the direction of the angular velocity vector.

Since the rotational angular momentum points in the direction of $\vec{\omega}$, we have the following for the angular momentum relative to the center of mass, even if the center of mass is moving in some complicated way:

ROTATIONAL ANGULAR MOMENTUM

$$\vec{L}_{rot} = I\vec{\omega}$$

This way of evaluating rotational angular momentum is often easier to use than the more basic expression in terms of cross products $\vec{r} \times \vec{p}$, especially because the moments of inertia for common objects can be looked up in reference books. For example, the moment of inertia about an axis passing through the center of a uniform solid disk of radius R is $\frac{1}{2}MR^2$ (see Problem 9.3 (page 330)), and the moment of inertia about an axis passing through the center of a uniform solid sphere of radius R is $\frac{2}{5}MR^2$.

Ex. 9.7 What is the moment of inertia of a nitrogen molecule around its center of mass if the mass of each atom is M and the distance between nuclei is d? (Note that the electrons hardly contribute to the moment of inertia because their mass is so small.)

Ex. 9.8 A bicycle wheel has almost all its mass M located in the outer rim at radius R. If it rotates on a stationary axle with angular speed ω, what is the angular momentum of the wheel? (Hint: It's helpful to think of dividing the wheel into the atoms it is made of.)

9.2.2 Rotational kinetic energy

The kinetic energy of the system shown in Figure 9.19, which is a rotating rigid object, is $\frac{1}{2}m_1 v_1^2 + \frac{1}{2}m_2 v_2^2 + \frac{1}{2}m_3 v_3^2 + \dots$. Since the speed of each mass is $v = \omega r_\perp$, we can write the rotational kinetic energy like this:

$$K_{rot} = \frac{1}{2}[m_1(r_{\perp 1}\omega)^2 + m_2(r_{\perp 2}\omega)^2 + m_3(r_{\perp 3}\omega)^2 + \dots]$$

$$= \frac{1}{2}[m_1 r_{\perp 1}^2 + m_2 r_{\perp 2}^2 + m_3 r_{\perp 3}^2 + \dots]\omega^2$$

We again see the moment of inertia showing up in the bracket, so we have the following calculation of rotational kinetic energy:

ROTATIONAL KINETIC ENERGY

$$K_{rot} = \frac{1}{2}I\omega^2 = \frac{1}{2}\frac{(I\omega)^2}{I} = \frac{L_{rot}^2}{2I}$$

Ex. 9.9 A disk whose mass is 10 kg and radius is 0.4 m makes one complete rotation every 0.2 s. What is its kinetic energy? (The moment of inertia of a disk about its center is $\frac{1}{2}MR^2$.)

Complications

We will deal with situations where the axis of rotation of a rigid body passes through the center of mass, and the object is symmetric or nearly so, in which case we can calculate the rotational angular momentum and rotational kinetic energy using the formulas $\vec{L}_{rot} = I\vec{\omega}$ and $K_{rot} = \frac{1}{2}I\omega^2$. In Section 9.11 on page 325 we discuss some complications that apply in the more general case. We will avoid these more complex rotational situations, which are treated in more advanced courses.

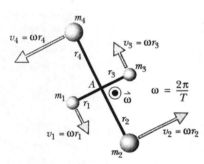

Figure 9.19 A case of rigid rotation about an axis with angular speed ω. Note the different speeds, with $v_i = \omega r_i$.

9.2.3 More exercises on angular momentum

The following exercises constitute an important worksheet that gives you practice in calculating angular momentum. Be sure to do all of these exercises and check your answers at the end of the chapter.

Ex. 9.10 A comet orbits the Sun (Figure 9.20). When it is at location 1 it is a distance d_1 from the Sun, and has momentum $\vec{p}_1$. Location A is at the center of the Sun.

(a) When the comet is at location 1, what is the direction of $\vec{L}_A$?

(b) When the comet is at location 1, what is the magnitude of $\vec{L}_A$?

When the comet is at location 2, it is a distance d_2 from the Sun, and has momentum $\vec{p}_2$.

(c) When the comet is at location 2, what is the direction of $\vec{L}_A$?

(d) When the comet is at location 2, what is the magnitude of $\vec{L}_A$?

Ex. 9.11 A barbell spins around a pivot at its center at A (Figure 9.21). The barbell consists of two small balls, each with mass m, at the ends of a very low mass rod of length d. The barbell spins clockwise with angular speed ω_0 radians/s.

(a) Calculate $\vec{L}_{\text{trans},1,A}$ (both direction and magnitude).

(b) Calculate $\vec{L}_{\text{tot},A}$ (both direction and magnitude).

(c) Calculate I for the barbell.

(d) What is the direction of $\vec{\omega}_0$?

(e) Calculate $\vec{L}_{\text{rot}}$ (both direction and magnitude).

(f) How does $\vec{L}_{\text{rot}}$ compare to $\vec{L}_{\text{tot},A}$?

(g) Calculate K_{rot}.

Ex. 9.12 The barbell in the previous exercise is mounted on the end of a low mass rigid rod of length b (Figure 9.22). The apparatus is started in such a way that although the rod rotates clockwise with angular speed ω_1 rad/s, the barbell maintains its vertical orientation.

(a) Calculate $\vec{L}_{\text{rot}}$ (both direction and magnitude).

(b) Calculate $\vec{L}_{\text{trans},B}$ (both direction and magnitude).

(c) Calculate $\vec{L}_{\text{tot},B}$ (both direction and magnitude).

Ex. 9.13 The apparatus in the previous exercise is restarted in such a way that it again rotates clockwise with angular speed ω_1 rad/s, but in addition, the barbell rotates clockwise about its center, with an angular speed ω_2 rad/s (Figure 9.23).

(a) Calculate $\vec{L}_{\text{rot}}$ (both direction and magnitude).

(b) Calculate $\vec{L}_{\text{trans},B}$ (both direction and magnitude).

(c) Calculate $\vec{L}_{\text{tot},B}$ (both direction and magnitude).

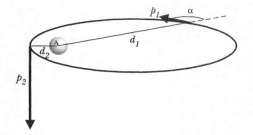

Figure 9.20 A comet orbits the Sun. Location A is at the center of the Sun.

Figure 9.21 A rotating barbell.

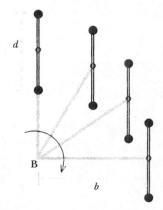

Figure 9.22 A barbell pivoted on a low-mass rotating rod. The barbell does not rotate.

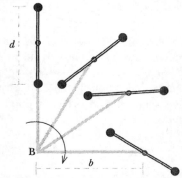

Figure 9.23 A barbell pivoted on a low-mass rotating rod. The barbell rotates.

9.3 Atomic and nuclear angular momentum

Many elementary particles have rotational angular momentum (or, if you wish to be very cautious, these particles behave as though they have rotational angular momentum, which comes down to essentially the same thing). Electrons bound in an atom may have translational angular momentum relative to the nucleus as well as their own rotational angular momentum. Atoms as a whole may have angular momentum, as do many nuclei. The total angular momentum is as expected the sum of the translational angular momenta plus the rotational angular momenta.

The surprising thing about angular momentum at the atomic and subatomic level is that it is quantized. The basic quantum of angular momentum is this:

The angular momentum quantum $= \hbar = \dfrac{h}{2\pi} = 1.05 \times 10^{-34}$ joule $\cdot$ second

where h is "Planck's constant," 6.63×10^{-34} joule-second. Whenever you measure a vector component of the angular momentum (that is, along the x or y or z axis), you get either a half-integer or integer multiple of this quantum of angular momentum.

Ex. 9.14 Show that $\hbar$ and angular momentum have the same units.

9.3.1 The Bohr model of the hydrogen atom

As discussed in the previous chapter, in 1911 Rutherford and his group discovered the nucleus in atoms. Stimulated by this discovery, in 1913 the Danish physicist Niels Bohr made a bold conjecture that a hydrogen atom could be modeled as an electron going around a proton in circular orbits, but only in those orbits whose translational angular momentum is an integer multiple of $\hbar$. Consequently, these orbits would have only certain radii, and only certain values of energy. The differences in energies between these quantized orbits match the observed energies of photons emitted by atomic hydrogen (Figure 9.24).

Bohr's basic hypothesis—that the angular momentum of electrons in a atom is quantized—has proven to be an important insight, and has been retained as a fundamental tenet of the quantum mechanical model of an atom. As this model has been refined, a probabilistic view of the motion of electrons has been adopted. In more sophisticated models electrons do not have precise trajectories, but only a "probability density"—a probability of being found at a particular location. These more complex quantum mechanical models explain a much wider range of atomic and molecular phenomena than does Bohr's original model.

The simple Bohr model does, however, predict correctly the allowed electronic energy levels for atomic hydrogen (as determined from the emission and absorption spectra of hydrogen atoms). We will work with this model to see how quantization of angular momentum leads to the prediction of electronic energy levels in a hydrogen atom.

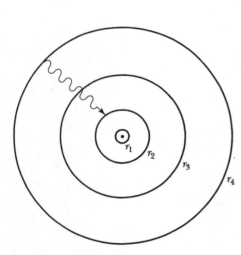

Figure 9.24 In the Bohr model of the hydrogen atom, electrons can be in only those circular orbits for which angular momentum is a multiple of $\hbar$. In a transition between allowed orbits, a photon is emitted.

Allowed radii of electron orbits

Because the proton has much more mass than an electron (about 2000 times as much), we'll make the approximation that the proton is at rest, with the electron in a circular orbit around the proton.

? If the electron momentum is p, and the radius of the circular orbit is r, what is the translational angular momentum?

The translational angular momentum of the electron relative to the location of the proton is $L_{\text{trans}, C} = rp\sin(90°) = rp$ (Figure 9.25). Bohr proposed that the only possible states of the hydrogen atom are those where the electron is in a circular orbit whose translational angular momentum is an integer multiple of $\hbar$:

ANGULAR MOMENTUM IS QUANTIZED

$$\left|\vec{L}_{\text{trans}, C}\right| = rp = N\hbar, \text{ where } N \text{ is an integer } (1, 2, 3,...)$$

The electric force and a circular orbit

? What is the magnitude of the electric force that the proton exerts on the electron?

$$F_{\text{el}} = \frac{1}{4\pi\varepsilon_0}\frac{e^2}{r^2}$$

where $+e$ is the electric charge on the proton ($-e$ for the electron).

? Use the momentum principle to relate this force to the circular motion of the electron.

In a circular orbit $\vec{p}$ is a rotating vector, so we know that $dp/dt = \omega p$, where ω is the angular speed of the electron in orbit, so we have this:

$$\omega p = F = \frac{1}{4\pi\varepsilon_0}\frac{e^2}{r^2}$$

Solving for r

We are looking for the allowed values of the orbit radius r, so we look for ways to express ω and p in terms of r. Bohr's angular momentum condition gives us p in terms of r:

$$p = \frac{N\hbar}{r}$$

We can write ω in terms of r, in the approximation that $v \ll c$, where m is the mass of the electron:

$$\omega = \frac{v}{r} = \frac{mv}{mr} = \frac{p}{mr} = \frac{N\hbar}{mr^2}$$

Putting these relations into $\omega p = F$, we have

$$\left(\frac{N\hbar}{mr^2}\right)\left(\frac{N\hbar}{r}\right) = \frac{1}{4\pi\varepsilon_0}\frac{e^2}{r^2}$$

Solving the latter relation for r, we obtain the following result (Figure 9.26):

ALLOWED BOHR RADII FOR ELECTRON ORBITS

$$r = N^2\frac{\hbar^2}{\left(\dfrac{1}{4\pi\varepsilon_0}\right)e^2 m}$$

Let's evaluate this result numerically, so that later we can compare with measurements of the spectrum of light emitted by excited atomic hydrogen. Using

$\hbar = 1.05\times10^{-34}$ J·s,

$1/4\pi\varepsilon_0 = 8.99\times10^9$ N·m^2/coulomb^2,

$e = 1.60\times10^{-19}$ coulomb,

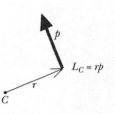

Figure 9.25 The angular momentum of the electron in a circular Bohr orbit relative to the proton. L_C is out of the page.

Figure 9.26 Bohr radii for N=1 through 5. The nucleus is not shown.

and the mass of the electron $m = 9.11 \times 10^{-31}$, we find this:

$$r = N^2 (0.53 \times 10^{-10} \text{ m})$$

This is a striking result. The simple Bohr model predicts that the smallest permissible electron radius ($n = 1$) is 0.53×10^{-10} m, and atoms are in fact observed to have radii of approximately this size.

Energy for a circular orbit

We cannot observe directly the radius of the orbit of an electron in an atom. What we can observe are photons emitted by excited atoms, which tell us the differences in energies between various energy levels of the atom. Using our model we can predict the energies of these energy levels, and see if the differences between these levels match the energies of photons observed in an atomic spectrum.

Given a set of possible values for r, we can calculate the possible values for energy of the electron. We rewrite the momentum principle for this case:

$$\frac{dp}{dt} = \omega p = \left(\frac{v}{r}\right) mv = \frac{mv^2}{r} = \frac{1}{4\pi\varepsilon_0}\frac{e^2}{r^2}$$

$$mv^2 = \frac{1}{4\pi\varepsilon_0}\frac{e^2}{r}$$

? Now write a formula for the kinetic energy plus electric potential energy of the hydrogen atom, using the Bohr model.

Kinetic energy:

$$K = \frac{1}{2}mv^2 = \frac{1}{2}\left(\frac{1}{4\pi\varepsilon_0}\frac{e^2}{r}\right) \qquad \text{(from previous equation)}$$

Electric potential energy:

$$U_{\text{el}} = -\frac{1}{4\pi\varepsilon_0}\frac{e^2}{r}$$

So the energy is this (omitting the rest energies):

$$E = K + U_{\text{el}} = -\frac{1}{2}\left(\frac{1}{4\pi\varepsilon_0}\frac{e^2}{r}\right)$$

According to the Bohr model only certain radii will actually occur. Inserting our previous expression for r, we get:

BOHR MODEL ENERGY LEVELS

$$E = -\frac{\left(\dfrac{1}{4\pi\varepsilon_0}\right)^2 e^4 m}{2N^2\hbar^2}$$

Evaluating this expression we get

$$E = -\frac{2.17 \times 10^{-18} \text{ J}}{N^2}$$

Converting to electron-volts (1 eV = 1.6×10^{-19} J) we have (Figure 9.27)

$$E = -\frac{13.6 \text{ eV}}{N^2}, \ N = 1, 2, 3, \dots$$

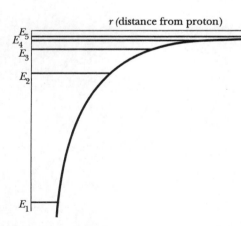

Figure 9.27 Bohr model prediction of electronic energy levels for a hydrogen atom. The potential energy curve for the electron–proton system is also shown.

This prediction for the quantized energies of atomic hydrogen agrees well with the observed electronic spectrum of hydrogen. We quoted this result for the quantized energy levels in Chapter 6. The energies are negative, corresponding to bound states.

Photons emitted by atomic hydrogen

Bohr proposed that electromagnetic radiation would be emitted when there was a sudden change from a higher energy level to a lower one. We saw in Chapter 6 that the Bohr energy formula does correctly predict the energies of photons emitted by excited atomic hydrogen.

? In the Bohr model, does photon emission correspond to the radius of the circular orbit getting larger or smaller? Does the quantum number N increase or decrease?

The kinetic plus potential energy is negative, corresponding to a bound state, so a higher energy is one that is less negative and therefore corresponds to a larger radius (since the energy is proportional to $-1/r$). Also, higher energy corresponds to a larger value of N, which is also associated with a larger radius. Therefore photon emission is associated with a decrease in r and a decrease in N (Figure 9.28). Photon absorption involves an increase in r and N.

Refinements of the Bohr model

The Bohr model was an important predecessor to current quantum mechanical models of the atom. It predicts the main aspects of the hydrogen energy levels correctly (neglecting various small effects). The idea that photon emission is associated with a drop from a higher energy level to a lower energy level, with the photon energy equal to the difference in the hydrogen energy levels, is fundamental to the extended quantum-mechanical treatment.

The notion of deterministic circular orbits has been abandoned and replaced with a picture of a probabilistic electron "cloud," but even in more sophisticated models the average radius of this electron cloud is larger for higher energies, just as in the Bohr model. From your study of chemistry you are probably familiar with the shapes of the "probability density" functions or "orbitals" describing the electron's location in the atom in different energy states, some of which are spherical, others dumbbell shaped.

The idea that angular momentum is quantized has far-reaching consequences and is a major element in modern quantum mechanics. However, it has become clear that the actual quantization rules in a hydrogen atom are more complex than those that are assumed in the Bohr model. For example, the translational angular momentum in the ground state ($N = 1$) is actually zero, not $\hbar$, and for the next higher state ($N = 2$), the z component of translational angular momentum can be either 0 or $\hbar$.

9.3.2 Particle spin

An important example of angular momentum quantization is the observation that many elementary particles have rotational angular momentum, and it is quantized. We know this in part because of observations of the interaction of these particles with other particles having nonzero angular momentum; in these interactions total angular momentum is constant.

For example, the electron, muon, and neutrino are said to have spin $1/2$, because measurement of a component of their rotational angular momentum always yields $\pm(1/2)\hbar$. Every quark has spin $1/2$, and particles built out of three quarks, such as protons and neutrons, necessarily have half-integral spin. The spin of the proton and neutron is $1/2$, as though two of the quarks have their spins opposed ($\Uparrow\Downarrow\Uparrow$), with no translational angular momentum of the quarks. Some short-lived three-quark particles have spin $3/2$. One way to get this is for all the quarks to have their spins aligned ($\Uparrow\Uparrow\Uparrow$), or there can be some translational angular momentum of the

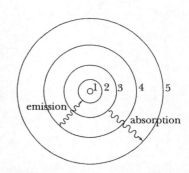

Figure 9.28 A photon is emitted in a transition from a higher to a lower energy level. A photon is absorbed in a transition from a lower to a higher energy level.

quarks. While rotational angular momentum can be half-integral, translational angular momentum is always integral (0, 1, 2, etc.).

Particles made out of one quark and one antiquark, called mesons, have integral spin. For example, the pion spin is zero (quark and antiquark spins opposed, ⇑⇓), and the spin of the "rho" meson is 1 (quark and antiquark spins aligned, ⇑⇑).

Presumably the angular momentum of a macroscopic object such as a baseball is also quantized, but the angular momentum of a baseball is huge compared to Planck's constant, and you don't notice the quantization, which is on an exceedingly fine scale (10^{-34} J·s!). Only at the atomic level are the effects of quantization really evident.

9.3.3 Consequences of angular momentum quantization

The quantization of angular momentum has far-reaching consequences. Angular momentum quantization plays a major role in determining the structure of atoms, and the nature of the chemical periodic table. The two lowest-energy-state electrons in an atom always have zero translational angular momentum and zero rotational angular momentum (because the two rotational angular momenta are always oppositely aligned and add up exactly to zero).

There is a very deep connection between angular momentum and the statistical behavior of particles. Particles such as the electron and the proton that have half-integral spin are called "fermions" and exhibit the peculiar behavior that two fermions cannot be in the exact same quantum state. This is why only two electrons can be in the lowest energy state of an atom, with their spins opposed. Their spin directions are forbidden to be the same, because then their energy and angular momentum would be exactly the same, which is forbidden for fermions. This prohibition is called the "Pauli exclusion principle."

On the other hand, particles with integral spin, called "bosons," are not subject to the Pauli exclusion principle, and there is no limit to the number of bosons that can be in the same energy state. In fact, there is a special state of matter called a "Bose-Einstein condensation" in which very large numbers of particles end up in exactly the same quantum state. This state of matter was predicted long before it was actually created and observed in 1995.

In a diatomic gas molecule such as oxygen (O_2), the kinetic energy associated with rotation of the molecule is of course $L_{rot}^2/(2I)$. But the rotational angular momentum L_{rot} is quantized, so the rotational energies of oxygen are quantized. The phenomenon of angular momentum quantization affects the specific heat of diatomic gases at low temperatures, as will be discussed in a later chapter. (You can of course refer to the angular momentum of a oxygen molecule as translational angular momentum of the two atoms instead of rotational angular momentum of the molecule.)

It is an important prediction of quantum mechanics that the x, y, or z component of the angular momentum (L_x, L_y, or L_z) can only be an integer or half-integer multiple of $\hbar$, whereas the square magnitude of the angular momentum has the quantized values $l(l+1)\hbar^2$, where l has integer or half-integer values, depending on whether a component has integer or half-integer values:

QUANTIZED VALUES OF L^2

$$L^2 = l(l+1)\hbar^2, \text{ where } l \text{ is integer or half-integer}$$

All nuclei that have an even number of protons and an even number of neutrons ("even-even" nuclei) have a total angular momentum of zero. Examples include carbon-12 (6 protons and 6 neutrons) and oxygen-16 (8 protons and 8 neutrons). The spins of protons and neutrons in even-even nuclei are paired up in the lowest-energy state of a nucleus with each other in such a way as to produce zero net angular momentum. This would be an exceedingly unlikely arrangement if angular momentum weren't quantized.

9.4 The angular momentum principle

So far we have concentrated on defining and calculating angular momentum. Next we will show that an off-center force that applies a twist to a system can change the angular momentum of the system.

For a single particle, the momentum principle says that the time derivative of the momentum is the net force: $d\vec{p}/dt = \vec{F}_{net}$. Let's see what the time derivative of angular momentum turns out to be. By applying the product rule for derivatives, we have

$$\frac{d(\vec{r}_A \times \vec{p})}{dt} = \frac{d\vec{r}_A}{dt} \times \vec{p} + \vec{r}_A \times \frac{d\vec{p}}{dt}$$

Since $\dfrac{d\vec{r}_A}{dt} = \vec{v}$ and $\vec{p} = \dfrac{m\vec{v}}{\sqrt{1-(v^2/c^2)}}$, we have

$$\frac{d\vec{r}_A}{dt} \times \vec{p} = \frac{\vec{v} \times m\vec{v}}{\sqrt{1-(v^2/c^2)}} = 0$$

since $|\vec{v} \times \vec{v}| = v^2 \sin(0°) = 0$. Therefore we have proved that the following is true for a single particle:

$$\frac{d(\vec{r}_A \times \vec{p})}{dt} = \vec{r}_A \times \vec{F}_{net}$$

We have previously defined the quantity $\vec{L}_A = \vec{r}_A \times \vec{p}$ to be the angular momentum of a particle relative to location A. We call the quantity $\vec{\tau}_A = \vec{r}_A \times \vec{F}_{net}$ the "torque" exerted about location A (Figure 9.29). Torque is usually written as τ, Greek lowercase letter tau. Torque is a word of Latin origin and simply means "twist." Twisting a system changes its angular momentum.

We can read the equation as saying "the rate of change of the angular momentum of a particle relative to location A is equal to the torque applied to the particle about location A."

THE ANGULAR MOMENTUM PRINCIPLE FOR A POINT PARTICLE

$$\frac{d\vec{L}_A}{dt} = \vec{r}_A \times \vec{F}_{net} = \vec{\tau}_A$$

The angular momentum principle is similar to the momentum principle $d\vec{p}/dt = \vec{F}_{net}$. We will apply this new principle after practicing how to calculate torques.

9.4.1 Calculating torque

Torque and angular momentum are both calculated by using the cross product. In this section we offer some additional ways to think about cross products, in the context of practice in calculating torques.

? In Figure 9.30, which way of pushing on a wrench is more effective in twisting the nut?

You can see that a force parallel to the handle is completely ineffective in twisting the nut. Evidently it is only the perpendicular component of a force that is effective in twisting an object.

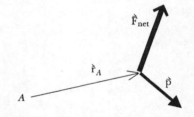

Figure 9.29 Torque relative to location A is defined as $\vec{\tau}_A = \vec{r}_A \times \vec{F}_{net}$.

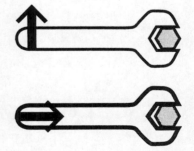

Figure 9.30 Twist a nut, or push straight at it?

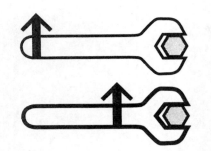

Figure 9.31 Push at the end, or near the nut?

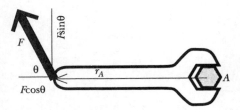

Figure 9.32 Push on the wrench at an angle.

? In Figure 9.31, where should you push on the wrench—at the end of the wrench or near the nut?

The farther away from the nut we push, the more effective we are in twisting the nut. The advantage of a long wrench is to provide more leverage in twisting.

Suppose we apply a force at an angle θ to the radius (Figure 9.32). Evidently the component of the force that is effective in twisting is $F\sin\theta$, the perpendicular component. Moreover, the bigger the lever arm (r_A, the distance from the nut at location A), the more effective we'll be. To capture both effects, we define the magnitude of "torque" (which means "twist") exerted by a force $\vec{F}$ relative to a location A as follows:

MAGNITUDE OF TORQUE RELATIVE TO LOCATION *A*

$$\text{torque} = r_A F \sin\theta$$

For a purely perpendicular force, $\theta = 90°$, $\sin\theta = 1$, and the torque is $r_A F$. For a force that is parallel to the lever arm, $\theta = 0°$, $\sin\theta = 0$, and the torque is zero. See Figure 9.30 for these extreme examples.

Direction of torque

The direction of the torque is given by the cross product and associated right-hand rule.

? In Figure 9.32, is the torque vector into the page or out of the page? Use the right-hand rule.

With your right hand at the location of the nut, point your fingers toward the point of application of the force (at the end of the wrench). Next curl your fingers toward the direction of the force vector. The only way you can do this results in your extended right thumb pointing into the page. If you try to keep your extended right thumb pointing out of the page, you will find that you can't curl your fingers toward the direction of the force.

Most nuts and bolts have what are called "right-handed threads." In Figure 9.32 a right-handed nut will advance in the direction of the torque vector, into the page. If you reverse the direction of the force, so the torque vector points out of the page, a right-handed nut will advance out of the page. (Exceptions include left-handed threads on gas cylinders containing explosive gases; the unexpected behavior alerts users to danger.)

The fact that a physical right-handed nut advances in the direction of the torque vector is another justification for drawing torque cross product vectors along an axis of rotation, perpendicular to the plane of the motion.

Ex. 9.15 If $r_A = 3$ m, $F = 4$ N, and $\theta = 30°$, what is the magnitude of the torque about location A, including units?

Ex. 9.16 If the force in Figure 9.32 were perpendicular to $\vec{r}_A$, but gave the same torque as in the preceding exercise, what would its magnitude be?

Ex. 9.17 At $t = 15$ s, a particle has angular momentum $\langle 3, 5, -2 \rangle$ kg·m^2/s relative to location A. A constant torque $\langle 10, -12, 20 \rangle$ N·m relative to location A acts on the particle. At $t = 15.1$ s, what is the angular momentum of the particle?

9.4.2 Example: Falling object

Let's compare the momentum principle and the angular momentum principle in a simple situation. Consider a mass m falling near the Earth (Figure 9.33). We choose a location A off to the side, on the ground.

Neglecting air resistance, the momentum principle gives $dp/dt = mg$, yielding $dv/dt = g$ (nonrelativistic). Next we see what the angular momentum principle gives.

? What is the direction of the torque on the mass relative to location A? What is its magnitude?

On the diagram you can see that $x = r_\perp$, so the torque on the mass relative to point A is $r_\perp mg = xmg$. The right-hand rule shows that the direction of the torque is into the page ($\otimes$).

? What is the direction of the angular momentum of the mass relative to point A? What is its magnitude?

The angular momentum of the mass relative to point A is $r_\perp mv = xmv$, since $x = r_\perp$. The right-hand rule shows that the direction of the angular momentum is into the page ($\otimes$). The angular momentum principle has components into the page of $d(xp)/dt = xmg$. Since x is a fixed value, this yields $dp/dt = mg$ as expected. So the angular momentum principle for a point particle is equivalent to the momentum principle. This isn't a big surprise, since we derived the new equation starting from the momentum principle.

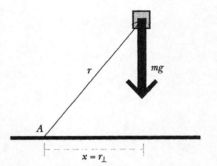

Figure 9.33 Analyze a falling mass using force and momentum, and using torque and angular momentum.

9.5 Conservation of angular momentum

Since we just found that the angular momentum principle didn't tell us anything new about a falling object, what's the point of all this? An important point is that the angular momentum is constant under certain circumstances. If the torque around a location A is zero, the angular momentum about that location does not change. There may be forces acting, and causing changes in the linear momentum, but if these forces don't exert any torque, the rate of change of angular momentum is zero. A major reason for defining torque is to be able to identify situations in which the torque about a location A is zero, in which case angular momentum is constant.

Angular momentum is a conserved quantity:

CONSERVATION OF ANGULAR MOMENTUM

$$\Delta \vec{L}_{A, \text{system}} + \Delta \vec{L}_{A, \text{surroundings}} = 0$$

If a system gains angular momentum, the surroundings lose that amount. An important special case that merits separate attention is the case of no torque acting on the system, in which case the angular momentum does not change. Comet orbits provide a nice example. Most comets have very long elliptical orbits around the Sun (Figure 9.34). We can treat the comet as a point particle, because it is tiny compared to the distances involved.

Ex. 9.18 Relative to the center of the Sun, explain why the torque exerted by the Sun's gravitational force on the comet is zero at every point along the orbit. What does that say about the angular momentum of the comet relative to the center of the Sun?

Ex. 9.19 In terms of the comet's mass m and speed v_1 when nearest the Sun (distance r_1 from the Sun), what is the comet's angular momentum relative to the center of the Sun (direction and

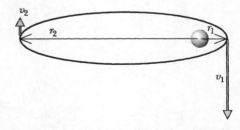

Figure 9.34 A long elliptical orbit of a comet around the Sun.

magnitude)? What is the comet's angular momentum relative to the center of the Sun (direction and magnitude) when it is farthest from the Sun (distance r_2 from the Sun) and traveling at speed v_2? What can you conclude about the relationship between v_1 and v_2?

Comments about comets

Some comets go far beyond Pluto and spend most of their time there, because they're traveling very slowly, as you have just shown. We see them when they come near the Sun, but only for a few months, because they're now traveling fast.

The result $r_1 v_1 = r_2 v_2$ at the closest and farthest points in the orbit is correct no matter what kind of "central" force is involved. For any force that acts along a line connecting two objects, there is no torque about a point at the center of one of the objects, so angular momentum is constant about that location. In the particular case of the gravitational force, we also have an energy equation relating v_1 and v_2:

$$\tfrac{1}{2} mv_1^2 - GMm/r_1 = \tfrac{1}{2} mv_2^2 - GMm/r_2$$

Historical note: Kepler and elliptical orbits

Based on careful, accurate naked-eye measurements made by Tycho Brahe before the invention of the telescope, Johannes Kepler in 1609 announced his discovery that the planets follow elliptical orbits around the Sun.

Kepler also stated that he had found that "a radius vector joining any planet to the Sun sweeps out equal areas in equal lengths of time." This is equivalent to conservation of angular momentum, as can be seen with the aid of Figure 9.35. The area swept out in a time Δt is the area of a triangle whose base is $v\Delta t$ and whose altitude is $r\sin\theta$. This area is $\tfrac{1}{2}(rv\sin\theta)\Delta t$, which is proportional to the angular momentum $rmv\sin\theta$.

In 1618 Kepler announced his discovery that the square of the time it takes a planet to go around the Sun is proportional to the cube of the mean distance from the Sun (a result you derived in Chapter 2 for the simpler case of circular orbits).

Later in the 1600's Newton explained all three of these discoveries as derivable from his second law of motion plus his universal law of gravitation. Kepler's insights provided important tests for Newton's theories.

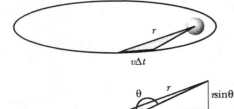

Figure 9.35 "Equal area in equal time."

9.6 Multiparticle systems

The angular momentum principle has already given us new insight into the motion of a single particle acted on by a torque due to a single force. The equation becomes a really powerful tool when we extend it to apply to multiparticle systems acted on by multiple torques. For example, we will be able to understand the counterintuitive behavior of spinning tops and gyroscopes.

We will now derive a multiparticle version of the angular momentum principle, following closely the derivation in Chapter 2 of a multiparticle version of the momentum principle.

To be as concrete as possible, we'll consider a system consisting of just three particles. It is easy to see how this generalizes to larger systems, including a block consisting of an astronomically huge number of atoms.

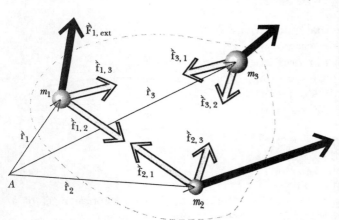

Figure 9.36 The angular momentum principle in a multiparticle system.

In Figure 9.36 we show all of the forces acting on each particle, where the lower case $\vec{f}$'s are internal forces, and the upper case $\vec{F}$'s are external forces, such as the gravitational attraction of the Earth. We write the angular momentum principle applied to each of the three particles. We measure angular momenta and torques relative to location A, but to avoid clutter we don't write the subscript A in any of these equations until the end. ($\vec{L}_1$ is the angular momentum of m_1 relative to A.)

$$\frac{d\vec{L}_1}{dt} = \vec{r}_1 \times (\vec{F}_{ext,1} + \vec{f}_{1,2} + \vec{f}_{1,3})$$

$$\frac{d\vec{L}_2}{dt} = \vec{r}_2 \times (\vec{F}_{ext,2} + \vec{f}_{2,1} + \vec{f}_{2,3})$$

$$\frac{d\vec{L}_3}{dt} = \vec{r}_3 \times (\vec{F}_{ext,3} + \vec{f}_{3,1} + \vec{f}_{3,2})$$

Nothing new so far. But now we add up these three equations. That is, we create a new equation by adding up all the terms on the left sides of the three equations, and adding up all the terms on the right sides, and setting them equal to each other. In doing so, we take into account that many of these terms cancel. By the reciprocity principle (Newton's third law), we have

$$\vec{f}_{1,2} = -\vec{f}_{2,1}$$

$$\vec{f}_{1,3} = -\vec{f}_{3,1}$$

$$\vec{f}_{2,3} = -\vec{f}_{3,2}$$

Therefore the torques of the internal forces cancel. For example, consider this piece of the sum:

$$\vec{r}_1 \times \vec{f}_{1,2} + \vec{r}_2 \times \vec{f}_{2,1} = \vec{r}_1 \times \vec{f}_{1,2} - \vec{r}_2 \times \vec{f}_{1,2} = (\vec{r}_1 - \vec{r}_2) \times \vec{f}_{1,2}$$

The vector $\vec{r}_1 - \vec{r}_2$ points from m_2 toward m_1, so the angle between it and $\vec{f}_{1,2}$ is either 0° or 180°, and $\sin\theta$ is zero. Consequently all of the torques associated with the internal forces cancel, and all that remains is this:

$$\frac{d(\vec{L}_1 + \vec{L}_2 + \vec{L}_3)}{dt} = \vec{r}_1 \times \vec{F}_{ext,1} + \vec{r}_2 \times \vec{F}_{ext,2} + \vec{r}_3 \times \vec{F}_{ext,3}$$

The right side of this equation represents the net torque due to external forces, $\vec{\tau}_{net,ext,A}$. The great importance of this equation is that the reciprocity of gravitational and electric forces has allowed us to get rid of all the torques due to internal forces, the torques that the particles in the system exert on each other. We rewrite our result like this:

THE ANGULAR MOMENTUM PRINCIPLE
FOR A MULTIPARTICLE SYSTEM

$$\frac{d\vec{L}_{tot,A}}{dt} = \vec{\tau}_{net,ext,A}$$

In words, the rate of change of the "total angular momentum" of a system relative to a location A, $\vec{L}_{tot,A} = \vec{L}_{1,A} + \vec{L}_{2,A} + \vec{L}_{3,A} + \dots$, is equal to the net torque due to external forces exerted on that system, relative to location A. The internal forces and torques do not appear in this multiparticle version of the angular momentum principle.

Often it pays to choose location A to be at the place where the center of mass happens to be at that instant. In that case, the angular momentum principle simplifies:

$$\frac{d\vec{L}_{cm}}{dt} = \frac{d}{dt}[(\vec{r}_{cm,cm} \times \vec{P}_{tot}) + \vec{L}_{rot}] = \frac{d\vec{L}_{rot}}{dt}, \text{ since } \vec{r}_{cm,cm} = 0$$

That is, the position of the center of mass, relative to the center of mass itself, is of course a zero vector. Therefore we have the important and useful special case for the motion relative to the center of mass

THE ANGULAR MOMENTUM PRINCIPLE
RELATIVE TO THE CENTER OF MASS

$$\frac{d\vec{L}_{rot}}{dt} = \vec{\tau}_{net,cm}$$

The child and the playground ride

We began the chapter with a child running and jumping onto a playground ride. The axle about which the disk turns may exert a force on the disk, but this force has a very small lever arm measured from the center of the axle. If we consider the combined multiparticle system of child plus disk, we expect that the angular momentum to be (nearly) constant. The initial angular momentum of the child should be equal to the sum of the final angular momentum of the child (on the now rotating disk) and the angular momentum of the disk.

Ex. 9.20 Consider a rotating star far from other objects. Its rate of spin stays constant, and its axis of rotation keeps pointing in the same direction. Why?

Ex. 9.21 Because the Earth is nearly perfectly spherical, gravitational forces act on it effectively through its center. Explain why the Earth's axis points at the North Star all year long. Also explain why the Earth's rotation speed stays the same throughout the year (one rotation per 24 hours). In your analysis, does it matter that the Earth is going around the Sun?

In actual fact, the Earth is not perfectly spherical. It bulges out a bit at the equator, and tides tend to pile up water at one side of the ocean. As a result, there are small torques exerted on the Earth by other bodies, mainly the Sun and the Moon. Over many thousands of years there are changes in what portion of sky the Earth's axis points toward (change of direction of rotational angular momentum), and changes in the length of a day (change of magnitude of rotational angular momentum). See page 324.

9.6.1 Summary: Angular momentum and torque

At last we have all the tools we need to investigate a variety of situations involving angular momentum. Let's summarize these new tools and concepts:
- A force exerts a torque relative to a chosen location A: $\vec{\tau}_A = \vec{r}_A \times \vec{F}$.
- Torque causes changes in angular momentum; the rate of change of the angular momentum is equal to the net torque:

$$\frac{d\vec{L}_A}{dt} = \vec{\tau}_{net,ext,A}$$

- If the net torque relative to A is zero, the angular momentum relative to A does not change. (And if we know the angular momentum isn't changing about some location, as in static equilibrium, we know that the net torque must be zero about that location.)

- The angular momentum can be divided into two pieces: translational angular momentum of the system as a whole (as though all the mass were concentrated at the center-of-mass point), plus rotational angular momentum relative to the center of mass: $\vec{L}_A = \vec{L}_{\text{trans},A} + \vec{L}_{\text{rot}}$.

- The rate of change of the rotational angular momentum is equal to the torque relative to the center of mass:

$$\frac{d\vec{L}_{\text{rot}}}{dt} = \vec{\tau}_{\text{net,ext,cm}}$$

- Rotational angular momentum is given by moment of inertia times angular velocity: $\vec{L}_{\text{rot}} = I\vec{\omega}$, where $I = m_1 r_{\perp 1}^2 + m_2 r_{\perp 2}^2 + m_3 r_{\perp 3}^2 + \dots$

- Rotational kinetic energy about the center of mass can be calculated as

$$K_{\text{rot}} = \tfrac{1}{2}I\omega^2 = \frac{1}{2}\frac{(I\omega)^2}{I} = \frac{L_{\text{rot}}^2}{2I}$$

9.7 The three fundamental principles of mechanics

We now have three fundamental principles that relate a change in some property of a system to the interactions of the system with external objects. In the special case in which there are no external interactions, these principles can be phrased as "conservation laws."

Momentum	Angular momentum	Energy
$\dfrac{d\vec{P}}{dt} = \vec{F}_{\text{net,ext}}$	$\dfrac{d\vec{L}_A}{dt} = \vec{\tau}_{\text{net,ext,A}}$	$\Delta E = W + Q$
If external forces: momentum changes.	If external torques: angular momentum changes.	If energy inputs: energy of system changes.
If no external forces: the momentum of a system is constant.	If no external torques: the angular momentum of a system is constant.	If no energy inputs: the energy of a system is constant.
Location of object does not matter.	Location of object relative to point A is important.	Location of object does not matter.

9.7.1 More exercises on the angular momentum principle

The following exercises constitute an important worksheet that gives you practice in using the angular momentum principle. Be sure to do all of these exercises and check your answers at the end of the chapter.

Ex. 9.22 A stationary bicycle wheel of radius R is mounted in the vertical plane on a horizontal frictionless axle (Figure 9.37). The wheel has mass M, all concentrated in the rim (the spokes have negligible mass). A lump of clay with mass m falls and sticks to the outer edge of the wheel at the location shown. Just before the impact the clay has a speed v.

(a) Just before the impact, what is the angular momentum of the combined system of wheel plus clay about the center C?

(b) Just after the impact, what is the angular momentum of the combined system of wheel plus clay about the center C?

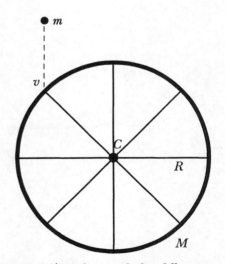

Figure 9.37 A lump of clay falls onto a wheel mounted on a frictionless axle.

(c) Just after the impact, what are the magnitude and direction of the angular velocity of the wheel?

(d) Qualitatively, what happens to the linear momentum of the combined system? Why?

Ex. 9.23 A disk of radius 8 cm is pulled along a frictionless surface with a force of 10 N by a string wrapped around the edge (Figure 9.38). 24 cm of string has unwound off the disk.

What are the magnitude and direction of the torque exerted about the center of the disk at this instant?

Ex. 9.24 In Figure 9.38, the uniform solid disk has mass 0.4 kg (moment of inertia $I = \frac{1}{2}MR^2$). At the instant shown, the angular velocity is 20 radians/s into the page.

(a) At this instant, what are the magnitude and direction of the angular momentum about the center of the disk?

(b) At a time 0.2 s later, what are the magnitude and direction of the angular momentum about the center of the disk?

(c) At this later time, what are the magnitude and direction of the angular velocity?

Ex. 9.25 A barbell mounted on a nearly frictionless axle through its center (Figure 9.39). At this instant, there are two forces of equal magnitude applied to the system as shown, with the directions indicated, and at this instant the angular velocity is 60 radians/s, counterclockwise. In the next 0.001 s, the angular momentum relative to the center increases by an amount 2.5×10^{-4} kg·m²/s.

What is the magnitude of each force?

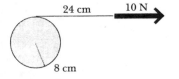
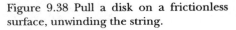

Figure 9.38 Pull a disk on a frictionless surface, unwinding the string.

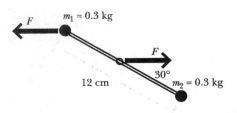

Figure 9.39 Forces applied to a barbell.

9.8 Applications to systems with no torques

We will apply our tools to two situations where no torques are applied.

9.8.1 An ice skater's spin

You may have seen an ice skater spin vertically on the tip of one skate, with her arms and one leg outstretched, then pull her leg in and bring her arms to a vertical position above her head. She then spins much faster.

What's going on? There is some frictional force of the ice on the tip of the skate, but this force is applied so close to the axis of rotation that the torque is small.

? During the short time when the skater quickly changes her configuration, what can you say about the rotational angular momentum? Why?

Small torque implies small change in angular momentum per unit time, so in a short time the angular momentum of the skater will hardly change.

? But if the rotational angular momentum hardly changes, how can the skater spin faster?

When she changes her configuration, the moment of inertia decreases, because some parts of her body are now closer to the axis of rotation. For the rotational angular momentum not to change, the angular velocity must increase to compensate for the decreased moment of inertia: $I_1\omega_1 = I_2\omega_2$, and therefore $\omega_2 = I_1\omega_1/I_2$.

Now you may quite legitimately be puzzled about how this actually works! *Why* does moving her arms and leg closer to the spin axis increase her angular speed? At one level of discussion, you can close your eyes (and maybe hold your nose) and say, "Well, we did all that general analysis of the effect of torques on the angular momentum of multiparticle systems, so if that's what we get when we apply these general principles, I guess that's that." But at another level it would be very nice to get a better sense of the detailed mechanisms involved in this odd phenomenon.

One approach is to analyze the changes in energy involved in this skating maneuver. The angular momentum L_{rot} of the skater doesn't change, so as her moment of inertia decreases, the kinetic energy of the skater $L_{rot}^2/(2I)$ actually increases.

? Where does this increased energy come from?

Evidently the skater has to expend chemical energy in order to increase her kinetic energy. In fact, at high spin rates it takes a noticeable effort to pull her arms and leg toward the spin axis. This is even more dramatic when holding heavy weights.

A popular physics demonstration is to sit on a rotating stool holding a dumbbell in each hand. Start spinning slowly with the arms held out, then pull your hands in toward your chest. It requires considerable effort to pull the dumbbells in. After all, the dumbbells would tend to move in straight lines, and you must exert a radial force just to keep turning them in a circle (though in that case you do no mechanical work, because the motion of the dumbbells has no radial component in the direction of the force you apply). To move the dumbbells into a smaller radius requires applying an even larger force which does work on the dumbbells, because there is now a radial component of the motion, in the direction of the radial force you exert.

9.8.2 A high dive

You may have seen a skilled diver leap off a high board, tuck himself into a tight ball (holding onto his ankles), and rotate quite fast for a few turns. Then he straightens out and enters the water like a knife, hardly ruffling the surface.

? What can you say about the diver's rotational angular momentum while he is in the air? Why?

We can probably neglect air resistance, because the diver is very far from reaching terminal speed. Then the only significant force acting is the Earth's gravitational force, which effectively acts through the center of mass. Therefore there is negligible torque about the center of mass, and the rotational angular momentum does not change.

? Early in the dive the diver is spinning rapidly. If rotational angular momentum is constant during the dive, where did that rapid spin come from?

When the diver jumps off the board, he must thrust with his feet in such a way that the force of the board on his feet has a sizable lever arm about his center of mass, so that by the time his feet lose contact with the board, the diver already has acquired a sizable rotational angular momentum.

? When the diver pulls out of the tucked position, why does he stop spinning rapidly?

There is a large increase in the moment of inertia, because many atoms are now much farther from the center of mass than they were when the diver was in the tucked position. Larger moment of inertia implies smaller angular speed, since the rotational angular momentum $I\omega$ does not change.

? Can his body go straight from then on?

His body can't really go completely straight, because he still has rotational angular momentum. But his angular speed may be so small that you hardly notice it, especially in comparison with the very rapid spin in the preceding tucked position. Moreover, his body could approximately follow the curving path of his center of mass, with the body rotating just enough to stay tangent to the trajectory. This enhances the illusion of straight motion. The most important aspect for good form is to arrange that the body rotate into the vertical position at the time of entering the water, so as to make little splash.

See Problem 9.24, which asks you to analyze a diver's motion.

9.9 Applications to systems with nonzero torques

In the remainder of this chapter we will analyze systems where there are external torques applied to the system, which can change the angular momentum of the system.

9.9.1 A meter stick on the ice

Consider a meter stick whose mass is 300 grams and which lies on ice (in Figure 9.40 we're looking down on the meter stick). You pull at one end of the meter stick, at right angles to the stick, with a force of 6 newtons. Assume that friction with the ice is negligible.

? The ensuing motion of the meter stick is quite complicated, but what is the initial rate of change dv_{cm}/dt of the center-of-mass velocity v_{cm} of the stick, at the instant when you first apply the force?

Using the momentum principle, $d\vec{P}/dt = \vec{F}_{net}$, we find that $dv_{cm}/dt = (6 \text{ N})/(0.3 \text{ kg}) = 20 \text{ m/s}^2$.

Next we'll determine the initial rate of change $d\omega/dt$ of the angular speed ω about the center of mass. We'll use the angular momentum principle about the center of mass, $d\vec{L}_{rot}/dt = \vec{\tau}_{net,ext,cm}$, with $L_{rot} = I\omega$.

? What is the torque about the center of mass? State the direction as well as the magnitude.

By the right-hand rule, the torque points into the page. The magnitude is $(0.5 \text{ m})(6 \text{ N})\sin 90° = 3 \text{ N·m}$.

? The moment of inertia I around the center of mass of a uniform rod of mass M and length L can be shown to be $ML^2/12$ (or looked up in a handbook). What is the initial rate of change $d\omega/dt$ of the angular speed ω?

We find that $d\omega/dt = (3 \text{ N·m})/[(0.3 \text{ kg·m}^2)/12] = 120 \text{ radians/s}^2$. In vector terms, $d\vec{\omega}/dt$ points into the page, corresponding to the fact that the angular velocity points into the page and is increasing.

Alternative analysis—taking torques around the end of the stick

It is interesting to analyze the motion of the meter stick by calculating torque and angular momentum about the end of the stick where the force is applied, rather than about the center of mass. To be cautious and correct, we should say that we are taking torques about a location fixed in the ice

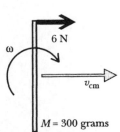

Figure 9.40 Pull one end of a meter stick that is lying on ice (negligible friction).

6 N

ω

v_{cm}

$M = 300 \text{ grams}$

next to the place where the end of the stick is momentarily located. This is a fixed location, not tied to the moving stick.

? About that location (by the end of the stick), what is the magnitude of the net external torque?

The torque is zero, because the distance is zero from that location to the point where the force is exerted.

? Therefore, what do you know about the angular momentum measured about that same location?

The angular momentum about that location is constant, since there is no torque.

? After a very short time interval the stick has a center-of-mass speed v_{cm} and some angular speed ω. At that instant, what is the angular momentum about the end of the stick where the force is applied?

The angular momentum is the sum of the translational angular momentum and the rotational angular momentum. The translational angular momentum is $(L/2)Mv_{cm}$, where L is one meter, and its direction is out of the page or up out of the ice (using the right-hand rule). The rotational angular momentum is $I\omega = (ML^2/12)\omega$, and its direction is into the page or down into the ice. Taking the net direction to be into the page, we have this:

$$\frac{d}{dt}\left[\frac{ML^2}{12}\omega - \frac{L}{2}Mv_{cm}\right] = 0$$

$$\frac{d\omega}{dt} = \frac{6}{L}\frac{dv_{cm}}{dt}\ ,\ \text{where } L = 1 \text{ m}.$$

Using the momentum principle we determined that $dv_{cm}/dt = 20 \text{ m/s}^2$. Evidently $d\omega/dt$ should equal $(6/1 \text{ m})(20 \text{ m/s}^2) = 120$ radians/s^2. This agrees with our analysis where we took torques about the center of mass of the stick.

9.9.2 A puck with string wound around it

Wrap a string around the outside of a hockey puck. Then pull on the string with a constant tension F_T (Figure 9.41). The puck has mass M and radius R. Assume that friction with the ice rink is negligible. Evidently $dv_{cm}/dt = F_T/M$.

? The moment of inertia of a uniform solid puck of mass M and radius R can be shown to be $MR^2/2$ (see Problem 9.3; or this result can be looked up in a handbook). What is the initial rate of change $d\omega/dt$ of the angular speed ω?

The torque about the center of mass is RF_T, into the page (down into the ice). The rotational angular momentum is $(MR^2/2)\omega$, into the page (down into the ice). Therefore $(MR^2/2)d\omega/dt = RF_T$, and we have $d\omega/dt = (2F_T)/(MR)$.

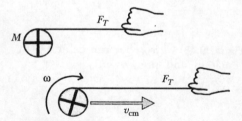

Figure 9.41 Wrap a string around a hockey puck, then pull along ice with negligible friction, with a constant tension F_T.

Ex. 9.26 Redo the analysis, calculating torque and angular momentum relative to a fixed location in the ice anywhere underneath the string (similar to the analysis of the meter stick around one end). Show that the two analyses of the puck are consistent with each other.

9.9.3 Static equilibrium

If the net torque on a system is zero, its angular momentum is constant. Conversely, if we know that the angular momentum is not changing, we can conclude that the net torque must be zero. For example, if a system is in static equilibrium, not only must the net force on the system be zero; the net torque must also be zero. This allows us to make conclusions about the individual torques whose vector sum is zero.

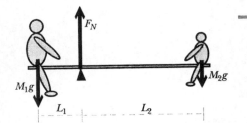

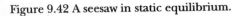

Figure 9.42 A seesaw in static equilibrium.

Ex. 9.27 As a simple application, consider a seesaw that is not rotating around its frictionless support, so that the angular momentum is not changing and is in fact zero (Figure 9.42). Write a force equation, and a torque equation around the support, assuming that the board itself has negligible mass compared to the people sitting on it.

Ex. 9.28 Write another torque equation around a location fixed in space where the person on the left is sitting, and show that it is in fact equivalent to the torque equation around the axle, when you take the force equation into consideration.

It is often convenient to choose your fixed location so that some of the forces create no torque around that location and therefore don't appear in the angular momentum principle.

9.10 *Gyroscopes

A gyroscope is a fascinating device, and its unusual properties have been exploited to stabilize ships and spacecraft. Its behavior provides a good model and analogy for some important aspects of the quantum mechanical behavior of atoms and nuclei. Magnetic resonance imaging (MRI) is based on the gyroscope-like behavior of spinning nuclei.

A gyroscope has a spinning disk mounted on an axle. If you place one end of the axle on a vertical support, you may observe a very complex motion (Figure 9.43). The gyroscope rises and falls ("nutation") as it revolves around the support ("precession").

One fruitful approach in modeling a complex physical system is to try to analyze the simplest motion that the system is capable of. For example, we can observe a pure precession of a gyroscope, with no nutation? It is possible with care to start the gyroscope with special initial conditions so that it merely precesses, without bobbing up and down. We'll try to analyze this special kind of gyroscope motion. In fact, we'll start with a particularly simple form of precession, with the rotational axis horizontal (Figure 9.44).

Figure 9.43 A gyroscope can exhibit both "nutation" and "precession."

We define some variables to make it easier to describe and discuss the situation. The gyroscope disk rotates on its axis with some spin angular speed that we'll call ω, and the disk has a moment of inertia I. The rotational angular momentum $\vec{L}_{rot}$ of the disk always points horizontally but is continually changing direction as the gyroscope precesses (if we neglect friction, the angular speed ω is constant). The gyroscope revolves around the support with angular speed Ω (Greek uppercase omega), constant in magnitude and direction (if we neglect friction).

Figure 9.44 A gyroscope which is precessing horizontally, with no nutation.

? What forces act on the gyroscope? What relationships are there among these forces?

The support pushes up with a force we'll call F_N, and the Earth pulls down with a force Mg through the center of mass (Figure 9.45). Since the center of mass stays at the same height all the time in the case of simple precession, it must be that the vertical component of the net force must be zero. Therefore $F_N = Mg$.

This isn't quite the whole story, though. Note that the center of mass of the gyroscope is moving in a circle. That requires that there be a radially inward force to make the momentum vector rotate (with angular speed Ω). The only object that can exert this force is the support. There must be a small horizontal frictional force f such that $\left| d\vec{P}/dt \right| = \Omega P = f$, where $P = Mv_{\text{cm}} = M(\Omega R)$; R is the radius of the circle traveled by the center of mass. This force, $f = MR\Omega^2$, is small if the precession rate Ω is small.

It is easy to observe with a toy gyroscope that as the spin of the disk slows down due to friction (smaller ω), the precession actually speeds up (larger Ω). If we could predict the precession speed for a given spin, we would largely "understand" this simple example of gyroscope motion. What can we use to attempt a prediction?

We've already used the momentum principle to conclude something about the forces F_N and f. What about energy? If the gyroscope precesses in the horizontal plane, there's no change in gravitational energy, because the center of mass remains always at the same height. There's no change in kinetic energy, either. So energy considerations don't seem to tell us anything interesting about precession (though they would be useful in analyzing the more complex rise and fall of nutation). The remaining principle to apply is the angular momentum principle (no wonder we left gyroscopes until this chapter).

? What is the magnitude of the rotational angular momentum of the gyroscope? What is the magnitude of the translational angular momentum of the gyroscope around the support?

We said that the rotating disk has moment of inertia I and angular speed ω, so the rotational angular momentum is simply $L_{\text{rot}} = I\omega$. The magnitude of the translational angular momentum is this:

$$L_{\text{support}} = \left| \vec{R} \times \vec{P} \right| = RP$$

Evidently the magnitudes of both the rotational and translational angular momenta are constant, not changing with time. But what about the directions of these angular momenta? Do they change with time?

? How does the direction of the translational angular momentum change with time? How does the direction of the rotational angular momentum change with time?

Neither the magnitude nor the direction of the translational angular momentum changes (neglecting friction): it has constant magnitude and points vertically upward at all times. But the direction of the rotational angular momentum is constantly changing as the gyroscope precesses. We need to calculate the rate at which the rotational angular momentum vector is changing, $d\vec{L}_{\text{rot}}/dt$. Since the rotational angular momentum vector is a rotating vector, which rotates with the precession angular speed Ω, we have this:

$$\left| \frac{d\vec{L}_{\text{rot}}}{dt} \right| = \Omega L_{\text{rot}}$$

? Therefore, what is the angular momentum principle for the gyroscope?

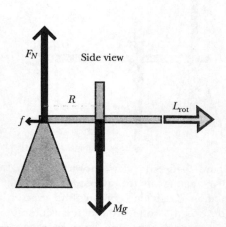

Figure 9.45 Side view showing external forces acting on the gyroscope, and the rotational angular momentum vector.

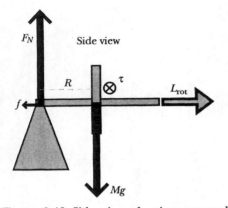

Figure 9.46 Side view showing external forces acting on the gyroscope, and the rotational angular momentum vector. There is a torque around the center of mass, into the page, due to the F_N force.

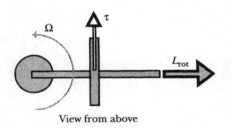

Figure 9.47 View from above, showing the torque vector and the rotational angular momentum vector.

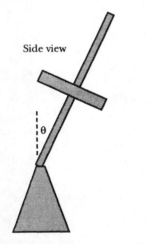

Figure 9.48 What is the precession rate when the spin axis is at an angle θ to the vertical?

The remaining element we need for the angular momentum principle is the torque that acts around the center of mass, at the center of the spinning disk. Look again at the force diagram (Figure 9.46), and you see that the magnitude of the torque about the center of mass is $RF_N = RMg$, and the direction of the torque is into the page. Seen from above, the torque points at a right angle to the rotational angular momentum (Figure 9.47). Therefore the angular momentum principle yields

$$\left| \frac{d\vec{L}_{\text{rot}}}{dt} \right| = \Omega L_{\text{rot}} = \tau_{\text{cm}}, \text{ where } L_{\text{rot}} = I\omega, \text{ and } \tau_{\text{cm}} = RMg$$

Solving for the precession angular speed Ω, we have

$$\Omega = \frac{\tau_{\text{cm}}}{L_{\text{rot}}} = \frac{RMg}{I\omega}$$

This is a surprisingly simple result for such a complicated system., and it agrees at least qualitatively with observations. A smaller spin ω is associated with a larger precession rate Ω. Conversely, a gyroscope which spins very fast precesses very slowly. If possible, your instructor will provide an opportunity to test the theory quantitatively by observing an actual gyroscope whose spin and precession rates can be measured.

Reflection

Take some time to reflect on how we arrived at this result. Look over the steps involved. The key point is that the rotational angular momentum varies in direction without varying in magnitude, and the magnitude of the rate of change of the rotational angular momentum is simply Ω times L_{rot}. This rate of change of the rotational angular momentum is equal to the net torque acting on the system (around the center of mass).

Ex. 9.29 To complete this reflection, determine the relationship between ω and Ω for the case of pure precession, but with the spin axis at an arbitrary angle θ to the vertical (Figure 9.48; $\theta = 90°$ is the case we treated of horizontal precession). If you have the opportunity, see whether this relationship holds for a real gyroscope.

Uses of gyroscopes

The precession we calculated is the result of a nonzero torque acting around the center of mass. Through clever mounting of the gyroscope it is possible to make this torque vanishingly small, in which case the axis does not precess but always maintains the same direction, which has been useful in navigation. The satellite-based Global Positioning System (GPS) can tell you where your airplane is; gyroscopically based instruments can tell you the orientation and attitude of your airplane at this known location.

Large gyroscopes have been used to stabilize ships against rolling in the sea. A gyroscope with very high rotational angular momentum (large moment of inertia and large angular speed) is hard to turn quickly, and this can provide mechanical stability. Gyroscopes are also used to stabilize the orientation of spacecraft and satellites.

Friction

We deliberately neglected friction in the calculation. Friction has two main effects on a gyroscope. Friction in the spin axis makes ω decrease with time, which leads to an increased precession rate Ω. Friction on the top of the support slows down the precession, which means that Ω is no longer equal to RMg/L_{rot} as it should be for pure precession. As a result of these effects, what you observe with a gyroscope that starts out in pure precession is that the spin axis eventually starts to tip lower, accompanied by faster precession. The gyroscope starts out with a stately, dignified, slow precession, but toward the end gives the impression of motion that is more and more frantic.

9.10.1 *Magnetic Resonance Imaging (MRI)

Electrically charged atomic and subatomic objects that have angular momentum act like little bar magnets with north and south poles along the angular momentum axis (we say that they have "magnetic moments"). It is as though the spinning charged particle constituted little loops of current, and current loops produce a magnetic field. Even some electrically neutral particles such as the neutron have magnetic moments, because they are built out of electrically charged quarks that have magnetic moments.

In the presence of an applied magnetic field (typically produced by other current-carrying coils of wire), a bar magnet tends to twist to line up with the applied magnetic field. That is, a magnetic field exerts a torque on a bar magnet. A well-known example is that a compass needle aligns with the Earth's magnetic field, which is useful for navigation. (The compass needle is magnetized and is itself a little bar magnet.)

Since a magnetic field applies a torque to a bar magnet, or to an atomic or subatomic particles that has angular momentum, such particles precess in the presence of an applied magnetic field (Figure 9.49). This phenomenon is exploited in a technique important in physics, chemistry, and biology, called NMR—nuclear magnetic resonance. A particularly useful application of NMR is in magnetic resonance imaging (MRI).

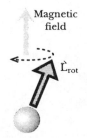

Figure 9.49 Precession of proton spin in an external magnetic field.

In MRI, a large, steady, uniform magnetic field created by large direct-current-carrying coils of wire surrounding the patient makes protons in the person's body precess. Many of the most common nuclei in the body are even-even nuclei (in particular, the most common isotopes of carbon and oxygen) and have zero angular momentum and therefore no magnetic moment, and these do not precess. But hydrogen nuclei (protons) are very common in the body, have rotational angular momentum, and do precess. The stronger the magnetic field, the larger the torque acting on the proton nuclear magnets, and the faster their precession. (Note that for a gyroscope, $\Omega = \tau/L_{rot}$; the precession rate is proportional to the applied torque τ.)

With the protons precessing in the presence of the large steady magnetic field, a small high-frequency (time-varying) magnetic field tuned to match exactly the precession frequency of the protons will flip the spins upside down. This is called a "resonance" phenomenon. The high-frequency signal is turned off, and the protons revert back to being aligned with the steady magnetic field. The act of flipping back emits radiation (at the precession frequency) that can be detected by coils connected to a receiver, and the strength of the signal indicates how many protons were affected.

That's the basic physical mechanism, but this by itself would not yield spatial detail about the interior of the body. The trick is to superimpose on the large steady magnetic field a small nonuniform magnetic field, which has the effect of establishing slightly different torques on protons at different locations in the body. As a result, when the protons flip back into alignment they radiate signals whose frequencies indicate their locations. A computer algorithm calculates how much of each frequency is present in the signal, and this indicates how many protons are at each location. In this way a very detailed image is built up of the interior of the body.

9.10.2 *Precession of spin axes in astronomy

As was mentioned earlier, the Earth is subject to small torques due to gravitational forces of the Sun and Moon acting on its nonspherical, "oblate" shape. There is a torque on the Earth's equatorial bulge, perpendicular to the spin rotation axis, and this causes the Earth's axis to precess very slowly, once around about every 26000 years, so that the "North Star" hasn't always been the star we call Polaris. This effect is called the "precession of the equinoxes" because it leads to a change in what month the spring and fall equinoxes occur. To see why there is such an effect, consider Figure 9.50.

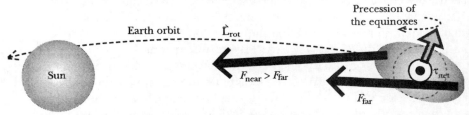

Figure 9.50 Forces on the Earth due to the Sun. The diagram is not to scale; the asymmetry of the Earth is greatly exaggerated. The net torque about the center of mass of the Earth is out of the page.

The Sun (or Moon) exerts a slightly larger force on the closer bulge, due to the $1/r^2$ dependence of the force, so there is a nonzero net torque around the center of mass of the Earth, pointing toward you, out of the page. This makes the rotational angular momentum $\vec{L}_{rot}$ of the Earth precess in a "retrograde" way—that is, in the opposite direction to most rotations in our Solar System. Slowly the direction of the axis of the Earth changes due to the torque. (The size of the equatorial bulge, and the differences in F_{near} and F_{far}, have been greatly exaggerated in Figure 9.50).

Since the Earth's axis is tipped 23° away from perpendicular to the Earth-Sun orbit, the precession of the axis makes a big change in the location of the "North Star." 13000 years from now the "North Star" will be a star that is 46° away from Polaris in the night sky.

The Moon going around the Earth is a kind of gyroscope. The Earth-Moon orbit is inclined a few degrees to the plane of the Earth-Sun orbit, and the Sun exerts a nonzero torque perpendicular to the angular momentum of the Earth-Moon system. To see why, consider a simpler case where the Earth and Moon have equal mass, and you see the same effect as with the equatorial bulge of the Earth (Figure 9.51).

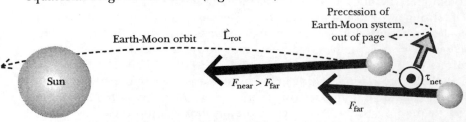

Figure 9.51 Forces on the Earth-Moon system due to the Sun. The diagram is not to scale; the Earth and Moon are shown as if their masses were equal. The net torque on the Earth-Moon system about its center of mass is out of the page.

This torque varies in magnitude during a month, but the averaged effect is that the Earth-Moon system precesses once around in about 18 years (this precession is also "retrograde"). This has an effect on the timing of eclipses, which can occur only when the Moon in its orbit is passing through the plane of the Earth-Sun orbit.

Tidal torques

If the Earth did not rotate, the Moon would create tides in the oceans that would pile up in line with the Moon. However, the rotating Earth drags these tidal bulges so that they are no longer in line with the Moon, and the Moon exerts a small torque (Figure 9.52). This torque is directed along the axis and acts to slow down the spin rate, so that the day is getting longer than 24 hours. This effect is called "tidal friction." This interaction with the Moon has the effect that as the rotational angular momentum of the Earth decreases, the translational angular momentum of the Earth-Moon system increases (conservation of total angular momentum), with the result that the Earth and Moon are getting farther apart. (These tidal forces have a small net tangential component that acts on both Earth and Moon to increase their translational angular momenta.)

Something like this presumably happened in the past to the Moon, when it was molten. The rotational angular momentum of the Moon decreased due to tidal torques exerted by the Earth, which are much larger than the tidal torques that the low-mass Moon exerts on the Earth. When the Moon's spin angular speed had decreased to be the same as its translational angular speed (currently about 2π radians or $360°$ per month), the process terminated. That is why the Moon now always displays nearly the same face to the Earth.

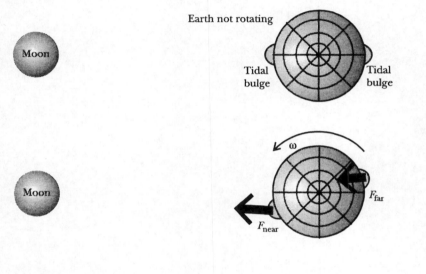

Figure 9.52 Tidal torques due to the Moon, viewed from a point above the North Pole. In the top image the Earth is not rotating; in the bottom image it is rotating. Not to scale; tidal effects greatly exaggerated.

9.11 *More complex rotational situations

In more advanced courses you may study nonsymmetric rotational situations like that in Figure 9.53. The angular velocity points along the axis of rotation, but the (translational) angular momentum of each mass is $\hat{\mathbf{r}}_A \times \vec{\mathbf{p}}$, so the total angular momentum $\vec{\mathbf{L}}_A$ does *not* point along the axis! Moreover, the angular momentum vector continually changes direction.

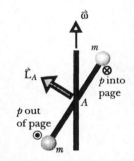

Figure 9.53 If the object lacks axial symmetry, the angular momentum and angular velocity may have different directions!

A symmetric object like that shown in Figure 9.54 has an angular momentum that does point along the axis, because the perpendicular components of the angular momenta cancel each other. This is the simpler kind of situation we have dealt with in this chapter. In this symmetric case it is true that $I = m_1 r_{\perp1}^2 + m_2 r_{\perp2}^2 + m_3 r_{\perp3}^2 + m_4 r_{\perp4}^2 + ...$ about the axis of rotation, and the simple formulas $\vec{\mathbf{L}}_{\text{rot}} = I\vec{\omega}$ and $K_{\text{rot}} = \frac{1}{2}I\omega^2$ are valid.

We calculated moment of inertia as a sum (integral) over $r_{\perp}^2$, the perpendicular distance to an axis. To deal with rotational motion in general requires expressing the moment of inertia as a "tensor," a 3 by 3 array of numbers representing integrals over x^2, xy, xz, y^2, yx, yz, z^2, zx, and zy. Matrix multiplication of this tensor times the angular velocity vector $\vec{\omega}$ yields an angular momentum vector that need not point in the same direction as $\vec{\omega}$. This is beyond the scope of this introductory course.

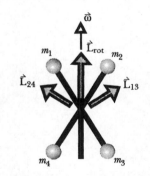

Figure 9.54 If the object is symmetric, the angular momentum is in the direction of the angular velocity.

9.12 *Vector rate of change of a rotating vector

A minor comment: The magnitude of the time rate of change of a rotating vector $\vec{\mathbf{X}}$ is $\left|d\vec{\mathbf{X}}/dt\right| = \omega X$. If you look at a diagram involving the angular velocity vector $\vec{\omega}$ you can see that the direction of $d\vec{\mathbf{X}}/dt$ is given by the cross product $\vec{\omega} \times \vec{\mathbf{X}}$. It can be shown rather easily that

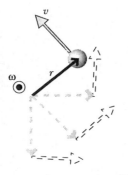

$$\frac{d\vec{X}}{dt} = \vec{\omega} \times \vec{X}$$

gives both the magnitude and direction of a rotating vector (Figure 9.55). This is another illustration of the usefulness of cross products.

Figure 9.55 In circular motion, both $\hat{r}$ and $\hat{v}$ are rotating vectors. $\vec{\omega}$ is out of the page. At any instant, $\vec{\omega} \times \hat{r}$ is in the direction of $\hat{v}$, while $\vec{\omega} \times \hat{v}$ is toward the center, in the direction of $\vec{a}$, as predicted by the relation $d\vec{X}/dt = \vec{\omega} \times \vec{X}$.

9.13 Summary

Fundamental Physical Principles

The angular momentum principle for a system:

$$\frac{d\vec{L}_{\text{tot,A}}}{dt} = \vec{\tau}_{\text{net,ext,A}} \quad \text{or} \quad \Delta\vec{L}_{\text{tot,A}} = \vec{\tau}_{\text{net,ext,A}}\Delta t$$

New concepts

Angular momentum of a particle about location A (Figure 9.56):

$$\vec{L}_A = \vec{r}_A \times \vec{p}$$

$$L_A = r_\perp p = r_A p \sin\theta \quad, \text{direction given by right-hand rule}$$

$$\vec{L}_A = <(yp_z - zp_y), (zp_x - xp_z), (xp_y - yp_x)>$$

Torque is a measure of the twist imparted by a force about location A:

$$\vec{\tau}_A = \vec{r}_A \times \vec{F}$$

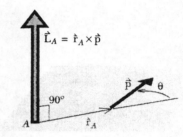

Figure 9.56 Angular momentum of a particle relative to location A.

Moment of inertia about an axis of rotation: $I = m_1 r_{\perp 1}^2 + m_2 r_{\perp 2}^2 + m_3 r_{\perp 3}^2 + ...$

Angular momentum is quantized in microscopic systems:
A component of angular momentum is an integer or half-integer multiple of $\hbar$.
The square magnitude of angular momentum has quantized values $L^2 = l(l+1)\hbar^2$, where l has integer or half-integer values.

Results

Angular momentum for a multiparticle system about location A can be divided into translational and rotational angular momentum (Figure 9.57):

$$\vec{L}_A = \vec{L}_{\text{trans},A} + \vec{L}_{\text{rot}}$$

$$\vec{L}_{\text{trans},A} = \vec{r}_{\text{cm},A} \times \vec{P}_{\text{tot}} \quad, \quad \vec{L}_{\text{rot}} = \vec{r}_1 \times \vec{p}_1 + \vec{r}_2 \times \vec{p}_2 + \vec{r}_3 \times \vec{p}_3$$

Rotational angular momentum in terms of moment of inertia: $\vec{L}_{\text{rot}} = I\vec{\omega}$
Rate of change of rotational angular momentum:

$$\frac{d\vec{L}_{\text{rot}}}{dt} = \vec{\tau}_{\text{net,ext,cm}}$$

Kinetic energy relative to center of mass in terms of moment of inertia and rotational angular momentum:

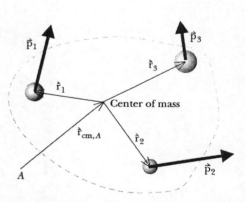

Figure 9.57 Calculating translational and rotational angular momentum.

$$K_{\text{rot}} = \frac{1}{2}I\omega^2 = \frac{1}{2}\frac{(I\omega)^2}{I} = \frac{L_{\text{rot}}^2}{2I}$$

Angular speed of precession for a gyroscope: $\Omega = \dfrac{\tau}{L_{\text{rot}}} = \dfrac{RMg}{I\omega}$

Bohr model of hydrogen: $E = -\dfrac{13.6 \text{ eV}}{N^2}$, $N = 1, 2, 3,...$

$I_{\text{disk}} = I_{\text{cylinder}} = \frac{1}{2}MR^2$, $I_{\text{sphere}} = \frac{2}{5}MR^2$ for axis passing through center

Uniform solid cylinder of length L, radius R, about axis perpendicular to cylinder, through center of cylinder: $I = \frac{1}{12}ML^2 + \frac{1}{4}MR^2$

9.14 Example problem: A satellite with solar panels

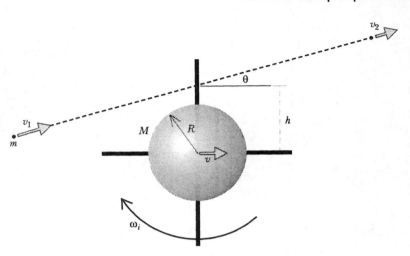

A satellite has four low-mass solar panels sticking out as shown. The satellite can be considered to be approximately a uniform solid sphere. Originally it is traveling to the right with speed v and rotating clockwise with angular speed ω_i.

A tiny meteor traveling at high speed v_1 rips through one of the solar panels and continues in the same direction but at reduced speed v_2. Afterwards, calculate the v_x and v_y components of the center-of-mass velocity of the satellite, and its angular velocity ω_f (magnitude and direction). Additional data are provided on the diagram.

Solution

What fundamental principles are useful starting points? The momentum principle and the angular momentum principle. The energy principle would not be a useful starting point, because we don't know how much thermal energy increase is associated with the meteor ripping through the panel. Take the satellite and meteor together as the system of interest, in which case the momentum and the angular momentum will remain constant, there being no external forces or torques.

Momentum principle, from just before to just after collision:

$$<(Mv + mv_1\cos\theta),\ m_1v_1\sin\theta,\ 0> =$$

$$<(Mv_x + mv_2\cos\theta),\ (Mv_y + mv_2\sin\theta),\ 0>$$

Angular momentum principle, about the original center of the satellite, component into the page, from just before to just after collision:

$$I\omega_i + hmv_1\cos\theta = I\omega_f + hmv_2\cos\theta,\ \text{where } I = \tfrac{2}{5}MR^2$$

Calculating just before and just after the collision makes it particularly easy to see how to calculate the translational angular momentum of the meteor, since it is directly above the center of the satellite at those instants. Note that we need $\cos\theta$, not $\sin\theta$, in calculating the translational angular momentum of the meteor. In the right-hand-rule, your fingers rotate through an angle that is $(90° - \theta)$.

Solving for the unknown final quantities:

$$v_x = v + \frac{m}{M}(v_1 - v_2)\cos\theta$$

$$v_y = \frac{m}{M}(v_1 - v_2)\sin\theta$$

$$\omega_f = \omega_i + \frac{hm}{(\tfrac{2}{5}MR^2)}(v_1 - v_2)\cos\theta,\ \text{clockwise, into the page}$$

9.15 Review questions

In addition to these review questions, note that the exercises on page 303 and page 315 constitute important worksheets that gives you practice in calculating angular momentum and torque.

Angular momentum

RQ 9.1 Give an example of a situation where an object is traveling in a straight line, yet has nonzero angular momentum.

RQ 9.2 Under what circumstances is angular momentum constant? Give an example of a situation where the x component of angular momentum is constant, but the y component isn't.

Translational and rotational angular momentum

RQ 9.3 Give examples of translational angular momentum and rotational angular momentum in our Solar System.

RQ 9.4 Give an example of a physical situation in which the angular momentum is zero yet the translational and rotational angular momenta are both nonzero.

The Bohr model of hydrogen

RQ 9.5 What features of the Bohr model of hydrogen are consistent with the later, full quantum mechanical analysis? What features of the Bohr model had to be abandoned?

Particle spin

RQ 9.6 Give some examples of atomic or nuclear phenomena associated with particle spin.

Torque

RQ 9.7 Under what conditions is the torque about some location equal to zero?

RQ 9.8 Make a sketch showing a situation where the torque due to a single force about some location is 20 N·m in the positive z direction, whereas about another location the torque is 10 N·m in the negative z direction.

Static equilibrium

RQ 9.9 What must be true for a system to be in static equilibrium (that is, not moving)?

Gyroscopes

RQ 9.10 Two gyroscopes are made exactly alike except that the spinning disk in one is made of low-density aluminum, whereas the disk in the other is made of high-density lead. If they have the same spin angular speeds and the same torque is applied to both, which gyroscope precesses faster?

9.16 Homework problems

Problem 9.1 Angular momentum in an elliptical orbit

In Chapter 2 you wrote a program to model the motion of a planet going around a fixed star. In this problem you will build on that program.

(a) Use initial conditions that produce an elliptical orbit. At each step calculate the translational angular momentum $\vec{L}_{trans,A}$ of the planet with respect to a location A chosen to be in the orbital plane but outside the orbit. Display this in two ways (i and ii below), and briefly describe in words what you observe:

(i) Display $\vec{L}_{trans,A}$ as an arrow with its tail at location A, throughout the orbit. Since the magnitude of $\vec{L}_{trans,A}$ is quite different from the magnitudes of the distances involved, you will need to scale the arrow by some factor to fit it on the screen.

(ii) Graph the component of $\vec{L}_{trans,A}$ perpendicular to the orbital plane as a function of time.

(b) Repeat part (a), but this time choose a different location B at the center of the fixed star, and calculate and display $\vec{L}_{trans,B}$ relative to that location B. As in part (a), display $\vec{L}_{trans,B}$ as an arrow (scaled appropriately), and also graph the component of $\vec{L}_{trans,B}$ perpendicular to the orbital plane, as a function of time, and briefly describe in words what you observe.

(c) Choose a location C which is not in the orbital plane, and calculate and display $\vec{L}_{trans,C}$ as an arrow throughout the orbit. (You do not need to make a graph.) Briefly describe in words what you observe.

Problem 9.2 Playground ride

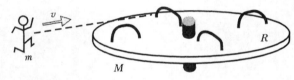

Figure 9.58 A child runs and jumps on a playground ride (Problem 9.2).

A playground ride consists of a disk of mass M and radius R mounted on a low-friction axle (Figure 9.58). A child of mass m runs at speed v on a line tangential to the disk and jumps onto the outer edge of the disk.

(a) If the disk was initially at rest, now how fast is it rotating? (The moment of inertia of a uniform disk is $\frac{1}{2}MR^2$.)

(b) If you were to do a lot of algebra to calculate the kinetic energies in part (a), you would find that $K_f < K_i$. Where has this energy gone?

(c) Calculate the change in *linear* momentum of the system consisting of the child plus the disk (but not including the axle), from just before to just after impact. What caused this change in the linear momentum?

(d) The child moves to a location a distance $R/2$ from the axle. Now what is the angular speed?

(e) If you were to do a lot of algebra to calculate the kinetic energies in part (d), you would find that $K_f > K_i$. Where has this energy come from?

(f) In part (a), estimate numerical values for all of the quantities and determine a value for the spin rate ω. Is your result reasonable?

Problem 9.3 Moment of inertia of a disk

Show that the moment of inertia of a disk of mass M and radius R is $\frac{1}{2}MR^2$. Divide the disk into narrow rings each of radius r and width dr. The contribution to I by one of these rings is simply $r^2 dm$, where dm is the amount of mass contained in that particular ring. The mass of any ring is the mass per unit area times the area of the ring. The area of this ring is approximately $2\pi r\, dr$. Use integral calculus to add up all the contributions.

Problem 9.4 Rotating disk

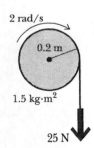

Figure 9.59 A disk rotates on a nearly frictionless axle (Problem 9.4).

A disk of radius 0.2 m and moment of inertia 1.5 kg·m^2 is mounted on a nearly frictionless axle (Figure 9.59). A string is wrapped tightly around the disk, and you pull on the string with a constant force of 25 N. After a while the disk has reached an angular speed of 2 radians/s. What is its angular speed 0.1 seconds later? Explain briefly.

Problem 9.5 Momentum and angular momentum

Two small objects each of mass m are connected by a lightweight rod of length d (Figure 9.60). At a particular instant they have velocities as shown and are subjected to external forces as shown. The system is moving in outer space.

In the following questions involving vectors, give components along the axes shown, and state which axis you're using for each component.

(a) What is the total (linear) momentum $\vec{P}_{total}$ of this system?

(b) What is the velocity $\vec{v}_{cm}$ of the center of mass?

(c) What is the total angular momentum $\vec{L}_A$ of the system relative to point A?

(d) What is the rotational angular momentum $\vec{L}_{rot}$ of the system?

(e) What is the translational angular momentum $\vec{L}_{trans}$ of the system relative to point A?

After a short time interval Δt,

(f) What is the total (linear) momentum $\vec{P}_{total}$ of the system?

(g) What is the rotational angular momentum $\vec{L}_{rot}$ of the system?

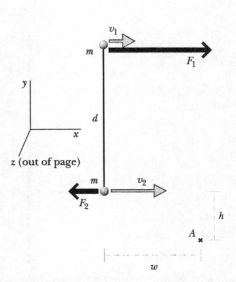

Figure 9.60 Two masses connected by a lightweight rod (Problem 9.5).

Problem 9.6 Ice skater

An ice skater whirls with her arms and one leg stuck out as shown, making one complete turn in one second (Figure 9.61). Then she quickly moves her arms up above her head and pulls her leg in as shown.

(a) Estimate how long it now takes for her to make one complete turn. Explain your calculations, and state clearly what approximations and estimates you make.

(b) Estimate the minimum amount of chemical energy she must expend to change her configuration.

Problem 9.7 Rotating stool and barbells

You sit on a rotating stool and hold barbells in both hands with your arms fully extended horizontally. You make one complete turn in 2 seconds. You then pull the barbells in close to your body.

(a) Estimate how long it now takes you to make one complete turn. Be clear and explicit about the principles you apply and about your assumptions and approximations.

(b) About how much energy did you expend?

Figure 9.61 A spinning skater pulls in one leg (Problem 9.6).

Problem 9.8 An asteroid collision

A spherical non-spinning asteroid of mass M and radius R moving with speed v_1 to the right collides with a similar non-spinning asteroid moving with speed v_2, to the left, and they stick together (Figure 9.62). The impact parameter is d. Note that $I_{sphere} = \frac{2}{5}MR^2$.

After the collision, what is the velocity v_{cm} of the center of mass and the angular velocity ω about the center of mass? (Note that each asteroid rotates about its own center with this same ω.)

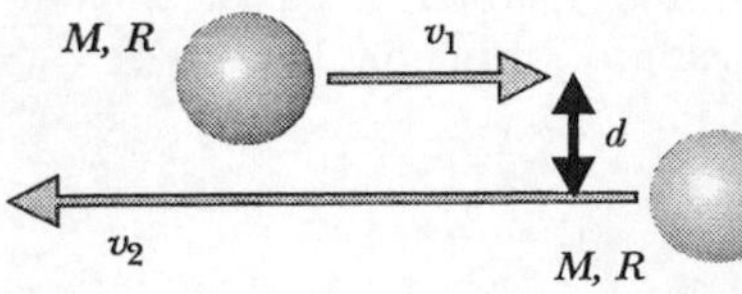

Figure 9.62 Two asteroids collide and stick together (Problem 9.8).

Just before collision

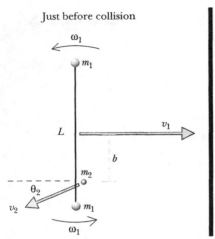

Figure 9.63 A spinning rod is struck by a small object (Problem 9.9).

Just after collision

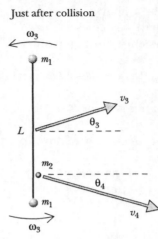

Problem 9.9 A collision with a rod with masses on the ends

Two small objects each of mass m_1 are connected by a lightweight rod of length L (Figure 9.63). At a particular instant the center of mass speed is v_1 as shown, and the object is rotating counterclockwise with angular speed ω_1. A small object of mass m_2 traveling with speed v_2 collides with the rod at an angle θ_2 as shown, at a distance b from the center of the rod. After being struck, the mass m_2 is observed to move with speed v_4, at angle θ_4. All the quantities are positive magnitudes. This all takes place in outer space.

For the object consisting of the rod with the two masses, write equations that, in principle, could be solved for the center of mass speed v_3, direction θ_3, and angular speed ω_3 in terms of the given quantities. State clearly what physical principles you use to obtain your equations.

Don't attempt to solve the equations; just set them up.

Problem 9.10 Space junk

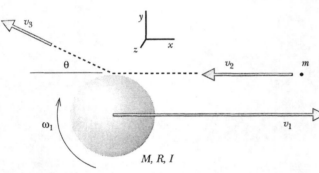

Figure 9.64 Space junk strikes a satellite (Problem 9.10).

A tiny piece of space junk of mass m strikes a glancing blow to a spherical satellite (Figure 9.64). After the collision the space junk is traveling in a new direction and moving more slowly. The space junk had negligible rotation both before and after the collision. The velocities of the space junk before and after the collision are shown in the diagram. The center of mass of the satellite is at its geometrical center. The satellite has mass M, radius R, and moment of inertia I about its center. Before the collision the satellite was moving and rotating as shown in the diagram.

(a) Just after the collision, what are the components of the center-of-mass velocity of the satellite (v_x and v_y) and its rotational speed ω?

(b) Calculate the rise in the thermal energy of the satellite and space junk combined. You do *not* need to substitute in the values for quantities you already calculated in part (a).

Problem 9.11 Space station

A space station has the form of a hoop of radius R, with mass M (Figure 9.65). Initially its center of mass is not moving, but it is spinning with angular speed ω_0. Then a small package of mass m is thrown by a spring-loaded gun toward a nearby spacecraft as shown; the package has a speed v after launch. Calculate the center-of-mass velocity of the space station (v_x and v_y) and its rotational speed ω after the launch.

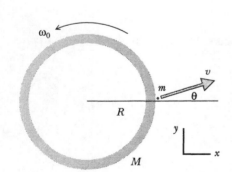

Figure 9.65 Launch a package from a space station (Problem 9.11).

Problem 9.12 Rotational spectrum of a diatomic molecule

(a) Calculate the energies of the quantized rotational energy levels for O_2. The most common oxygen nucleus contains 8 protons and 8 neutrons. Estimate any quantities you need. See discussion of diatomic molecules on page 308; the parameter l has values 0, 1, 2, 3...

(b) Describe the emission spectrum for electromagnetic radiation emitted in transitions among the rotational O_2 energy levels. Include a calculation of the lowest-energy emission in electron volts (1.6×10^{-19} J).

(c) It is transitions among "electronic" states of atoms that produce visible light, with photon energies on the order of a couple of electron-volts. Each electronic energy level has quantized rotational and vibrational (harmonic oscillator) energy sublevels. Explain why this leads to a visible spectrum that contains "bands" rather than individual energies.

Problem 9.13 The Bohr model

The Bohr model correctly predicts the main energy levels not only for atomic hydrogen but also for other "one-electron" atoms where all but one of the atomic electrons has been removed, such as in He$^+$ (one electron removed) or Li^{++} (two electrons removed).

(a) Predict the energy levels for a system consisting of a nucleus containing Z protons and just one electron. You need not recapitulate the entire derivation for the Bohr model, but do explain the changes you have to make to take into account the factor Z.

(b) The negative muon (μ^-) behaves like a heavy electron, with the same charge as the electron but with a mass 207 times as large as the electron mass. As a moving μ^- comes to rest in matter, it tends to knock electrons out of atoms and settle down onto a nucleus to form a "one-muon" atom. For a system consisting of a lead nucleus (Pb^{208} has 82 protons and 126 neutrons) and just one negative muon, predict the energy of a photon emitted in a transition from the first excited state to the ground state. The high-energy photons emitted by transitions between energy levels in such "muonic atoms" are easily observed in experiments with muons.

(c) Calculate the radius of the smallest Bohr orbit for a μ^- bound to a lead nucleus (Pb^{208} has 82 protons and 126 neutrons). Compare with the approximate radius of the lead nucleus (remember that the radius of a proton or neutron is about 10^{-15} m, and the nucleons are packed closely together in the nucleus).

Comments: This analysis in terms of the simple Bohr model hints at the result of a full quantum-mechanical analysis, which shows that in the ground state of the lead-muon system there is a rather high probability for finding the muon *inside* the lead nucleus. Nothing in quantum mechanics forbids this penetration, especially since the muon does not participate in the strong interaction.

The eventual fate of the μ^- in a muonic atom is that it either decays into an electron, neutrino, and antineutrino, or it reacts through the weak interaction with a proton in the nucleus to produce a neutron and a neutrino. This "muon capture" reaction is more likely if the probability is high for the muon to be found inside the nucleus, as is the case with heavy nuclei such as lead.

Problem 9.14 Nuclear gamma ray

The nucleus dysprosium-160 (containing 160 nucleons) acts like a spinning object with quantized angular momentum, $L^2 = l(l+1)\hbar^2$, and for this nucleus it turns out that l must be an even integer (0, 2, 4...). When a Dy-160 nucleus drops from the $l=2$ state to the $l=0$ state, it emits an 87 keV photon (87×10^3 eV).

(a) What is the moment of inertia of the Dy-160 nucleus?

(b) Given your result from part (a), find the approximate radius of the Dy-160 nucleus, assuming it is spherical. (In fact, these and similar experimental observations have shown that some nuclei are not quite spherical.)

(c) The radius of a (spherical) nucleus is given approximately by $(1.3\times10^{-15}$ m$)A^{1/3}$, where A is the total number of protons and neutrons. Compare this prediction with your result in part (b).

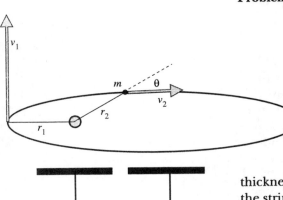

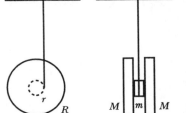

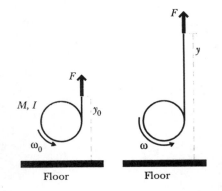

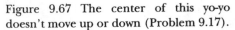

Figure 9.66 A yo-yo is released and moves downward (Problem 9.15).

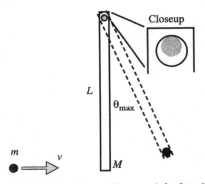

Figure 9.67 The center of this yo-yo doesn't move up or down (Problem 9.17).

Figure 9.68 A bullet strikes a stick that is suspended from an axle (Problem 9.15).

Problem 9.15 A comet

A certain comet of mass m at its closest approach to the Sun is observed to be at a distance r_1 from the center of the Sun, moving with speed v_1. At a later time the comet is observed to be at a distance r_2 from the center of the Sun, and the angle between $\hat{r}_2$ and the velocity vector is measured to be θ. What is v_2? Explain briefly.

Problem 9.16 Yo-yo

A yo-yo is constructed of three disks (Figure 9.66): two outer disks of mass M, radius R, and thickness d, and an inner disk (around which the string is wrapped) of mass m, radius r, and thickness d. The yo-yo is suspended from the ceiling and then released with the string vertical.

Calculate the tension in the string as the yo-yo falls. Note that when the center of the yo-yo moves down a distance y, the yo-yo turns through an angle y/r, which in turn means that the angular speed ω is equal to v_{cm}/r. The moment of inertia of a uniform disk is $\frac{1}{2}MR^2$.

Problem 9.17 A hovering yo-yo

String is wrapped around an object of mass M and moment of inertia I. You pull the string with your hand straight up with some constant force F such that the center of the object does not move up or down, but the object spins faster and faster (Figure 9.67). This is like a yo-yo; nothing but the vertical string touches the object.

When your hand is a height y_0 above the floor, the object has an angular speed ω_0. When your hand has risen to a height y above the floor, what is the angular speed ω of the object?

Your result should not contain F nor the (unknown) radius of the object. Explain what physics principles you are using.

Problem 9.18 Bullet and stick

A stick of length L and mass M hangs from a low-friction axle (Figure 9.68). A bullet of mass m traveling at a high speed v strikes near the bottom of the stick and quickly buries itself in the stick.

(a) During the brief impact, is the linear momentum of the stick+bullet system constant? Explain why or why not. Include in your explanation a sketch of how the stick shifts on the axle during the impact.

(b) During the brief impact, around what point does the angular momentum of the stick+bullet system remain constant?

(c) Just after the impact, what is the angular speed ω of the stick (with the bullet embedded in it)? (Note that the center of mass of the stick has a speed $\omega L/2$. The moment of inertia of a uniform rod about its center of mass is $\frac{1}{12}ML^2$.)

(d) Calculate the change in kinetic energy from just before to just after the impact. Where has this energy gone?

(e) The stick (with the bullet embedded in it) swings through a maximum angle θ_{max} after the impact, then swings back. Calculate θ_{max}.

Problem 9.19 An object rolls down a ramp

A solid object of uniform density with mass M, radius R, and moment of inertia I rolls without slipping down a ramp at an angle θ to the horizontal. The object could be a hoop, a disk, a sphere, etc.

(a) Carefully follow the complete analysis procedure explained in Chapter 3, but with the addition of the angular momentum principle about the center of mass. Note that in your force diagram you must include a small frictional force f that points *up* the ramp. Without that force the object will slip. Also note that the condition of nonslipping implies that the instantaneous velocity of the atoms of the object that are momentarily in contact with the ramp is zero, so $f < \mu F_N$ (no slipping). This zero-velocity condition also implies that $v_{cm} = \omega R$, where ω is the angular speed of the object, since the instantaneous speed of the contact point is $v_{cm} - \omega R$.

(b) The moment of inertia about the center of mass of a uniform hoop is MR^2, for a uniform disk it is $(1/2)MR^2$, and for a uniform sphere it is $(2/5)MR^2$. Calculate the acceleration dv_{cm}/dt for each of these objects.

(c) If two hoops of different mass are started from rest at the same time and the same height on a ramp, which will reach the bottom first? If a hoop, a disk, and a sphere of the same mass are started from rest at the same time and the same height on a ramp, which will reach the bottom first?

(d) Write the energy equation for the object rolling down the ramp, and for the point-particle system. Show that the time derivatives of these equations are compatible with the force and torque analyses.

Problem 9.20 Pulling a rotating device

A string is wrapped around a uniform disk of mass M and radius R. Attached to the disk are four low-mass rods of radius b, each with a small mass m at the end (Figure 9.69).

The apparatus is initially at rest on a nearly frictionless surface. Then you pull the string with a constant force F. At the instant when the center of the disk has moved a distance d, a length w of string has unwound off the disk.

(a) At this instant, what is the speed of the center of the apparatus? Explain your approach.

(b) At this instant, what is the angular speed of the apparatus? Explain your approach.

(c) You keep pulling with constant force F for an additional time Δt. By how much ($\Delta \omega$) does the angular speed of the apparatus increase in this time interval Δt?

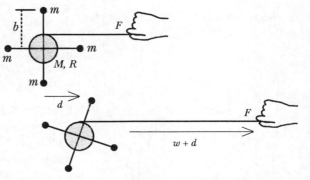

Figure 9.69 A rotating disk with four masses attached (Problem 9.20).

Problem 9.21 A rotating disk with masses sliding on a rod

A rod of length L and negligible mass is attached to a uniform disk of mass M and radius R (Figure 9.70). A string is wrapped around the disk, and you pull on the string with a constant force F. Two small balls each of mass m slide along the rod with negligible friction. The apparatus starts from rest, and when the center of the disk has moved a distance d, a length of string s has come off the disk, and the balls have collided with the ends of the rod and stuck there. The apparatus slides on a nearly frictionless table. Here is a view from above:

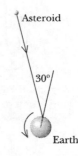

Figure 9.70 The masses slide on the rod (Problem 9.21).

(a) At this instant, what is the speed v of the center of the disk?

(b) At this instant the angular speed of the disk is ω. How much thermal energy has been produced?

(c) In the next short amount of time Δt, by how much will the angular speed change ($\Delta \omega$)?

Problem 9.22 Asteroid collision

Suppose an asteroid of mass 2×10^{21} kg is nearly at rest outside the solar system, far beyond Pluto. It falls toward the Sun and crashes into the Earth at the equator, coming in at an angle of 30 degrees to the vertical as shown, against the direction of rotation of the Earth (Figure 9.71). It is so large that its motion is barely affected by the atmosphere.

(a) Calculate the impact speed.

(b) Calculate the change in the length of a day due to the impact.

Figure 9.71 An asteroid crashes into the Earth (Problem 9.22; not to scale).

View from above

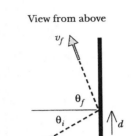

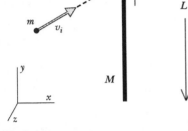

Figure 9.72 A stick lying on ice is struck by a small mass (Problem 9.23).

Figure 9.73 Tucked position (side view).

Figure 9.74 Entering water (front view).

Problem 9.23 Stick on ice

A stick of mass M and length L is lying on ice (Figure 9.72). A small mass m traveling at high speed v_i strikes the stick a distance d from the center and bounces off with speed v_f as shown in the diagram, which is a top view of the situation. The magnitudes of the initial and final angles to the x axis of the small mass's velocity are θ_i and θ_f All of the symbols in the diagram represent positive numbers.

(a) Afterwards, what are the velocity components v_x and v_y of the center of the stick? Explain briefly.

(b) Afterwards, what is the magnitude and direction of the angular velocity ω of the stick?

(c) What is the increase in thermal energy of the objects? You can leave your expression in terms of the initial quantities and v_x, v_y, and ω.

Problem 9.24 Diver

A diver dives from a high platform. When he leaves the platform, he tucks tightly (Figure 9.73) and performs three complete revolutions in the air, then straightens out with his body fully extended before entering the water (Figure 9.74). He is in the air for a total time of 1.4 seconds. What is his angular speed ω just as he enters the water? Give a numerical answer.

Be explicit about the details of your model, and include (brief) explanations. You will need to estimate some quantities.

Problem 9.25 Gyroscope experiment—qualitative

This problem requires that you have a toy gyroscope available. The purpose of this problem is to make as concrete as possible the unusual motions of a gyroscope and their analysis in terms of fundamental principles. In all of the following studies, the effects are most dramatic if you give the gyroscope as large a spin angular speed as possible.

(a) Hold the spinning gyroscope firmly in your hand, and try to rotate the spin axis quickly to point in a new direction. Explain qualitatively why this feels "funny." Also explain why you *don't* feel anything odd when you move the spinning gyroscope in any direction *without* changing the direction of the spin axis.

(b) Support one end of the spinning gyroscope (on a pedestal or in an open loop of the string) so that the gyroscope precesses *counterclockwise* as seen from above. Explain this counterclockwise precession direction; include sketches of top and side views of the gyroscope.

(c) Again support one end of the spinning gyroscope so that the gyroscope precesses *clockwise* as seen from above. Explain this clockwise precession direction; include sketches of top and side views of the gyroscope.

Problem 9.26 Gyroscope experiment—quantitative

This problem requires that you have a toy gyroscope available. The purpose of this problem is to make as concrete as possible the unusual motions of a gyroscope and their analysis in terms of fundamental principles. In all of the following studies, the effects are most dramatic if you give the gyroscope as large a spin angular speed as possible.

(a) If you knew the spin angular speed of your gyroscope, you could predict the precession rate. Invent an appropriate experimental technique and determine the spin angular speed approximately. Explain your experimental method and your calculations. Then predict the corresponding precession rate, and compare with your measurement of the precession rate. You will have to measure and estimate some properties of the gyroscope and how it is constructed.

(b) Make a *quick* measurement of the precession rate with the spin axis horizontal, then make another quick measurement of the precession rate with the spin axis nearly vertical. (It you make quick measurements, friction on the spin axis doesn't have much time to change the spin angular speed.) Repeat, this time with the spin axis initially nearly vertical, then horizontal. Making all four of these measurements gives you some indication of how much the spin unavoidably changes due to friction while you are quickly changing the angle. What do you conclude about the dependence of the precession rate on the angle, assuming the same spin rate at these different angles? What is the theoretical prediction for the dependence of the precession rate on angle (for the same spin rate)?

Problem 9.27 Wood or steel top

(a) A solid wood top spins at high speed on the floor, with a spin direction shown in Figure 9.75. Using appropriately labeled diagrams, explain the direction of motion of the top (you do not need to explain the magnitude).

(b) How would the motion change if the top had a higher spin rate? Explain briefly.

(c) If the top were made of solid steel instead of wood, explain how this would affect the motion (for the same spin rate).

Problem 9.28 Bicycle wheel

A bicycle wheel with a heavy rim is mounted on a lightweight axle, and one end of the axle rests on top of a post. The wheel is observed to precess in the horizontal plane. With the spin direction shown, does the wheel precess clockwise or counterclockwise? Explain in detail, including appropriate diagrams.

Problem 9.29 Gyroscope

The axis of a gyroscope is tilted at an angle of 30° to the vertical. The rotor has a radius of 15 cm, mass 3 kg, moment of inertia 0.06 kg·m^2, and spins on its axis at 30 radians/s. It is supported in a cage (not shown) in such a way that without an added weight it does not precess. Then a mass of 0.2 kg is hung from the axis at a distance of 18 cm from the center of the rotor.

(a) Viewed from above, does the gyroscope precess in a 1) clockwise or a 2) counterclockwise direction? That is, does the top end of the axis move 1) out of the page or 2) into the page in the next instant? Explain your reasoning.

(b) How long does it take for the gyroscope to make one complete precession?

Problem 9.30 Angular momentum of the Earth

(a) Calculate the magnitude of the translational angular momentum of the Earth relative to the center of the Sun. See data on inside back cover.

(b) Calculate the magnitude of the rotational angular momentum of the Earth. How does this compare to your result in part (a)?

(The angular momentum of the Earth relative to the center of the Sun is the sum of the translational and rotational angular momenta. The rotational axis of the Earth is tipped 23.5° away from a perpendicular to the plane of its orbit.)

Figure 9.75 A wood or steel top (Problem 9.27).

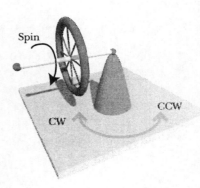

Figure 9.76 A bicycle wheel on a pivot (Problem 9.28).

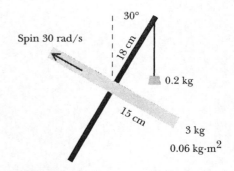

Figure 9.77 A gyroscope (Problem 9.29).

9.17 Answers to exercises

9.1 (page 298)   A: 30 kg·m^2/s, into page ($\otimes$)

B: 30 kg·m^2/s, into page ($\otimes$)

C: 0

D: 50 kg·m^2/s, out of page ($\odot$)

E: 50 kg·m^2/s, out of page ($\odot$)

F: 50 kg·m^2/s, out of page ($\odot$)

G: 0

H: 30 kg·m^2/s, into page ($\otimes$)

9.2 (page 298)   0.38 kg·m^2/s, into page ($\otimes$)

9.4 (page 300)   Nonzero; into the page; zero; (no direction)

9.5 (page 300)   No; nonzero; into the page; nonzero; into the page

9.6 (page 301)   Nonzero; up (toward the sky); nonzero; up

9.7 (page 302) $2M\left(\dfrac{d}{2}\right)^2 = \frac{1}{2}Md^2$

9.8 (page 302) $MR^2\omega$

9.9 (page 302)   395 J

9.10 (page 303)   (a) up; (b) $d_1 p_1 \sin\alpha$; (c) up; (d) $d_2 p_2$

9.11 (page 303)   (a) $(d/2)p = (d/2)[m(d/2)\omega_0]$ into page

(b) $2m(d/2)^2\omega_0$ into page ($\otimes$)

(c) $2m(d/2)^2$

(d) into page

(e) $2m(d/2)^2\omega_0$ into page ($\otimes$)

(f) they are equal

(g) $m(d/2)^2\omega_0^2$

9.12 (page 303)   (a) 0

(b) $b(2m)(b\omega_1)$ into page ($\otimes$)

(c) $b(2m)(b\omega_1)$ into page ($\otimes$)

9.13 (page 303)   (a) $2m(d/2)^2\omega_2$ into page ($\otimes$)

(b) $b(2m)(b\omega_1)$ into page ($\otimes$)

(c) $2m[b^2\omega_1 + (d/2)^2\omega_2]$ into page ($\otimes$)

9.15 (page 310)   6 N·m

9.16 (page 310)   2 N

9.17 (page 310)   (4,3.8,0) kg·m^2/s

9.18 (page 311)   Force directed toward center of Sun, so zero torque about the location of the Sun.

Angular momentum about center of Sun *constant.*

9.19 (page 311) $r_1 m v_1$, into page; $r_2 m v_2$, into page; $r_1 v_1 = r_2 v_2$

9.20 (page 314)   Negligible external forces means negligible external torques.

Therefore magnitude and direction of angular momentum constant.

9.21 (page 314)   No torques around center of mass means no change in rotational angular momentum, so rotational angular momentum stays constant in magnitude (which determines length of day) and direction (which determines what "North Star" the axis points at). Doesn't matter that Earth is going around Sun; rotational angular momentum affected solely by torque around center of mass.

9.22 (page 315) (a) $R\cos(45°)mv$ out of page (⊙)

(b) $R\cos(45°)mv$ out of page (⊙)

(c) $\dfrac{R\cos(45°)mv}{(M+m)R^2}$ out of page (⊙)

(d) There is a change of linear momentum of the clay (but not of the wheel), caused by a force applied by the axle to the combined system.

9.23 (page 316) 0.8 N·m, into the page, or <0,0,–0.8> N·m

9.24 (page 316) (a) 0.0256 kg·m^2/s, into page (⊗), or

   <0,0,–0.0256> kg·m^2/s

(b) 0.1856 kg·m^2/s, into page (⊗), or

   <0,0,–0.1856> kg·m^2/s

(c) 145 rad/s, into page (⊗), or

   <0,0,–145> rad/s

9.25 (page 316) 8.33 N

9.26 (page 319) $\dfrac{dL_{\text{string},z}}{dt} = \dfrac{d}{dt}\left[RMv_{cm} - \dfrac{1}{2}MR^2\omega\right] = 0$, since no torque

here. Differentiating, we get $\dfrac{d\omega}{dt} = \dfrac{2}{R}\dfrac{dv_{cm}}{dt} = \dfrac{2F_T}{MR}$, as

before.

9.27 (page 320) $\dfrac{dP_z}{dt} = F_N - M_1g - M_2g = 0$

$\dfrac{dL_{\text{support},z}}{dt} = L_1M_1g - L_2M_2g = 0$ (+z out of page)

9.28 (page 320) $\dfrac{dL_{\text{left person},z}}{dt} = L_1F_N - (L_1 + L_2)M_2g = 0$

Substitute $F_N = (M_1 + M_2)g$ and get

$L_1M_1g - L_2M_2g = 0$

9.29 (page 322) Same as in horizontal case: $\Omega = \dfrac{RMg}{I\omega}$

Chapter 10

Entropy:
More Limits on the Possible

Chapter 10

Entropy: More Limits on the Possible

We have now studied in some detail the three fundamental principles of classical mechanics: the energy principle, the momentum principle, and the angular momentum principle. These three powerful principles, together with the statistical concepts to be introduced in this chapter, are sufficient to allow us to model quite large and complex systems.

In this chapter we go deeply into the statistical nature of the behavior of systems that include a huge number of atoms. Our goal is to explain some of the macroscopic thermal properties of matter by reasoning about the behavior of individual atoms. A theme that will run through all of the topics in this chapter is the question of how the total energy of a system is divided up among the various components of the system.

10.1 Statistical issues

Temperature

We have repeatedly encountered a connection between the temperature of an object and the average motion of the atoms that make up the object, but our understanding of the meaning and role of temperature is incomplete. What is the quantitative relationship between temperature and microscopic energy?

Direction of thermal energy flow

When two blocks of different temperatures are placed in contact with each other, we observe that energy flows from the hotter block to the colder block (Figure 10.1). We can understand this in a rough way. The average energy of atoms in the hotter block is greater than the average energy of atoms in the colder block. In the interface region, where atoms of the two blocks are in contact with each other, it seems more likely that an atom in the hotter block will lose energy rather than gain energy when it collides with an atom in the colder block.

But couldn't energy flow the other way? After all, sometimes an individual atom in the hotter block might happen to have a lot less than its average energy when it collides with an individual atom in the colder block that happens to have a lot more than its average energy. In that case energy would be transferred from the colder block to the hotter block (Figure 10.2).

This "uphill" movement of energy seems unlikely, but how unlikely is it? Should we occasionally observe a hot block get hotter when we put it in contact with a cold block? If you put an ice cube in your drink, could your drink get warmer and the ice cube get colder? This would not violate conservation of energy, because the total energy of the two objects would still be the same. So what physical principle would be violated?

Reversibility

A related question, and a very deep one, has to do with the "reversibility" of processes. The fundamental physical interactions seem to be completely "reversible" in the following technical, physics sense: Make a movie of an alpha particle scattering off a gold nucleus, then run the movie forward or reversed. Both views look entirely reasonable. In contrast, make a movie of a

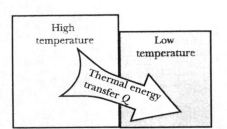

Figure 10.1 We observe that energy flows from a hotter object to a cooler one.

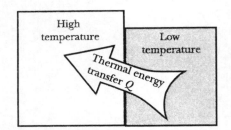

Figure 10.2 Could energy be transferred from the cooler object to the hotter one?

ball bouncing on the floor with decreasing height on each bounce, then run the movie forward or reversed. The reversed movie looks silly: you see a ball bouncing higher and higher with each bounce! Would such a motion violate conservation of energy? Not necessarily. It could be that the atoms in the floor happen on average to give energy to the ball, and the floor is getting colder as a result of this continuing loss of energy. So what physical principle is violated by a process represented by the reversed movie?

Statistical models

This chapter addresses these questions by applying probabilistic and statistical ideas to the behavior of systems. This subject is called "statistical mechanics." We will first apply statistical mechanics to a simple model of a solid, and we'll find that we can go surprisingly far with just a few new concepts. The main concepts that we will develop for the specific case of a solid apply in general to a wide variety of systems.

How will we know whether our statistical model explains anything? Our criterion for understanding is whether the predictions of our microscopic model agree with measurements of macroscopic systems, such as measurements of heat capacity (the amount of energy required to increase the temperature of an object by one degree).

10.2 A statistical model of solids

We have noted in previous chapters that the interactions between atoms in a solid are electric in nature, but that a detailed model of them involves quantum mechanics. The interatomic potential energy function encountered in Chapter 4 provides a reasonably accurate description of these interactions. Because of the similarity of this function to the potential energy curve for a harmonic oscillator (a mass on a spring), in previous chapters we have modeled a solid as a large number of tiny masses (the atoms) connected to their neighbors by springs (the interatomic bonds), as shown in Figure 10.3. This model has allowed us to understand qualitatively how solids interact with other objects. We would now like to use this model to ask detailed quantitative questions about the distribution of energy in a solid.

In this chapter we focus on calculating how probable a particular distribution of speeds or energies in a solid would be. We do this for a solid because it happens to be easier to calculate probabilities for a simple model of a solid than it is for a gas or a liquid. The atoms in a solid are nearly fixed in position, so we don't have to consider how likely different spatial arrangements might be (unlike the situation for a gas). Reasoning about the energy distribution in a solid will also enable us to draw conclusions about the transfer of thermal energy from one solid object to another.

10.2.1 The Einstein model of a solid

Because our goal is to calculate the probability of particular distributions of energy among atoms, we can simplify our model of a solid even further. We can think of each atom as moving independently of its neighbors, as though it were connected to rigid walls rather than to other atoms (Figure 10.4). Of course in a real solid energy is exchanged with neighboring atoms. However, to address our current questions we do not need to worry about the mechanism of energy exchange between atoms, since we will focus on calculating probabilities for various distributions of energy among the atoms of the solid. This is a different kind of model from those we have so far constructed, because this very simple model would not allow us to predict in detail the dynamic motion of each atom in the solid, nor to ask how long it would take for energy to flow from one end of a solid to the other. It will not

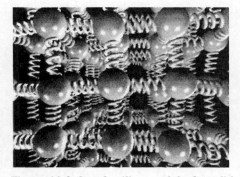

Figure 10.3 Our familiar model of a solid as massive balls connected by springs.

Figure 10.4 A single atom can oscillate in three dimensions. We simplify our model by assuming that each atom moves independently of the surrounding atoms—in effect, that it is connected to rigid walls instead of moving atoms.

shed light on the details of a process, but it does allow us to ask questions about initial and final states. Since at the moment the questions we want to address involve initial and final states, this model is useful, and it is mathematically much simpler than the connected-atoms model.

Einstein proposed this simple model in 1907, and he found that some basic properties of solids such as heat capacity could in fact be understood using this model. This model also allows us to understand in detail the statistical nature of energy transfer between a hot object and a cold object, and why two objects come to "thermal equilibrium" (the same final temperature). This simple model of a solid will help us gain a more sophisticated and powerful understanding of the meaning of temperature.

A three-dimensional oscillator—three one-dimensional oscillators

We will consider each atom in a solid to be connected by springs to immovable walls. Each isolated atom is a three-dimensional spring-mass system, with $\vec{s}$ representing the three-dimensional vector displacement away from a fixed equilibrium position (because we are ignoring collective motions of groups of atoms). Since $p^2 = p_x^2 + p_y^2 + p_z^2$ and $s^2 = s_x^2 + s_y^2 + s_z^2$, we can write the energy of a three-dimensional classical oscillator as consisting of three parts, corresponding to the x, y, and z oscillations:

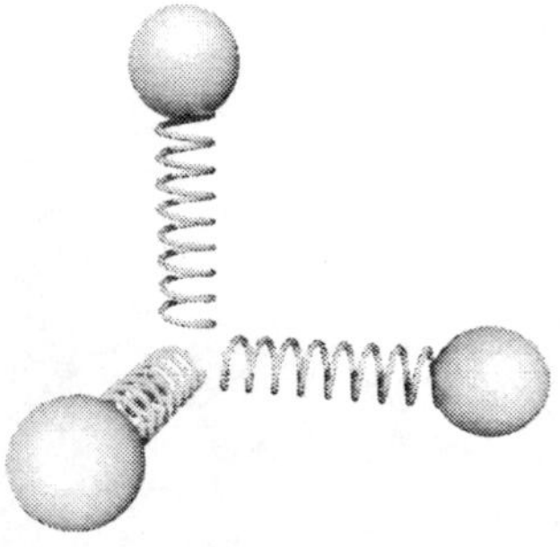

Figure 10.5 Since we are considering the atoms to be independent, in our model we can replace a single three-dimensional oscillator (one atom) by three independent one-dimensional oscillators.

$$K_{\text{vib}} + U_s = \left(\frac{p_x^2}{2m} + \frac{1}{2}k_s s_x^2 + U_0\right) + \left(\frac{p_y^2}{2m} + \frac{1}{2}k_s s_y^2 + U_0\right) + \left(\frac{p_z^2}{2m} + \frac{1}{2}k_s s_z^2 + U_0\right)$$

Recall from chapter 6 that a quantized oscillator (a quantum mechanical "ball and spring") has evenly spaced energy levels. A complete quantum mechanical analysis of an oscillator that is free to oscillate in three dimensions rather than just one dimension leads to the conclusion that the motion of a 3-D oscillator can be separated into x, y, and z components, each of which has the same energy level structure as the familiar one-dimensional oscillator. This is mathematically equivalent to replacing each three-dimensional oscillator (an atom) with three ordinary one-dimensional oscillators, which we will do to simplify our model. We will think of a block as containing N one-dimensional oscillators, corresponding to $N/3$ atoms (Figure 10.5).

In Chapter 6 we discussed energy quantization in atomic spring-mass systems. As predicted by quantum mechanics and abundantly confirmed by experiments, energy can be added to a one-dimensional atomic oscillator only in multiples of one "quantum" of energy $\hbar\omega_0$, where $\omega_0 = \sqrt{k_s/m}$ (Figure 10.6):

$$\text{Added energy can be } 0,\ 1\hbar\omega_0,\ 2\hbar\omega_0,\ 3\hbar\omega_0,\ \text{etc.}$$

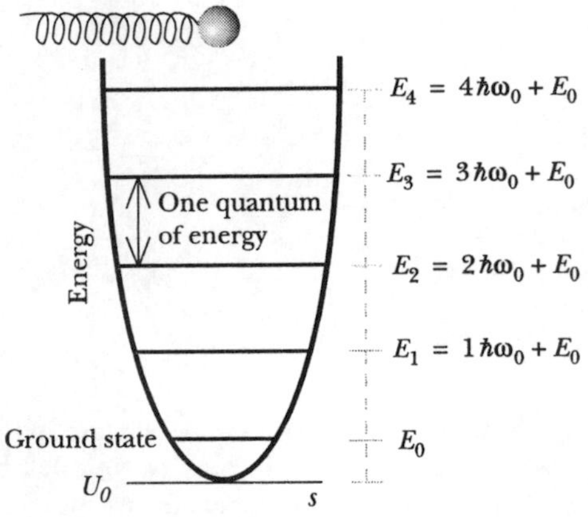

$E_4 = 4\hbar\omega_0 + E_0$
$E_3 = 3\hbar\omega_0 + E_0$
One quantum of energy
Energy
$E_2 = 2\hbar\omega_0 + E_0$
$E_1 = 1\hbar\omega_0 + E_0$
Ground state
E_0
U_0
s

Figure 10.6 The quantized energy levels of a one-dimensional atomic oscillator.

Here $\hbar = h/(2\pi) = 1.05\times10^{-34}$ joule · second, k_s is the spring stiffness, and m is the mass of the atom. The "ground state" (lowest energy level) of the quantum oscillator is not U_0 but $\frac{1}{2}\hbar\omega_0 + U_0$. This offset doesn't matter for what we are going to do; it just sets the baseline energy level, and we will measure energies starting from the ground state.

10.2.2 Distributing energy among objects

When two identical blocks are brought into contact, their total energy is shared among all the oscillators (atoms) in both blocks. It seems plausible that the most likely outcome would be that the total thermal energy be shared equally between identical blocks. However, might there be some probability that the first block would have more energy than the other, or even have all of it? With a large number of atoms the probability of even a small deviation from the most probable distribution turns out to be hugely unlikely. In order to look at this question in detail, we need to find a way to calculate the probability of various possible distributions of the energy. If we can find the most probable energy distribution, we can predict the eventual equilibrium distribution of energy between two objects brought into thermal contact with each other.

Distributing energy: A single atom inside a solid

Consider a single atom inside a solid, whose energy we model in terms of the energy of three one-dimensional oscillators (x, y, z), neglecting interactions with the neighbors. Each of these three oscillators can have 0, 1, 2, etc. number of "quanta" of vibrational energy $(\hbar\omega_0)$ added to its ground state.

Suppose the total vibrational energy added to the atom is 4 quanta, and we ask how we might distribute this energy among the three oscillators. We will soon see that enumerating all the possible ways of distributing the energy among the oscillators leads to a deeper understanding of the statistical behavior of a solid. We could give all the energy to the first one-dimensional oscillator and none to the others, or all to the second, or all to the third, as shown in Figure 10.7.

Or we could give three quanta to one oscillator, and give the remaining quantum to one of the others, as shown in Figure 10.8.

Or we could give two quanta to one oscillator, and distribute the other two to the others, as shown in Figure 10.9.

That's it—there aren't any more ways to distribute the energy. (Check to make sure that you can't think of any other arrangements.) By explicitly listing all the possible arrangements, we see that there are 15 different ways that the four energy quanta could be distributed among the three one-dimensional oscillators.

In each of these 15 cases the total energy of the three-oscillator atom is exactly the same. When we make macroscopic measurements of the energy of a block, we don't know and we usually don't care exactly how the energy is distributed among the atoms that make up the block, because the internal energy of the block is the same for all of these distributions. However, the number of different arrangements affects the probability of certain processes occurring, as we shall see.

"Microstate" and "macrostate"

Some terminology: We say that there are 15 "microstates," such as the microstate with 3 quanta in the first oscillator, 0 in the second, and 1 in the third (the second microstate in Figure 10.8). The 15 microstates correspond to one "macrostate," which is characterized by the total energy being equal to 4 quanta of energy, no matter how distributed.

THE FUNDAMENTAL ASSUMPTION OF STATISTICAL MECHANICS

A fundamental assumption of statistical mechanics is that in our state of microscopic ignorance, each microstate (microscopic distribution of energy) corresponding to a given macrostate (total energy) is equally probable.

That is, we assume that if we could observe the detailed arrangement whenever these three oscillators (corresponding to one atom) have a total energy of 4 quanta, we would find on the average that each of these arrangements would occur 1/15th of the times that we looked. This fundamental assumption is plausible, but ultimately it is justified by the fact that deductions based on this assumption do agree with experimental observations.

Two interacting atoms

We can now consider the case of two blocks in thermal contact (Figure 10.10). The importance of counting arrangements ("microstates") can be

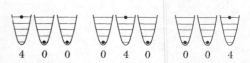

Figure 10.7 Three ways of distributing four quanta of vibrational energy among three one-dimensional oscillators.

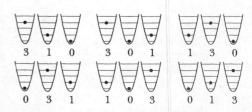

Figure 10.8 Six more ways of distributing four quanta of vibrational energy among three one-dimensional oscillators.

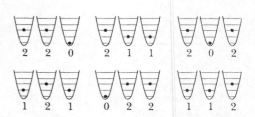

Figure 10.9 Still more ways of distributing four quanta of vibrational energy among three one-dimensional oscillators.

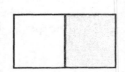

Figure 10.10 Two small blocks in thermal contact.

atom 1	atom 2	# ways
4	0	15·1
0	4	1·15

Figure 10.11 If we give all four quanta to one atom or the other, there are 30 ways of distributing four quanta of energy between the two atoms (i.e. among six independent oscillators).

atom 1	atom 2	# ways
3	1	10·3
1	3	3·10

Figure 10.12 We find 60 more ways to distribute four quanta among two atoms (six oscillators) if we give three quanta to one atom and one quantum to the other.

atom 1	atom 2	# ways
2	2	6·6

Figure 10.13 We find 36 ways of distributing four quanta of energy by giving two quanta to each atom.

q_1	q_2	# Ways$_1$	# Ways$_2$	# Ways$_1$ · # Ways$_2$
0	4	1	15	15
1	3	3	10	30
2	2	6	6	36
3	1	10	3	30
4	0	15	1	15

Figure 10.14 Summary of the 126 different ways to distribute four quanta of vibrational energy between two atoms (each consisting of three independent oscillators).

seen when we have systems of oscillators interacting with each other. Consider the smallest possible "blocks," two neighboring atoms (a total of 6 one-dimensional oscillators), and suppose 4 quanta of vibrational energy, $4\hbar\omega_0$, are distributed among these 6 oscillators. We already know that there are 15 ways to distribute the 4 quanta among the three one-dimensional oscillators of the first atom, and in that case there is only one way to distribute zero quanta to the second atom. Similarly, in Figure 10.11 we see that there are 15 ways to distribute the 4 quanta among the three one-dimensional oscillators of the second atom, and in that case there is only one way to distribute zero quanta to the first atom.

But since the two atoms are in contact with each other, the 4 quanta could also be shared between the atoms. For example, the first atom might have only 3 quanta, with the second atom having the other quantum.

? How many ways are there to distribute 3 quanta among 3 one-dimensional oscillators? Try to list all the possibilities, as we did for the case of 4 quanta.

You should have found ten ways, which we could list as 300 (that is, 3 quanta on the first oscillator, 0 on the second, and 0 on the third), 030, 003, 201, 210, 021, 120, 012, 102, and 111.

? For each of these arrangements, how many ways are there to distribute the one remaining quantum among the other 3 one-dimensional oscillators?

There are just three ways to arrange one quantum among the other 3 oscillators: 100, 010, and 001.

The product of these two numbers is the total number of ways of distributing the 4 quanta in such a way that there are 3 quanta on the first atom and 1 on the other. This product is $10 \cdot 3 = 30$ ways. These results are summarized in Figure 10.12.

Now let's give just 2 quanta to the first atom (leaving 2 quanta for the other atom).

? How many ways are there to distribute 2 quanta among 3 one-dimensional oscillators?

You should find that there are six ways: 200, 020, 002, 101, 110, and 011.

? For each of these arrangements, how many ways are there to distribute the 2 remaining quanta among the other 3 one-dimensional oscillators?

Clearly there are also six ways to arrange the remaining 2 quanta among the other 3 oscillators. The total number of ways of distributing the 4 quanta in such a way that there are 2 quanta on the first atom and 2 on the other is $6 \cdot 6 = 36$ (Figure 10.13)

Now we have carried out enough calculations to be able to make a table of all the ways of distributing 4 quanta between two atoms (that is, among 6 one-dimensional oscillators). In this table, q_1 is the number of quanta (0 to 4) assigned to the first atom, and q_2 is the number of quanta assigned to the second (with q_1+q_2 necessarily equal to 4). In Figure 10.14 we show the number of ways to arrange the quanta among the first three one-dimensional oscillators, among the second three oscillators, and the product, which is the total number of ways of distributing q_1 and q_2 quanta.

Figure 10.15 is a graph of the total number of ways of distributing the quanta vs. q_1. Remember that the basic assumption of statistical mechanics is that all of these 126 different arrangements ($15 + 30 + 36 + 30 + 15 = 126$ microstates) with a total energy of 4 quanta are equally probable.

? Therefore, if two atoms share 4 quanta, what is the most likely division of the energy between the two? What is the probability at some instant that the energy is equally divided? What is the probability at some instant that the first atom has all the energy?

Evidently the most probable division is 2 and 2, since there are 36 ways for this to be done, whereas all other divisions have fewer ways for this to happen. The probability for a 2-2 division is 36/126, or 29%, so even though this is the most probable division, it happens less than one-third of the time. The probability that the first atom has all the energy is the number of ways for this to happen (15) divided by 126: 15/126 = 12%.

This is a microcosm of what happens when two identical blocks are brought into contact: the most probable division of the thermal energy is that it is shared equally. In order to understand in detail how this works in a real macroscopic system, we need to consider large numbers of atoms and large numbers of quanta. For numbers of atoms or quanta even slightly larger than the few we've been considering, it becomes practically impossible to figure out the number of arrangements of quanta by explicitly listing all the possibilities as we did up to now. We need a formula for calculating the number of arrangements.

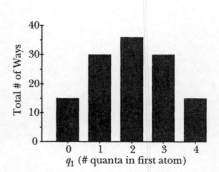

Figure 10.15 Histogram showing the total number of ways of distributing four quanta of vibrational energy between two atoms (six one-dimensional oscillators).

Ex. 10.1 For practice in counting microstates, determine how many ways there are to arrange 3 quanta among 4 one-dimensional oscillators. (This would represent one-and-a-third atoms, so this doesn't make physical sense.)

10.2.3 One system or many?

The results for distributing 4 quanta among 2 atoms (6 oscillators) can be thought of in two complementary ways. You can say, "I will make frequent observations of my two-atom system, and I expect that in 29% of these observations I'll find the energy split evenly (2-2)." Alternatively, you can set up 100 of these two-atom systems and say, "Whenever I look at my 100 systems, I expect that 29 of them will have the energy split evenly (2-2)." Sometimes the "one system, many observations" view is particularly helpful, and sometimes the "many systems, one observation" view is the more useful way to think about the statistical nature of a phenomenon.

10.2.4 A formula for the number of arrangements of quanta

Clearly it would be not only tedious but impractical to keep on counting states as we did above for systems involving very large numbers of atoms. We need a formula for calculating how many different ways a set of objects can be arranged. We can develop this formula in a general way, using concepts of probability, and then apply it to solids made up of atomic oscillators.

Suppose you have five billiard balls in a bag, numbered 1 through 5. You draw them out of the bag, one at a time, and record the sequence of numbers, such as 34152 (Figure 10.16). Then you put them back in the bag and repeat. How many different number sequences are possible?

- The first ball might be any one of the five balls.
- For each of these 5 possibilities, there are four possibilities for the second ball, or 5·4 = 20 choices so far.
- For each of the 5·4 = 20 choices of the first two balls, there are three different possible choices for the third ball, or 5·4·3 choices so far.

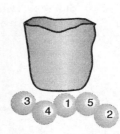

Figure 10.16 Five numbered billiard balls are taken one at a time from a bag. How many different number sequences are possible?

- For each of the $5 \cdot 4 \cdot 3 = 60$ choices of the first three balls, there are 2 possible choices for the fourth ball, for a total of $5 \cdot 4 \cdot 3 \cdot 2 \cdot 1 = 120$ possibilities, since there is only one remaining ball to choose.

Evidently there are 120 different "permutations" of the five integers. It would be exceedingly tedious to list all these different arrangements, but we have a simple formula to calculate how many there are: 5!, which is the standard notation for "5 factorial," meaning $5 \cdot 4 \cdot 3 \cdot 2 \cdot 1$. Fortunately, for analyzing a solid all we care about is how many arrangements (microstates) there are for a particular total energy, not the details about how much of the energy is assigned to which oscillators. So all we need from a formula is the number of arrangements.

? We can easily check this factorial formula for the case of three balls. Explicitly list all possible permutations of the numbers 1, 2, and 3, and verify that there are indeed $3! = 3 \cdot 2 \cdot 1 = 6$ possible arrangements.

Making fewer distinctions

Now suppose that of the five balls, three are red and two are green, and we ask how many different arrangements of the colors are possible, such as RGRRG or GGRRR. We know that there are $5! = 120$ different arrangements of the numbered balls, but if we're only interested in the color sequence, the numerical order of the red balls is irrelevant, as is the numerical order of the green balls.

There are $3! = 6$ permutations of the red balls among each other, and $2! = 2$ permutations of the green balls among each other. Therefore there are many fewer than 120 distinctively different color sequences, and we need to correct for this by dividing by the extra permutations:

$$\text{\# of color sequences} = \frac{5!}{3!2!} = \frac{120}{(6)(2)} = 10$$

? Check this result by listing all the different ways of ordering 3 R's and 2 G's.

Arranging quanta among oscillators

By an appropriate choice of visual representation, we can convert our problem of calculating the number of arrangements of q quanta among N one-dimensional oscillators ($N/3$ atoms) into the problem we just solved, for which we have a formula. Consider again the specific case of $q = 4$ quanta distributed among $N = 3$ one-dimensional oscillators. We'll represent a quantum of energy by the symbol •, and a (fictitious) boundary between oscillators by a vertical bar |. Here is a picture of a particular situation where the first oscillator has 2 quanta, the second oscillator has 1 quantum, and the third oscillator has 1 quantum; together with the boundaries between oscillators we have 6 objects arranged in a particular sequence:

$$• \, • \, | \, • \, | \, •$$

This sequence represents a total energy of 4 quanta and 2 boundaries (a total of 6 things, if we consider a boundary to be a "thing"). Figure 10.17 illustrates two such sequences. Notice that we need $N - 1 = 2$ vertical bars to be able to indicate which oscillators have how much energy. How many such sequences like this are there? Rearranging quanta and boundaries is equivalent to moving quanta between oscillators. In this pictorial form, the problem is like having 4 red balls and 2 green ones. Therefore we can write a formula for the number of different arrangements (number of different microstates):

$$\frac{6!}{4!2!} = 15 \text{ ways of arranging 4 quanta among 3 oscillators}$$

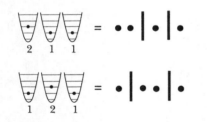

Figure 10.17 Two different representations of number of quanta in three oscillators.

This agrees with our earlier calculations. Generalizing to arbitrary numbers of quanta distributed among arbitrary numbers of oscillators, we have this important result:

WAYS TO ARRANGE q QUANTA AMONG N ONE-DIMENSIONAL OSCILLATORS

$$\frac{(q+N-1)!}{q!(N-1)!}$$

Ex. 10.2 Verify that this formula gives the correct number of ways to arrange 0, 1, 2, 3, or 4 quanta among 3 one-dimensional oscillators, given in the table on page 346.

Very big numbers

As you increase q and/or N, this formula gets very big very fast. For example, the number of ways to distribute 100 quanta among 300 oscillators (100 atoms) is about 1.7×10^{96}, which is 17 followed by 95 zeros!

In striking contrast, there is only one way to arrange to have all 100 energy quanta be placed on just one particular oscillator out of all the 300 oscillators. Although this arrangement would satisfy the requirement that the total energy of the system be 100 quanta, this is extremely improbable. The fundamental assumption of statistical mechanics is that all microstates are equally probable, so the odds that the actual microstate is the one with all the energy given to just one particular oscillator of your choice is 1 in 1.7×10^{96}, which makes this unlikely event essentially impossible.

A typical macroscopic object such as a block of ordinary size contains 10^{23} or more atoms, not a mere hundred, and the number of quanta is even larger. How likely is it that all the energy will be found concentrated on one particular atom? The mind boggles at the astronomically huge odds against this ever happening. Could it happen? Yes. Is any human ever likely to observe it? No.

Ex. 10.3 Suppose you look once every second at a system with 300 oscillators and 100 energy quanta, to see whether your favorite oscillator happens to have all the energy (all 100 quanta) at the instant when you look. You expect that just once out of 1.7×10^{96} times you will find all of the energy concentrated on your favorite oscillator. On the average, about how many years will you have to wait? Compare this to the age of the Universe, which is thought to be about 10^{10} years. (1 year $\approx \pi \times 10^{7}$ seconds.)

10.3 Thermal equilibrium of two blocks in contact

We now have the tools necessary for analyzing in some detail what will happen when two blocks are brought into contact and approach thermal equilibrium. We'll choose two blocks made of the same material, so a quantum of energy is the same for the atomic oscillators in both blocks (same atomic mass m; same interatomic forces, so same "spring stiffness" k_s). We'll make the analysis somewhat general by choosing blocks of different sizes. One block contains N_1 one-dimensional oscillators ($N_1/3$ atoms) and initially contains q_1 quanta of energy, and the other block contains N_2 one-dimensional oscillators ($N_2/3$ atoms) and initially contains q_2 quanta of energy.

? We want to treat the simple case where the total energy $q_1 + q_2$ of the two-block system remains fixed at all times. What simplifying assumption should we make about the situation?

During the entire process we assume that there is little energy transferred into or out of the surrounding air or the supports for the blocks, so that the total energy of the two blocks doesn't change. However, the number of quanta in each block, q_1 or q_2, need not stay fixed, since energy can flow back and forth between the two blocks.

Consider a very concrete example. Suppose $N_1 = 300$ (100 atoms), $N_2 = 200$ (about 67 atoms; or choose 201 oscillators if you wish to be exact), and there is a total energy distributed throughout the two blocks corresponding to $q_1 + q_2 = 100$ quanta (Figure 10.18). We use a computer to calculate the number of ways that q_1 quanta can be distributed among the $N_1 = 300$ oscillators of the first block (using the formula we developed earlier), and we multiply this number times the number of ways that $q_2 = (100 - q_1)$ quanta can be distributed among the $N_2 = 200$ oscillators of the second block.

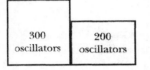

Figure 10.18 A total of 100 quanta of vibrational energy are available to be distributed between two systems, one consisting of 300 oscillators (100 atoms), the other consisting of 200 oscillators (about 67 atoms).

The product of these two calculations is the number of ways that we can arrange the 100 quanta so that the first block has q_1 quanta and the second block has $q_2 = (100 - q_1)$ quanta, for a total energy shared between the two blocks of 100 quanta. We have the computer do this calculation for $q_1 = 0$, 1, 2, 3,...99, 100 quanta, which corresponds to $q_2 = 100, 99, 98,...1, 0$ quanta. The following table shows the first few results, where the number of microstates is denoted by Ω (Greek uppercase omega):

q_1	$q_2 =$ $(100 - q_1)$	$\Omega_1 = \dfrac{(q_1 + 300 - 1)!}{q_1!(300 - 1)!}$	$\Omega_2 = \dfrac{(q_2 + 200 - 1)!}{q_2!(200 - 1)!}$	Total # of Ways $\Omega_1 \Omega_2$
0	100	1	2.772 E+81	2.772 E+81
1	99	300	9.271 E+80	2.781 E+83
2	98	4.515 E+04	3.080 E+80	1.391 E+85
3	97	4.545 E+06	1.016 E+80	4.619 E+86
4	96	3.443 E+08	3.331 E+79	1.147 E+88
...	...	...	...	...

We see in the last column that there is an enormous number of ways of arranging 100 quanta among 500 oscillators, and that this number is growing rapidly in the table. In Figure 10.19 the possible number of arrangements of quanta is plotted on the y axis, and q_1, the number of quanta assigned to the first block, is plotted on the x axis.

The most probable arrangement (indicated by the highest point of the peak on the graph) is that 60 of the 100 quanta will be found in the first block, which contains 300 oscillators out of the total of 500 oscillators. This seems reasonable, since it does seem most probable that the energy would be distributed uniformly throughout the two blocks (since they're made of the same material), and that would give 3/5 (60%) of the energy to the first block. It is gratifying that our statistical analysis leads to this plausible result.

It appears from the width of the peak shown in the graph that we shouldn't be too surprised if occasionally we would find that the first block contains anywhere between 40 to 80 of the 100 quanta, but it appears that it

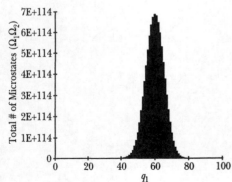

Figure 10.19 The number of ways of distributing 100 quanta of vibrational energy between two blocks having 300 and 200 oscillators, respectively. q_1 is the number of quanta in the first (larger) block.

is very unlikely to find fewer than 40 or more than 80 of the quanta in the first block.

Relatively speaking, how likely is it for *none* of the energy to be in the first block? In that case, $q_1 = 0$, and from the table we see that the number of ways to arrange the 100 quanta this way is 2.772 E+81 (2.772×10^{81}). That's an awfully big number, but how big is it compared to the most probable arrangement, where $q_1 = 60$, for which there are about 7×10^{114} ways according to the graph? Evidently it is less likely by about a factor of 10^{33} (!) to find the energy split 0-100 rather than 60-40. Is it possible according to the laws of physics for none of the energy to be in the first block? Yes. Is it likely that we would ever observe such an unusual distribution? Most emphatically not!

Note that the number of ways to arrange the quanta 0-100 (2.772×10^{81}) isn't actually visible on the graph, because it is 10^{33} times smaller than the peak of the graph. Values of q_1 outside the 40 to 80 range are invisible on the graph, because they are relatively so very small compared to the peak.

Problem 10.1 Probability distribution

(a) Model a system consisting of two atoms (three oscillators each), among which 4 quanta of energy are to be distributed. Write a program to display a histogram showing the total number of possible microstates of the two-atom system *vs.* the number of quanta assigned to atom 1. Compare your calculations and histogram to the one on page 347 (you should get the same distribution).

(b) Model a system consisting of two solid blocks, block 1 containing 300 oscillators and block 2 containing 200 oscillators. Find the possible distributions of 100 quanta among these blocks, and plot number of microstates *vs.* number of quanta assigned to block 1. Compare your calculations and histogram to those on page 350. Determine the distribution of quanta for which the probability is half as large as the most probable 60-40 distribution.

(c) Do a series of calculations distributing 100 quanta between two blocks whose total number of oscillators is 500, but whose relative number of atoms varies. For example, consider equal numbers of oscillators, and ratios of 2:1, 5:1, etc. Describe your observations.

Width of the distribution

Compared to the graph on page 347 of our earliest calculation (6 oscillators and 4 quanta), the peak shown in Figure 10.19 occupies a much narrower range of the graph, reflecting the sharply decreased relative probability of observing extreme distributions of the energy.

For further comparison, Figure 10.20 is a graph of the number of ways of distributing 1000 quanta among two blocks containing 3000 and 2000 oscillators. This peak is much narrower than the peak in the previous graph (where there were only 100 quanta distributed among only 300 and 200 oscillators). This is a general trend. The larger the number of quanta and oscillators, the narrower is the peak around the most probable distribution.

The fractional width of the peak is the width of the peak at half height divided by the value of q_1 that gives the maximum probability (in the present case, the width divided by 600). It can be shown that the fractional width of the peak is proportional to $1/\sqrt{q}$ or to $1/\sqrt{N}$, whichever is larger. That is, if you quadruple the number of quanta or the number of oscillators, the fractional width of the peak decreases by a factor of 2.

? Consider two blocks that are of ordinary macroscopic size, containing 3×10^{23} atoms and 2×10^{23} atoms, and many quanta per atom. Qualitatively, what would you expect about the width of the

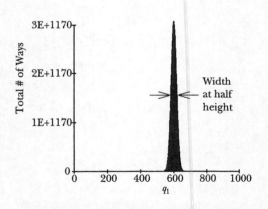

Figure 10.20 The number of ways of distributing 1000 quanta between two blocks containing 3000 and 2000 oscillators, respectively.

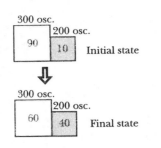

Figure 10.21 If the initial energy distribution between two systems in thermal contact is not the most probable energy distribution, energy will be exchanged until the most probable distribution is reached.

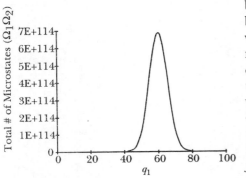

Figure 10.22 Ways of distributing 100 quanta of vibrational energy between a system of 300 oscillators and a system of 200 oscillators.

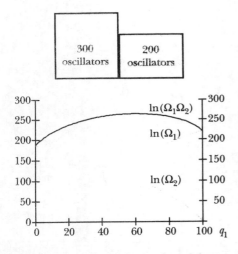

Figure 10.23 A logarithmic plot of the ways of distributing 100 quanta of vibrational energy between two systems. The number of quanta in system 1 is plotted on the x axis, and the natural log of the number of arrangements on the y axis.

peak, if you could calculate it? How likely is it that you would ever observe a significant fluctuation away from the most probable 60-40 split?

You would expect the peak to be extremely narrow, and hence the probability of significant fluctuations would be very low.

The most probable is the only real possibility

These considerations show that in the world of macroscopic objects such as ordinary-sized blocks, the most probable arrangement is essentially the *only* arrangement that is ever observed. That is why you do not see a block suddenly leap up from the table when all of the thermal energy of the table floods into the block. On the other hand, at the microscopic level we should not be surprised if the energy of one of the atoms in a block varies a lot, since the most probable distribution includes many different arrangements of the quanta, with varying numbers of quanta on one particular atom.

10.3.1 Entropy and equilibrium

We now have a good statistical description of the thermal equilibrium of two blocks in contact. At equilibrium, energy is distributed between the two blocks in the most probable manner, based on having the largest number of ways of achieving this distribution (largest number of microstates for the macrostate of given total energy). We have been studying very small systems consisting of a hundred or more atoms, and we found that in small systems there is some significant probability of finding the energy distributed somewhat differently than the most probable 60-40 division. (Objects consisting of only a few hundred atoms are called "nanoparticles" and are currently the subject of intense research, because many of their properties and behaviors are intermediate between those of atoms and those of large-scale objects.) For large macroscopic objects containing 10^{20} atoms or more, the most probable distribution of the energy is essentially the only energy distribution we will ever observe, because the probability of distributions that are only slightly different is very small.

Next we will study more deeply the details of why a particular thermal equilibrium becomes established. Suppose for example that the first block (a nanoparticle with 300 oscillators, or 100 atoms) starts out with 90 quanta and the second (200 oscillators, or about 67 atoms) with only 10 quanta (Figure 10.21). When we put them together, we expect this to shift toward a 60-40 distribution (Figure 10.22). Studying the details of why this shift occurs will lead us to a deep understanding of the concepts of temperature and something called "entropy."

Since we start from a non-equilibrium energy distribution, it would be nice if we could make some kind of graph where we could see the total number of ways to arrange the quanta even when we are far from equilibrium. As we have seen, the graph of the total number of microstates is so strongly peaked that we don't see anything outside the peak. For example, on the graph in Figure 10.22 where we can see 10^{114} we can't see the relatively much smaller value of 10^{81}. A way to get around this problem is to plot the logarithm of the data, which makes the data visible across the whole range of energy distributions.

In statistical mechanics it is standard practice to use the base-*e* or "natural" logarithm ("ln"). For the case of distributing 100 quanta among 300 and 200 one-dimensional oscillators, which we studied before, Figure 10.23 shows a plot of the natural logarithms of the number of ways Ω_1 to arrange q_1 quanta in block 1, the number of ways Ω_2 to arrange $q_2 = (100 - q_1)$ quanta in block 2, and the total number of ways $\Omega_1\Omega_2$ to arrange the 100 quanta among the 500 oscillators.

Because we are considering discrete microstates, each of these curves is really a set of 100 dots, but we connect the dots and make continuous curves. This is particularly appropriate when we deal with macroscopic objects, where an increment of one quantum along the x axis is practically an infinitesimal fraction of the axis.

By taking (natural) logarithms, we have converted a highly-peaked graph into a slowly varying one. To make sure you understand how this graph is related to the peaked graph, do the following:

? For this situation, we calculated on page 350 that there is just 1 way to arrange 0 quanta in block 1, and 2.772×10^{81} ways to arrange 100 quanta in block 2. Take the natural logarithm of this number and verify that all three curves on the graph make sense to you at $q_1 = 0$ ($q_2 = 100$). Note that at the right end of the graph, at $q_1 = 100$, the curve for $\ln(\Omega_1)$ goes higher than $\ln(2.772 \times 10^{81})$, because block 1 has a larger number of oscillators among which to distribute the 100 quanta (300 oscillators compared with 200).

Problem 10.2 Natural logarithm (ln) of ways to arrange energy

Start with your solution to Problem 10.1 (b). For the same system of two blocks, with $N_1 = 300$ oscillators and $N_2 = 200$ oscillators, plot $\ln(\Omega_1)$, $\ln(\Omega_2)$, and $\ln(\Omega_1\Omega_2)$, for q_1 running from 0 to 100 quanta. Your graph should look like the one in Figure 10.23. Determine the maximum value of $\ln(\Omega_1\Omega_2)$ and the value of q_1 where this maximum occurs. What is the significance of this value of q_1?

Definition of entropy

We will deal repeatedly with the natural logarithm of the number of ways to arrange energy among a group of atoms (the number of microstates corresponding to a particular macrostate of specified energy). This quantity, $\ln(\Omega)$, when multiplied by the Boltzmann constant k, is called the "entropy" of the object and is denoted by the letter "S" (the triple equal sign means "is defined as"):

DEFINITION OF ENTROPY S

$$S \equiv k \ln \Omega$$

The Boltzmann constant is $k = 1.4 \times 10^{-23}$ J/K, so entropy has units of joules per kelvin. The Boltzmann constant is included in the definition for consistency with an older, macroscopically based definition of entropy. Figure 10.23 is a plot of S/k, or you can think of it as a plot of S measured in units of k, just as we have been measuring energy in units (quanta) of $\hbar\sqrt{k_s/m}$, where k_s is the effective spring stiffness. Alternatively, you could make a graph of S ($= k \ln\Omega$) instead of $\ln\Omega$.

Since $k\ln(\Omega_1\Omega_2) = k\ln(\Omega_1) + k\ln(\Omega_2)$ (this is a property of logarithms), the entropy S of the two-block system is equal to the entropy S_1 of the first block plus the entropy S_2 of the second block. A consequence of defining entropy in terms of a logarithm is that we can consider entropy as describing a property of a system, and when there is more than one object in a system we get the total entropy simply by adding up the individual entropies, $S_{tot} = S_1 + S_2$, as is the case with energy ($E_{tot} = E_1 + E_2$).

Since both $k\ln(\Omega_1\Omega_2)$ and $\Omega_1\Omega_2$ go through a maximum at the same value of q_1, we can state the condition for thermal equilibrium of the two blocks, in terms of entropy:

In equilibrium the most probable energy distribution is that which maximizes the total entropy $S_{tot} = S_1 + S_2$ of the two blocks.

10.4 The second law of thermodynamics

The "first law of thermodynamics" is another name for the energy principle $\Delta E = W + Q$, which by now should be very familiar to you. The "second law of thermodynamics" is however something new. It can be stated in a number of equivalent forms, but there is a particularly useful formulation in terms of entropy:

THE SECOND LAW OF THERMODYNAMICS

If a closed system is not in equilibrium, the most probable consequence is that the entropy will increase.

To say it differently, a closed system will tend toward maximum entropy.

As a specific example of the second law of thermodynamics, consider our two blocks. The most probable energy distribution is the one for which the total entropy is a maximum, and if initially the energy distribution is something else, it is highly likely that the entropy will increase. For nanoparticles there can be significant fluctuations away from this state of maximum entropy (with accompanying decrease in total entropy), but for ordinary-sized systems these fluctuations are extremely small, and for practical purposes the entropy of a closed macroscopic system never decreases.

Loosely, one can restate the second law of thermodynamics like this: "A closed system tends toward increasing disorder." This is overly vague without precise definitions of "order" and "disorder," and it is the definition of entropy as $k\ln\Omega$ that provides the needed precision. A closed system tends toward increasing entropy, which is a measure of the number of microstates corresponding to a particular macrostate of given energy. An extreme example of high "order" is a situation in which all the energy is concentrated on a single oscillator, for there is only one way to arrange the energy like this. In this extreme case, not only do we correctly perceive the situation to be one of high order and low disorder, but the quantitative measurement of disorder (the entropy) is $k\ln(1)$, which equals 0, the lowest possible value.

Irreversibility

Any process in which the entropy of the Universe increases is "irreversible" in the technical, physics sense: a reversed movie of the process looks odd. But any process in which the entropy of the Universe doesn't change is in principle "reversible," and there do exist processes that are approximately or nearly reversible. For example, a steel ball bearing that bounces vertically on a steel plate rebounds almost to the same height from which it was dropped, and this represents a (nearly) reversible process, because the system returns (nearly) to its original state.

However, if we watch for a while, the ball bearing returns to lower and lower heights and eventually settles down to rest on the plate. If the ball bearing kept returning to its original height, this would not be a violation of energy conservation. It is the second law of thermodynamics that says that we cannot expect the process to be completely reversible. There is a larger number of ways to share the energy between the ball and the plate than the number of ways for the ball to keep all the energy.

Forward

Reverse

Figure 10.24 If the height of a bouncing ball increased with time instead of decreasing with time, this would not necessarily violate conservation of energy.

Suppose we make a movie of the ball bearing bouncing lower and lower and coming to rest on the plate. Then we run it backwards, and our friends see the ball starting to bounce, and bouncing higher and higher (Figure 10.24). This needn't violate energy conservation: energy in the plate could be flowing into the ball. But this reversed movie certainly looks odd, presumably because we have an instinctive sense, based on lots of experience, that the entropy of the Universe increases rather than decreases.

REVERSIBLE PROCESS: *S* OF THE UNIVERSE DOESN'T CHANGE

IRREVERSIBLE PROCESS: *S* OF THE UNIVERSE INCREASES

Ex. 10.4 You see a movie in which a shallow puddle of water coalesces into a perfectly cubical ice cube. How do you know the movie is being played backwards? Otherwise, what physical principle would be violated?

10.5 What is temperature?

Until now, we have associated temperature with the average energy of a molecule. The concept of entropy makes possible a deeper connection between our macroscopic measurements of temperature and a fundamental, atomic, statistical view of matter and energy. We will develop a statistically based definition of temperature.

Consider again our two blocks. On the graph in Figure 10.25 we plot entropy $S = k\ln\Omega$ rather than $\ln\Omega$. Remember that we are approximating a set of dots by a smooth curve, since the dots are very close together (tiny energy spacing). At the maximum of the total entropy $S = S_1 + S_2$, the total entropy curve is horizontal (slope = 0), and the following relationship is true at equilibrium:

$$\frac{dS}{dq_1} = \frac{dS_1}{dq_1} + \frac{dS_2}{dq_1} = 0$$

Since $q_2 = (100 - q_1)$, this leads to

$$\frac{dS_1}{dq_1} - \frac{dS_2}{dq_2} = 0 \text{ (at equilibrium)}$$

dS_2/dq_2 is the slope moving from right to left and is positive, since q_2 increases from right to left. Therefore we obtain the following result:

$$\frac{dS_1}{dq_1} = \frac{dS_2}{dq_2} \text{ (at equilibrium)}$$

dS_1/dq_1 is a measure of the state of the first block, and dS_2/dq_2 is a measure of the state of the second block.

? What physical property can these derivatives represent? What physical property of the blocks is the same when the two blocks reach thermal equilibrium?

The temperature. It must be that the derivative of the entropy with respect to the energy is somehow related to temperature.

To see what this relationship is, suppose that when the two blocks were initially brought into contact, the first block contained $q_1 = 90$ quanta and the second block only $q_2 = 100 - 90 = 10$ quanta, as illustrated in Figure 10.26. Note that in Figure 10.26 the initial slope of the entropy curve for block 2 is steeper than the initial slope of the entropy curve for block 1. (Remember: dS_2/dq_2 is the slope moving from right to left and is positive, since q_2 increases from right to left.) Therefore, if we remove one quantum

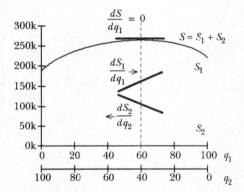

Figure 10.25 Entropy *vs.* number of quanta of energy in system 1.

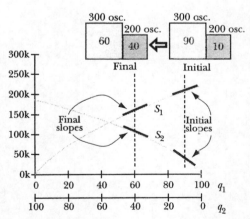

Figure 10.26 Location of the initial and final states of the two-block system on a plot of entropy *vs.* number of energy quanta in system 1.

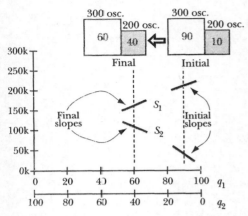

Figure 10.27 Location of the initial and final states of the two-block system on a plot of entropy *vs.* number of energy quanta in system 1.

of energy from block 1 and give it to block 2, we'll increase the entropy in block 2 more than we'll decrease the entropy in block 1.

? Will this result in a net increase or a net decrease in the total entropy S? (See Figure 10.27, a copy of Figure 10.26.)

Since the entropy of block 2 increases more than the entropy of block 1 decreases, there is a net increase of the entropy of the two blocks together. We already established that the state of the two blocks will evolve toward greater total entropy, not less total entropy.

? Consider which block gives up energy, and which block takes it up. Which block is initially at a higher temperature?

Evidently block 1 must be at the higher temperature, because on average it will give energy to block 2 in order that the total entropy increase.

? Does a steeper or a less steep slope of entropy vs. energy correspond to higher temperature?

Initially, block 1 has the higher temperature, and S_1 vs. q_1 has the smaller slope. Initially, block 2 has the lower temperature, and S_2 vs. q_2 has the larger slope.

? Therefore, which of the following looks like a better guess for relating temperature to entropy: $T_1 \propto dS_1/dq_1$? Or $1/T_1 \propto dS_1/dq_1$?

Since block 1 has the higher temperature and the smaller slope (smaller dS/dq), it seems possible that $1/T_1 \propto dS_1/dq_1$.

10.5.1 Definition of temperature

For these reasons, we *define* temperature in terms of dS/dq. But the magnitude of one quantum of energy varies for different systems, so we can't compare different systems if we use q. So instead of q we use E_{int}, the internal energy above the ground state of a system measured in joules. For a group of oscillators, E_{int} is the number of quanta q times the energy per quantum: $E_{int} = q\hbar\sqrt{k_s/m}$ for a system of harmonic oscillators, where k_s is the spring stiffness.

DEFINITION OF TEMPERATURE *T*

$$\frac{1}{T} \equiv \frac{dS}{dE_{int}}$$

The Boltzmann constant k in the definition $S \equiv k\ln\Omega$ makes the units come out right. With the internal energy E_{int} of a system measured in joules (J), and entropy S measured in joules per kelvin (J/K), temperature T is measured in kelvins (K). (Remember that we're approximating a set of closely-spaced dots by a curve, so taking a derivative makes sense.)

This is a highly sophisticated and abstract way of defining temperature. How does this relate to the temperature that is measured by an ordinary thermometer? We will see in Chapter 11 that the temperature defined by $1/T = dS/dE_{int}$ is the same as the "absolute" temperature, which appears for example in the ideal gas law $PV = N_{moles}RT$, which you have probably encountered in previous chemistry or physics courses. On this temperature scale, ice melts at a temperature of +273 K (0° C), and water boils at +373 K (100° C). To put it another way, absolute zero Kelvin is at −273° Celsius.

Review of temperature definition

To recapitulate, the greater the dependence of the entropy on the energy for an object (the steeper the slope of S *vs.* E_{int}), the more "eager" the object is to take in energy to contribute to increasing the total entropy of the total system. It will do this if it can take energy from another object that has a smaller dependence of entropy on energy (smaller slope), since the second object's entropy will decrease less than the first object's entropy will increase. An object with a large dS/dE_{int} has a low temperature, since it is low-temperature objects that take in energy from high-temperature objects.

Conversely, the smaller the dependence of the entropy on the energy, the less reluctant the object is to give up energy and decrease its own entropy, if it can give the energy to an object with a greater dependence of entropy on energy. An object with a small dS/dE_{int} has a high temperature, since it is high-temperature objects that donate energy to low-temperature objects.

Work, thermal energy transfer, and entropy

We should really write $1/T = \partial S/\partial E_{int}$, involving a "partial derivative," which means that we hold everything but S and E constant when we take the derivative. In particular, we hold the volume constant, which means that we do no work on the system, so we should write $1/T = (\partial S/\partial E_{int})_V$, where the subscript V means "take the (partial) derivative holding the volume constant."

One might ask whether there is any entropy change of the surroundings associated with energy exchange in the form of work. The answer turns out to be no for processes that are reversible (there do exist irreversible processes in which work can be associated with an increase in entropy). Thermal energy transfer (associated with a temperature difference between objects in contact) is "disorganized" energy transfer and is associated with entropy change. Work is "organized" energy transfer and in many situations doesn't affect the entropy. It is beyond the scope of this course to prove this rigorously, but we can illustrate the basic issues in terms of an Einstein solid.

If you mechanically squeeze or compress a block, doing work on it to raise its energy, it can be shown that you shift the energy levels upward without changing which state an oscillator is in, as shown in Figure 10.28. The "spring" stiffnesses don't change, but there is more energy stored in the "springs." On the other hand, if there is thermal energy transfer into the system (due to a temperature difference), there is no change in the energy level, but an oscillator will jump to a higher level. This jump from one level to another affects the probability calculations and the entropy, which is the natural log of the number of ways to arrange the energy quanta.

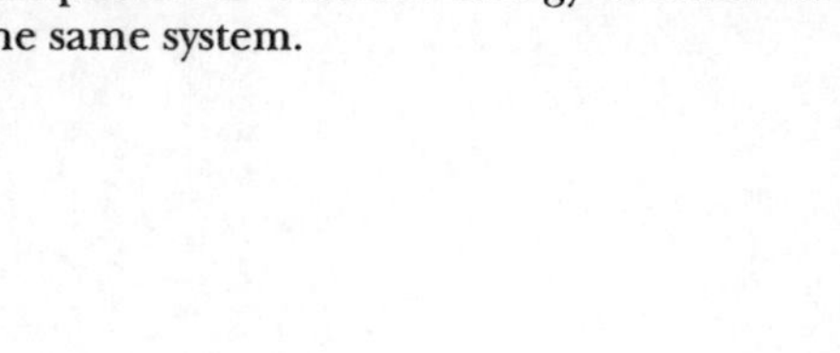

Roughly speaking, thermal energy transfer Q (due to a temperature difference) alters which state the system is in and affects the entropy. Work W alters the energy levels without changing which level the system is in, and doesn't affect the entropy. A small amount of thermal energy transfer into a system of amount Q, with $\Delta E_{int} = Q$, leads to an entropy change of that system.

Since $\dfrac{1}{T} = \dfrac{\Delta S}{\Delta E_{int}} = \dfrac{\Delta S}{Q}$, we conclude the following:

ENTROPY CHANGE ASSOCIATED WITH SMALL *Q*

$$\Delta S = \frac{Q}{T} \text{ (small } Q, \text{ associated with nearly constant } T)$$

If there is a great deal of thermal energy transfer, causing a large temperature change, we must add up (integrate) the contributions to the entropy change.

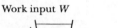
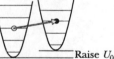

Work input W Thermal energy transfer Q

Raise U_0

Figure 10.28 Mechanical work (compression or stretching of an oscillator system) compared to thermal energy transfer to the same system.

Ex. 10.5 There was thermal energy transfer of 5000 J into a system, and the entropy increased by 10 J/K. What was the approximate temperature of the system?

Ex. 10.6 It takes about 335 joules to melt one gram of ice. During the melting, the temperature stays constant. Which has higher entropy, a gram of liquid water at 0° C or a gram of ice at 0° C? Does this make sense? How large is the entropy difference?

Problem 10.3 Temperature

Modify your calculations from Problem 10.2 (page 353), to plot the temperature in kelvins of block 1 as a function of the number of quanta q_1 present in the first block. On the same graph plot the temperature in kelvins of block 2 as a function of q_1 (of course, $q_2 = q_{tot} - q_1$).

In order to plot the temperature in kelvins, you must determine the values of ΔE and ΔS that correspond to a one-quantum change in energy. Consider the model we are using. The energy of one quantum, in joules, is $\Delta E = \hbar \sqrt{k_s / m}$. The increment in entropy corresponding to this increment in energy is $\Delta S = k\Delta(\ln\Omega)$. Assume that the blocks are made of aluminum. In Problem 3.2 (page 82) you made an estimate of the interatomic spring constant k_s for aluminum. We show later, on page 360, that the effective k_s for oscillations in the Einstein solid is expected to be about 4 times the value obtained from measuring Young's modulus.

What is the significance of the value of q_1 (and of q_2) where the temperature curves for the two blocks cross (the temperatures are equal)?

10.5.2 Can the entropy of an object decrease?

? We analyzed two blocks that initially had a 90-10 energy distribution. In the process of changing to a 60-40 distribution, was it possible for the entropy of *one* of the blocks to decrease significantly? Is this a violation of the second law of thermodynamics?

The entropy of block 1 decreased, but the entropy of block 2 increased even more. We believe that the Universe is a closed system, and that the total entropy of the Universe continually increases and never decreases. However, some portions of the Universe may experience a decrease of entropy, as long as there is at least as much increase elsewhere.

Irreversibility again

We started with a 90-10 distribution of the energy but found the most probable final distribution to be 60-40. This change is effectively irreversible. The 60-40 distribution is so enormously more probable than the 90-10 arrangement that an observer will essentially *never* see a 90-10 distribution spontaneously recur. Closed macroscopic systems (such as two blocks inside an insulating box) always evolve toward the most probable arrangement, and stay there (or so close that you can hardly tell the difference).

10.6 Heat capacity of a solid

In Problem 10.3 (page 358) you calculated the relationship between the temperature of a block and its energy. How can we compare these of calculations with experiment, when we don't have a good way to measure the total energy in a solid block? It is, however, possible to measure *changes* in energy and temperature. Suppose that we add a known amount of energy

ΔE to a system, and we measure the resulting rise in temperature ΔT. The ratio of the energy input to the temperature rise on a per-atom basis is called the heat capacity on a per-atom basis, C:

$$C = \frac{\Delta E_{\text{atom}}}{\Delta T}, \text{ where } \Delta E_{\text{atom}} = \frac{\Delta E_{\text{system}}}{N_{\text{atoms}}}$$

The heat capacity can also be defined on a per-gram or per-kilogram basis, or on a per-mole basis ("molar heat capacity").

Different materials have different heat capacities. For example, at room temperature the heat capacity of water (on a per-gram basis) is 4.2 J/K/gram, while the heat capacity of iron at room temperature is 0.84 J/K/gram. Heat capacity is an important property of a material for two reasons. It has practical consequences in science and engineering because it, along with the thermal conductivity, determines the thermal interactions that the material has with other objects. Also, heat capacity is an experimentally measurable quantity that we can compare with theoretical calculations, to test the validity of our models.

How would we measure the heat capacity of a block? One way is to enclose the block inside an insulating box and use an electric heater inside the box to raise the temperature of the block (Figure 10.28). The electric power (in watts) times the amount of time the current runs (in seconds) gives us the input energy ΔE (in joules), and we can use a thermometer to measure the temperature rise ΔT of the block. The heat capacity (per atom) is $\Delta E_{\text{atom}}/\Delta T = (\Delta E_{\text{block}}/\Delta T)/N_{\text{atoms}}$. The heater itself should have a small mass compared to the sample of material whose heat capacity we're measuring, so that the heater isn't a significant part of the material that is being warmed up.

The following exercise illustrates another way to measure heat capacity:

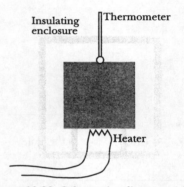

Figure 10.29 Schematic diagram of an apparatus for measuring the heat capacity of a solid.

Ex. 10.7 A 100 gram block of metal at a temperature of 20° C is placed into an insulated container with 400 grams of water at a temperature of 0° C. The temperature of the metal and water ends up at 2° C. What is the heat capacity of this metal, per gram? (The heat capacity of water is 4.2 joules/K/gram.)

Experimental results for heat capacity

Measurements of the heat capacity of aluminum (circles) and lead (squares) are displayed in Figure 10.30. They are plotted as heat capacity (J/K) per atom to facilitate comparison. Note that at low temperatures the heat capacity of both substances changes dramatically with temperature. It is not initially obvious why this should be the case, and in fact this temperature dependence was not predicted by the classical theory, developed in the 1800's. Note also that at room temperature and above, the heat capacity (per atom) of both materials is about the same.

Will our simple statistical model of a solid be able to predict the temperature dependence of the heat capacity of a solid? In Problem 10.3 (page 358) you determined the relationship between energy and temperature for a block (in fact, you did the calculation for two different blocks). This calculation can easily be extended to predict the heat capacity per oscillator, which is $\Delta E_{\text{osc}}/\Delta T$, where E_{osc} is the average energy per oscillator. We can compare this prediction with experimental data by noting that the heat capacity per atom is 3 times the heat capacity per oscillator.

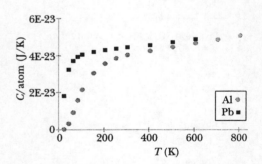

Figure 10.30 Measured heat capacities of aluminum and lead, shown on a per-atom basis.

Effective spring stiffness

In Chapter 3 (page 81) we determined the interatomic spring stiffness k_s of aluminum and lead by relating it to the macroscopic stress-strain relationship (Young's modulus). In the following computation (Problem 10.4) you will use the Einstein independent-oscillator model of a solid to predict the heat capacity of aluminum and lead. We need to use a value for k_s in this model, in order to convert the number of quanta to a value for energy in joules. Presumably the spring stiffness k_s that we use in $\hbar\sqrt{k_s/m}$ should be related to the spring stiffness k_s determined from Young's modulus. However, we shouldn't expect exact agreement, because our model ignores the fact that the atoms are not actually isolated from each other, which is a significant simplification.

Moreover, the effective spring stiffness for oscillations in the x, y, or z directions can be expected to be larger than the value of k_s estimated in Chapter 3 for two reasons. When an atomic core is displaced a distance s_x from its equilibrium position, the spring to the left and the spring to the right each exert a force of $k_s s_x$, so the combined force is $2k_s s_x$. This implies that the effective spring stiffness for oscillations would be $2k_s$.

In addition, in the Einstein model we divide the metal into cubes surrounding each independent atomic oscillator, so each of the two springs to the left and right of an atomic core is effectively half the length of a full interatomic spring, and each half-spring would have a spring stiffness of $2k_s$ (see Ex. 3.5 on page 78).

These considerations suggest that the effective spring stiffness for thermal oscillations in our model might be a factor of 4 larger than the spring stiffness we determined in Chapter 3 from stretching a wire.

Problem 10.4 Heat capacity

Modify your analysis of Problem 10.3 to determine the heat capacity as a function of temperature for a single block of metal. In order to see all of the important effects, consider a single block of 35 atoms (105 oscillators) with up to 300 quanta of energy. Note that in this analysis you are calculating quantities for a *single* block, not two blocks in contact.

To make specific comparison with experimental data, consider the cases of aluminum (Al) and lead (Pb). For each metal, plot the theoretical heat capacity C per atom *vs.* T (K), with dots showing the actual experimental data (converted to the same basis, J/K per atom).

Adjust the interatomic spring stiffness k_s until your calculations approximately fit the experimental data given in the adjoining table. (You will need to convert to a per-atom value.) What value of k_s gives a good fit? Compare with the estimated values of k_s obtained in Problem 3.2 (page 82), but remember that those estimated values are probably 4 times smaller than the effective spring constant for oscillations.

T (K)	C, Al (J/K/mole)	C, Pb (J/K/mole)
20	0.23	11.01
40	2.09	19.57
60	5.77	22.43
80	9.65	23.69
100	13.04	24.43
150	18.52	25.27
200	21.58	25.87
250	23.25	26.36
300	24.32	26.82
400	25.61	27.45

Problem 10.4 Heat capacity data.

Figure 10.31 displays the result of a similar calculation, shown with heat capacity data for copper. Our simple model of a solid as a collection of independent harmonic oscillators does a surprisingly good job of fitting experimental data over a wide range of temperatures. If we were able to include more oscillators in our calculations (by using double precision arithmetic to handle larger numbers, or by using mathematical techniques such as Stirling's approximation for factorials), we could extend our predictions to lower temperatures.

The deviation of the experimental data from the prediction at very high temperatures suggests that at these temperatures a simple harmonic oscilla-

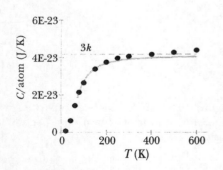

Figure 10.31 Computed heat capacity of copper, compared with experimental values.

tor is not a good model of the atoms in this solid. At high quantum levels, the harmonic oscillator potential energy curve is a poor approximation to the actual potential energy curve describing interatomic interactions.

Note an interesting aspect of the graph: at room temperatures, the predicted heat capacity per atom approaches a constant value of $3k$ (three times the Boltzmann constant, 1.4×10^{-23}). This value agrees quite well with the measured heat capacity of a variety of substances at ordinary temperatures.

Energy quantization and heat capacity

The key difference between our model of solids (the Einstein model) and earlier classical models that did not predict a temperature dependence of heat capacity lies in the quantization of energy in atomic oscillators. Statistical mechanics was originally developed in the 1800's, before the beginning of quantum theory. The classical theory did predict that at high temperatures the energy of each atom in a solid would be $3kT$ (kT per nonquantized one-dimensional oscillator), so that the heat capacity per atom would be $3k$.

If you examine your calculations, you'll find that the high-temperature limit (heat capacity per atom = $3k$) corresponds to temperatures high enough that kT is significantly larger than one quantum of energy. In a situation where the average energy per one-dimensional oscillator is about kT and is large compared to one quantum of energy, the quantization of the energy doesn't make much difference in the analysis. That is, if energy quanta are small compared to the energies of interest, mathematical analysis can be carried out adequately by considering energy to be continuous rather than discrete. In such cases pre-quantum and quantum calculations will give the same results, as is the case here at high temperatures.

On the other hand, at low temperatures the average energy per one-dimensional oscillator is comparable to or smaller than one quantum of energy, and the continuous, nonquantum calculations are not valid. The classical theory provided no explanation for the discovery that as materials were cooled down to very low temperatures, the heat capacity decreased with decreasing temperature. In 1907 Einstein carried through the analysis we have just done and predicted the curve we have just plotted. The good agreement with both low-temperature *and* high-temperature measurements of heat capacity was strong additional evidence for the hypothesis that the energy of oscillators is indeed quantized. (At extremely low temperatures the model of a solid must be refined, taking into account the electrons in the metal, among other things, to achieve full agreement between theory and experiment.)

The fact that the heat capacity for all materials decreases at low temperatures has practical consequences. For example, it makes it difficult to cool a sample to a very low temperature. Cooling a sample depends essentially on putting the sample in contact with a "sink," a large object that is already at a lower temperature, so that there is thermal energy transfer (Q) out of the sample into the sink, lowering the temperature of the sample and not raising the temperature of the sink very much. But at very low temperatures the sink has a low heat capacity, so this is difficult to achieve.

Ex. 10.8 In an insulated container a 100-watt electric heating element of small mass warms up a 300 gram sample of copper for 6 seconds. The initial temperature of the copper was 20° C (room temperature). Predict the final temperature of the copper.

Ex. 10.9 Since $\Delta T = \Delta E / C$, what will happen at low temperatures to the temperature of the sink when some energy ΔE is transferred to it from the sample? Why is this unfortunate?

10.6.1 Which of our results are general?

We have analyzed simple models of solid matter. Nevertheless the basic conclusions are quite general. For example, if our two model blocks were made of different materials, so that the energy quanta were of different size in the two blocks, this would complicate the procedures for evaluating the number of ways Ω to arrange the energy, but the basic conclusion would remain, that the entropy will increase to a maximum.

10.7 The Boltzmann distribution

So far we have mainly been concerned with the thermal equilibrium of two blocks, and how thermal equilibrium arises as that particular distribution of energy between the two blocks that has (by far) the largest number of ways to arrange the quanta. What can we say about the probability of observing a particular amount of energy associated with one particular atom (or one particular one-dimensional oscillator)? Addressing this question will lead us to the "Boltzmann distribution," which provides insight into the behavior of a very wide variety of physical, chemical, and biological phenomena.

Consider a collection of 300 one-dimensional oscillators (100 atoms; object 1) in contact with just a single one-dimensional oscillator (object 2), as shown in Figure 10.32. There are 100 quanta of energy distributed between the two objects.

No matter how many quanta are in object 2, there is only one way to distribute them in that single oscillator. Therefore the number of ways of arranging q_2 quanta in the one oscillator (object 2) is just $\Omega_2 = 1$, so in our calculations of $\Omega = \Omega_1\Omega_2$ we can focus on Ω_1, which is equal to Ω.

In Figure 10.33 we show the familiar calculation of the entropy $S = k\ln\Omega$ for the larger object 1, as a function of the energy E_1 (q_1 times one quantum of energy in joules). On an expanded scale we show a portion of a plot of the entropy *vs.* E_1 near the maximum possible value of E_1, which is the total energy E_{tot} of the combined system (object 1 plus object 2).

As usual, the total entropy of the two systems does tend to maximize, but in this case the maximum is at the right end of the scale, and the slope is nonzero there. We see that $\ln(\Omega)$ is falling approximately *linearly* with increasing energy in the single oscillator (decreasing energy in the large object 1), so the number of microstates Ω must be falling *exponentially* with increasing energy $\Delta E = (E_{tot} - E_1)$ in the single oscillator. We can be quantitative about this.

What we would like to know is the probability for the single oscillator to have a certain amount of the total energy, so we need to evaluate Ω for small values of $\Delta E = (E_{tot} - E_1)$. We can obtain this probability by the following argument. The straight line on the graph has a slope dS/dE_1, so in this region, S can be represented by

$$S = A - \frac{dS}{dE_1}\Delta E$$

where A is some constant, so that the entropy decreases with increasing energy ΔE in the small system.

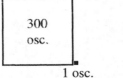

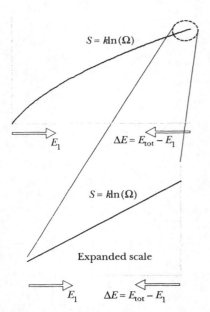

Figure 10.32 A single oscillator in contact with a system of 300 oscillators (100 atoms).

$S = k\ln(\Omega)$

E_1 $\Delta E = E_{tot} - E_1$

$S = k\ln(\Omega)$

Expanded scale

E_1 $\Delta E = E_{tot} - E_1$

Figure 10.33 Entropy (in units of k) as a function of E_1, for the larger object.

But $\dfrac{dS}{dE_1} = \dfrac{1}{T}$, so we have

$$S = k\ln\Omega = A - \frac{\Delta E}{T}, \text{ and } \ln\Omega = \frac{A}{k} - \frac{\Delta E}{kT}$$

Take the exponential of this equation—that is, use both sides of the equation as powers of *e:*

$$e^{\ln\Omega} = e^{\frac{A}{k} - \frac{\Delta E}{kT}}$$

Since *e* raised to the natural logarithm of some quantity is that quantity, we have the following, where *B* is some other constant:

$$\Omega = Be^{-\frac{\Delta E}{kT}}$$

This is what we were looking for—the "Boltzmann distribution" of energy to be found in a microscopic system. We have calculated the number of ways (number of microstates) associated with finding an amount ΔE of energy in a single oscillator. This is proportional to the probability of finding this energy in the single oscillator. By ΔE we mean the amount of energy in a small object in thermal equilibrium with a much larger object. We're measuring ΔE from the ground state (minimum-energy state) of the small object.

Taking ratios of probabilities, we have this:

$$\frac{\text{Prob. oscillator has energy } \Delta E \text{ above ground state}}{\text{Prob. oscillator is in ground state}} = \frac{Be^{-\frac{\Delta E}{kT}}}{Be^{-\frac{0}{kT}}} = e^{-\frac{\Delta E}{kT}}$$

Here is a memorable way to state our result, where again ΔE is the energy above the ground state:

THE BOLTZMANN DISTRIBUTION
The probability of finding a microscopic system to be in a state with energy ΔE above the ground state is proportional to

$$e^{-\frac{\Delta E}{kT}}$$

Although we have derived the Boltzmann distribution by considering a particular situation (a single quantum oscillator in thermal equilibrium with a large number of oscillators), the result is actually very general and applies to a very wide variety of phenomena.

? What is the most likely value of ΔE, the energy to be found in a microscopic system that is in thermal equilibrium with a large system? Are you likely to find an energy that is much larger than kT?

The exponential is largest for $\Delta E = 0$, so the most likely energy for the microscopic system is zero. That is, you're most likely to find the system in its ground state. Since e^{-1} is 0.37, the probability of the energy being much larger than $\Delta E = kT$ is rather small.

If the temperature is so low that kT is small compared to the energy E_1 of the first excited state, you are unlikely ever to find the system in one of its excited states. The system is thermally inert, because the probability of taking in *any* amount of energy is very small. In contrast, if the temperature is high enough that $kT > E_1$, then you will sometimes find the system in one of its excited states, although it is still true that the *most* likely situation is that it will be in its ground state (Figure 10.34).

The Boltzmann distribution has far-reaching consequences. For example, chemical and biochemical reaction rates typically depend strongly on

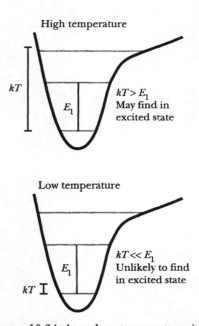

Figure 10.34 At a low temperature it is unlikely to find a system in an excited state, but at a high temperature the likelihood increases.

temperature because with higher temperature the reactants are moving faster and may be found in excited states. Physical reaction rates are also affected. At the very high temperatures found in the interior of our Sun, kinetic energies are high enough to overcome the electric repulsion between nuclei and to allow the nuclei to come in contact, so that they can undergo thermonuclear fusion reactions. A gas becomes a plasma if the temperature is so high that kT is comparable to the ionization energy.

Ex. 10.10 A certain atom has its first and second excited states E_1 and E_2 above the ground state energy. Calculate the relative probability of finding the atom in the second excited state compared to finding it in the first excited state.

Ex. 10.11 At room temperature, show that $kT \approx \frac{1}{40}$ eV. It is useful to memorize this result, because it tells a lot about what phenomena are likely to occur at room temperature.

10.7.1 One microscopic system *vs.* many

We have found that one microscopic system in contact with a large system has an exponential energy distribution. The meaning of this result is this: You make repeated observations of the microscopic system, and you count the number of times you find the system in its ground state, how many times you find it in the first excited state, etc. You will find that the probabilities for finding the system in each of these states are proportional to the Boltzmann distribution.

But we have also determined the energy distribution in a macroscopic system! Take an energy "snapshot" of a macroscopic system. That is, suppose you can make a simultaneous measurement of the energy contained in each of the microscopic components of this large system. What fraction of the microscopic systems will be in the ground state, or the first excited state? The fraction given by the Boltzmann distribution! This follows because we could analyze any one of these microscopic systems as being a small system in thermal contact with a large system.

These two views of the Boltzmann distribution, one microscopic system observed repeatedly, or a large number of microscopic systems observed once, complement each other. Sometimes one view is more helpful, sometimes the other.

10.8 Application: The Boltzmann distribution in a gas

The Boltzmann distribution applies to any kind of system—not just a solid. As a major application of the Boltzmann distribution, we will study a gas consisting of molecules which don't interact much with each other. Examples are the so-called "ideal gas" (with no interactions at all), and any real gas at sufficiently low density that the molecules seldom come near each other.

In order to apply the Boltzmann distribution, we need a formula for the energy of a molecule in the gas. We will omit rest energy, and we will also omit nuclear energy and electronic energy from our total, because at ordinary temperatures there is not enough energy available in the surroundings to raise the molecule to an excited nuclear state or an excited electronic state. Also, for simplicity, instead of writing ΔE_{vib} to represent an amount of vibrational energy above the ground vibrational state, we will simply write E_{vib}.

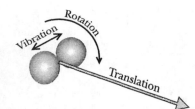

Figure 10.35 Energy of an oxygen molecule.

The energy of a single gas molecule in the gravitational field near the Earth's surface (excluding rest energy, nuclear energy, and electronic energy) is this:

$$K_{\text{trans}} + E_{\text{vib}} + E_{\text{rot}} + Mgy_{\text{cm}}$$

where E_{vib} and E_{rot} are the vibrational and rotational energies relative to the center of mass (if the molecule contains more than one atom, as in Figure 10.35) and y_{cm} is the height of the molecule's center of mass above the Earth's surface. The mass of the molecule is M. For brevity we will speak of "the energy of the gas molecule" but of course the gravitational potential energy really applies to the system of gas molecule plus Earth.

? For this expression for the energy to be accurate, why must the gas be an "ideal" gas (or a real gas at low density)?

In a real gas at somewhat high density we may not be able to neglect the potential energy associated with the intermolecular forces, in which case this expression for the energy would not be adequate.

If the temperature T is the same everywhere in the (ideal) gas, the probability that a particular molecule will have a certain amount of energy is proportional to this, where P stands for probability:

$$P(E) \propto e^{-\frac{[K_{\text{trans}} + E_{\text{vib}} + E_{\text{rot}} + Mgy_{\text{cm}}]}{kT}}$$

Again, we mentally divide the gas into two systems: one particular molecule of interest, and all the rest of the molecules. These two systems are in thermal equilibrium with each other, because the gas molecules are continually colliding with each other and can share energy. The energy for the one particular molecule is then expected to follow the Boltzmann distribution.

We know that vibration and rotation are quantized. When the gas is confined inside a finite container the momentum (or velocity) and height are also quantized, but under almost all conditions the size of the energy quantum associated with momentum (or velocity) and position is so small compared to kT that it is appropriate to take a nonquantum approach for these variables. We speak of the probability that the gas molecule has an x component of velocity within the range between some v_x and $v_x + dv_x$, and similarly for v_y and v_z. Here dv_x is considered to be a very small amount, small compared to the average speed of the molecules. We also speak of the probability that the x component of the gas molecule's position lies between some x and $x+dx$, and similarly for y and z. Here dx is considered to be a very short distance, small compared to the size of the container, but large compared to the size of a molecule.

(Note: To avoid excess subscripts, in the following discussion we will simply write v for the center-of-mass speed v_{cm}, and y for the center-of-mass height y_{cm}.)

Given these considerations, the formula

$$P(E) \propto e^{-\frac{[K_{\text{trans}} + E_{\text{vib}} + E_{\text{rot}} + Mgy]}{kT}} \, dx \, dy \, dz \, dv_x \, dv_y \, dv_z$$

is proportional to the probability that one particular molecule of interest will have the specified energy, with its position and velocity components within dx, dv_x, etc. We will use this formula to explore various properties of a gas.

Separating the various factors

It is useful to group the terms of the distribution formula like this, where the velocity and position distributions are treated in a classical (non-quantum) way:

$$P(E) \propto \left[e^{-\frac{K_{trans}}{kT}} dv_x dv_y dv_z \right]\left[e^{-\frac{Mgy}{kT}} dx dy dz \right]\left[e^{-\frac{E_{vib}}{kT}} \right]\left[e^{-\frac{E_{rot}}{kT}} \right]$$

The first bracket gives us the distribution of velocities, the second the distribution of positions, the third the distribution of vibrational energy, and the fourth the distribution of rotational energy. We will discuss each of these individually before putting it all together.

10.8.1 Height distribution in a gas

A striking property of Earth's atmosphere is that in high mountains the air density and pressure are significantly lower than at sea level. The air density is so low on top of Mount Everest that most climbers carry oxygen tanks. Can we explain this? In order to get at the main issue, the variation of density with height, we make the rough approximation that the temperature is constant—the same in the mountains as at sea level.

? How bad an approximation is it to consider the temperature to be the same at all altitudes?

Even in high mountains the temperature is typically above –29° C (–20° F), which is 244 K, and this is only 17% lower than room temperature of 293 K (20° C or 68° F). So maybe this approximation isn't too bad. Focus just on that part of the Boltzmann distribution that deals with position:

$$P(y) \propto e^{-\frac{Mgy}{kT}} dx dy dz$$

This is proportional to the probability of finding one particular molecule between x and $x+dx$, y and $y+dy$, and z and $z+dz$. Evidently there's nothing very interesting about x and z (directions parallel to the ground). But in the vertical direction there is an exponential fall-off with increasing height for the probability of finding a particular molecule at height y (Figure 10.36).

Looked at another way, our exponential formula tells us how the number density of the atmosphere depends on height, because in telling us about the behavior of one representative molecule, the formula also tells us something about all the molecules.

We shifted gears from thinking about one molecule to thinking about many. Think again about one single air molecule. Suppose you place it on a table, which is at room temperature. It is not a very unusual event for a thermally agitated atom in the table to hit our air molecule hard enough to send it kilometers high into the air! Actually, our particular molecule will very soon run into another air molecule and not make it very high in one great leap, but on average we do find lots of air molecules very high above sea level rather than finding them all lying on the ground.

Notice again that kT is the important factor in understanding the statistical behavior of matter. Here it sets the scale for the variation with height of the atmosphere's number density.

At a height where the gravitational potential energy $U_g = mgy$ is equal to kT, the number density has dropped by a factor of $e^{-1} = 1/e = 0.37$. Dry air at sea level is 78% nitrogen (one mole N_2 = 28 grams), 21% oxygen (one mole O_2 = 32 grams), about 1% argon, and 0.03% CO_2. An average mass of 29 grams per mole is good enough for most calculational purposes.

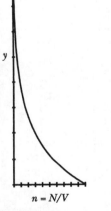

$n = N/V$

Figure 10.36 Number density vs. height in a constant-temperature atmosphere.

Ex. 10.12 Suppose you put one air molecule on your desk, so that it is in thermal equilibrium with the desk at room temperature. Suppose that there is no atmosphere to get in the way of this one molecule bouncing up and down on the desk. Calculate the typical height that the air molecule will be above your desk, so that $Mgy \approx kT$.

Ex. 10.13 Marbles of mass $M = 10$ grams are lying on the floor. They are of course in thermal equilibrium with their surroundings. What is a typical height above the floor for one of these marbles? That is, for what value of y is $Mgy \approx kT$?

Ex. 10.14 Approximately what fraction of the sea-level air density is found at the top of Mount Everest, a height of 8848 meters above sea level?

10.8.2 Distribution of velocities in a gas

Next we look at the distribution of molecular speeds. Since we're explicitly interested in v_{cm}, it will be useful to express translational kinetic energy as $\frac{1}{2}Mv_{cm}^2$ rather than $p^2/(2M)$ in this discussion. Since $v^2 = v_x^2 + v_y^2 + v_z^2$, we can rewrite the formula for the distribution of velocity in a gas (remember that we are simply writing v for the center-of-mass speed v_{cm}):

$$P(v_x, v_y, v_z) \propto \left[e^{-\frac{\frac{1}{2}Mv_x^2}{kT}} \, dv_x \right]\left[e^{-\frac{\frac{1}{2}Mv_y^2}{kT}} \, dv_y \right]\left[e^{-\frac{\frac{1}{2}Mv_z^2}{kT}} \, dv_z \right]$$

The distribution for each velocity component is a bell-shaped curve, called a "Gaussian." In Figure 10.37 we show the distribution of the x component of velocity for helium at room temperature.

? Judging from this graph, what is the average value for v_x? What are the average values of v_y and v_z? Why are these results plausible?

Evidently the average value for the x component of the velocity is zero. This is reassuring, because a molecule is as likely to be headed to the right as it is to be headed to the left. Similarly, the average values for the y and z components of velocity are zero.

It would not be particularly surprising to find a gas molecule with an x component of velocity such that $\frac{1}{2}Mv_x^2 \approx kT$. On the other hand, it would be very surprising to find a gas molecule with a value of v_x many times larger. As usual, kT sets the scale for thermal phenomena.

Distribution of speeds in a gas

By converting from rectangular coordinates to "spherical" coordinates, the velocity-component distribution can be converted into a speed distribution. We won't go into the details, but the main idea is that the volume element in rectangular coordinates in "velocity space" $dv_x dv_y dv_z$ turns into the volume of a spherical shell with surface area $4\pi v^2$, and thickness dv, as shown in Figure 10.38.

Transferring to spherical coordinates, we have the following:

$$e^{-\frac{\frac{1}{2}Mv^2}{kT}} \, dv_x dv_y dv_z = e^{-\frac{\frac{1}{2}Mv^2}{kT}} \, 4\pi v^2 dv$$

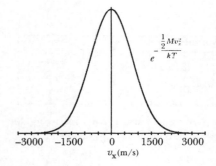

Figure 10.37 Distribution of the x component of velocity for helium at room temperature.

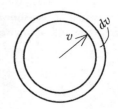

Figure 10.38 The volume of a shell with radius v and thickness dv is $4\pi v^2 dv$.

The only thing missing is a "normalization" factor in front to make the integral over all speeds from 0 to infinity be equal to 1.0 (since our one molecule must have a speed somewhere in that range). Here is the "Maxwell-Boltzmann" distribution for a low-density gas:

MAXWELL-BOLTZMANN SPEED DISTRIBUTION (LOW-DENSITY GAS)

$$P(v) = 4\pi\left(\frac{M}{2\pi kT}\right)^{\frac{3}{2}} e^{-\frac{\frac{1}{2}Mv^2}{kT}} v^2$$

The probability that a molecule of a gas have a center-of-mass speed within the range v to $v+dv$ is given by $P(v)dv$. The distribution function for helium is shown for two different temperatures in Figure 10.39. You can see that the average speed of a helium atom is predicted to be about 1200 m/s at room temperature (20° Celsius, which is 293 Kelvin). The prediction was first made by Maxwell, in the mid-1800's, long before the development of quantum mechanics. In retrospect, this worked because the quantum levels of the translational kinetic energy are so close together as to be almost a continuum, so the classical model is a good approximation.

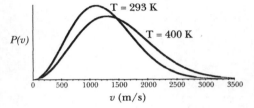

Figure 10.39 Distribution of speeds of helium atoms at two temperatures.

The average number of helium atoms in a container that have speeds in the range of 415.4 m/s to 415.7 m/s can be calculated by evaluating Maxwell's formula with $v = 415.4$ m/s and multiplying by $dv = 0.3$ m/s, which gives the probability of finding one molecule in this speed range. If there are N atoms in the container, multiplying by N gives the total number of molecules likely to be in that speed range at any given instant. The area of the vertical slice shown in Figure 10.40 represents the fraction of helium atoms that are likely to have speeds between 500 m/s and 600 m/s. At higher temperatures the distribution shifts to higher speeds (Figure 10.39).

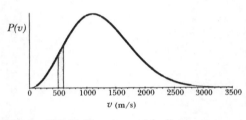

Figure 10.40 Fraction of helium atoms at 293 K with speeds between 500 and 600 m/s.

10.8.3 Measuring the distribution of speeds in a gas

The actual distribution of speeds of molecules in any gas can be measured by an ingenious experiment. Make a tiny hole in a container of gas and let the molecules leak out into a vacuum (that is, a region from which the air is continually pumped out). The moving molecules run through collimating holes and then through a slot in a rapidly rotating drum and strike a row of devices at the other side of the drum that can detect gas molecules (Figure 10.41). Such measurements confirm the Maxwell prediction.

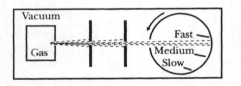

Figure 10.41 Apparatus for measuring the distribution of speeds of gas molecules.

Fast molecules strike soonest, followed by medium-speed molecules, and finally slow molecules, so molecules with different speeds strike at different locations along the rotating drum. The number of molecules striking various locations along the drum is a direct indication of the distribution of speeds of molecules in the gas. (A correction has to be applied to these data to obtain the speed distribution inside the gas, because high-speed molecules emerge through the hole more frequently than low-speed molecules do, even if their numbers are the same in the interior of the gas.)

How might we display the results of this experiment? We could present the data as a graph (Figure 10.42) of the fraction $\Delta N/N$ of the N helium atoms that have speeds between 0 and 500 m/s, between 500 and 1000 m/s, between 1000 and 1500 m/s, etc.

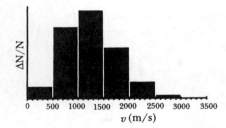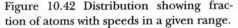

Figure 10.42 Distribution showing fraction of atoms with speeds in a given range.

10.8.4 Average translational kinetic energy in a gas

For a gas confined inside a container, quantum mechanics predicts that the kinetic energy $\frac{1}{2}Mv^2$ is quantized, but if the container is of ordinary macroscopic size, the energy quanta are extremely small compared to the average kinetic energy, even at very low temperatures. Most gases liquefy before the temperature drops so low as to invalidate the pre-quantum analysis for the velocity distribution.

Whenever the energy has a term containing a square of a position or momentum component, such as x^2 or p_x^2, pre-quantum theory predicts an associated average energy of $\frac{1}{2}kT$. This result is valid in quantum mechanics for high temperatures, where kT is large compared to the quantum energy spacing. This result follows from using an integral to take the average value of a quadratic term when the distribution is a Gaussian involving that quadratic term. That is, using integral tables (or integrating by parts), you can show that the average value of a quadratic term is $\frac{1}{2}kT$:

$$\overline{w^2} = \frac{\displaystyle\int_0^\infty w^2 e^{-\frac{w^2}{kT}} dw}{\displaystyle\int_0^\infty e^{-\frac{w^2}{kT}} dw} = \tfrac{1}{2}kT$$

The average value is obtained by weighting the values of w^2 by the probability of finding that value; the denominator takes care of normalizing the distribution properly.

As an example, when we did quantum-based calculations for the Einstein solid, we found that at high temperatures the heat capacity per atom was $3k$, which implies an average energy of $3kT$. We modeled an atomic oscillator as three one-dimensional oscillators, each having an average energy of $\frac{1}{2}kT$ corresponding to kinetic energy $p^2/(2M)$ and another $\frac{1}{2}kT$ for the potential energy $\frac{1}{2}k_s x^2$. This is a total of six quadratic terms, implying an average energy per atom of $6(\frac{1}{2}kT)$, or $3kT$. We therefore expect the heat capacity per atom in a solid at high temperature to be $3k$, which is indeed what is observed.

HIGH-TEMPERATURE AVERAGE ENERGY

If $kT \gg$ the energy quantum, the average energy associated with a quadratic energy term is $\frac{1}{2}kT$.

The number of quadratic terms in the expression for the energy is often called the "degrees of freedom."

? What is the average translational kinetic energy $\frac{1}{2}M\overline{v^2}$ in terms of kT? (A bar over a quantity means "average value.") What is the associated contribution to the heat capacity of the gas?

Since there are three quadratic terms in the translational kinetic energy, the average value of K_{trans} should be three times $\frac{1}{2}kT$. We have the following important result:

AVERAGE K_{trans} FOR AN IDEAL GAS

$$\overline{K}_{\text{trans}} = \tfrac{1}{2}M\overline{v^2} = \tfrac{3}{2}kT$$

The contribution to the heat capacity is $\frac{3}{2}k$.

Root-mean-square speed

The square root of $\overline{v^2}$ is called the "root-mean-square" or "rms" speed:

$$v_{\text{rms}} \equiv \sqrt{\overline{v^2}}$$

With this definition of v_{rms}, we write $\frac{1}{2}m\overline{v^2} = \frac{1}{2}mv_{\text{rms}}^2 = \frac{3}{2}kT$.

Ex. 10.15 Calculate v_{rms} for a helium atom in the room you're in (whose temperature is probably about 293 K).

Ex. 10.16 Calculate v_{rms} for a nitrogen molecule (N_2; molecular mass 28) in the room you're in (whose temperature is probably about 293 K). Air is about 80% nitrogen.

10.8.5 Average speed vs. rms speed

The way you find the average speed of molecules in a gas is to weight the speed by the number of molecules that have that speed, and divide by the total number of molecules:

$$\bar{v} = \frac{N_1 v_1 + N_2 v_2 + \text{etc.}}{N} = \frac{N_1}{N} v_1 + \frac{N_2}{N} v_2 + \text{etc.}$$

For a continuous distribution of speeds this turns into an integral, where the weighting factors are given by the Maxwell speed distribution, which gives the probability that a molecule have a center-of-mass speed in the range v to $v+dv$. Using integral tables one finds this:

$$\bar{v} = \int_0^\infty 4\pi \left(\frac{m}{2\pi kT}\right)^{\frac{3}{2}} e^{-\frac{\frac{1}{2}mv^2}{kT}} (v) v^2 \, dv = \sqrt{\frac{8}{3\pi}} \sqrt{\frac{3kT}{m}} = 0.92 \, v_{rms}$$

This shows that the average speed is smaller than the rms speed ($0.92 v_{rms}$). This same calculational scheme can be used to determine any average. For example, if you calculate the average value of v^2 by this method, you find $\overline{v^2} = 3kT/m$, as expected.

On page 368 we commented that the average speed $\bar{v}$ of helium atoms at room temperature is about 1200 m/s. In Ex. 10.15 on page 370 you found that v_{rms} at room temperature is 1360 m/s. This is an example of the fact that the rms speed is higher than the average speed.

Ex. 10.17 The rms speed is somewhat higher than the average speed due to the averaging of squared speeds. Calculate the average of the numbers 1, 2, 3, and 4, then calculate the rms average (the square root of the average of their squares), and show that the rms average is larger than the simple average.

10.8.6 Application: Retaining a gas in the atmosphere

Since $v_{rms} = \sqrt{3k(T/m)}$, helium atoms typically travel much faster than nitrogen molecules in our atmosphere, due to the small mass of the helium atoms. Some few helium atoms will be going much faster than the average and may attain a high enough speed to escape from the Earth entirely (escape velocity from the Earth is about 1.1×10^4 m/s). This leads to a continuous leakage of high-speed helium atoms and explains why there is essentially no helium in our atmosphere, and no hydrogen either (Figure 10.43).

? Why doesn't the Moon have any atmosphere at all?

The Moon's gravitational field is so weak that escape speed is quite small, and all common gases can escape, not just hydrogen and helium.

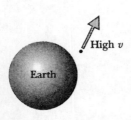

Figure 10.43 Low mass atoms or molecules in our atmosphere typically have speeds high enough to escape from the Earth entirely.

Ex. 10.18 Calculate the escape speed from the Moon and compare with typical speeds of gas molecules. The mass of the Moon is 7×10^{22} kg, and its radius is 1.75×10^{6} m.

10.8.7 Application: Speed of sound

Sound waves in a gas consist of propagation of variations in density, and the fundamental mechanism for this kind of wave propagation involves collisions between neighboring molecules, whose speeds are proportional to v_{rms} (and roughly comparable to the speed of sound). For example, compare the v_{rms} for nitrogen, 496 m/s, which you calculated in Ex. 10.16 (page 370) with the speed of sound in air (which is mostly nitrogen) at 293 K, which is measured to be 344 m/s. (You may know the approximate rule that a 1-second delay between lightning and thunder indicates a distance of about 1000 feet, which is about 300 m.)

High gas density   Low gas density v

Figure 10.44 Periodic variations in gas density make a sound wave, which travels through a gas with speed v (the speed of sound).

Ex. 10.19 Should the speed of sound in air increase or decrease with increasing temperature? What percentage change would result from doubling the absolute temperature? (This effect is readily observed by measuring the speed of sound in a gas as a function of absolute temperature. The excellent agreement between theory and experiment provides additional evidence for our understanding of gases.)

10.8.8 Vibrational energy in a diatomic gas molecule

We repeat that the distribution of velocity, position, vibrational energy, and rotational energy is this:

$$P(E) \propto \left[e^{-\frac{K_{trans}}{kT}} dv_x dv_y dv_z \right]\left[e^{-\frac{Mgy}{kT}} dxdydz \right]\left[e^{-\frac{E_{vib}}{kT}} \right]\left[e^{-\frac{E_{rot}}{kT}} \right]$$

We have treated the distribution of velocity and position. Next we discuss the distribution of vibrational energy:

$$P(E_{vib}) \propto e^{-\frac{E_{vib}}{kT}}$$

For a monatomic gas such as helium or argon, there is no vibrational energy term. But for a diatomic molecule such as N_2 or O_2 or HCl, the vibrational energy is

$$E_{vib} = \frac{p_1^2}{2 m_1} + \frac{p_2^2}{2 m_2} + \frac{1}{2} k_s s^2$$

where k_s is the effective spring stiffness corresponding to the interatomic electric force (not to be confused with k, the Boltzmann constant), and s is

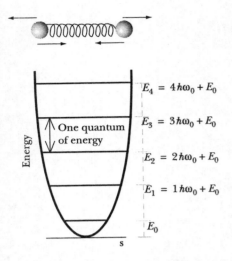

Figure 10.45 The quantized vibrational energy levels of a diatomic oscillator.

the stretch of the interatomic bond. This formula for the vibrational energy is essentially the same as the formula for a one-dimensional spring-mass oscillator, which we have studied in detail (Figure 10.45).

Actually, since $p_1 = p_2$ for momenta p_1 and p_2 relative to the center of mass, we can write

$$\frac{p_2^2}{2m_2} = \frac{p_1^2}{2m_2} = \left(\frac{m_1}{m_2}\right)\left(\frac{p_1^2}{2m_1}\right)$$

$$\frac{p_1^2}{2m_1} + \left(\frac{m_1}{m_2}\right)\frac{p_1^2}{2m_1} + \frac{1}{2}k_s s^2 = \left(1 + \frac{m_1}{m_2}\right)\frac{p_1^2}{2m_1} + \frac{1}{2}k_s s^2$$

Note that this equation has a form exactly like that for a spring-mass oscillator, with a different mass.

Earlier we modeled a solid as a large number of isolated three-dimensional atomic oscillators (each corresponding to three one-dimensional oscillators, because there are spring-like interatomic forces on an atom from neighboring atoms in all directions). This is an overly simplified model of a solid, because in a solid the atoms interact with each other. For example, if an atom moves to the left this affects the atoms to the right and to the left.

In a low-density gas however the vibrational oscillators really are independent of each other, because the gas molecules aren't even in contact with each other except when they happen to collide. So the analysis we carried out for the Einstein model of a solid applies even better to the vibrational portion of the energy in a real gas than it does to a real solid.

Among the results that apply immediately are that the heat capacity associated with the vibrational energy of one oscillator is k at high temperatures. The heat capacity decreases at very low temperatures where kT is small compared to the energy spacing of the quantized oscillator energies. Just as it was a surprise when the heat capacity of metals was found to decrease at low temperatures, there was a similar surprise in the measurements of the heat capacity of diatomic gases at low temperatures, because the contribution of the vibrational motion vanished.

? There are two quadratic terms in the vibrational energy (kinetic and spring). Therefore, at high temperatures what is the average vibrational energy in terms of kT? What is the associated contribution to the heat capacity of the gas?

At high temperatures the vibrational motion of a diatomic molecule (a one-dimensional oscillator) is expected to have an average energy that is approximately $2(\frac{1}{2}kT) = kT$, and contributes k to the heat capacity.

10.8.9 Rotational energy in a diatomic gas molecule

Finally we consider the rotational-energy portion of the distribution formula:

$$P(E_{rot}) \propto e^{-\frac{E_{rot}}{kT}}$$

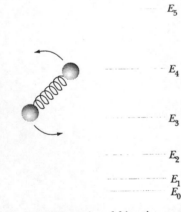

Figure 10.46 Rotational kinetic energy levels of a diatomic molecule.

For a monatomic gas there is no rotational term. But for a diatomic gas there is rotational kinetic energy associated with rotation of the "dumbbell" consisting of the two nuclei. The rotational kinetic energy is this (Figure 10.46):

$$\frac{1}{2}I\omega_x^2 + \frac{1}{2}I\omega_y^2 = \frac{L_{rot,x}^2}{2I} + \frac{L_{rot,y}^2}{2I}$$

This corresponds to rotational angular momenta $L_{rot,x}$ and $L_{rot,y}$ about the x axis and about the y axis. Note that only the nuclei contribute significantly to the rotational kinetic energy, because the electrons have much less mass.

Rotation about the z axis connecting the two nuclei is irrelevant, for a somewhat subtle reason. The angular momentum is quantized, which leads to energy quantization. For rotations around the z axis the energy is $L_{rot,z}^2/(2I)$, where $L_{rot,z}$ is the z component of the rotational angular momentum. Since I for the tiny nuclei is extremely small compared with the moment of inertia about the x and y axes of the diatomic molecule, the rotational energy of the associated first excited state is enormous, and this state is not excited at ordinary temperatures (Figure 10.47).

Since there are two quadratic energy terms associated with rotation, we conclude that at high temperatures the rotational motion of a diatomic molecule has an average energy that is approximately $2(\frac{1}{2}kT) = kT$, and contributes k to the heat capacity.

The spacing between quantized rotational energies for a diatomic molecule is even smaller than the vibrational energy quantum. As a result, the gas must be cooled to a very low temperature before the pre-quantum results become invalid. Many gases liquefy before this low-temperature regime is reached.

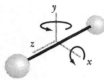

Figure 10.47 A diatomic molecule can rotate around the x or the y axis. We say there are two rotational degrees of freedom.

10.8.10 Heat capacity of a gas

We are now in a position to discuss the heat capacity of a gas as a function of temperature. This property is important in calculating the thermal interactions of a gas. Historically, measurements of the heat capacity of gases were important in testing theories of statistical mechanics. We'll concentrate on calculating c_v, the heat capacity at constant volume (meaning no mechanical work is done on the gas). The associated experiment would be to add a known amount of energy to a gas in a rigid container and measure the temperature rise of the gas.

Consider the average energy of a diatomic molecule such as N_2 or HCl, consisting of the translational kinetic energy associated with the motion of the center of mass, plus the vibrational energy, plus the rotational energy, plus the gravitational energy of the molecule plus the Earth:

$$\left\{\frac{\overline{p_x^2}}{2m} + \frac{\overline{p_y^2}}{2m} + \frac{\overline{p_z^2}}{2m}\right\} + \left\{\left(1 + \frac{m_1}{m_2}\right)\frac{\overline{p_1^2}}{2m_1} + \frac{1}{2}k_s\overline{s^2}\right\} + \left\{\frac{\overline{L_{rot,x}^2}}{2I} + \frac{\overline{L_{rot,y}^2}}{2I}\right\} + Mg\overline{y}_{cm}$$

$$\underbrace{\qquad\qquad}_{\text{Translation}} \qquad \underbrace{\qquad\qquad}_{\text{Vibration}} \qquad \underbrace{\qquad}_{\text{Rotation}} \underbrace{\quad}_{\text{Gravitation}}$$

? At high temperature, what should the heat capacity at constant volume be?

There are seven quadratic terms in the expression for the energy, so if the temperature is very high, we expect an average energy of $7(\frac{1}{2}kT)$ and a heat capacity at constant volume $c_v = \frac{7}{2}k$.

The discrete energy levels for a diatomic molecule include electronic, vibrational, and rotational energy levels (Figure 10.48). The electronic energy levels correspond to particular configurations of the electron clouds, and the spacing between these levels is typically 1 eV or more. Since at room temperature kT is about $\frac{1}{40}$ eV, the electronic levels do not contribute to the heat capacity at ordinary temperatures. The vibrational energy spacing is much smaller, and the rotational energy spacing is smaller still. A diatomic gas at high temperature has a band-like spectrum, since transitions between electronic levels may be accompanied by one or more vibrational or rotational quanta.

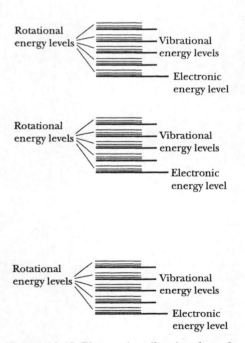

Figure 10.48 Electronic, vibrational, and rotational energy levels for a diatomic molecule.

What about *Mgy*?

What about the Mgy_{cm} term in the energy? Remember that we are trying to predict the heat capacity of a gas at constant volume (so that no mechanical work is done on the system). We could measure c_v by having a known amount of thermal energy transfer Q into a closed container of gas and observing the temperature rise of the gas. During this process the height y_{cm} of the center of mass of the gas in the container hardly changes, so the Mgy term doesn't contribute to the heat capacity.

Actually there is an extremely small increase in y_{cm} of the gas in the container due to the temperature dependence of the height distribution (page 366), but the associated contribution to the heat capacity is negligible.

Heat capacity *vs.* temperature

Figure 10.49 is a graph of the heat capacity (at constant volume) of a diatomic gas as a function of temperature. In the following exercises, see whether you can explain these values of the heat capacity in terms of our analyses of the various contributions to the energy of a diatomic gas molecule.

Remember that kT is about $\frac{1}{40}$ eV at room temperature.

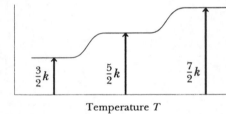

Temperature T

Figure 10.49 Heat capacity c_v of a diatomic gas.

Ex. 10.20 What is the heat capacity per molecule (at constant volume) of a monatomic gas such as helium or neon? Why doesn't the heat capacity depend on temperature?

Ex. 10.21 At high temperatures, what is the heat capacity (at constant volume) per molecule of a diatomic gas such as oxygen or nitrogen?

Ex. 10.22 Suppose we lower the temperature of a diatomic gas to a point where kT is small compared to the energy of the first excited vibrational state, but still large compared to the energy of the first excited rotational state. Now what is the heat capacity (at constant volume)?

Ex. 10.23 Lower the temperature even more, so that kT is small compared to the energy of the first excited rotational state (some gases liquefy before this low temperature is reached). What is the heat capacity (at constant volume) at this low temperature?

Ex. 10.24 As the temperature is decreased, the heat capacity (at constant volume) for H_2 eventually decreases to $\frac{3}{2}k$, before the gas liquefies. Does this transition to $c_v = \frac{3}{2}k$ happen at a higher or lower temperature for D_2 (deuterium, whose nuclei each contain a proton plus a neutron)? Why? (Hint: Remember that it is the rotational angular momentum that is quantized.)

10.9 Summary

Fundamental physical principles

Fundamental assumption of statistical mechanics: Each arrangement of the same total energy within a system is equally probable.

Two blocks in thermal contact evolve to that division of energy that has associated with it the largest number of ways $\Omega = \Omega_1\Omega_2$ of arranging the total energy among the atoms (largest number of microstates for a macrostate of given energy).

Second law of thermodynamics: The entropy of a closed system never decreases. Only in a reversible process does the entropy of a closed system stay constant.

New concepts

Entropy: $S \equiv k\ln\Omega$, where $k = 1.4\times10^{-23}$ J/K

Temperature: $\dfrac{1}{T} = \dfrac{\partial S}{\partial E}$

Adding small Q to a system raises entropy by $\Delta S = \dfrac{Q}{T}$

Heat capacity per atom: $C = \dfrac{\Delta E_{\text{atom}}}{\Delta T}$, where $\Delta E_{\text{atom}} = \dfrac{\Delta E_{\text{system}}}{N_{\text{atoms}}}$

Results

WAYS TO ARRANGE *q* QUANTA AMONG *N* ONE-DIMENSIONAL OSCILLATORS

$$\Omega = \frac{(q+N-1)!}{q!(N-1)!}$$

The Boltzmann distribution for the energy of a microscopic system in thermal contact with a macroscopic system:

$$e^{-\frac{\Delta E}{kT}}$$

For each energy term involving a quadratic term such as x^2 or p_x^2 ("degree of freedom"), if the average energy is large compared to kT the average energy is $\frac{1}{2}kT$ and the contribution to the heat capacity is $\frac{1}{2}k$. This contribution decreases at low temperatures. At high temperatures, heat capacity per atom in a solid $\approx 3k$.

Distribution of energy for a diatomic gas:

$$P(E) \propto \left[e^{-\frac{K_{\text{trans}}}{kT}}\,dv_x dv_y dv_z\right]\left[e^{-\frac{Mgy}{kT}}\,dxdydz\right]\left[e^{-\frac{E_{\text{vib}}}{kT}}\right]\left[e^{-\frac{E_{\text{rot}}}{kT}}\right]$$

Speed distribution for a gas molecule:

$$P(v) = 4\pi\left(\frac{M}{2\pi kT}\right)^{\frac{3}{2}}e^{-\frac{\frac{1}{2}Mv^2}{kT}}v^2$$

10.10 Example problem: A lead nanoparticle

In Chapter 3, from Young's modulus for lead we found that the effective interatomic "spring" stiffness was about 5 N/m.

(a) For a nanoparticle consisting of 3 lead atoms, what is the approximate temperature when there are 5 quanta of energy in the nanoparticle?

(b) What is the approximate specific heat per lead atom at this temperature? Compare with the approximate high-temperature specific heat per atom for lead.

Solution

We'll use the Einstein model of a solid, though with only three atoms our conclusions will be very approximate.

(a) We model the three atoms as 9 independent quantized oscillators. One quantum of energy in one of these oscillators is this many joules:

$$\hbar\omega_0 = \hbar\sqrt{k_s/m} = (1.05\times10^{-34}\text{ J·s})\sqrt{\frac{4(5\text{ N/m})}{(207\times1.7\times10^{-27}\text{ kg})}} = 7.92\times10^{-22}\text{ J}$$

We need to calculate the number of ways to arrange q quanta in the neighborhood of $q = 5$, and the associated entropy (Figure 10.50).

Since $1/T = \partial S/\partial E$, $T \approx \Delta E/\Delta S$. Take differences from $q = 4$ to $q = 6$, in order to approximate the slope at $q = 5$; energy increase is 2 quanta:

$$T \approx \frac{N\hbar\omega_0}{k\Delta(\ln\Omega)} = \frac{2(7.92\times10^{-22}\text{ J})}{(1.4\times10^{-23}\text{ J/K})(8.01-6.20)} \approx 62.5\text{ K}$$

(b) Find T_1 at midpoint of 4 to 5 interval, T_2 at midpoint of 5 to 6 interval, energy increase from T_1 to T_2 is one quantum:

$$T_1 \approx \frac{(7.92\times10^{-22}\text{ J})}{(1.4\times10^{-23}\text{ J/K})(7.16-6.20)} \approx 58.9\text{ K}$$

$$T_2 \approx \frac{(7.92\times10^{-22}\text{ J})}{(1.4\times10^{-23}\text{ J/K})(8.01-7.16)} \approx 66.6\text{ K}$$

As expected, these two temperatures bracket the temperature of 62.5 K at $q = 5$. We want the heat capacity on a per-atom basis, and there are 3 atoms:

$$C_{\text{per atom}} = \frac{1}{3}\frac{\Delta E}{\Delta T} \approx \frac{1}{3}\frac{(7.92\times10^{-22}\text{ J})}{(66.6-58.9)\text{ K}} \approx 3.4\times10^{-23}\text{ J/K}$$

The approximate high-temperature heat capacity for a solid is $3k$ per atom, which is $3(1.4\times10^{-23}\text{ J/K}) = 4.2\times10^{-23}\text{ J/K}$. This suggests that at a temperature of about 62.5 K, lead is not quite at the high-temperature limit.

The actual experimental value for lead at 60 K (page 360) is 22.43 J/K/mole, which on a per-atom basis is

$$(22.43\text{ J/K/mole})\left(\frac{1\text{ mole}}{6\times10^{23}\text{ atoms}}\right) = 3.7\times10^{-23}\text{ J/K}$$

Our 3-atom calculation is not a very accurate model of a macroscopic block of lead, in part because with only three atoms it is not a very good model to say that each atom is connected by spring-like bonds to six neighboring atoms. Despite the failings of our simple model, our result is rather close to the actual value.

q	Ω	$\ln\Omega$
4	$\frac{12!}{4!8!} = 495$	6.20
5	$\frac{13!}{5!8!} = 1287$	7.16
6	$\frac{14!}{6!8!} = 3003$	8.01

Figure 10.50 Calculation of entropy $(S = k\ln\Omega)$.

10.11 Review questions

Arranging things

RQ 10.1 List explicitly all the ways of arranging 2 quanta among 4 one-dimensional oscillators.

A formula for the number of arrangements

RQ 10.2 How many different ways are there to get 5 heads in 10 throws of a true coin? How many different ways are there to get no heads in 10 throws of a true coin?

RQ 10.3 How many different ways are there to arrange 4 quanta among 3 atoms in a solid?

Thermal equilibrium

RQ 10.4 Energy conservation for two blocks in contact with each other is satisfied if all the energy is in one block and none in the other. Would you expect to observe this distribution in practice? Why or why not?

RQ 10.5 What is the advantage of plotting the (natural) logarithm of the number of ways of arranging the energy among the many atoms (natural logarithm of the number of microstates)?

RQ 10.6 Which has a higher temperature, a system whose entropy changes rapidly with increasing energy, or one whose entropy changes little with increasing energy?

Heat capacity

RQ 10.7 Explain why it is a disadvantage for some purposes that the heat capacity of all materials decreases at low temperatures.

RQ 10.8 Sketch and label graphs of heat capacity *vs.* temperature for hydrogen gas (H_2) and oxygen gas (O_2), using the same temperature scale. Explain briefly.

Second law of thermodynamics

RQ 10.9 Two blocks with different temperatures had entropies of 10 J/K and 35 J/K before they were brought in contact. What can you say about the entropy of the combined system after the two came in contact with each other?

The Boltzmann distribution

RQ 10.10 Explain qualitatively the basis for the Boltzmann distribution. Never mind the details of the math for the moment. Focus on the trade-offs involved with giving energy to a single oscillator *vs.* giving that energy to a large object.

RQ 10.11 Many chemical reactions proceed at rates that depend on the temperature. Discuss this from the point of view of the Boltzmann distribution.

Heat capacity

RQ 10.12 Which has more internal energy at room temperature—a mole of helium or a mole of air?

Distributions

RQ 10.13 Figure 10.51 shows the distribution of speeds of atoms in a particular gas at a particular temperature. Approximately what is the average speed? Is the rms (root-mean-square) speed bigger or smaller than this? Approximately what fraction of the molecules have speeds greater than 1000 m/s?

Figure 10.51 Distribution of speeds of atoms in a gas at a particular temperature (RQ 10.13).

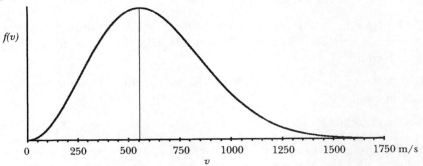

10.12 Homework problems

Problem 10.1 (page 351) Probability distribution

(a) Model a system consisting of two atoms (three oscillators each), among which 4 quanta of energy are to be distributed. Write a program to display a histogram showing the total number of possible microstates of the two-atom system *vs.* the number of quanta assigned to atom 1. Compare your calculations and histogram to the one on page 347 (you should get the same distribution).

(b) Model a system consisting of two solid blocks, block 1 containing 300 oscillators and block 2 containing 200 oscillators. Find the possible distributions of 100 quanta among these blocks, and plot number of microstates *vs.* number of quanta assigned to block 1. Compare your calculations and histogram to those on page 350. Determine the distribution of quanta for which the probability is half as large as the most probable 60-40 distribution.

(c) Do a series of calculations distributing 100 quanta between two blocks whose total number of oscillators is 500, but whose relative number of atoms varies. For example, consider equal numbers of oscillators, and ratios of 2:1, 5:1, etc. Describe your observations.

Problem 10.2 (page 353) Natural logarithm (ln) of ways to arrange energy

Start with your solution to Problem 10.1 (b). For the same system of two blocks, with $N_1 = 300$ oscillators and $N_2 = 200$ oscillators, plot $\ln(\Omega_1)$, $\ln(\Omega_2)$, and $\ln(\Omega_1\Omega_2)$, for q_1 running from 0 to 100 quanta. Your graph should look like the one in Figure 10.23. Determine the maximum value of $\ln(\Omega_1\Omega_2)$ and the value of q_1 where this maximum occurs. What is the significance of this value of q_1?

Problem 10.3 (page 358) Temperature

Modify your calculations from Problem 10.2 (page 353), to plot the temperature in kelvins of block 1 as a function of the number of quanta q_1 present in the first block. On the same graph plot the temperature in kelvins of block 2 as a function of q_1 (of course, $q_2 = q_{tot} - q_1$).

In order to plot the temperature in kelvins, you must determine the values of ΔE and ΔS that correspond to a one-quantum change in energy. Consider the model we are using. The energy of one quantum, in joules, is $\Delta E = \hbar\sqrt{k_s/m}$. The increment in entropy corresponding to this increment in energy is $\Delta S = k\Delta(\ln\Omega)$. Assume that the blocks are made of aluminum. In Problem 3.2 (page 82) you made an estimate of the interatomic spring constant k_s for aluminum. We show later, on page 360, that the effective k_s for oscillations in the Einstein solid is expected to be about 4 times the value obtained from measuring Young's modulus.

What is the significance of the value of q_1 (and of q_2) where the temperature curves for the two blocks cross (the temperatures are equal)?

Problem 10.4 (page 360) Heat capacity

Modify your analysis of Problem 10.3 to determine the heat capacity as a function of temperature for a single block of metal. In order to see all of the important effects, consider a single block of 35 atoms (105 oscillators) with up to 300 quanta of energy. Note that in this analysis you are calculating quantities for a *single* block, not two blocks in contact.

To make specific comparison with experimental data, consider the cases of aluminum (Al) and lead (Pb). For each metal, plot the theoretical heat capacity C per atom *vs.* T (K), with dots showing the actual experimental data (converted to the same basis, J/K per atom).

Adjust the interatomic spring stiffness k_s until your calculations approximately fit the experimental data given in the adjoining table. (You will need to convert to a per-atom value.) What value of k_s gives a good fit? Compare with the estimated values of k_s obtained in Problem 3.2 (page 82), but remember that those estimated values are probably 4 times smaller than the effective spring constant for oscillations.

10.12.1 Additional problems

Problem 10.5 Square root of N

For some examples of your choice, demonstrate by carrying out actual computer calculations that the "square-root" rule holds true for the fractional width of the peak representing the most probable arrangements of the energy. Warning: Check to see what is the largest number you can use in your computations; some programs or programming environments won't handle numbers bigger than 10^{307}, for example, and larger numbers are treated as "infinite."

Problem 10.6 Copper and aluminum blocks

A 50-gram block of copper (one mole has a mass of 63.5 grams) at a temperature of 35° C is put in contact with a 100-gram block of aluminum (molar mass 27 grams) at a temperature of 20° C. The blocks are inside an insulated enclosure, with little contact with the walls. At these temperatures, the high-temperature limit is valid for the heat capacity. Calculate the final temperature of the two blocks.

Problem 10.7 Low-temperature heat capacity of copper

Young's modulus for copper is measured by stretching a copper wire to be about 1.2×10^{11} N/m^2. The density of copper is about 9 grams/cm^3, and the mass of a mole is 63.5 grams. Starting from a very low temperature, use these data to estimate roughly the temperature T at which we expect the heat capacity for copper to approach $3k$. Compare your estimate with the data shown on page 360.

Problem 10.8 Experiment: Measurement of the heat capacity of water

The goal of this experiment is to understand, in a concrete way, what heat capacity is and how it can be measured. You will need a microwave oven, a styrofoam coffee cup, and a clock or watch.

In the range of temperature where water is a liquid (0 C to 100 C), it is approximately true that it takes 4.2 J of energy (1 calorie) to raise the temperature of 1 gram of water through 1 degree Kelvin. To measure this heat capacity of water, we need some way to raise a known mass of water from a

T (K)	C, Al (J/K/mole)	C, Pb (J/K/mole)
20	0.23	11.01
40	2.09	19.57
60	5.77	22.43
80	9.65	23.69
100	13.04	24.43
150	18.52	25.27
200	21.58	25.87
250	23.25	26.36
300	24.32	26.82
400	25.61	27.45

Problem 10.4　Heat capacity data.

known initial temperature to a final temperature which can also be measured, while we keep track of the energy supplied to the water. One way to do this, as discussed in the text, is to warm up water in a well-insulated container within which a heater, whose power output is known, warms up the water. In this experiment, instead of a well-insulated box with a heater, we will use microwave power which preferentially warms up water by exciting rotational modes of the water molecules, as opposed to burners or heaters which warm up water in a pan by first warming up the pan.

Use a styrofoam coffee cup of known volume in which water can be warmed up. The density of water is 1 gram/cm^3.

> It is a good idea not to fill the cup completely full, because this makes it more likely to spill.

One method of recording the initial temperature of the water is to get water from the faucet and wait for it to equilibrate with room temperature (which can either be read off a thermostat or estimated based on past experience). After waiting about a half hour for this to happen, place the cup in the microwave oven and turn on the oven at maximum power. The cup needs to be watched as it warms up, so that when the water starts to boil, the elapsed time can be noted accurately.

> BE CAREFUL! A styrofoam cup full of hot liquid can buckle if you hold it near the rim. Hold the cup near the bottom. If the cup is full, do not attempt to move the cup while the water is hot. A spill can cause a painful burn.

On the back of the microwave oven (or inside the front door), there is usually a sticker with specifications which says "Output Power = ...Watts" which can be used to calculate the energy supplied. If there is no indication, use a typical value of 600 watts for a standard microwave oven. Using all the quantities measured above and knowing the temperature interval over which you have warmed up the water, you can calculate the heat capacity of water.

(a) Show and explain all your data and calculations, and compare with the accepted value for water (4.2 J/K per gram).

(b) Discuss why your result might be expected to differ from the accepted value. For each effect that you consider, state whether this effect would lead to a result that is larger or smaller than the accepted value.

Problem 10.9 Atmosphere at very high altitudes

In studying a voyage to the Moon in Chapter 2 we somewhat arbitrarily started at a height of 50 kilometers above the surface of the Earth.

(a) At this altitude, what is the density of the air as a fraction of the density at sea-level?

(b) Approximately how many air molecules are there in one cubic centimeter at this altitude?

(c) At what altitude is the air density one-millionth (10^{-6}) that at sea level?

Problem 10.10 Two nanoparticles of iron

A nanoparticle consisting of four iron atoms (object 1) initially has 1 quantum of energy. It is brought into contact with a nanoparticle consisting of two iron atoms (object 2) which initially has 2 quanta of energy. The mass of one mole of iron is 56 grams.

(a) Using the Einstein model of a solid, calculate and plot $\ln \Omega_1$ *vs.* q_1 (the number of quanta in object 1), $\ln \Omega_2$ *vs.* q_1, and $\ln \Omega_{total}$ *vs.* q_1 (put all three plots on the same graph). Show your work and explain briefly.

(b) Calculate the approximate temperature of the objects at equilibrium. State what assumptions or approximations you made.

Problem 10.11 Creating a plasma

At sufficiently high temperatures, the thermal speeds of gas molecules may be high enough that collisions may ionize a molecule (that is, remove an outer electron). An ionized gas in which each molecule has lost an electron is called a "plasma." Determine approximately the temperature at which air becomes a plasma.

Problem 10.12 Heat capacity

100 joules of thermal energy transfer are given to air in a 50-liter rigid container and to helium in a 50-liter rigid container, both initially at STP (standard temperature and pressure).

(a) Which gas experiences a greater temperature rise?

(b) What is the temperature rise of the helium gas?

Problem 10.13 Heat capacity of hydrogen

(a) Below about 80 K the heat capacity at constant volume for hydrogen gas (H_2) is $\frac{3}{2}k$ per molecule, but at higher temperatures the heat capacity increases to $\frac{5}{2}k$ per molecule due to contributions from rotational energy states. Use these observations to estimate the distance between the hydrogen nuclei in an H_2 molecule.

(b) At about 2000 K the heat capacity at constant volume for hydrogen gas (H_2) increases to $\frac{7}{2}k$ per molecule due to contributions from vibrational energy states. Use these observations to estimate the stiffness of the "spring" that approximately represents the interatomic force.

Problem 10.14 Five microscopic objects

Figure 10.52 shows a one-dimensional row of 5 microscopic objects each of mass

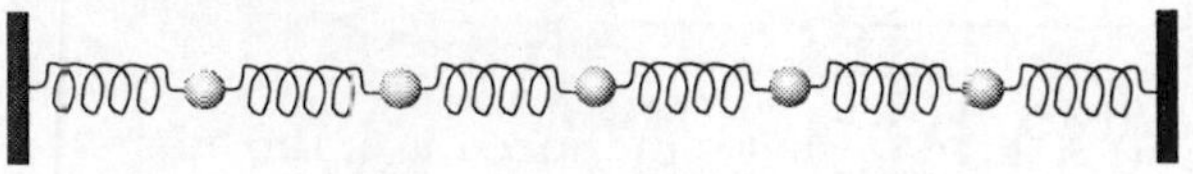

Figure 10.52 A one-dimensional row of microscopic masses and springs (Problem 10.14).

4×10^{-26} kg, connected by forces that can be modeled by springs of stiffness 15 N/m. These objects can move only along the x axis.

(a) Using the Einstein model, calculate the approximate entropy of this system for total energy of 0, 1, 2, 3, 4, and 5 quanta.

(b) Calculate the approximate temperature of the system when the total energy is 4 quanta.

(c) Calculate the approximate heat capacity per object when the total energy is 4 quanta.

(d) If the temperature is raised very high, what is the approximate heat capacity per object? Give a numerical value and compare with your result in part (c).

Problem 10.15 Pluto's atmosphere

In 1988 telescopes viewed Pluto as it crossed in front of a distant star. As the star emerged from behind the planet, light from the star was slightly dimmed as it went through Pluto's atmosphere. The observations indicated that the atmospheric density at a height of 50 km above the surface of Pluto is about one-third the density at the surface. The mass of Pluto is known to be about 1.5×10^{22} kg, and its radius is about 1200 km. Spectroscopic data indicate that the atmosphere is mostly nitrogen (N_2). Estimate the temperature of Pluto's atmosphere. State what approximations and/or simplifying assumptions you made.

Problem 10.16 Buckminsterfullerene

Buckminsterfullerene, C_{60}, is a large molecule consisting of sixty carbon atoms connected to form a hollow sphere. The diameter of a C_{60} molecule is about 7×10^{-10} m. It has been hypothesized that C_{60} molecules might be found in clouds of interstellar dust, which often contain interesting chemical compounds. The temperature of an interstellar dust cloud may be very low, around 3 K. Suppose you are planning to try to detect the presence of C_{60} in such a cold dust cloud by detecting photons emitted when molecules undergo transitions from one rotational energy state to another.

Approximately, what is the highest-numbered rotational level from which you would expect to observe emissions? The rotational levels are $l = 0, 1, 2, 3, \ldots$

Problem 10.17 A disk and brake

A box contains a uniform disk of mass M and radius R which is pivoted on a low-friction axle through its center. A block of mass m is pressed against the disk by a spring, so that the block acts like a brake, making the disk hard to turn. The box and the spring have negligible mass. A string is wrapped around the disk (out of the way of the brake) and passes through a hole in the box. A force of constant magnitude F acts on the end of the string. The motion takes place in outer space. At time t_i the speed of the box is v_i, and the rotational speed of the disk is ω_i. At time t_f the box has moved a distance x, and the end of the string has moved a longer distance d, as shown in Figure 10.53.

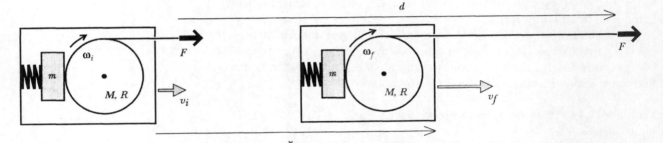

Figure 10.53 The box in Problem 10.17 shown both at its initial position, and after it has moved a distance x.

(a) At time t_f, what is the speed v_f of the box?

(b) During this process, the brake exerts a tangential friction force of magnitude f. At time t_f, what is the angular speed ω_f of the disk?

(c) At time t_f, assume you know (from part b) the rotational speed ω_f of the disk. From time t_i to time t_f, what is the increase in thermal energy of the apparatus?

(d) Suppose the numerical result for the increase in thermal energy in part c is 8×10^4 J. The disk and brake are made of iron, and their total mass is 1.2 kg. At time t_i their temperature was 350 K. At time t_f, what is their approximate temperature?

Problem 10.18 Coin tosses

The reasoning developed for counting microstates applies to many other situations involving probability. For example, if you flip a coin 5 times, how many different sequences of 3 heads and 2 tails are possible? Answer: 10 different sequences, such as HTHHT or TTHHH. In contrast, how many different sequences of 5 heads and 0 tails are possible? Obviously only one, HHHHH, and our formula gives $5!/[5!0!] = 1$, using the standard definition that 0! is defined to equal 1.

If the coin is equally likely on a single throw to come up heads or tails, any specific sequence like HTHHT or HHHHH is equally likely. But there is only one way to get HHHHH, while there are 10 ways to get 3 heads and 2 tails, so this is 10 times more probable than getting all heads.

Use the formula $5!/[N!(5-N)!]$ to calculate the number of ways to get 0 heads, 1 head, 2 heads, 3 heads, 4 heads, or 5 heads in a sequence of 5 coin tosses. Make a graph of the number of ways *vs.* the number of heads.

10.13 Answers to exercises

10.1 (page 347)   20

10.3 (page 349) 5.4×10^{88} years; 5.4×10^{78} times the lifetime of the Universe!

10.4 (page 355)   Looks *very* improbable; second law of thermodynamics being violated.

10.5 (page 358)   500 K

10.6 (page 358)   Liquid; yes, liquid more disordered; 1.23 J/K (note that 0° C is 273 K)

10.7 (page 359)   1.9 J/K/gram

10.8 (page 361)   25° C

10.9 (page 362)   Large rise in temperature of sink; no longer as cold as before

10.10 (page 364) $e^{-(E_2 - E_1)/kT}$

10.12 (page 367)   About 8.4 kilometers! (about 5 miles)

10.13 (page 367)   About 4×10^{-20} m !

10.14 (page 367)   About 1/3

10.15 (page 370)   1360 m/s

10.16 (page 370)   496 m/s

10.17 (page 370)   2.50; 2.74

10.18 (page 371)   2300 m/s; not very large compared with 500 m/s

10.19 (page 371)   Increase; factor of $\sqrt{2}$, so increase about 40%

10.20 (page 374) $\frac{3}{2}k$; translational energy quanta are very small.

10.21 (page 374) $\frac{7}{2}k$

10.22 (page 374) $\frac{5}{2}k$

10.23 (page 374) $\frac{3}{2}k$

10.24 (page 374)   Lower temperature. Rotational energy is $\frac{S_z^2}{2I}$, so energy of rotational first excited state for D_2 is smaller (larger nuclear mass means larger moment of inertia, since internuclear distance about the same).

Chapter 11

The Kinetic Theory of Gases

Chapter 11

The Kinetic Theory of Gases

In this chapter we will extend our model of gases. The emphasis is on description and analysis at the atomic level, and how the atomic picture relates to large-scale, macroscopic observations of the behavior of gases. Building on our prior studies, we will address the following issues:

- The rate at which gas molecules escape through a small hole.
- How far gas molecules go between collisions.
- The dependence of pressure on temperature, and the ideal gas law.
- Energy relations in processes involving gases.

This variety of applications illustrates the usefulness of an atomic-level approach, which is known as "the kinetic theory of gases."

11.1 Gases, solids, and liquids

Remember that a gas has no fixed structure, unlike a solid. The molecules are not bound to each other but move around very freely, which is why a gas does not have a well-defined shape of its own; it fills whatever container you put it in (Figure 11.1). Think for example of the constantly shifting shape of a cloud, or the deformability of a balloon, in contrast with the rigidity of a block of aluminum. On average, gas molecules are sufficiently far apart that most of the time they hardly interact with each other. This low level of interaction is what makes it feasible to model a gas in some detail, using relatively simple concepts.

The molecular motion must be sufficiently violent that molecules can't stay stuck together. At high enough temperatures, any molecules that do manage to bind to each other temporarily soon get knocked apart again by high-speed collisions with other molecules. At a low enough temperature however, molecules move sufficiently slowly that collisions are no longer violent enough to break intermolecular bonds. Rather, more and more molecules stick to each other in a growing mass as the gas turns into a liquid or, at still lower temperatures, a solid.

Why we don't study liquids

For completeness, we should say why we don't study liquids in this course. A liquid is intermediate between a solid and a gas. The molecules are sufficiently attracted to each other that the liquid doesn't fly apart like a gas (Figure 11.2), yet the attraction is not strong enough to keep each molecule near a fixed equilibrium position. The molecules can slide past each other, giving liquids their special property of fluid flow (unlike solids) with fixed volume (unlike gases).

The analysis of liquids in terms of atomic, microscopic models is quite difficult compared with gases, where the molecules only rarely come in contact with other atoms, or compared with solids, where the atoms never move very far away from their equilibrium positions. For this reason, in this introductory course with its emphasis on atomic-level description and analysis we concentrate mostly on understanding gases and solids.

Figure 11.1 Molecules in a gas.

Figure 11.2 Molecules in a liquid.

11.2 Gas leaks through a hole

An example of a phenomenon that we can analyze using simple ideas about the speeds of gas molecules is the leakage rate of a gas through a small hole in a container filled with the gas. First we'll consider a simplified one-directional example, in order to understand the basic issues before stating the results for a real three-dimensional gas.

One-directional gas

The chain of reasoning that we follow is basically geometric. Consider a situation where many gas molecules are all traveling to the right inside a tube. For the moment, temporarily assume that they all have the same speed v (Figure 11.3). The cross-sectional area of the tube is A (Figure 11.4). There are N gas molecules in the tube of volume V, and we will frequently use the symbol $n = N/V$ to stand for the number of molecules per cubic meter:

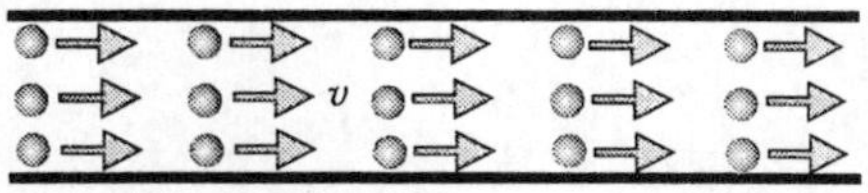

Figure 11.3 Side view of molecules all traveling to the right with speed v inside a tube. There are n molecules per cubic meter inside the tube.

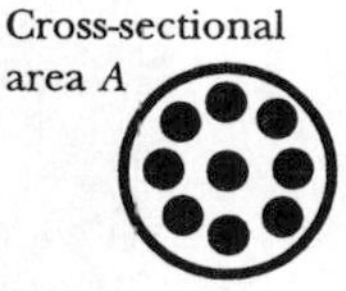

NUMBER DENSITY: NUMBER PER CUBIC METER

$$\text{Definition: } n \equiv \frac{N}{V} \, \frac{\text{number of molecules}}{\text{m}^3}$$

Because we eventually want to be able to calculate how fast a gas will leak through a hole in a container, we will calculate how many molecules leave this tube in a short time interval Δt. For a molecule that is traveling at speed v to be able to reach the right end of the tube in this time interval Δt, it must be within a distance $v\Delta t$ of the end (Figure 11.5).

Since the cross-sectional area of the tube is A, the volume V of the tube that contains just those molecules that will leave the tube during the time interval Δt is simply $A(v\Delta t)$; see Figure 11.6.

There are n molecules per m^3 inside the tube, and $N = nV$ molecules inside the volume V. So the number N of molecules that will leave in the time interval Δt is

$$N = nV = n(Av\Delta t)$$

Dividing by the time interval Δt, we find the following important result:

OF MOLECULES CROSSING AREA *A* PER SECOND

nAv (one-directional case; all molecules have same speed)

? Does this formula make sense? What would you expect if you increased the number of molecules per cubic meter, or the cross-sectional area, or the speed?

The formula does make sense. The more molecules per unit volume ($n = N/V$), the more molecules will reach the right end of the tube per unit time. The bigger the cross-sectional area (A), the more molecules that will pass through that area per unit time. The faster the molecules are moving (v), the more molecules from farther away that can reach the end of the tube in a given time interval. The units are right: $(\#/\text{m}^3)(\text{m}^2)(\text{m/s}) = \#/\text{s}$.

Effect of different speeds

We need to account for the fact that the gas molecules don't all have the same speed v. Suppose n_1 molecules per unit volume have speeds of approximately v_1, n_2 molecules per unit volume have speeds of approximately v_2, etc. The number of molecules crossing an area A per second is

$$n_1 A v_1 + n_2 A v_2 + \text{etc.}$$

The average speed of all the molecules is by definition the following, where we weight each different speed by the number of molecules per unit volume that have that approximate speed:

$$\bar{v} = \frac{n_1 v_1 + n_2 v_2 + \text{etc.}}{n}$$

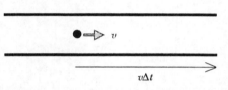

Figure 11.4 End view of molecules all traveling the same direction in a tube of cross-sectional area A.

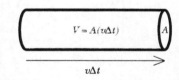

Figure 11.5 A molecule that will leave the tube in a time Δt must be within a distance $v\Delta t$ from the end.

Figure 11.6 Volume containing the molecules that will leave the tube in time Δt.

Therefore,

$$n_1 A v_1 + n_2 A v_2 + \text{etc.} = nA\bar{v}$$

A horizontal bar over a symbol is a standard notation for "average," and $\bar{v}$ means average speed.

Finally we have a valid formula for the number of molecules in a one-directional flow leaving the right end of the tube, even in the situation where they have a distribution of different speeds:

OF MOLECULES CROSSING AREA *A* PER SECOND

$nA\bar{v}$ (one-directional case; molecules have various speeds)

A "one-directional" gas may sound a bit silly, but this formula does apply to real one-directional flows such as the flow of water or gas through a pipe or the flow of electrons through a copper wire in an electric circuit.

A three-dimensional gas

A more realistic model of a gas would have approximately equal numbers of molecules heading to the left as well as to the right in the tube, in which case the number of molecules leaving the right end of the tube would be only $\frac{1}{2}nA\bar{v}$. In a real three-dimensional gas, molecules are moving in all directions. Only those molecules that are headed in the $+x$ direction can pass through a hole located to the right, not those moving in the $-x$ direction nor in the $\pm y$ or $\pm z$ directions. The molecules are headed randomly in all six directions, so we might expect our formula would have a factor of $1/6$.

However, the actual factor is $1/4$, which comes from detailed averaging over all directions and is related to our use of the average speed (magnitude of velocity), rather than averages of velocity components v_x or v_y or v_z. We don't want to get bogged down in the rather heavyweight mathematics required to prove this, so we just state that the factor is $1/4$ rather than $1/6$.

OF MOLECULES CROSSING AREA *A* PER SECOND

$\frac{1}{4}nA\bar{v}$ (three-dimensional case; molecules have various speeds)

$$\text{Remember that } n = \frac{N}{V}\frac{\text{\# of molecules}}{\text{m}^3}$$

Let's see what this formula predicts for the rate at which helium will escape from a balloon through a hole. You probably learned in chemistry that at "standard temperature and pressure" (STP, which means a temperature of $0°$ Celsius or 273 Kelvin, and a pressure of 1 atmosphere), one mole of a gas occupies 22.4 liters (a liter is 1000 cubic centimeters). As we saw on page 368, the average speed $\bar{v}$ of helium atoms at ordinary temperatures is about 1200 m/s.

Ex. 11.1 Suppose we make a circular hole 1 millimeter in diameter in a balloon. Calculate the initial rate at which helium escapes through the hole (at $0°$ C), in number of helium atoms leaving the balloon per second.

11.2.1 Cooling of the gas

There is an interesting and important effect of gas escaping through a hole. If you look back over the derivations of the formulas, you can see that faster molecules escape disproportionately to their numbers. Faster molecules can be farther away from the hole than is true for slower molecules and still escape through the hole in the next short time interval. As a result, the distribution of speeds of molecules inside the container becomes somewhat depleted of high speeds.

Ex. 11.2 What can you say about the temperature of the gas inside the container as the gas escapes? Why?

11.3 Mean free path

There is another important property of a gas that we can understand at this point in terms of our atomic model. Suppose we place a special molecule somewhere in the helium gas, one whose movements we can trace. For example, it might be a molecule of perfume. On average, how far does this molecule go before it runs into a molecule of the gas? The average distance between collisions is called the "mean free path." It plays an important role in many phenomena, including the creation of electric sparks in air (discussed in Volume II of this textbook).

The calculation of the mean free path depends on a simple geometrical argument. Draw a cylinder along the direction of motion of the special molecule, with length d and radius $R+r$, where R is the radius of the molecule, and r is the radius of a gas molecule, as shown in Figure 11.7 and Figure 11.8.

The geometrical significance of this cylinder is that if the path of the special molecule comes within one molecular radius of a gas molecule, there will be a collision, so any gas molecule whose center is inside the cylinder will be hit. How long should the cylinder be for there to be a collision?

We define d to be the average distance the special molecule will travel before colliding with another molecule, so the cylinder drawn in Figure 11.7 should contain on average about one gas molecule. The cross-sectional area of the cylinder is $A \approx \pi(R + r)^2$ and the volume of the cylinder is Ad. If n stands for the number of gas molecules per cubic meter, we can write a formula involving the mean free path:

MEAN FREE PATH

$$nAd \approx 1 \quad \text{where} \quad A \approx \pi(R + r)^2 \quad (\text{and } n = N/V)$$

We are ignoring some subtle effects in this calculation, but this analysis gives us the main picture: the mean free path is shorter for higher density or larger molecular size.

To get an idea of the order of magnitude of a typical mean free path, assume that an N_2 molecule in the air has a radius of approximately 2×10^{-10} m (the radius of one of the N atoms being about 10^{-10} m), and let's calculate approximately the mean free path d of an N_2 molecule moving through air. At "standard temperature and pressure" (STP, 0° Celsius, 1 atmosphere pressure) one mole of air (6×10^{23} molecules) has a volume of 22.4 liters (22.4×10^3 cm^3).

$$n = \frac{6 \times 10^{23} \text{ molecules}}{(22.4 \times 10^3 \text{ cm}^3)(10^{-6} \text{ m}^3/\text{cm}^3)} = 2.7 \times 10^{25} \frac{\text{molecules}}{\text{m}^3}$$

$$A \approx \pi(2 \times 2 \times 10^{-10} \text{ m})^2 = 5 \times 10^{-19} \text{ m}^2$$

$$d \approx \frac{1}{nA} \approx \frac{1}{\left(2.7 \times 10^{25} \frac{\text{molecules}}{\text{m}^3}\right)(5 \times 10^{-19} \text{ m}^2)} = 7 \times 10^{-8} \text{ m}$$

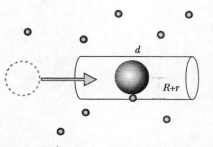

Figure 11.7 If a "special molecule" (dashed circle) enters a cylinder of length d and radius $R+r$, it will collide with a gas molecule (dark).

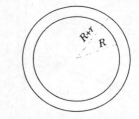

Figure 11.8 End view of cylinder.

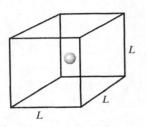

Figure 11.9 The volume of air occupied on average by one molecule.

Ex. 11.3 It is interesting to compare the mean free path of about 7×10^{-8} m to the average spacing L between air molecules, which is the cube root of the volume occupied on average by one molecule (Figure 11.9). Calculate L. You may be surprised to find that the mean free path d is much larger than the average molecular spacing L. The molecules represent rather small targets.

11.4 Pressure and temperature

Next we will use statistical ideas to relate what we know about the motion of molecules in a gas to the pressure that the gas exerts on its container.

In a party balloon, helium atoms are continually hitting the rubber walls of the balloon. In Ex. 11.1 on page 388 you found that on average, at STP (standard temperature and pressure) about 6×10^{21} helium atoms hit a 1-mm-diameter section of the balloon every second. The time-and space-averaged effect of this bombardment is an average force exerted on every square millimeter of the balloon. This average force per unit area is called the "pressure" P and is measured in N/m^2, also called a "pascal." We are going to calculate how big the pressure is in terms of the average speed of the helium atoms, thus building a link between the microscopic behavior of the helium atoms and the macroscopic time- and space-averaged pressure.

When the velocity distribution is stable, on average a helium atom bounces off the wall with no change of kinetic energy. We emphasize that this is the average behavior. Any individual atom may happen to gain or lose energy in the collision with a vibrating atom in a rubber molecule, but if the balloon is not being warmed up or cooled down, the velocity distribution in the gas is stable, and on average the helium atoms rebound from the wall with the same kinetic energy they had just before hitting the wall (Figure 11.10).

? If the kinetic energy doesn't change when an atom bounces off the wall, is there any change in the atom's momentum?

Unlike speed or kinetic energy, momentum is a vector quantity, and there is a large change in the x component of momentum, from $+p_x$ to $-p_x$ (There is no change in the y component.) Therefore the momentum change of the helium atom is $\Delta p_{x, \text{helium}} = -2p_x$.

? What caused this change in the momentum of the helium atom?

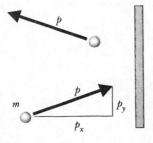

Figure 11.10 On average an atom bounces off the wall without changing speed.

A force is required to change the momentum of an object. In this case the force was applied by the wall of the container (or in more detail, by an atom in a rubber molecule in the balloon). By the principle of reciprocity (Newton's third law), the helium atom must have applied an equal and opposite force to the wall (or more precisely, to that atom in the rubber molecule). Therefore the wall must have acquired an amount of momentum $\Delta p_{x, \text{wall}} = 2p_x$.

If we could calculate the average time Δt between collisions of helium atoms with a small area A of the wall, we could express the pressure (force per unit area) as follows, since $d\vec{p}/dt = \vec{F}$:

$$P = \frac{F}{A} = \frac{1}{A}\frac{\Delta p_x}{\Delta t}$$

Assuming one direction and one speed

We already know that if there are n_{right} helium atoms per unit volume that have an x component of velocity equal to $+v_x$ (that is, moving to the right toward the wall), the number of such atoms that hit an area A to their right in a time Δt is $nAv_x\Delta t$. Therefore the average time between collisions is

$$\Delta t = 1/(n_{\text{right}}Av_x)$$

and we find

$$P = \frac{F}{A} = \frac{1}{A}\frac{\Delta p_x}{\left(\dfrac{1}{n_{\text{right}}Av_x}\right)} = (n_{\text{right}}v_x)(2p_x) = 2n_{\text{right}}\left(\frac{p_x}{m}\right)p_x = 2n_{\text{right}}\frac{p_x^2}{m}$$

An alternative argument

Alternatively, we can say that $n_{\text{right}}Av_x$ helium atoms with this x component of velocity hit the area A every second, and each of them delivers $2p_x$ of momentum to the wall, so that the total momentum transfer to the wall per second is

(# hits per second) (momentum transfer per hit)

$$(n_{\text{right}}Av_x)(2p_x)$$

Divide by A to get the force per unit area, and you again get

$$P = (n_{\text{right}}v_x)(2p_x) = 2n_{\text{right}}\frac{p_x^2}{m}$$

Taking the speed distribution into account

Actually, this is just the contribution to the pressure made by those atoms that happened to have this particular value of v_x to the right. On average only half the atoms are headed to the right, so we replace n_{right} by $n/2$, where n is the number of atoms per unit volume (going in either direction).

$$P = 2n_{\text{right}}\frac{p_x^2}{m} = n\frac{p_x^2}{m}$$

We also need to average over the slow and fast atoms:

$$P = n\frac{\overline{p_x^2}}{m}$$

We'll explain what we mean by this average. Suppose n_1 atoms per unit volume have x components of momentum of approximately p_{x1}, n_2 atoms per unit volume have x components of momentum of approximately p_{x2}, etc. The pressure is

$$P = n_1\left(\frac{p_{x1}^2}{m}\right) + n_2\left(\frac{p_{x2}^2}{m}\right) + \dots$$

The average value of p_x^2 for all the atoms is by definition the following, where we weight each different value of p_x^2 by the number of atoms per unit volume that have that approximate value:

$$\overline{p_x^2} = \frac{n_1 p_{x1}^2 + n_2 p_{x2}^2 + \text{etc.}}{n}, \text{ so we have } n_1 p_{x1}^2 + n_2 p_{x2}^2 + \text{etc.} = n\overline{p_x^2}$$

Therefore we can write this:

$$P = n_1\left(\frac{p_{x1}^2}{m}\right) + n_2\left(\frac{p_{x2}^2}{m}\right) + \dots = n\frac{\overline{p_x^2}}{m}$$

Taking direction into account

We can re-express the pressure in terms of momentum p rather than p_x by the following steps:

First, note that $p = \sqrt{p_x^2 + p_y^2 + p_z^2}$, or $p^2 = p_x^2 + p_y^2 + p_z^2$.

Second, since the atoms are flying around in random directions, there should be no difference in averaging in the x, y, or z direction, so $\overline{p_x^2} = \overline{p_y^2} = \overline{p_z^2}$, which implies that $\overline{p^2} = 3\overline{p_x^2}$.

Taking these factors into account, instead of $P = n\dfrac{\overline{p_x^2}}{m}$ we can write the following important result:

GAS PRESSURE IN TERMS OF ATOMIC QUANTITIES

$$P = \tfrac{1}{3} n \frac{\overline{p^2}}{m}, \text{ where } n = N/V$$

? Look back over the line of reasoning that led us to this result, and reflect on the nature of the argument. Stripped of the details, try to summarize the major steps leading to this result.

We reached this result by combining two effects: the number of molecules hitting an area per second is proportional to v, and the momentum transfer is also proportional to v. Hence the force per unit area is proportional to v^2.

11.4.1 The ideal gas law

It is useful to rewrite our result for the pressure, factoring out the term $\overline{p^2}/(2m)$, which is the average translational kinetic energy K_{trans} of a molecule of mass m:

$$P = \tfrac{2}{3} n \left(\frac{\overline{p^2}}{2m} \right) = \tfrac{2}{3} n \overline{K}_{\text{trans}}$$

The pressure of a gas is proportional to the number density (number of molecules per cubic meter, N/V), and proportional to the average translational kinetic energy of the gas molecules. This gives us a connection to temperature, because we found in Chapter 10 that $\overline{K}_{\text{trans}} = \tfrac{3}{2} kT$.

Substituting into the pressure formula, $P = \tfrac{2}{3} n \overline{K}_{\text{trans}} = \tfrac{2}{3} n (\tfrac{3}{2} kT) = nkT$, which relates pressure to temperature and is called the "ideal gas law":

MOLECULAR VERSION OF THE IDEAL GAS LAW

$$P = nkT$$

$n = N/V$ = number of molecules per m^3

$k = R/6\times10^{23}$ is the "Boltzmann constant" (1.4×10^{-23} J/K)

Ex. 11.4 A mole of a gas (6×10^{23} molecules) occupies 22.4 liters under standard conditions (temperature of 0° Celsius and one atmosphere pressure). Calculate the value of one atmosphere pressure in units of N/m^2.

11.4.2 The macroscopic ideal gas law

We can compare our molecular version of the gas law with experiments that measure the macroscopic properties of gases. For gases with fairly low densities, measurements of the pressure for all gases (helium, oxygen, nitrogen, carbon dioxide, etc.) are well summarized by the macroscopic "ideal gas law" which is probably familiar to you from chemistry, and which describes the observed behavior of any low-density gas:

$$P = \frac{(\text{\# of moles})RT}{V} \quad \text{(mole version of the gas law)}$$

Here R is the "gas constant" (8.3 J/K/mole), T is the absolute temperature in Kelvins, and V is the volume in cubic meters. To compare with our microscopic prediction, we can convert this macroscopic version of the ideal gas law to a version involving microscopic quantities. Note the following:

$$\text{\# of moles} = \frac{(\text{\# of molecules})}{(6\times10^{23} \text{ molecules/mole})} = \frac{N}{6\times10^{23}}$$

where N is the number of molecules in the gas. Therefore we have

$$P = \left(\frac{N}{6\times10^{23}}\right)\frac{RT}{V} = \left(\frac{N}{V}\right)\left(\frac{R}{6\times10^{23}}\right)T$$

But this is simply $P = nkT$, where $n = N/V$ and $R/6\times10^{23}$ is the Boltzmann constant k:

$$k = \frac{R}{6\times10^{23}} = \frac{8.3 \text{ J/K/mole}}{6\times10^{23} \text{ molecules/mole}} = 1.4\times10^{-23} \text{ J/K}$$

Warning about the meaning of "n"

You may be familiar with the gas law written in the form $PV = nRT$, where n is the number of moles, whereas we mean something else by "n"—the number of molecules per unit volume, N/V. On the rare occasions when we need to refer to the number of moles, we'll write it out as "# of moles."

11.4.3 Temperature from entropy or from the ideal gas law

In Chapter 10 we found the relationship $\overline{K}_{\text{trans}} = \frac{3}{2}kT$, based on the Boltzmann distribution as applied to a low-density gas. The Boltzmann distribution in turn was based on the statistical mechanics definition of temperature in terms of entropy as $1/T = \partial E/\partial S$.

When we inserted $\overline{K}_{\text{trans}} = \frac{3}{2}kT$ into the kinetic theory result for pressure, $P = \frac{1}{3}n(\overline{p^2}/m)$, we obtained the molecular version of the ideal gas law, $P = nkT$, which we showed was equivalent to the macroscopic version of the ideal gas law, $P = (\text{\# of moles})RT/V$.

Low-density gases are described well by the ideal gas law, and for that reason gases are used to make accurate thermometers. You measure the pressure P and volume V of a known number of moles of a low-density gas and determine the temperature from the macroscopic gas law:

$$T = \frac{PV}{(\text{\# of moles})R} \quad \text{(gas thermometer)}$$

The fact that we could start from $1/T = \partial E/\partial S$ and derive the ideal gas law proves that the temperature measured by a gas thermometer is exactly the same as the "thermodynamic temperature" defined in terms of entropy.

11.4.4 Real gases

Our analysis works well for low-density gases. For high-density gases, there are two major complications. First, the molecules themselves take up a considerable fraction of the space, so the effective volume is less than the geometrical volume, and V in the gas law must be replaced by a smaller value.

Second, at high densities the short-range electric forces between molecules have some effect. In this context these intermolecular forces are called "van der Waals" forces, which are the gradient of the interatomic potential energy discussed in Chapter 4. In a low-density gas almost all of the energy is kinetic energy, but in a high-density gas some of the energy goes into configurational energy associated with the interatomic potential energy. The effect is that in the gas law P must be replaced by a smaller value.

When these two effects are taken into effect, the resulting "van der Waals" equation fits the experimental data quite well for all densities of a gas, although the corrections are different for different gases due to differences in molecular sizes and intermolecular forces.

11.4.5 Energy of a diatomic gas

The formula for the average translational kinetic energy $\overline{K}_{trans} = \frac{3}{2}kT$ is valid even for a gas with multiatom molecules such as nitrogen (N_2) or oxygen (O_2). But, as you will recall from Chapter 7, this translational kinetic energy is only a part of the total energy of the molecule. In addition to translational energy, a diatomic molecule can have rotational and vibrational energy relative to the center of mass (Figure 11.11), so the energy contains additional terms. We write the energy of a diatomic molecule in the following way:

$$K_{trans} + E_{vib} + E_{rot} + Mgy_{cm}$$

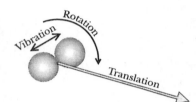

Figure 11.11 Energy of an oxygen molecule.

The key point, as we discussed in Chapter 10, is that the average energy of a diatomic molecule is greater than the average energy of an atom in a monatomic gas at the same temperature. For a monatomic gas such as helium the average total energy per molecule is just $\overline{K}_{trans} = \frac{3}{2}kT$.

11.4.6 Application: Weight of a gas in a box

The exponential dependence of number density ($n \propto e^{-\frac{mgy}{kT}}$) and therefore of pressure on height makes it possible to understand something that otherwise might be rather odd about weighing a box that contains a gas (Figure 11.12). Let's do the weighing in a vacuum so we don't have to worry about buoyancy forces due to surrounding air. The box definitely weighs more on the scales if there is gas in it than if there is a vacuum in it. In fact, if the mass of the gas is M, the additional weight is Mg, because the momentum principle for multiparticle systems refers to the net force on a system, and the gravitational contribution to the net force is the sum of the gravitational forces on each individual gas molecule in the box, Mg.

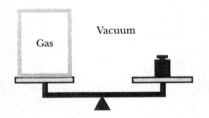

Figure 11.12 Weighing a box containing a gas.

That seems reasonable until you think about the details of what is going on inside the box. At any given instant, the vast majority of the gas molecules are not touching the box! And some of the gas molecules are colliding with the top of the box, exerting an upward force on the box. How can these molecules possibly contribute to the weight measured by the scales? What is the mechanism for the momentum principle working out correctly for this multiparticle system?

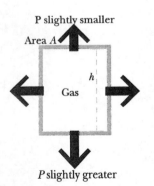

Figure 11.13 The force on the top of the box differs slightly from the force on the bottom of the box.

The number density n at the bottom of the box is slightly larger than the number density at the top of the box, if all the gas is at the same temperature, which is a good assumption in most situations. Therefore the pressure, $P = nkT$, is larger at the bottom than at the top. Let's calculate the y components of the forces associated with these slightly different pressures, where A is the area of the top (and bottom) of the box, and h is the height of the box (Figure 11.13).

$$F_{net,y} = P_{top}A - P_{bottom}A = A\Delta P$$

where ΔP is the pressure difference from bottom to top inside the box. We can calculate ΔP directly, by starting from the fact that for a small height change Δy, we have this:

$$\frac{\Delta P}{\Delta y} \approx \frac{dP}{dy}, \text{ so } \Delta P \approx \frac{dP}{dy}\Delta y$$

If we can evaluate dP/dy, we can determine the small pressure difference ΔP. Since $P = nkT$, the height dependence of the pressure is the same as for the number density (for constant temperature), and we can write this:

$$\frac{dP}{dy} = \frac{d}{dy}\left[P_{\text{bottom}} e^{-\frac{mgy}{kT}}\right]$$

$$\frac{dP}{dy} = P_{\text{bottom}} e^{-\frac{mgy}{kT}}\left[-\frac{mg}{kT}\right] \approx -P_{\text{bottom}}\left[\frac{mg}{kT}\right]$$

because the factor $e^{-\frac{mgy}{kT}}$ is very close to 1. The result is negative because the force the gas exerts on the top of the box is smaller than the force on the bottom. Choosing $\Delta y = h$, and writing $P_{\text{bottom}} = nkT$, we have this result for ΔP:

$$\Delta P = \frac{dP}{dy}\Delta y \approx -nkT\left[\frac{mg}{kT}\right]h = -nmgh$$

Now that we know the pressure difference, we can calculate the y component of the net force:

$$F_{\text{net},y} = A\Delta P = -nmg(Ah)$$

? But Ah is the volume V of the box, and $n = N/V$ is the number of molecules per m^3 in the box, so what does this formula reduce to?

We have the striking result that the air inside the box pushes down on the box with a force equal to the combined weight Mg of all N molecules in the box:

$$F_y = -Nmg = -Mg \text{ !!}$$

We have shown that the difference in the time- and space-averaged momentum transfers by molecular collisions to the top and bottom of the box is equal to the weight of the gas in the box, as predicted by the momentum principle for a multiparticle system. The pressure difference is very slight, but then the weight of the gas is very small, for that matter. What is surprising is that any particular instant, relatively few of the molecules are actually in contact with the box, yet the effects of these relatively few molecules is the same as though they were all sitting on the bottom of the box.

You could think of a box full of water in the same way. The water pressure is larger at the bottom than at the top, but with water this difference is quite large, corresponding to the much higher density of water. In fact, a column of water only 10 meters high makes a pressure equal to that produced by the many kilometers of atmosphere (atmospheric pressure at sea level is about 10^5 N/m^2).

Ex. 11.5 In the preceding discussion the box was in vacuum, but under normal conditions the box we were weighing would have been surrounded by air. Suppose the box has thin walls and is initially open to the air. Then we close the lid, trapping air inside it. Consider all forces including buoyancy forces, and determine what the scales will read: the weight of the box alone or the box plus air?

Maximum height

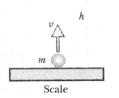

Scale

Figure 11.14 The ball has just hit the scale and has rebounded upward.

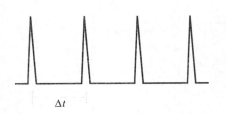

Δt

Figure 11.15 Impacts of the ball on the scale.

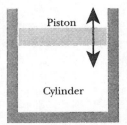

Piston

Cylinder

Figure 11.16 A cylinder with a piston that can move vertically.

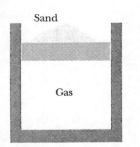

Sand

Gas

Figure 11.17 The piston can be loaded with varying amounts of sand.

11.4.7 Application: Weight of a bouncing molecule

There is a related calculation that is amusing. If you did a related homework problem in Chapter 8 on collisions, you already went through a similar calculation. Consider a single molecule (otherwise in vacuum) that bounces up and down on a scale without losing significant energy. Figure 11.14 shows a snapshot at the instant that the molecule is just bouncing up after hitting the scale.

What do the scales read? If the scales can respond quickly, we will see brief spikes each time the molecule strikes it (Figure 11.15). Let's determine what the time-averaged force is.

? How long does it take for the molecule to reach the top of its trajectory, h, starting with a speed v? What then is the time Δt between impacts? How much momentum transfer is there to the scale on each impact?

It takes a time interval v/g to go up (for the speed to decrease from v to 0 with acceleration $-g$) and another time interval v/g to come down, so the time between impacts is $\Delta t = 2v/g$. Each impact transfers an amount of momentum $\Delta P = 2mv$. Therefore the time-averaged force is

$$\frac{\Delta P}{\Delta t} = \frac{2mv}{2v/g} = mg \text{ !!}$$

If the scale is sluggish, and can't respond in a time as short as $\Delta t = 2v/g$, the scale will simply register the value mg, just as though the molecule were sitting quietly on the scale.

11.5 Energy transfers with the surroundings

The final topic in this chapter on gases is an analysis of energy transfers between a gas and its surroundings. We will address such questions as these: How much thermal energy transfer Q is required to raise the temperature of a gas by one degree (the heat capacity of a gas)? How does the temperature of a gas change when you compress it quickly? Do the answers to these questions depend on what kind of gas is involved?

We need a device that lets us control the flow of energy into and out of a gas, in the form of work or in the form of thermal energy transfer. We will use a system consisting of a cylindrical container containing gas that is enclosed by a piston that can move in and out of the cylinder with little friction but which fits tightly enough to keep the gas from leaking out (Figure 11.16). This is similar to a cylinder in an automobile engines, into which is sprayed a mixture of gasoline vapor and air. The mixture is ignited by a spark, and the chemical reactions raise the temperature and pressure very high very quickly. The piston is pushed outward, which turns a shaft which ultimately drives the wheels.

Force and pressure

Figure 11.17 shows a cylinder with a vertical-running piston on which we can load varying amounts of sand, in order to be able to control the pressure of the contained gas, and to be able to do controlled amounts of work on the gas. We will also put the cylinder in contact with hot or cold objects and allow thermal energy transfer into or out of the cylinder.

? Consider the piston plus sand as the system of interest for a moment and think about what forces act on this system.

A physics diagram for the piston+sand system (Figure 11.18) includes the downward gravitational forces on the piston (Mg) and on the sand (mg), an upward time- and space-averaged force due to the pressure of the enclosed gas on the lower surface of the piston, and a downward time- and space-averaged force due to the pressure of the outside atmosphere on the upper surface of the system.

Since pressure is force per unit area, the upward force is PA, where A is the surface area of the bottom of the piston. Similarly, the downward force contributed by the outside atmosphere is $P_{air}A$. These pressure-related forces are not actually continuous, since they are the result of collisions of individual gas molecules with the piston and sand. However, the rate of collisions is so extremely high over the area of the piston that the force seems essentially constant. For example, in Ex. 11.1 on page 388 you found that on average, at STP (standard temperature and pressure) about 6×10^{21} helium atoms strike a tiny 1-mm-diameter section every second.

? In mechanical equilibrium (velocity of piston not changing), solve for the gas pressure inside the cylinder.

The momentum principle tells us that in equilibrium the net force on the piston+sand system must be zero, from which we are able to deduce that the pressure of the gas inside the cylinder is

$$P = P_{air} + \frac{Mg + mg}{A}$$

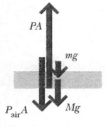

Figure 11.18 Forces on the piston.

A sudden change

By varying the amount of sand (m) we can vary the pressure of the gas under study.

? If you suddenly add a lot of sand, what happens?

If you suddenly add a lot of sand to the piston, there is suddenly a sizable nonzero net downward force on the piston+sand system: $F_{net} = AP_{air} + Mg + mg - AP \neq 0$. The piston starts to pick up speed downwards. As it does so, it runs into gas molecules and tends to increase their speeds.

? What happens to the temperature of the gas in the cylinder?

Since higher average speed means higher temperature, the gas temperature starts to increase, at the same time that the volume of the gas is decreasing. This is a double whammy: both increased temperature and decreased volume contribute to increased pressure, since $P = (N/V)kT$, where N is the number of gas molecules in the cylinder. Therefore the pressure in the gas quickly rises, and eventually there will be a new equilibrium with a lower piston (supporting more sand) and a higher gas pressure in the cylinder.

However, getting to that new equilibrium is pretty complicated. If there is no friction or other energy dissipation, the piston will oscillate down and up, with the gas pressure going up and down. It is even possible to determine an effective "spring constant" for the gas and calculate the frequency of the oscillation. However, in any real system there will be some friction, so we know that the system will eventually settle down to a new equilibrium configuration.

Quasistatic processes

To avoid these complicated (though interesting) transient effects, we will study what happens when we add sand very carefully and very slowly, one grain at a time, and we assume that the new equilibrium is established almost immediately. This is called a "quasistatic compression" because the system is at all times very nearly in static equilibrium (Figure 11.19). Similarly, if we slowly remove one grain at a time, we can carry out a "quasistatic ex-

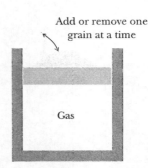

Figure 11.19 By adding or removing one grain of sand at a time we effect a "quasistatic" compression or expansion.

pansion." Note in particular that at no time does the piston have any significant amount of kinetic energy, and macroscopic kinetic energy is essentially zero at all times. Of course there is plenty of microscopic kinetic energy in the gas molecules and the outside air molecules, and in the thermal motion of atoms in the cylinder walls, piston, and sand.

Suppose we carry out a lengthy, time-consuming quasistatic compression by adding lots of sand, one grain at a time, very slowly. The piston goes down quite a ways, the pressure in the gas is a lot higher, and the volume of the gas is a lot smaller. If we know the pressure and volume, we can calculate the temperature by using the ideal gas law $P = (N/V)kT$ (assuming the gas density isn't too high to make this invalid).

But can we predict how low the piston will go for a given amount of added sand? Oddly enough, no—not without knowing something more about this device. There are two extreme cases that are both important in practice and calculable for an ideal gas. If the apparatus is made of metal (a very good thermal conductor) and is in good thermal contact with a large object at temperature T, the process will proceed at nearly constant temperature, with thermal energy transfer from the gas into the surroundings. If the apparatus is made of glass (a very poor thermal conductor), the temperature of the gas inside the cylinder will rise in a predictable way, with negligible energy transfer to the surroundings in the form of thermal energy transfer Q. We will analyze both kinds of processes—constant temperature processes, and no-Q processes.

11.5.1 Constant-temperature (isothermal) compression

Suppose the cylinder is made of metal (which is a very good thermal conductor) and is sitting in a very big tub of water whose temperature is T, as shown in Figure 11.20.

As we compress the gas, the temperature in the gas starts to increase. However, this will lead to energy flowing out of the gas into the water, because whenever the temperatures differ in two objects that are in thermal contact with each other, we have seen that there is a transfer of energy from the hotter object into the colder object. In fact, for many materials the rate of energy transfer is proportional to the temperature difference—double the temperature difference, double the rate at which energy transfers from the hotter object into the colder one.

The mechanism for thermal energy transfer is that atoms in the hotter object are on average moving faster than atoms in the colder object, so in collisions between atoms at the boundary between the two systems it is likely that energy will be gained by the colder object and lost by the hotter object.

Energy transfer out of the gas will lower the temperature of the gas, since the total energy of the gas is proportional to the temperature. Quickly the temperature of the gas will fall back to the temperature of the water. The temperature of the big tub of water on the other hand will hardly change as a result of the energy added to it from the gas, because there is such a large mass of water to warm up.

Therefore the entire quasistatic compression takes place essentially at the temperature of the water, and the final temperature of the gas is nearly the same as the initial temperature of the gas. This is a constant-temperature compression (also called an "isothermal" compression, which just means constant temperature). Similarly, we can slowly remove sand from the piston and carry out a quasistatic constant-temperature expansion.

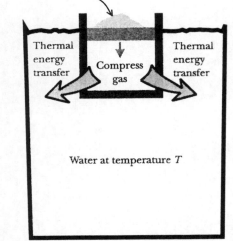

Figure 11.20 A cylinder of gas immersed in a large volume of water.

? Suppose the original gas pressure and volume were P_1 and V_1, and as the result of a constant-temperature process the final volume is V_2. What is the final pressure P_2?

Since the temperature hasn't changed, from the ideal gas law $P = (N/V)kT$ we can deduce that $P_1 V_1 = P_2 V_2$, and therefore $P_2 = P_1(V_1/V_2)$.

Work and thermal energy transfer in a constant-temperature compression

A more difficult question we can ask (and answer!) is this: How much thermal energy was added to the water in the constant-temperature compression?

? What energy inputs and outputs were made to the gas? What energy change occurred in the gas?

The piston did work on the gas, and there was thermal energy transfer out of the gas (and into the water). The Earth's gravitational force did work on the gas (since the center of mass of the gas went down), but this is negligibly small compared to the work done by the piston (the lowering of the heavy piston involves much more gravitational energy than the lowering of the low-mass gas).

? Did the total energy of the gas change (ignoring the small gravitational energy change)?

Since the total energy of an ideal gas is proportional to temperature (including rotational and vibrational energy if the gas is not monatomic), and we made sure that we kept the temperature constant, the total energy of the gas did not change. Therefore we have this energy equation for the open system that is the gas:

? Δ(energy of gas) = (W by piston) − ($|Q|$ that flowed into water) = 0

Symbolically, $\Delta E_{gas} = W + Q = 0$, where Q is negative (thermal energy transfer out of system). In thermal processes of this kind, the energy equation is called "the first law of thermodynamics."

If we can calculate the work done by the piston, we can equate that result to the amount of thermal energy transfer into the water. The piston exerts a (time- and space-averaged) force PA on the gas, where P is the pressure in the gas and A is the cross-sectional area of the piston (Figure 11.21).

? If the piston drops a distance h, is the work it does PAh?

As the piston drops, the pressure in the gas increases, so the force is not constant. We have to integrate the variable force through the distance h in order to determine the work. If we measure x downward from the initial piston position, at each step dx of the way the increment of work done is $PA dx$. But $A dx$ is an increment of volume (base area A time altitude dx), and the change in the volume is negative. Putting this all together, we have this:

WORK DONE BY A PISTON ON A GAS

$$W = -\int_{V_1}^{V_2} P dV$$

Check the sign: If the volume decreases, the integral will be negative, and the minus sign in front of the integral makes the work done on the gas be positive, which is correct. Conversely, in an expansion of a gas, the integral is positive, and the work is negative, because the gas is doing work on the piston rather than the other way round.

? In a constant-temperature (isothermal) compression, work is done on the gas. Where does this energy flow go? How does the temperature stay constant?

As the piston moves down, it increases the average energy of the gas molecules that run into it. Therefore the input work starts to raise the tempera-

Figure 11.21 Force on the gas by the piston.

ture of the gas (higher speeds), but this higher temperature causes thermal energy transfer out of the cylinder, into the (very slightly) lower-temperature water. The net effect is for energy to flow into the gas in the form of mechanical work, and out of the gas into the water in the form of thermal energy transfer. There is no change in the total energy of the gas, because the temperature of the gas didn't change. The outward thermal energy transfer brings the gas temperature back to what it was before the falling piston tried to increase the temperature.

Now we can calculate quantitatively the work done by the piston in the constant-temperature compression, which is equal to the thermal energy transfer from the gas to the surrounding water. We add lots of sand, one grain at a time (to maintain quasi-static equilibrium), and we compress the gas from an initial volume V_1 to a final volume V_2 (Figure 11.22).

Replace P by NkT/V, where T is a constant:

$$W = -\int_{V_1}^{V_2} P\,dV = -\int_{V_1}^{V_2} \frac{NkT}{V}\,dV$$

$$W = -NkT\int_{V_1}^{V_2}\frac{dV}{V} = -NkT[\ln V]_{V_1}^{V_2} = -NkT\ln\left(\frac{V_2}{V_1}\right)$$

Since the process was a compression, V_2 is smaller than V_1. Since $-\ln(V_2/V_1) = \ln(V_1/V_2)$, we have

$$W_{\text{by piston}} = |Q_{\text{into water}}| = NkT\ln\left(\frac{V_1}{V_2}\right)$$

Just as the symbol W is normally used for "work," so the symbol Q is normally used for "thermal energy transfer." Here Q is negative (energy output).

Graphical representation

There is a useful graphical representation of this constant-temperature process. We plot the pressure P vs. the volume V, as shown in Figure 11.23, and the process is represented by a curve of constant T (which is also a curve of constant PV, since $PV = NkT$). The area under the process curve represents the work W, which is also the thermal energy transfer Q that flows to the surroundings.

$$W = -\int_{V_1}^{V_2} P\,dV$$

The first law of thermodynamics

For historical reasons, in thermal processes the energy principle is called "the first law of thermodynamics."

THE FIRST LAW OF THERMODYNAMICS

$$\Delta E_{\text{sys}} = W + Q$$

In the particular case of constant-temperature (isothermal) compression of an ideal gas, we have

$$\Delta E_{\text{int}} = W + Q = 0$$

where E_{int} is the "internal" energy of the gas (the sum of the translational, rotational, vibrational, and other energy terms of all the molecules).

Initial volume
V_1

More sand

Final volume
V_2

Figure 11.22 The gas is compressed at a constant temperature T (the cylinder is immersed in a large tub of water at temperature T).

P

V

Figure 11.23 If T is kept constant, then the product PV is also constant.

Ex. 11.6 A mole of nitrogen is compressed (by piling lots of sand on the piston) to a volume of 12 liters at room temperature (293 K). The cylinder is placed on an electric heating element whose temperature is maintained at 293.001 K. A quasistatic expansion is carried out at constant temperature by very slowly removing grains of sand from the top of the piston, with the temperature of the gas staying constant at 293 K. When the volume is 18 liters, how much thermal energy transfer Q has gone from the heating element into the gas? How much work W has been done on the piston by the gas? How much has the energy of the gas changed?

(You must assume that there is no thermal energy transfer from the gas to the surrounding air, and no friction in the motion of the piston, all of which is pretty unrealistic in the real world! Nevertheless there are processes that can be approximated by a constant-temperature expansion. This problem is an idealization of a real process.)

11.5.2 Heat capacity

We were able to calculate the important properties of a constant-temperature compression, where the apparatus was in good thermal contact with its surroundings. Before analyzing the opposite extreme, where the apparatus is thermally insulated from its surroundings and no thermal energy transfer is involved, we will review the concept of heat capacity in this context.

Lock the piston in position so that the volume of the gas won't change (Figure 11.24). Put the apparatus in a big tub of water whose temperature is higher than the gas temperature, so that thermal energy transfer will go from the water to the gas. Allow an amount of thermal energy transfer Q to enter the gas and observe how much temperature rise ΔT has occurred in the gas. We define the "molecular heat capacity C_V at constant volume" in the following way, where N as usual is the total number of molecules in the gas:

$$Q = NC_V\Delta T \text{ (constant volume)}$$

The larger the heat capacity, the smaller the temperature rise ΔT for a given energy input Q.

? What is C_V for a monatomic gas such as helium?

Since the total energy in a monatomic gas is $N\overline{K}_{\text{trans}} = \frac{3}{2}NkT$, thermal energy transfer Q will increase the total energy of the gas by an amount $\frac{3}{2}Nk\Delta T$, so $C_V = \frac{3}{2}k$. Experimental measurements of the heat capacity of monatomic gases agree with this prediction.

For gases that are not monatomic, such as nitrogen (N_2), c_V can be larger than $\frac{3}{2}k$ because the total energy can be larger than $\frac{3}{2}NkT$ due to rotational and vibrational energy relative to the center of mass of the molecule. Before quantum mechanics, theoretical predictions for the contribution of the rotational and vibrational energies to the heat capacity of gases did not agree with experimental measurements. This was one of the puzzles that was eventually resolved by quantum mechanics, as we found in Chapter 10.

HEAT CAPACITY AT CONSTANT VOLUME, PER MOLECULE

$$C_V = \tfrac{3}{2}k \text{ for a monatomic gas (He, etc.)}$$

$$C_V \ge \tfrac{3}{2}k \text{ for other gases (}N_2\text{, etc.)}$$

Heat capacity at constant pressure

If we don't lock the piston but let the gas expand at constant pressure, the incoming thermal energy transfer Q not only raises the energy of the gas by an amount $NC_V\Delta T$ but also raises the piston, which involves an amount of

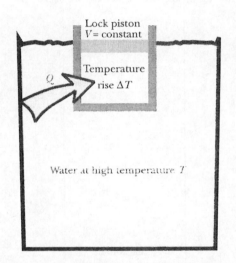

Figure 11.24 If the piston is locked so the volume of the gas cannot change, we can measure c_v of the gas.

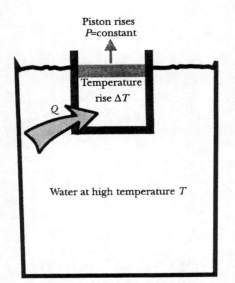

Piston rises
P=constant

Temperature
rise ΔT

Q

Water at high temperature T

Figure 11.25 If the piston is free to move, the pressure inside the cylinder remains the same, and we can measure c_p of the gas.

work W (Figure 11.25). We define the "molecular heat capacity C_P at constant pressure" as follows:

$$Q = NC_P\Delta T \text{ (constant pressure)}$$

Evidently C_P is bigger than C_V, because $Q = NC_P\Delta T = NC_V\Delta T + W$. We can calculate the work done on the piston by the gas as the gas expands, which is the negative of the work done by the piston on the gas:

$$W = \int_{V_1}^{V_2} PdV = P\int_{V_1}^{V_2} dV = PV_2 - PV_1 = NkT_2 - NkT_1 = Nk\Delta T$$

Therefore for an ideal gas there is a simple relationship between C_V and C_P:

$$Q = NC_P\Delta T = NC_V\Delta T + Nk\Delta T$$

$$C_P = C_V + k$$

Ex. 11.7 What is the molecular heat capacity at constant pressure C_P for helium?

Molar heat capacity

Heat capacity is often given on a per-mole basis rather than a per-molecule basis. The molar heat capacity at constant volume for an ideal gas is $C_V = \frac{3}{2}R$, where R is the molar gas constant (8.3 J/K), which is $6\times10^{23}k$. The molar heat capacity at constant pressure is $C_P = C_V + R$ for an ideal gas.

11.5.3 No-Q (adiabatic) compression

With the apparatus made of metal and sitting in a big tub of water, the temperature during the compression didn't change. Now we analyze the opposite extreme. We make the cylinder and piston out of glass or ceramic (which are poor conductors for thermal energy transfer) to minimize thermal energy transfer out of the gas as the gas gets hotter during the compression. In fact, we make the approximation that there is no thermal energy transfer at all (Figure 11.26). Such a no-Q process is also called "adiabatic" (which means "no thermal energy transfer").

How realistic is such a no-Q approximation? For many real situations this can be a rather good approximation if the compression or expansion is fast, so that there isn't enough time for significant thermal energy transfer. The flow of thermal energy from one object to another is a rather slow process. For example, a cup of coffee sitting on a table may stay quite hot for ten minutes or more.

But didn't we say that we were going to carry out compressions and expansions very slowly, "quasistatically"? Yes, so there is a contradiction. However, it is often the case that motion may be slow enough to be a good approximation to a quasistatic process and nevertheless may also be fast compared to the time required for significant thermal energy transfer. Again, think of how long it takes a cup of coffee to cool off.

We again consider the gas in the cylinder as the system of interest, and the work done by the piston is equal to the increase in energy of the gas:

$$W = -\int_{V_1}^{V_2} PdV = NC_V\Delta T$$

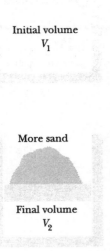

Initial volume
V_1

More sand

Final volume
V_2

Figure 11.26 If the cylinder and piston are made of insulating material, no thermal energy transfer can occur between the gas and its surroundings. This is called an adiabatic compression.

? Why did we use C_V in this equation? The volume isn't constant in this compression!

An ideal gas is unique among materials in that its total energy is entirely determined by the temperature, independent of volume or pressure. So the change of energy of the gas itself is just $NC_V\Delta T$, even when the volume is not constant. As a practical matter, real gases behave this way as long as the density is not too high. At high densities the energy of the gas includes a significant term associated with the interatomic forces.

We can work through the integration, starting from a "differential" form of the equation shown above:

$$-PdV = NC_V dT$$

Use the gas law $P = (N/V)kT$ to substitute for the pressure:

$$-NkT\frac{dV}{V} = NC_V dT$$

Divide through by kT and rearrange:

$$\left(\frac{C_V}{k}\right)\frac{dT}{T} + \frac{dV}{V} = 0$$

$$\left(\frac{C_V}{k}\right)\int\frac{dT}{T} + \int\frac{dV}{V} = 0$$

Now integrate:

$$\left(\frac{C_V}{k}\right)\ln T + \ln V = \text{constant}$$

Rewrite using the properties of logarithms:

$$\ln T^{\left(\frac{C_V}{k}\right)} + \ln V = \ln\left[T^{\left(\frac{C_V}{k}\right)}V\right] = \text{constant}$$

And therefore we have

$$T^{\left(\frac{C_V}{k}\right)}V = \text{constant}$$

Ex. 11.8 Use the ideal gas law to eliminate T from this expression, and show that $PV^\gamma = \text{constant}$ in a no-Q process, where the parameter γ is defined as the ratio of the constant-pressure heat capacity to the constant-volume heat capacity, $\gamma \equiv C_P/C_V$. (Since C_p is greater than C_v, γ is greater than 1.)

In particular, if the initial pressure and volume are P_1 and V_1, and the final pressure and volume are P_2 and V_2, then $P_2 V_2^\gamma = P_1 V_1^\gamma$. Alternatively, we also have $T_2^{(C_V/k)}V_2 = T_1^{(C_V/k)}V_1$.

Graphical representation

Again, there is a useful graphical representation of this no-Q process. In Figure 11.27 we plot the pressure P vs. the volume V, and the process is represented by a curve along which $PV^\gamma = \text{constant}$. The area under the process curve represents the work W. For comparison we also show the curve for a constant-temperature process (T and PV constant). The no-Q curve is much steeper than the constant-temperature curve.

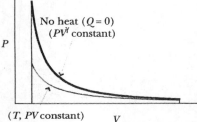

Figure 11.27 PV curves for a no-Q (adiabatic) process and a constant temperature (isothermal) process.

11.5.4 Work vs. thermal energy transfer

We have studied the response of a gas to energy inputs and outputs, both mechanical (work) and thermal (thermal energy transfer). Work represents organized, orderly, macroscopic energy input. Thermal energy transfer represents disorganized, disorderly, microscopic energy input. There is randomness at the atomic level in the collisions of atoms, which is the basic mechanism for thermal energy transfer. In Chapter 10 we saw deep consequences of the distinction between work W and thermal energy transfer Q.

11.6 *Application: A random walk

If you would find it useful to study one more application of the basic statistical concepts underlying our analysis of a gas, here is an interesting one. If we could watch one special molecule wandering around in the air, colliding frequently with air molecules, its path would look something like Figure 11.28. This kind of motion is called a "random walk." It may surprise you to find that despite the random nature of this motion we can calculate something significant about the motion, using simple statistical reasoning.

One dimension

For simplicity, first we'll consider just the x component of the motion. Pick the origin of the x axis to be at the original position of the special molecule. The first thing that happens is that the special molecule moves until it collides with an air molecule. We call this first x component of the displacement Δx_1. This component of the displacement may be to the right ($+x$ direction) or to the left ($-x$ direction). As a result of the collision, the special molecule may change direction, and change speed. The next displacement before another collision we call Δx_2, etc. After N collisions, the net displacement Δx away from the origin is

$$\Delta x = \Delta x_1 + \Delta x_2 + \Delta x_3 + \Delta x_4 + ... + \Delta x_N$$

After N collisions, what is the average (most probable) position of the special molecule? On average, it is just as likely that it has moved to the right as moved to the left, so the average x component of displacement ought to be zero. It is easy to see that this will be the case, by taking the average value:

$$\overline{\Delta x} = \overline{\Delta x_1} + \overline{\Delta x_2} + \overline{\Delta x_3} + \overline{\Delta x_4} + ... + \overline{\Delta x_N}$$

Each of the individual x displacements are equally likely to be to the left or the right, so the average value of the nth x displacement (for $n = 1, 2, 3, ...N$) is zero. Therefore the average value of the net x component of displacement is also zero.

On average, after N collisions the special molecule ends up to the left of the origin as often as it ends up to the right. But we can ask the question, "On average, how far away from the origin (along the x axis) does the special molecule get, no matter whether it ends up to the left or the right?" A good indicator of this distance is the average value of the square of the net x-displacement, $(\Delta x)^2$, because that's always a positive number. We can calculate this average.

Since $\Delta x = \Delta x_1 + \Delta x_2 + \Delta x_3 + \Delta x_4 + ... + \Delta x_N$, we have

$$(\Delta x)^2 = (\Delta x_1)^2 + (\Delta x_2)^2 + ... + (\Delta x_N)^2 + 2\Delta x_1 \Delta x_2 + 2\Delta x_1 \Delta x_3 + 2\Delta x_1 \Delta x_4 + ...$$

In this square of the net x displacement, there are two kinds of terms: squares of individual x displacements such as $(\Delta x_2)^2$ and "cross terms" like $2\Delta x_1 \Delta x_3$. Given the random nature of the process, we expect that one

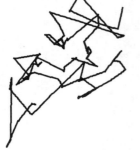

Figure 11.28 A random walk, generated by a computer program, using random numbers.

fourth of the time both Δx_1 and Δx_3 are positive (product is positive), one-fourth of the time they're both negative (product is positive), one-fourth of the time Δx_1 is positive with Δx_3 negative (product is negative), and one-fourth of the time Δx_1 is negative with Δx_3 positive (product is negative). Therefore, the average value of each cross term $2\Delta x_i \Delta x_j$ $(i \neq j)$ is zero.

As for the other terms, the squares of individual x-displacements such as $(\Delta x_2)^2$, the average value of each of these terms is some number d_x^2, related to d^2, the square of the mean free path that we discussed in Section 11.3 on page 389.

? Count up how many of these square terms there are for N collisions and calculate the average value of the square of the net displacement:

$$\overline{(\Delta x)^2} = ?$$

There are N of these terms, and the square root of this quantity is the "root-mean-square" or "rms" value of the net displacement:

$$\Delta x_{rms} = \sqrt{\overline{(\Delta x)^2}} = (\sqrt{N})d_x$$

We can also write this formula in terms of time. If we let v be the average speed of the special molecule between collisions, and we let T be the average time between collisions, we have $v = d/T$. Also, the total time t for the N collisions is $t = NT$.

? Rewrite the formula for the rms displacement in terms just of d_x, v, and t (that is, eliminate N and T): $\Delta x_{rms} = ?$

We find that $\Delta x_{rms} = \sqrt{Nd_x d_x} = \sqrt{N(vT)d_x} = \sqrt{vd_x}\sqrt{t}$. This is a somewhat curious result. For ordinary motion at constant speed, displacement increases proportional to time: double the time, double the displacement. In contrast, the rms displacement in a random walk grows with the square root of the time: on average, the rms displacement doubles when the time quadruples.

This calculation was done for the x component of the motion, but we can generalize our result to real three-dimensional motion in a gas. In Figure 11.29 is a three-dimensional displacement involving Δx, Δy, and Δz.

You can see in Figure 11.29 a three-dimensional version of the Pythagorean theorem for triangles. We can assume that the motion in each dimension is independent of the motions in the other two dimensions, and we have the following result:

$$\Delta r_{rms} = \sqrt{\overline{(\Delta x)^2} + \overline{(\Delta y)^2} + \overline{(\Delta z)^2}} = \sqrt{Nd_x^2 + Nd_y^2 + Nd_z^2} = (\sqrt{N})d$$

(Here, d_x is the component in the x direction of the three-dimensional mean free path d.)

Writing the rms displacement in terms of the average speed v, the time t, and the mean-free path d, we have

$$\Delta r_{rms} = \sqrt{vd}\sqrt{t}$$

This result is somewhat unusual, because it predicts (correctly, it turns out) that for this random process the displacement from the starting location is proportional to the square root of the time rather than to the time.

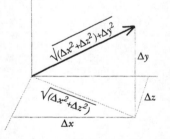

Figure 11.29 Pythagorean theorem in three dimensions.

Ex. 11.9 As you have calculated, the average speed of an air molecule at room temperature is about 500 m/s. What is Δr_{rms} for an air molecule after one second?

11.7 Summary

Fundamental physical principles

No new fundamental physical principles; application of older principles plus statistical reasoning.

New concepts

Statistical reasoning about systems containing large number of atoms.

Mean free path d: $n[\pi(R+r)^2(d)] \approx 1$

Root-mean-square speed: $v_{rms} \equiv \sqrt{\overline{v^2}}$

Work done on a gas: $W = -\int_{V_1}^{V_2} P dV$

First law of thermodynamics: $\Delta E_{sys} = W + Q$

Molecular heat capacity at constant volume C_v: $Q = NC_V \Delta T$

Molecular heat capacity at constant pressure C_p: $Q = NC_P \Delta T$

Random walk root-mean-square displacement:

$$\Delta r_{rms} = (\sqrt{N})d = \sqrt{v}d\sqrt{t}$$

Results

of gas molecules hitting an area A per second $= \frac{1}{4} nA\bar{v}$

$$P = \frac{2}{3} n\left(\frac{\overline{p^2}}{2m}\right) = nkT, \text{ where } n = N/V \text{ (\# of molecules per cubic meter)}$$

For a multiatom gas molecule, $\Delta \bar{E}_{tot} = \Delta(\frac{3}{2}kT + \bar{E}_{rot} + \bar{E}_{vib})$

$C_V = \frac{3}{2}k$ for a monatomic gas (He, etc.)

$C_V \geq \frac{3}{2}k$ for other gases (N_2, etc.)

$C_P = C_V + k$ (molecular heat capacity);

$C_P = C_V + R$ (molar heat capacity)

In a constant-temperature (isothermal) compression,

$$W_{\text{by piston}} = Q_{\text{into surroundings}} = NkT\ln\left(\frac{V_1}{V_2}\right)$$

In a no-Q (adiabatic) compression,

$$PV^\gamma = \text{constant}, \text{ where } \gamma = C_p/C_v, \text{ and also } T^{\left(\frac{C_v}{k}\right)}V = \text{constant}$$

$k = 1.4\times10^{-23}$ J/K $R = (6\times10^{23})k = 8.3$ J/K

At "standard temperature and pressure" (STP, which means a temperature of 0° Celsius or 273 Kelvin, and a pressure of 1 atmosphere), one mole of a gas occupies 22.4 liters (a liter is 1000 cubic centimeters).

Sea-level pressure (1 atm) is 10^5 N/m^2

11.8 Example problem: Leakage from a balloon

A party balloon is about one foot in diameter (about 30 centimeters). The volume of a sphere is $\frac{4}{3}\pi r^3$, where r is the radius of the sphere. Refer to the calculation you made in Exercise 11.1 on page 388 for the initial leak rate through a circular hole 1 millimeter in diameter.

(a) If the helium were to escape at a constant rate equal to the initial rate, about how long would it take for all the helium to leak out?

(b) As the helium escapes, the balloon shrinks, and the number of helium atoms per cubic meter n will stay roughly constant, in which case our analysis is pretty good. However, remember that the temperature of the gas drops due to a preferential loss of high-speed atoms. Would this effect make the amount of time to empty be more or less than the value you calculated in part (a)?

Solution

(a) The leak rate was 6.3×10^{21} atoms per second. Calculate number of helium atoms in the balloon originally:

$$\frac{4}{3}\pi(0.3 \text{ m})^3 \left(\frac{6 \times 10^{23} \text{ atoms}}{22.4 \times 10^3 \text{ cm}^3} \right) \left(\frac{10^6 \text{ cm}^3}{\text{m}^3} \right) = 3 \times 10^{24} \text{ atoms}$$

Assuming constant rate:

$$\frac{3 \times 10^{24} \text{ atoms}}{6.3 \times 10^{21} \text{ atoms/s}} = 500 \text{ s}$$

Note that when you blow up an ordinary rubber balloon the pressure is higher than one atmosphere, which means higher number density n but also a correspondingly higher leak rate, which is proportional to n. To a first approximation the density doesn't matter in this estimate of the time to empty. Also note that often when you puncture a balloon the balloon rips, creating a large opening; this is not the case we are considering.

(b) With the preferential loss of high-speed atoms, the speed distribution inside the balloon shifts to lower speeds (corresponding to a lower temperature). If the average speed is lower, the leak rate is lower, and it should take longer for the balloon to empty than we calculated in part (a), where we assumed a constant leak rate.

We've implicitly done the analysis in vacuum. If the balloon is in air, air molecules enter the balloon through the same hole.

11.9 Review questions

Gas leaks

RQ 11.1 Gas leaks at a rate L (in molecules per second) through a small circular hole. If the density of the gas is doubled, and the Kelvin temperature is doubled, and the radius of the hole is doubled, what is the new leak rate?

Mean free path

RQ 11.2 How does the mean free path of an atom in a gas change if the temperature is increased, with the volume kept constant?

RQ 11.3 What is the approximate time between collisions for one particular air molecule?

Speed of sound

RQ 11.4 How does the speed of sound in a gas change when you raise the temperature from $0°$ C to $20°$ C? Explain briefly.

Compressions and expansions

RQ 11.5 If you expand the volume of a gas containing N molecules to twice its original volume, while maintaining a constant temperature, how much thermal energy transfer is there from the surroundings?

RQ 11.6 If you compress a volume of helium containing N atoms to half the original volume in a well-insulated cylinder, what is the ratio of the final pressure to the original pressure?

11.10 Homework problems

Problem 11.1 Leakage from a tank

In the example problem (page 407) we considered leakage from a flexible party balloon. If the leakage is from a rigid container (a metal storage tank, for example), the number of atoms per unit volume, n, will decrease with time t. If the total volume of the tank is V, there are $N = nV$ atoms in the tank at any instant, so we can write the following "differential" equation (that is, an equation that involves derivatives):

$$\frac{d}{dt}(nV) = -\tfrac{1}{4}nA\bar{v}$$

This says "the rate of change of the number of atoms in the tank is equal to the (negative) of the rate at which atoms are leaving the tank."

(a) Show that the differential equation is satisfied if $n = n_{\text{initial}}e^{-\frac{A\bar{v}}{4V}t}$, where t is the time elapsed since the hole was made. Just plug this function of n, and its derivative, into the equation and show that the two sides of the equation are equal for all values of the time t. Also show that the initial particle density n is equal to n_{initial}.

(b) Despite the fancy math, this solution is really only approximate, because the average speed isn't a constant but is decreasing. Suppose however that we use a heater to keep the container and the gas at a nearly constant temperature, so that the average speed does remain nearly constant. Suppose the (rigid) container is again a sphere 30 cm in diameter, with a circular hole 1 millimeter in diameter. About how long would it take for most of the helium to leak out? Explain your choice of what you mean by "most."

Problem 11.2 A helium leak

A rigid, thermally insulated container with a volume of 22.4 liters is filled with one mole of helium gas (4 grams per mole) at a temperature of 0 Celsius (273 K). The container is sitting in a room, surrounded by air at STP.

(a) Calculate the pressure inside the container in N/m^2.

(b) Calculate the root-mean-square average speed of the helium atoms.

(c) Now open a tiny square hole in the container, with area 10^{-8} m^2 (the hole is 0.1 mm on a side). After 5 seconds, how many helium atoms have left the container?

(d) Air molecules from the room enter the container through the hole during these 5 seconds. Which is greater, the number of air molecules that enter the container or the number of helium atoms that leave the container? Explain briefly.

(e) Does the pressure inside the container increase slightly, stay the same, or decrease slightly during these 5 seconds? Explain briefly. If you have to make any simplifying assumptions, state them clearly.

Problem 11.3 Gaseous diffusion

Natural uranium ore consists mostly of the isotope U-238 (92 protons and 146 neutrons), but 0.7% of the ore consists of the isotope U-235 (92 protons and 143 neutrons). Because only U-235 fissions in a reactor, industrial processes are used to enrich the uranium by enhancing the U-235 content.

One of the enrichment methods is "gaseous diffusion." The gas UF_6, uranium hexafluoride, is manufactured from supplies of natural uranium and fluorine (each of the six fluorine atoms has 9 protons and 10 neutrons). A container is filled with UF_6 gas. There are tiny holes in the container, and gas molecules leak through these holes into an adjoining container, where pumps sweep out the leaked gas.

(a) Explain why the gas that initially leaks into the second container has a slightly higher fraction of U-235 than is found in natural uranium.

(b) Estimate roughly the practical change in the concentration of U-235 that can be achieved in this single-stage separation process. Explain what approximations or simplifying assumptions you have made to obtain your estimate.

(c) A typical nuclear reactor requires uranium that has been enriched to the point where about 3% of the uranium is U-235. Estimate roughly the number of stages of gaseous diffusion required (that is, the number of times the gas must be allowed to leak from one container into another). Note that the effects are multiplicative.

This is why a practical gaseous diffusion plant has a large number of stages, each operating at high pressure, which makes this an expensive process. The first large gaseous diffusion plant was constructed during World War II at Oak Ridge, Tennessee, and used cheap Tennessee Valley Authority electricity.

Problem 11.4 Spacecraft emergency

You are on a spacecraft measuring 8 m by 3 m by 3 m when it is struck by a piece of space junk, leaving a circular hole of radius 4 mm, unfortunately in a place that can be reached only by making a time-consuming spacewalk. About how much time do you have to patch the leak? Explain what approximations you make in assessing the seriousness of the situation.

Problem 11.5 Meteor in air

A roughly spherical meteor made mainly of iron (density about 8 grams per cubic centimeter) is hurtling downward through the air at low altitude. At an instant when its speed is 10^4 m/s, calculate the approximate rate of change of the meteor's speed. Do the analysis for two different meteors— one with a radius of 10 meters and one with a radius of 100 meters.

Start from fundamental principles. Do not try to use some existing formula that applies to a very different situation. Follow the kind of *reasoning* used in this chapter, applied to the new situation, rather than trying to use the *results* of this chapter.

A major difference from our earlier analyses is that the meteor is traveling much faster than the average thermal speed of the air molecules, so it is a good approximation to consider the air molecules below the meteor to be essentially at rest, and to assume that no air molecules manage to catch up with the meteor and hit it from behind. The meteor drills a temporary hole in the atmosphere, a vacuum, that gets filled explosively by air rushing in after the meteor has passed.

There is good evidence that a very large meteor, perhaps 10 kilometers in diameter, hit the Earth near the Yucatan Peninsula in southern Mexico 65 million years ago and caused so much damage that the dinosaurs became extinct. See the excellent account in *T. rex and the Crater of Doom*, by Walter Alvarez (Princeton University Press, 1997). Alvarez is the geologist who made the first discoveries leading to our current understanding of this cataclysmic event.

Problem 11.6 Manipulating a gas

A horizontal cylinder 10 cm in diameter contains helium gas at room temperature and atmospheric pressure. A piston keeps the gas inside a region of the cylinder 20 cm long.

(a) If you *quickly* pull the piston outward a distance of 12 cm, what is the approximate temperature of the helium immediately afterwards? What approximations did you make?

(b) How much work did you do, including sign? (Note that you need to consider the outside air as well as the inside helium in calculating the amount of work *you* do.)

(c) Immediately after the pull, what force must you exert on the piston to hold it in position (with the helium enclosed in a volume that is 20+12 = 32 cm long)?

(d) You wait while the helium slowly returns to room temperature, maintaining the piston at its current location. After this wait, what force must you exert to hold the piston in position?

(e) Next, you very *slowly* allow the piston to move back into the cylinder, stopping when the region of helium gas is 25 cm long. What force must you exert to hold the piston at this position? What approximations did you make?

Problem 11.7 Pumping up a bicycle tire

Atmospheric pressure at sea level is about 10^5 N/m^2, which is about 15 pounds per square inch (psi). A bicycle tire typically is pumped up to 50 psi above atmospheric pressure (psi "gauge"), for an actual pressure of about 65 psi. In rapidly pumping up a bicycle tire, starting from atmospheric pressure, about how high does the temperature of the air rise? Explain what approximations and simplifying assumptions you make.

11.11 Answers to exercises

11.1 (page 388) 6.3×10^{21} atoms per second

11.2 (page 389) T decreases because average v decreases.

11.3 (page 390) $L \approx 0.3 \times 10^{-10}$ m

11.4 (page 392) 10^5 N/m^2

11.5 (page 395) Box alone; buoyancy force is Mg upward.

11.6 (page 400) 1000 J; 1000 J; 0

11.7 (page 402) $\frac{3}{2}k + k = \frac{5}{2}k$

11.8 (page 403) $\left(\dfrac{PV}{Nk}\right)^{c_v/k} V = \text{constant}$

$P^{c_v/k} V^{c_v/k+1} = \text{new constant}$

Take (c_v/k) root:

$PV^{1+k/c_v} = PV^{(c_v+k)/c_v} = PV^{c_p/c_v}$

11.9 (page 405) 6 millimeters

Chapter 12

The Efficiency of Engines

Chapter 12

The Efficiency of Engines

12.1 Fundamental limitations on efficiency

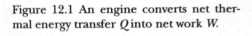

Net thermal energy transfer Q

Engine

Net work W

Figure 12.1 An engine converts net thermal energy transfer Q into net work W.

An important technology involves the conversion of thermal energy transfer Q into useful work W (Figure 12.1). For example, in a steam-powered electricity generating plant, coal is burned to warm up water that drives a steam engine to turn a generator, which converts the work done by the steam engine into electric energy. Energy conservation of course puts limits on how much useful work you can get from the burning of the coal. But there is a further limitation due to the second law of thermodynamics. It turns out that in a practical generating plant only about one-third of the energy of the coal can be turned into useful work! The fundamental problem is that thermal energy transfer is a disorderly energy transfer, and the second law of thermodynamics takes that into account.

We will find that the most efficient processes are "reversible" processes, that is, processes that produce no change in the entropy of the Universe (a reversed movie of such a process looks possible). In the next sections we discuss two processes, mechanical friction and finite-temperature-difference thermal energy transfer, which are major contributors to entropy production and whose effects must be minimized in order to obtain the most efficient performance in a mechanical system. Where we are headed is to establish limits on how efficiently thermal energy can be turned into mechanical energy in an engine, as a consequence of the second law of thermodynamics. This is a revealing example of the power of the second law of thermodynamics to set limits on phenomena.

12.1.1 Friction and entropy production

The second law of thermodynamics says that the entropy of the Universe never decreases. Portions of the Universe may experience a decrease of entropy, but only if the entropy of other portions increases at least as much (and typically more). An example of a process that increases the total entropy of the Universe is sliding friction. A block sliding across a table slows down as kinetic energy associated with the overall motion of the block turns into thermal energy inside the block, with an increase in entropy.

This friction process is irreversible. We would be astonished if after coming to rest the block suddenly started picking up speed back toward its starting position, although this would not violate conservation of energy. But the probability is vanishingly small that the thermal energy in the block could convert back into an orderly motion of the block.

12.1.2 Thermal energy transfer and entropy production

Mechanical friction contributes to increasing the entropy of the Universe. We will show that thermal energy transfer between two objects that have different temperatures also makes the total entropy of the Universe increase. This is then a process to avoid if possible in an efficient engine.

Connect a metal bar between a large block at a high temperature T_H and another large block at a low temperature T_L (Figure 12.2). We write the rate of thermal energy transfer from the high-temperature block (the "source") to the lower-temperature block (the "sink") as $\dot{Q}$. The dot over the letter Q means "rate" or "amount per second." This thermal energy transfer rate is

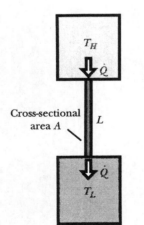

T_H

$\dot{Q}$

Cross-sectional area A L

$\dot{Q}$

T_L

Figure 12.2 Energy flows as thermal energy transfer at a rate $\dot{Q}$ from high temperature to low.

- proportional to the "thermal conductivity" σ of the bar (metals have higher thermal conductivity than glass or plastic),

- proportional to the temperature difference (twice the temperature difference, twice the rate of energy transfer),

- proportional to the cross-sectional area A of the bar (twice the cross-sectional area is like having two bars), and

- inversely proportional to the length L of the bar (twice the length of the bar, half the energy transfer rate):

RATE OF THERMAL ENERGY TRANSFER

$$\dot{Q} = \sigma A \frac{(T_H - T_L)}{L} \quad (J/s)$$

The quantity $(T_H - T_L)/L$ is called the "temperature gradient." The larger the gradient (the more rapidly the temperature changes with distance along the bar), the larger the number of joules per second of thermal energy transfer.

We want to show you that thermal energy transfer between different temperatures increases the total entropy of the Universe. Suppose the source and sink are so large that for a while the thermal energy transfer doesn't change the temperatures much. Since $1/T = (dS)/(dE)$, it follows that a small thermal energy input (dE) into a system, of amount Q, leads to an entropy change of that system:

$$\Delta S = \frac{Q}{T}$$

Suppose the thermal energy transfer *rate* is $\dot{Q}$. In a short time interval Δt, the high-temperature source at temperature T_H loses an amount of energy $\dot{Q}\Delta t$.

? In this short time interval Δt, does the entropy of the source increase or decrease? By how much?

Since $\Delta S_H = (\Delta E)/T_H$, and the high-temperature source loses energy, we have $\Delta S_H = -(\dot{Q}\Delta t)/T_H$.

In this same time interval, the low-temperature sink at temperature T_L gains the same amount of energy, $\dot{Q}\Delta t$.

? In this short time interval Δt, does the entropy of the sink increase or decrease? By how much?

Since the low-temperature sink gains energy, we have $\Delta S_L = +(\dot{Q}\Delta t)/T_L$

Here is a crucial and perhaps not entirely obvious point: Does the entropy of the bar change during this time interval? No. The blocks are so large that their temperatures don't change much in a short time interval, so the temperatures at the ends of the bar, and all along the bar, are not changing. Every second, energy enters the bar, but the same amount of energy leaves the bar, so the energy of the bar isn't changing. Nothing about the bar is changing. It is merely a conduit for the thermal energy transfer, but its state is not changing.

Therefore the change in the total entropy of the Universe in a time interval Δt is

$$\Delta S_{\text{source}} + \Delta S_{\text{sink}} + \Delta S_{\text{bar}} = -\frac{\dot{Q}\Delta t}{T_H} + \frac{\dot{Q}\Delta t}{T_L} + 0 = \dot{Q}\Delta t\left(\frac{1}{T_L} - \frac{1}{T_H}\right)$$

? Is this quantity positive or negative? Is this consistent with the second law of thermodynamics?

Since $T_L < T_H$, $1/T_L > 1/T_H$, and the entropy change of the Universe in this process is positive. This is consistent with the second law of thermodynamics, which states that the entropy of the Universe should never decrease.

Reversible and irreversible processes

An increase in the total entropy of the Universe is associated with an irreversible process, because to return to the earlier state would require that the total entropy of the Universe actually decrease, which won't happen with macroscopic systems. We would be astonished if the thermal energy transfer suddenly reversed and ran from the colder block thermally back "uphill" to the hotter block!

There is only one way to carry out thermal energy transfer reversibly or nearly so (that is, with little or no change in the total entropy of the Universe)—do it between systems whose temperatures are very nearly equal to each other ($T_H \approx T_L$). But there's a practical problem.

? If the two temperatures are nearly equal (to avoid increasing the entropy of the Universe), at what rate is energy transferred from the (slightly) hotter source to the (slightly) cooler sink?

Thermal energy transfer will flow extremely slowly, because $\dot{Q}$ is proportional to the temperature difference (or more specifically to the temperature gradient). If there is hardly any temperature difference, the energy transfer rate will be very small. So we can carry out thermal energy transfers nearly reversibly, hardly changing the entropy of the Universe, but only if we do it so slowly as to be of little practical use in driving some kind of mechanical engine that converts thermal energy transfer into work.

> *Ex. 12.1* The thermal conductivity of copper (a good thermal conductor) is 400 watt/K/m (for comparison, the thermal conductivity of iron is about 70 watt/K/m, and that of glass is only about 1 watt/K/m). One end of a copper bar 1 cm on a side and 30 cm long is immersed in a large pot of boiling water (100° C), and the other end is embedded in a large block of ice (0° C). It takes about 335 joules to melt one gram of ice. How long does it take to melt one gram of ice? Does the entropy of the Universe increase, decrease, or stay the same?

12.2 A maximally efficient process

Despite the serious practical problem that reversible thermal energy transfer proceeds infinitesimally slowly, we will analyze reversible "engines" in which (nearly) reversible thermal energy input with tiny temperature differences is used (very slowly!) to do something mechanically useful, such as lift a weight or turn an electric generator. We will also assume that we can nearly eliminate sliding friction. The idea that we will pursue is to see how efficiently we can convert disorderly thermal energy into orderly mechanical energy (a lifted weight). We expect that reversible processes, which don't increase the total entropy of the Universe, should be the most efficient processes, though we recognize that these most efficient processes must proceed excruciatingly slowly.

Since we will use nearly reversible processes to do work, we could run the engine backwards, and such a backwards-running engine turns out to be a refrigerator—an engine in which work input to the engine can lead to extracting thermal energy from something to make it colder or to keep it cold.

The conception of a theoretically most efficient possible engine, and the recognition that this ideal engine would have to be reversible and not increase what we now call the total entropy of the Universe, was due to a young French engineer, Sadi Carnot, in 1824. His achievement is all the more remarkable because it came before the principle of energy conservation was established!

After using the second law of thermodynamics to determine the maximum possible efficiency of infinitely slow engines, in the last part of the chapter we will analyze engines that run at useful speeds. We will find that they are even less efficient than the infinitely slow engines.

12.2.1 A cyclic process of a reversible engine

Some of the first practical engines drove pumps to pump water out of deep mines in England, repeatedly lifting large amounts of water large distances. The energy came from burning coal, which boiled water to make steam in a cylinder which pushed up on a piston, and operated the pump. People started wondering how efficient such an engine could be. What limits the amount of useful work you can get from burning a ton of coal?

We describe a scheme for running an engine in a reversible way, which ought to be as efficient as possible. For concreteness, our engine is a device consisting of a cylinder containing an ideal gas, with a piston and some sand on the piston to adjust the pressure on the gas (Figure 12.3). This is a simple and familiar device which permits energy exchanges in the form of work and thermal energy transfer. As we'll see, the actual construction details of the engine don't matter for the theoretical question regarding the maximum possible efficiency, though they matter very much in the actual construction of a useful engine.

A constant-temperature (isothermal) expansion

In addition to the engine itself, we need a large high-temperature source, large enough that extracting some energy from it will cause only a negligible decrease in its temperature. Start by arranging that the engine (the gas cylinder with its piston) has a temperature very slightly lower than the temperature T_H of the high-temperature source. We connect the engine to the high-temperature source and perform a reversible constant-temperature (isothermal) expansion of the engine, doing some work on the surroundings in the process (Figure 12.3). We lift the piston by slowly shifting some sand sideways from the piston onto nearby shelves. This requires almost no work on our part but results in the lifting of some weight. We make sure that the temperature of the gas doesn't change during this process. There is thermal energy transfer into the engine because it is very slightly cooler, but the temperature difference is so slight that there is negligible change in the entropy of the Universe.

Entropy and energy in lifting the piston

Remember that the entropy change of a system when there is thermal energy transfer Q is $\Delta S = Q/T$. In lifting the piston the entropy of the high-temperature source has changed by an amount $\Delta S_H = -Q_H/T_H$, where Q_H is a positive quantity. The entropy of the gas has changed by an amount $\Delta S_{gas} = +Q_H/T_H$. The entropy of the Universe hasn't changed at all. (The increased entropy of the gas is associated with the fact that there are more ways to arrange the molecules in a large volume than in a small volume.)

Because the temperature hasn't changed, there is no change in the energy of the ideal gas. Recall that the energy of an ideal gas, or a low-density real gas, is a function solely of the temperature, not the volume. (A dense real gas is more complicated, because the electric potential energy for pairs

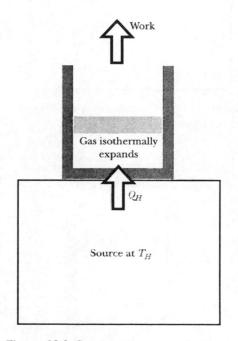

Figure 12.3 Constant-temperature expansion; gas temperature just slightly lower than T_H

of molecules depends on distance, and therefore the energy of the gas depends on volume as well as temperature.)

We have succeeded in converting all of the thermal energy transfer from the high-temperature source into useful work (lifting the piston), because none of the input energy went into changing the energy of the gas. This is 100% efficiency in converting thermal energy into useful work on the surroundings. What's the problem?

The need for a cycle

The problem is this: We need to be able to do this again, and again, and again. For example, we want to keep pumping water out of the deep mine, over and over. But to repeat the lifting process, we have to return to the original state, with the gas compressed. We could simply reverse the process, doing work on the gas (constant-temperature compression), with thermal energy transfer from the gas into the high-temperature source. But the net effect would be that there was zero net thermal energy transfer to the gas, and zero net work done on the surroundings.

We need to run our engine in a non-trivial repetitive "cycle," where we can lift the piston repeatedly. Each time we lift the piston and do useful work, we need to get the piston back down again without having to do the same amount of work to push it down. If we can get the gas back to its original state with net work done on the surroundings, we will have a useful device.

One possibility for bringing the piston down would be to let the gas cool down. But to lower the temperature we would have to place the gas cylinder in contact with a large object at some low temperature T_L, called the "sink" because, as we'll see, we will dump some thermal energy into it. We can't place the gas cylinder in contact with the sink immediately, because the gas is at a high temperature T_H, and placing the hot gas in contact with the cold sink would mean that there would be a large temperature difference, and there would be a large increase in the entropy of the Universe associated with the thermal energy transfer from the hot gas to the cold sink.

A no-Q (adiabatic) expansion

To avoid this large production of entropy, we need to bring the temperature of the gas down almost to T_L *before* making contact with the sink. To do this, we disconnect the gas cylinder from the high-temperature source and perform a reversible no-Q (adiabatic) expansion, which does some more work on the surroundings and lowers the temperature of the engine (Figure 12.4). This is accomplished by slowly removing some weight from the piston, allowing the gas to expand and the temperature to fall. We stop the expansion when the temperature of the gas is just slightly higher than the low temperature T_L of the sink.

A constant-temperature (isothermal) compression

We can now safely place the gas cylinder in contact with the low-temperature sink. This is okay as far as entropy production is concerned, because we made sure that the temperature of the gas is only very slightly higher than the sink temperature T_L. The piston is now quite high, and we need to bring it down. We do this by slowly adding some weight to the piston, performing a constant-temperature (isothermal) compression (Figure 12.5). There is thermal energy transfer into the sink. Because the temperature of the ideal gas is nearly T_L at all times, there is no energy change in the gas. Therefore the work that we do is equal to the thermal energy transfer Q_L into the sink.

There is an increase in the entropy of the sink $\Delta S_L = +Q_L/T_L$, and there is an entropy decrease in the entropy of the gas $\Delta S_{gas} = -Q_L/T_L$. The

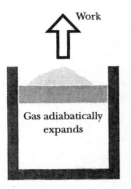

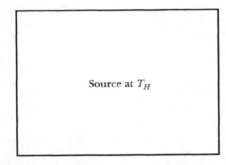

Figure 12.4 No-Q expansion; gas temperature drops to T_L.

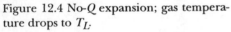

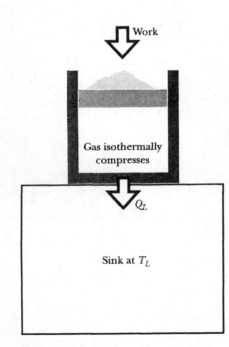

Figure 12.5 Constant-temperature compression; gas temperature just slightly higher than T_L.

entropy of the Universe doesn't change. (The decreased entropy of the gas is associated with the fact that there are fewer ways to arrange the molecules in a small volume than in a large volume.

A no-Q (adiabatic) compression

We stop the compression at a particular state of the gas chosen with care so that we can do the following: We disconnect the engine from the low-temperature sink and perform a no-Q compression that raises the temperature of the gas to a temperature just barely less than T_H, the temperature of the high-temperature source. We do some work to perform this compression (Figure 12.6).

This compression brings the system back to its original state (density, pressure, temperature), so we can repeat the cycle of four processes all over again. The possible usefulness of this engine is that we can run it repeatedly, over and over. What remains to be analyzed is how much work is done on the surroundings in one cycle, and how much thermal energy input we need to supply. Of course if we find that the net work is zero, the engine won't be useful. However, explicit calculations for an ideal gas show that at least in that case there is net work done on the surroundings. We will find that this is true for any reversible engine run in such a cycle.

This reversible cycle of two constant-temperature (isothermal) processes and two no-Q (adiabatic) processes running between a high-temperature source and a low-temperature sink is called a "Carnot cycle."

Summary of the cycle

Here is a summary of the four processes of the cycle:
- Isothermal expansion in contact with T_H
- Adiabatic expansion (no contact); temperature falls to T_L
- Isothermal compression in contact with T_L
- Adiabatic compression (no contact); temperature rises to T_H

The four processes return the gas to its original temperature and volume. The question is, did we get any net work out of this cycle?

12.2.2 Entropy in a cycle of a reversible engine

Let Q_H be the (absolute value of the) thermal energy extracted from the source at temperature T_H during the constant-temperature expansion, and let Q_L be the (absolute value of the) thermal energy dumped into the sink at temperature T_L during the constant-temperature compression. Remember that the entropy change of a system when there is thermal energy transfer Q is $\Delta S = Q/T$.

The entropy change of the high-temperature source in one cycle $= -\dfrac{Q_H}{T_H}$

The entropy change of the low-temperature sink in one cycle $= +\dfrac{Q_L}{T_L}$

? Is there any entropy change *in the engine* in one cycle?

The entropy change of the engine in one cycle is zero, because the state of the gas is brought back to its original state.

? Therefore, in one cycle, how much entropy change is there in the surroundings as a result of the work done on and by the surroundings?

Only the thermal energy transfer affects the entropy during one cycle (work doesn't contribute), and we calculated these through the formula $\Delta S = Q/T$. We were careful to run the engine in a reversible way, avoiding

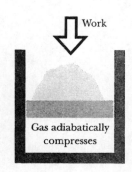

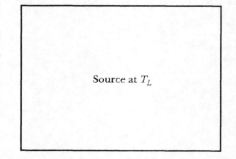

Figure 12.6 No-Q compression; gas temperature rises to T_H. Engine is now back to its original state, ready to begin another cycle.

making any increase in the total entropy of the Universe, so in one cycle we have

$$-\frac{Q_H}{T_H} + \frac{Q_L}{T_L} + 0 = 0$$

Therefore we can write this important relationship for one cycle of the engine:

ONE CYCLE OF A REVERSIBLE ENGINE

$$\frac{Q_H}{T_H} = \frac{Q_L}{T_L}$$

This is a surprisingly simple result for such a complicated process. What does this simple result depend on? Solely on the second law of thermodynamics, that statistically the total entropy of the Universe will essentially never decrease.

We illustrated the processes in the cycle with a cylinder filled with an ideal gas, but the result above doesn't actually depend at all on the details of how the engine is constructed. In particular, it doesn't matter whether the engine contains helium gas, or a block of rubber, or a liter of alcohol. There would be an easily visible difference between using a solid or liquid rather than a gas in the engine, because the distance through which the piston would move would be much smaller than with a gas. But the simple result shown above would remain the same.

12.2.3 Energy in a cycle of a reversible engine

Let's review the energy changes in the engine in one cycle. The source inputs an amount of energy Q_H. The sink removes an amount of energy Q_L. The working substance returns to its original state, so it undergoes no change of energy. Is there anything else? Yes, it may be that there has been net work W done on the surroundings. How can we tell? We can use conservation of energy of the engine for one cycle:

$$\Delta E_{engine} = Q_H - Q_L + W = 0$$

? Use the result that $Q_H/T_H = Q_L/T_L$ (because the total entropy of the Universe doesn't change), to determine whether W is positive or negative.

Because $Q_H = (T_H/T_L)Q_L > Q_L$, we have $W = Q_L - Q_H < 0$. The fact that W is negative means that our engine does net work on the surroundings in one cycle. Figure 12.7 outlines the basic scheme. In one cycle the net effect is that the engine takes in energy Q_H from a hot source, does some work on something (for example, turns an electric generator), and exhausts the remaining energy Q_L into a cold sink. The magnitude of the work done is $|W| = Q_H - Q_L$.

The crucial issue is that the engine will not run in repeatable cycles without exhausting some energy to the low-temperature sink. You *cannot* convert all of the thermal energy Q_H into useful work. We need the low-temperature block to allow us to begin the constant-temperature compression phase of the cycle in a way that avoids any thermal energy transfer with significant temperature difference, which would increase the total entropy of the Universe.

We have to pay for the high-temperature energy Q_H which we use, and the loss of some of the input energy in the form of Q_L is an unfortunate fact of life. We pay for coal or fuel oil or electricity to warm something up to the high temperature T_H, from which we can draw energy to do work for us.

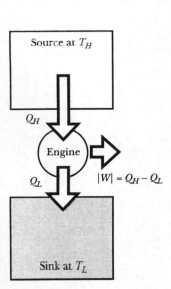

Figure 12.7 Conservation of energy lets us determine the net work output.

For example, in an old railroad steam engine a fire maintained water at a high temperature to constitute the high-temperature source. In an automobile engine we burn gasoline to create a high temperature and push the pistons. In an electricity generating station we burn coal or fuel oil, or use nuclear fission reactions, to create a high-temperature source from which to drive the generators. We'd like to get our full money's worth (Q_H), but we don't. We only get $|W| = Q_H - Q_L$, having "lost" some of the energy to a low-temperature sink, which typically is our ordinary surroundings at around 20° C (293 K).

Efficiency

This leads to the question, what fraction of the energy Q_H we pay for turns into useful work W? This fraction is called the "efficiency" of the engine:

$$\text{efficiency} \equiv \frac{|W|}{Q_H}$$

? Given this definition of efficiency, prove the following result:

THE EFFICIENCY OF A REVERSIBLE ENGINE

$$\text{Efficiency} = 1 - \frac{T_L}{T_H}$$

We have $|W|/Q_H = (Q_H - Q_L)/Q_H = 1 - Q_L/Q_H = 1 - T_L/T_H$. This may seem surprising. The efficiency of a reversible engine depends solely on the ratio of the absolute temperatures of the source and sink. We emphasize that it doesn't matter how the engine is designed, or what kind of materials it is made of. This is the highest efficiency we can ever achieve for converting thermal energy into useful work.

Ex. 12.2 What is the efficiency of a reversible engine if the source is a large container of boiling water and the sink is a large block of melting ice?

Ex. 12.3 An actual electricity generating plant is powered by a nuclear reactor, with a high temperature of 300° C and low temperature of 25° C (near room temperature). What is the efficiency, assuming that we can treat the processes as being reversible?

12.2.4 No other engine can be more efficient

We can show that no engine running between temperatures T_H and T_L can be more efficient than a reversible engine.

The proof is a "proof by contradiction." Suppose that an inventor claims to have invented some cleverly designed new kind of (cyclic) engine, with a higher efficiency than $(1 - T_L/T_H)$ when run between these same two temperatures. In that case for a given Q_H the new engine would exhaust less Q_L than is exhausted from a reversible engine, and the entropy change of the Universe with this new engine would be negative instead of zero:

$$-\frac{Q_H}{T_H} + \frac{Q_L}{T_L} < 0 \quad \text{(Impossible!)}$$

This would be a violation of the second law of thermodynamics, so it is impossible. We can be quite sure that the inventor's claims are not valid. Note carefully that the inventor's claims don't violate energy conservation. Nothing about energy conservation forbids converting 100% of the input energy Q_H into work W in a cycle. The impossibility lies rather in the massive improbability of seeing the total entropy of the Universe decrease.

We took great pains to make a reversible engine, for which the total entropy change of the Universe was zero. No other engine can be more efficient. In fact, any real engine will have some friction and some thermal energy transfer across differing temperatures, in which case the entropy change of the Universe will be positive. So the second law of thermodynamics leads to the following, where "= 0" applies only to ideal, reversible engines, and "> 0" applies to real engines:

ENTROPY CHANGE OF THE UNIVERSE FOR REAL ENGINES

$$\Delta S_{\text{Universe}} = -\frac{Q_H}{T_H} + \frac{Q_L}{T_L} \geq 0$$

We should not underestimate the need for the ingenuity of inventors, however. The reversible engine made with an ideal gas cylinder is pretty useless for practical purposes. The challenge of ingenuity is to design engines that are practical, and this involves many kinds of engineering design decisions and trade-offs. But no matter how ingenious the design, the second law of thermodynamics puts a rigid limitation on how efficient *any* engine can be.

Ex. 12.4 If $\Delta S_{\text{Universe}} > 0$ in one cycle of an engine, show that the efficiency is less than the ideal efficiency $(1 - T_L/T_H)$ obtained with a reversible engine.

12.2.5 Running the engine backwards: A refrigerator or heat pump

We could run our engine backwards, since we took care to make all aspects of the cycle reversible. Starting from our original starting point, we would disconnect from the hot block, perform a no-Q expansion to lower the temperature to that of the cold block, connect to the cold block, do a constant-temperature expansion, then disconnect from the cold block. Next we do a no-Q compression to raise the temperature to that of the hot block, connect to the hot block, and do a constant-temperature compression back to the original state.

The net effect in one cycle is that the low-temperature block gives some energy Q_L to the engine, the high-temperature block absorbs some energy Q_H from the engine, and we do some (positive) work on the system (instead of the engine doing work on the surroundings). Here is the formula for conservation of energy of the engine when we run this reversed engine:

$$\Delta E_{\text{engine}} = Q_L - Q_H + W = 0$$

? Use the result that $Q_H/T_H = Q_L/T_L$ (which is still valid for our reversed engine, as you can convince yourself) to determine whether W is positive or negative.

Since $Q_H = (T_H/T_L)Q_L > Q_L$, we have $W = Q_H - Q_L > 0$. This seems an odd kind of "engine": it doesn't do any work for us—we have to do work on it. Is that of any use? Yes! At the cost of doing some work, we extract energy from a low-temperature block and exhaust it into a high-temperature block. This is a refrigerator (Figure 12.8).

Consider how we keep food cold in an ordinary home refrigerator. A reversed engine maintains the inside of the refrigerator at a low temperature

T_L, while the room is at a higher temperature T_H. Although the door and walls are heavily insulated, some thermal energy transfer does leak through from the room, and this energy, Q_L, must be removed to keep the food at a constant low temperature T_L. We exhaust an amount of energy Q_H into the room at temperature T_H. It takes an amount of work W to achieve this effect of moving energy out of the cold region and into the warm region.

In the most favorable case (no increase in total entropy) we have the following two conditions stemming from entropy and energy considerations (the high-temperature exhaust energy must equal the sum of the low-temperature energy plus the work we put in to drive the cycle):

$$\frac{Q_H}{T_H} = \frac{Q_L}{T_L} \text{ and } Q_H = Q_L + W$$

Here our view of "efficiency" shifts a bit. What we care about is how much work W we have to do to remove an amount of leakage energy Q_L from the inside of the refrigerator.

? Show that the energy removal we get per amount of work done is as follows:

$$\frac{Q_L}{W} = \frac{1}{(T_H/T_L) - 1} > 1$$

Isn't this result somewhat surprising? Are we getting something for nothing when Q_L is bigger than the work W that we do? No, there's no violation of energy conservation. It's just that Q_L plus W equals the output energy Q_H, and we only have to add a modest amount of work to drive the energy "uphill."

Heat pump

A related device is the "heat pump" used in some areas to warm homes by driving low-temperature energy "uphill" into a higher-temperature house. Figure 12.9 shows a house whose interior is maintained at a temperature T_H despite continual leakage Q_H into the outside air. Energy Q_L at low temperature T_L is pumped out of the ground into the house, by the addition of some work W that we do in an engine called a heat pump. This is somewhat similar to the refrigerator situation, except that now what we care about is how much high-temperature energy Q_H we get per amount of work W done by the heat pump.

? Show that this ratio is given by the following formula for the heat pump:

$$\frac{Q_H}{W} = \frac{1}{1 - (T_L/T_H)} > 1$$

Again, this feels like we're getting something for nothing, since Q_H is greater than the work W that we do. After all, if you warm the house directly with gas or oil or electric energy, you have to pay for every joule of leakage, not some small fraction of the leakage. So why aren't heat pumps much more commonly used than they are? Basically because we've been calculating the best deal we can possibly obtain (the case of the reversible engine—that is, one whose operation leads to no increase in the total entropy of the Universe). Real engines running either forward or reversed don't attain the theoretical maximum performance, as we will see in the next sections.

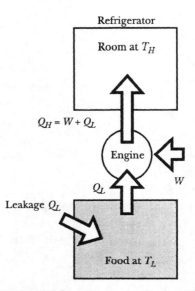

Figure 12.8 Run the engine backwards and you have a refrigerator.

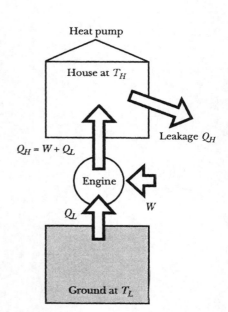

Figure 12.9 Run the engine backwards and you have a heat pump.

Ex. 12.5 Suppose there is a leakage rate of 50 watts through the insulation into a refrigerator, which we maintain at 3° C. What is the minimum electric power required to continually remove this leakage energy, to maintain the low temperature? In that case, what is the rate at which thermal energy is exhausted into the room? Room temperature is about 20° C.

Ex. 12.6 Suppose the temperature underground from where we draw low-temperature energy for a heat pump is about 5° C, and we want to keep the house at a temperature of 20° C. How many joules of work must our heat pump supply for every joule of leakage of energy there is out of the house?

12.3 *Why don't we attain the theoretical efficiency?

In Ex. 12.3 on page 421 you calculated the reversible-engine efficiency of an actual nuclear-powered electricity generating plant to be 48%, but the observed efficiency of this plant is only 30%. Real engines do not attain the efficiency predicted by the second law of thermodynamics for ideal reversible engines. In fact, the real efficiency is often only about half of the theoretical efficiency. The reason for less than optimum performance in real engines is partly due to mechanical friction, but this effect can be minimized by good design and proper lubrication. The main limitation on performance comes from the practical necessity of incorporating thermal energy transfers between parts of the system that are at significantly different temperatures, which leads to sizable increases in the total entropy of the Universe, and to much reduced efficiency.

The problem is speed. As we saw earlier in the chapter, the rate of thermal energy transfer $\dot{Q}$ in joules per second between two objects (such as the hot or cold block and the engine) at temperatures T_H and T_L connected by some conducting material of length L and cross-sectional area A, with thermal conductivity σ is this:

$$\dot{Q} = \sigma A \frac{(T_H - T_L)}{L}$$

? So what is the rate of thermal energy transfer in a reversible engine when in contact with the hot block or the cold block?

Alas, a reversible engine is totally impractical when it comes to getting anything done in a finite amount of time, because the rate of thermal energy transfer is zero. If an engine is perfectly reversible, a cycle takes an infinite amount of time!

Consider the portion of an engine's cycle where the working substance is in thermal contact with the high-temperature source. In order to carry out the expansion quickly, there must be a high rate of thermal energy transfer from the source into the working substance. That means we need a large contact area A, a short distance L, and a high thermal conductivity σ. Most significant of all, the temperature of the source (T_H) must be considerably higher than the temperature of the engine, leading to irreversibility and increase of the entropy of the Universe.

The efficiency of a nonreversible engine

It is actually possible to calculate the effect of such nonreversible thermal energy transfer, as was first pointed out in an article titled "Efficiency of a

Carnot engine at maximum power output," by F. L. Curzon and B. Ahlborn, *American Journal of Physics* volume **43**, pages 22-24 (January 1975).

Suppose that when the engine is in contact with the high-temperature block at temperature T_H, the engine is at a lower temperature T_1, so that there is a finite thermal energy transfer rate into the engine of $\dot{Q}_H = b(T_H - T_1)$ joules per second, where b is a constant that lumps together the factors of thermal conductivity, cross-sectional area, and length ($\sigma A/L$). Similarly, assume that when the engine is in contact with the low-temperature block at temperature T_L, the engine is at a higher temperature T_2, so that there is a finite thermal energy transfer rate into the working substance of $\dot{Q}_L = b(T_2 - T_L)$ joules per second. We're making the factor b be the same for both the high-temperature and the low-temperature contacts. This simplifies the algebra, and it turns out that using different b factors doesn't affect the final result anyway.

Now the energy flow diagram looks like Figure 12.10, with an ideal reversible engine running between the temperatures T_1 and T_2, but with irreversible finite-rate thermal energy transfer between this reversible engine and the source and sink at temperatures T_H and T_L.

Energy enters the engine from the high-temperature source at a rate $\dot{Q}_H$, work is done at a rate $\dot{W}$, and energy is dumped into the low-temperature sink at a rate $\dot{Q}_L$. All of these quantities are measured in watts (joules per second). From energy conservation we have

$$\dot{Q}_H = \dot{W} + \dot{Q}_L$$

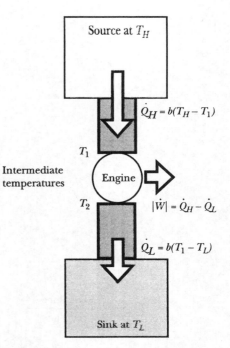

Figure 12.10 An engine that runs at a non-zero rate.

Curzon and Ahlborn pointed out that although this engine will necessarily be less efficient than a reversible engine (due to the irreversible thermal energy transfers), it might be useful to ask the question, "What is the *maximum* output power $\dot{W}$ per input power $\dot{Q}_H$?" This question comes down to the question of what value of b will maximize the output power. This is essentially an engineering design question, yet there is an unusually simple answer to this question about the maximum output power.

There must be a maximum power output

By considering two extreme designs (Figure 12.11) it is easy to see that there must be a maximum possible power output (per power input). One extreme set of operating conditions is the reversible engine we've already considered, in which all the energy flow rates are zero, including the output power $\dot{W}$.

Another extreme is to arrange conditions in such a way that $T_1 \approx T_2$, so that we have essentially connected the source and sink by a conducting bar. In that case the input power $\dot{Q}_H$ is equal to $\dot{Q}_L$, and the useful output power $\dot{W}$ is again zero.

Somewhere between these two extremes, for both of which there is zero output power, we expect to find a maximum output power. We now look for the conditions that maximize the output power. The search will be carried out by expressing the output power in terms of T_1, the higher of the two temperatures to which the engine is directly exposed in Figure 12.10. Then we will vary T_1 (by adjusting b in Figure 12.10) until we maximize the power output. The details are given in a derivation at the end of the chapter. We obtain the following surprisingly simple result:

THE EFFICIENCY OF A MAXIMUM-POWER ENGINE

$$\frac{\dot{W}}{\dot{Q}_H} \leq 1 - \sqrt{\frac{T_L}{T_H}}$$

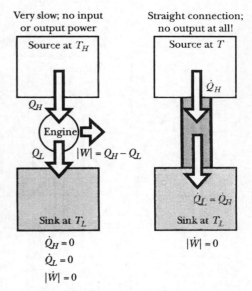

Figure 12.11 Two extreme designs where there is zero power output.

This gives us the upper bound on the fraction of the thermal input power that can end up as output mechanical power if we maximize the output power. The formula looks quite similar to the reversible-engine zero-power efficiency, but the square root makes a big difference.

Comparison with the real world

Curzon and Ahlborn give an example of a plant powered by a nuclear reactor, with high temperature 300° C and low temperature 25° C. Using the formula given above, the calculated efficiency is 28%, and the actual observed efficiency is 30% (see exercise at the end of this section). This confirms the estimate that Curzon and Ahlborn made that

- the main source of irreversibility is in finite-temperature-difference thermal energy transfer, and
- power plants will normally be run in such a way as to maximize output power.

There is an economic issue here. Suppose that you are responsible for building and operating an electricity generating plant, and you must supply one megawatt of power (one million joules per second). If the operating expense of buying fuel is extremely high but capital investment in generators is inexpensive, it makes sense to build lots of generators and run them very slowly, nearly reversibly, to make the required megawatt of electric power. If the operating expense of buying fuel is relatively low but capital investment in generators is expensive, it makes sense to build few generators and go for maximum power, even though the fuel is not used very efficiently in producing the megawatt of electric power.

Similar issues apply to refrigerators and heat pumps. Thermal energy leaks into a refrigerator at some rate (joules per second) and must be removed by "pushing it thermally uphill" to a higher temperature, the temperature of the room. But in order to have a nonzero rate of thermal energy flow from the food to the engine, the engine must reach an even lower temperature than the food, with entropy-producing thermal energy transfer from the food into the engine. Also, in order for there to be a nonzero rate of thermal energy flow from the engine to the room air, the engine must reach an even higher temperature than the air, with entropy-producing thermal energy transfer from the engine into the air. Schematically the situation looks like Figure 12.12.

Often a refrigerator or freezer has exposed coils (a "heat exchanger") where the thermal energy transfer occurs between the working substance circulating in the coils and the air. If you touch the coils, you find that they are indeed much hotter than room temperature, in order to drive a sufficiently high rate of thermal energy transfer into the air. Notice not only that the heat exchanger must be hotter than the air, but that it is a rather large and costly device because it has to have a large surface area to get a large rate of energy transfer.

In the case of heat pumps, where the heat pump picks up low-temperature energy underground, the heat pump must reach an even lower temperature in order to get a nonzero rate of thermal energy flow from the ground into the engine. Also, in order for there to be a nonzero rate of thermal energy flow into the house, the engine must reach an even higher temperature than the air, with entropy-producing thermal energy transfer from the engine into the air. The radiators (heat exchangers) in the house must be considerably hotter than the air. There is considerable expense in all the metal in the ground and in the house that enables adequate thermal energy transfer rate. Schematically the situation looks like Figure 12.13.

So although a heat pump does have a theoretical advantage in warming a house in part from energy in the cold ground, the expense of the heat exchangers and the problems of going to rather low temperatures in the

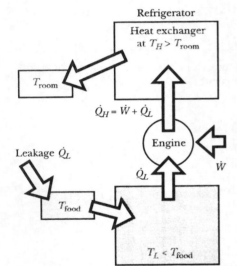

Refrigerator

Heat exchanger
at $T_H > T_{room}$

T_{room}

$\dot{Q}_H = \dot{W} + \dot{Q}_L$

Engine

Leakage $\dot{Q}_L$

$\dot{Q}_L$

$\dot{W}$

T_{food}

$T_L < T_{food}$

Figure 12.12 A refrigerator with nonzero cooling rate.

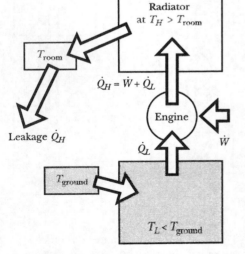

Heat pump

Radiator
at $T_H > T_{room}$

T_{room}

$\dot{Q}_H = \dot{W} + \dot{Q}_L$

Engine

Leakage $\dot{Q}_H$

$\dot{Q}_L$

$\dot{W}$

T_{ground}

$T_L < T_{ground}$

Figure 12.13 A heat pump with nonzero thermal energy transfer rate.

ground are practical limitations, especially in very cold climates. Heat pumps are more useful in climates where the winters are not too severe.

Ex. 12.7 Curzon and Ahlborn give an example of a plant powered by a nuclear reactor, with high temperature 300° C and low temperature 25° C. In Ex. 12.3 on page 421 you calculated the reversible-engine zero-power efficiency to be 48%. Now calculate the non-reversible-engine maximum-power efficiency for this power plant and compare with the observed efficiency, which is 30%.

12.4 *Derivation: Maximum-power efficiency

In this appendix we find the conditions under which the power output is maximized for an engine that connects to high- and low-temperature reservoirs through finite-temperature-difference connections.

Expressing the power in terms of T_1

We will express the power output in terms of T_1, the higher temperature to which the reversible engine is subjected. We assume that the engine, the part that runs between the temperatures T_1 and T_2, can be considered to be a reversible engine because there are no different-temperature thermal energy transfers within it, and we're assuming that we can make mechanical friction negligibly small.

? In terms of the temperatures T_1 and T_2, what does the second law of thermodynamics say about the relationship between $\dot{Q}_H$ and $\dot{Q}_L$?

Since we have a reversible engine, we know that $\dot{Q}_H / T_1 = \dot{Q}_L / T_2$. Using this relationship, we can write a formula for the thermal energy transfer into the low-temperature sink in two different ways:

$$\dot{Q}_L = b(T_2 - T_L), \text{ and also } \dot{Q}_L = \frac{T_2}{T_1}\dot{Q}_H = \frac{T_2}{T_1}b(T_H - T_1)$$

Therefore we have

$$b(T_2 - T_L) = \frac{T_2}{T_1}b(T_H - T_1)$$

With a bit of manipulation we can solve this equation for T_2 in terms of T_1:

$$T_2 = \frac{T_L T_1}{2T_1 - T_H}$$

? For the inner, reversible engine, the input power is of course $\dot{Q}_H = b(T_H - T_1)$, and using the result for the efficiency of a reversible engine, show that

$$\dot{W} = b(T_H - T_1)\left(1 - \frac{T_2}{T_1}\right)$$

Using our solution for T_2 in terms of T_1, this becomes

$$\dot{W} = b(T_H - T_1)\left(1 - \frac{T_L}{2T_1 - T_H}\right)$$

Varying the thermal conduction to maximize the power

We consider the source and sink temperatures T_H and T_L to be fixed, so the only temperature that is unspecified in this equation is T_1, which we will vary until we maximize the output power $\dot{W}$. Physically, this corresponds to varying the parameter b that determines the thermal energy transfer rate. As usual, we can maximize $\dot{W}$ by differentiating it with respect to T_1, then set this derivative equal to zero (corresponding to a maximum, where the slope is zero):

$$\frac{d\dot{W}}{dT_1} = b\left[-1 + \frac{T_L}{2T_1 - T_H} + (T_H - T_1)\frac{2T_L}{(2T_1 - T_H)^2}\right] = 0$$

Solving for T_1 and simplifying, we get a quadratic equation:

$$4T_1^2 - (4T_H)T_1 + (T_H^2 - T_L T_H) = 0$$

There are two solutions to the quadratic equation:

$$T_1 = \frac{T_H \pm \sqrt{T_L T_H}}{2}$$

Only the "+" sign makes physical sense in this situation, because if T_L were nearly as large as T_H, the solution with the "–" sign would make T_1 nearly 0 instead of lying between T_L and T_H.

We conclude that if $T_1 = (T_H + \sqrt{T_L T_H})/2$, the output power $\dot{W}$ will be maximized. Since $\dot{Q}_H = b(T_H - T_1)$, our value for T_1 determines the value of b $(= \sigma A/L)$ that will make T_1 come out right to maximize the power output for a given $\dot{Q}_H$.

Finally we can plug our value for T_1 back into the expression for $\dot{W}$ to find out what the efficiency at maximum power turns out to be:

$$\frac{\dot{W}}{\dot{Q}_H} = \frac{\dot{W}}{b(T_H - T_1)} = 1 - \frac{T_L}{2T_1 - T_H} = 1 - \frac{T_L}{\sqrt{T_L T_H}}$$

Here is the final result, where we write "≤" because we have calculated an upper bound on the efficiency (having taken into account issues of thermal energy transfer rate but not mechanical friction):

THE EFFICIENCY OF A MAXIMUM-POWER ENGINE

$$\frac{\dot{W}}{\dot{Q}_H} \leq 1 - \sqrt{\frac{T_L}{T_H}}$$

This gives us the fraction of the thermal input power that ends up as output mechanical power if we maximize the output power. The formula looks quite similar to the reversible-engine zero-power efficiency, but the square root makes a big difference.

12.5 Summary

Fundamental physical principles

No new physical principles.

New concepts

Efficiency of an engine.

Results

RATE OF THERMAL ENERGY TRANSFER

$$\dot{Q} = \sigma A \frac{(T_H - T_L)}{L} \ \text{(J/s)}$$

For a cyclic engine running between high temperature T_H and low temperature T_L we have

ENTROPY CHANGE OF THE UNIVERSE FOR REAL ENGINES

$$\Delta S_{\text{Universe}} = -\frac{Q_H}{T_H} + \frac{Q_L}{T_L} \geq 0$$

THE EFFICIENCY OF A REAL ENGINE

$$\text{Efficiency} = \frac{W}{Q_H} \leq 1 - \frac{T_L}{T_H}$$

The "=" sign applies only if the engine is reversible (extremely slow processes, no friction). A heat engine run in reverse is a refrigerator (or a heat pump).

If we maximize power output, we have

THE EFFICIENCY OF A MAXIMUM-POWER ENGINE

$$\frac{\dot{W}}{\dot{Q}_H} \leq 1 - \sqrt{\frac{T_L}{T_H}}$$

12.6 Review questions

Thermal conduction

RQ 12.1 An aluminum bar 30 cm long and 3 cm by 4 cm on its sides is connected between two large metal blocks at temperatures of 135° C and 20° C, and energy is transferred from the hotter block to the cooler block at a rate $\dot{Q}_1$. If instead the two blocks were connected by an aluminum bar that is 20 cm long and 3 cm by 2 cm on its sides, what would be the thermal energy transfer rate?

Engines

RQ 12.2 A (nearly) reversible engine is used to melt ice as well as do some useful work. If the engine does 1000 joules of work and dumps 400 joules into the ice, what is the temperature of the high-temperature source?

RQ 12.3 For engines where the high and low temperatures are not very different from each other, show that the efficiency of a maximum-power engine is about half of the efficiency of a reversible engine.

12.7 Homework problems

Problem 12.1 An engine containing an ideal gas

In one cycle of a reversible engine running between a high-temperature source and a low-temperature sink, a consequence of the second law of thermodynamics is that $Q_H/T_H = Q_L/T_L$. This result is independent of what kind of material the engine contains. It is instructive to check this general result for a specific model where we can calculate everything explicitly. Consider a reversible cycle of an ideal gas of N molecules, starting at high temperature T_H and volume V_1.

(1) Perform a constant-temperature (isothermal) expansion to volume V_2, and calculate the associated thermal energy transfer Q_H into the gas.

(2) Next perform a no-Q (adiabatic) expansion to volume V_3 and temperature T_L.

(3) Next perform a constant-temperature (isothermal) compression to volume V_4, and calculate the associated thermal energy transfer Q_L out of the gas.

(4) Finally perform a no-Q (adiabatic) compression back to the starting state, temperature T_H and volume V_1.

Show that $Q_H/T_H = Q_L/T_L$ for this cycle. (Hint: Determine the ratio of the final volume to the initial volume for each of the four processes.)

12.8 Answers to exercises

12.1 (page 416) 25 seconds; entropy of Universe increases

12.2 (page 421) 0.27

12.3 (page 421) 0.48

12.4 (page 422) $\Delta S_{\text{Universe}} > 0$ implies $\dfrac{Q_L}{T_L} > \dfrac{Q_H}{T_H}$, which implies

$\dfrac{Q_L}{Q_H} > \dfrac{T_L}{T_H}$, so

$$W = Q_H - Q_L = Q_H\left(1 - \frac{Q_L}{Q_H}\right) < Q_H\left(1 - \frac{T_L}{T_H}\right)$$

12.5 (page 424) 3 watts; 53 watts

12.6 (page 424) 0.05 J

12.7 (page 427) 28%, which is close to but less than the observed 30%. This implies that the power plant is not run at maximum power.

Appendix A

Vector Review

A.1 Review: Vectors and vector notation

Vectors play a very important role in the study of physics, because the systems under consideration are two or three dimensional, and the directions of forces and motion are critical.

A.1.1 Basic properties of vectors

Suppose you start at the origin and walk 4 meters along the x axis, then 2 meters parallel to the y axis, as in Figure A.1. We represent your new position as a

vector $\vec{r}$ = <4, 2> m

x component r_x = 4 m and y component r_y = 2 m

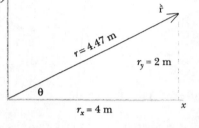

Figure A.1 A position vector and its x and y components.

We represent the position vector graphically in Figure A.1 by an arrow whose tail is at the origin and whose arrowhead is at your position.

From the Pythagorean theorem for right triangles, the net distance you have moved from the starting point is

$$\sqrt{(4 \text{ m})^2 + (2 \text{ m})^2} = \sqrt{20} \text{ m} = 4.47 \text{ m}.$$

We say that the magnitude r of the position vector $\vec{r}$ is

$$|\vec{r}| = r = 4.47 \text{ m}$$

The magnitude of a vector is written either with absolute-value bars around the vector as $|\vec{r}|$, or simply by writing the symbol for the vector without the little arrow above it, r.

Direction of a vector

One way to describe the direction of the vector is by stating the angle θ between the vector and the x axis, as shown in Figure A.1. Since the cosine of an angle is the adjacent side divided by the hypotenuse, we can read off the diagram that

$$\cos\theta = \frac{r_x}{|\vec{r}|} = \frac{r_x}{r} = \frac{4 \text{ m}}{\sqrt{20} \text{ m}} = 0.894$$

$$\theta = \arccos(0.894) = 26.6°$$

If on the other hand you know r and θ, you can find the x component:

$$r_x = r\cos\theta$$

Alternatively, since the sine of an angle is the opposite side divided by the hypotenuse, we can read off the diagram that

$$\sin\theta = \frac{r_y}{|\vec{r}|} = \frac{r_y}{r} = \frac{2 \text{ m}}{\sqrt{20} \text{ m}} = 0.447$$

$$\theta = \arcsin(0.447) = 26.6°$$

If you know r and θ, you can find the y component:

$$r_y = r\sin\theta$$

Another way to determine the angle θ is with the tangent, which is the opposite side divided by the adjacent side:

$$\tan\theta = \frac{r_y}{r_x} = \frac{2\ m}{4\ m} = 0.5$$

$$\theta = \arctan(0.5) = 26.6°$$

The tangent connects the x and y components:

$$r_y = r_x\tan\theta$$

Summary of vectors in two dimensions

$$\vec{r} = <r_x, r_y>\ \text{(Figure A.2)}$$

magnitude $|\vec{r}| = r = \sqrt{r_x^2 + r_y^2}$

$\cos\theta = r_x/r$ $\qquad\qquad r_x = r\cos\theta$

$\sin\theta = r_y/r$ $\qquad\qquad r_y = r\sin\theta$

$\tan\theta = r_y/r_x$ $\qquad\qquad r_y = r_x\tan\theta$

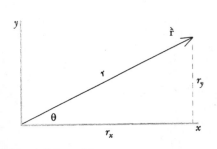

Figure A.2 A position vector.

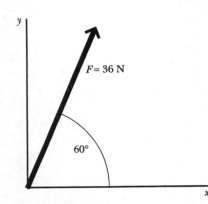

Figure A.3 Find the components (Ex. A2).

Ex. A.1 A velocity $\vec{v}$ has components $v_x = -8$ m/s and $v_y = 5$ m/s. What is the speed v (or $|\vec{v}|$)? What is the direction, as an angle relative to the x axis?

Ex. A.2 In Figure A.3, what are the x and y components of the force vector $\vec{F}$?

A.1.2 Vectors and scalars

In three dimensions a vector is a triple of numbers $<x, y, z>$. Quantities like the position of an object and the momentum of an object can be represented as vectors:

$$\vec{r} = <x, y, z>$$
$$\vec{p} = <p_x, p_y, p_z>$$

Each of the numbers in the triple is referred to as a "component" of the vector. The x component of the vector $\vec{p}$ is the number p_x. A componentsuch as p_x is not a vector, since it is only one number.

A scalar is a single number:

$$m = 5\ \text{kg}$$

Technically, although a component of a vector is a single number, it is not a scalar. If you rotate your coordinate axes, the x, y, and z components of a vector change, but a true scalar such as $m = 5$ kg doesn't change.

Vectors are mathematical entities that have their own vector algebra and vector calculus operations. For example, if a vector is multiplied by a scalar, each of the components of the vector is multiplied by the scalar:

$$a\vec{r} = <ax, ay, az>$$
$$b\vec{p} = <bp_x, bp_y, bp_z>$$

Multiplication by a scalar "scales" a vector, keeping its direction the same but making its magnitude larger or smaller.

Ex. A.3 What is the result of multiplying the vector $\vec{a}$ by the scalar f, where $\vec{a} = \langle 0.02, -1.7, 30.0 \rangle$ and $f = 2.0$?

A.1.3 The magnitude of a 3D vector

The magnitude of a vector is a scalar. If the vector represents a position in space, then the magnitude of the vector represents the distance between that position and the origin (Figure A.4). If the vector represents momentum, then the magnitude of the momentum vector represents speed times mass (for nonrelativistic momenta):

$$|\vec{r}| = r = \sqrt{(x^2 + y^2 + z^2)}$$

$$|\vec{p}| = p = \sqrt{(p^2_x + p^2_y + p^2_z)} \approx mv$$

Note that the magnitude of a vector is always positive—it can never be negative! It makes no sense to write $|\vec{r}| = -3 \text{ m}$.

In the VPython programming language, mag(any_vector) gives the magnitude of the vector.

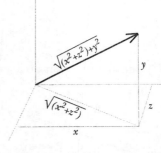

Figure A.4 The magnitude of a vector is the square root of the sum of the squares of its components (3D version of the Pythagorean theorem).

Ex. A.4 What is the magnitude of the vector $\vec{v}$, where $\vec{v} = \langle 8 \times 10^6, 0, -2 \times 10^7 \rangle$ m/s?

A.1.4 Adding and subtracting vectors

The sum of two vectors is another vector, obtained by adding the components of the vectors:

$$\vec{A} = \langle A_x, A_y, A_z \rangle$$

$$\vec{B} = \langle B_x, B_y, B_z \rangle$$

$$\vec{A} + \vec{B} = \langle (A_x + B_x), (A_y + B_y), (A_z + B_z) \rangle$$

Note that the magnitude of this vector is *not* in general equal to the sum of the magnitudes of the two original vectors! For example, the magnitude of the vector <3, 0, 0> is 3, and the magnitude of the vector <–2,0,0> is 2, but the magnitude of the vector (<3, 0, 0> + <–2,0,0>) is 1, not 5!

Here is a graphical interpretation. In Figure A.5 you first walk along displacement vector $\vec{A}$, followed by walking along displacement vector $\vec{B}$. What is your net displacement vector $\vec{C}$? The x component C_x of your net displacement is the sum of A_x and B_x. Similarly, the y component C_y of your net displacement is the sum of A_y and B_y.

To subtract one vector from another, we subtract the components of the second from the components of the first:

$$\vec{A} - \vec{B} = \langle (A_x - B_x), (A_y - B_y), (A_z - B_z) \rangle$$

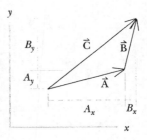

Figure A.5 Adding vectors.

Ex. A.5 What is the sum $\vec{F}_1 + \vec{F}_2$, if $\vec{F}_1 = \langle 3 \times 10^2, 0, -2 \times 10^2 \rangle$N and $\vec{F}_2 = \langle 1.5 \times 10^2, -3 \times 10^2, 0 \rangle$N?

Ex. A.6 What is the magnitude of $\vec{F}_1$? What is the magnitude of $\vec{F}_2$? What is the magnitude of $\vec{F}_1 + \vec{F}_2$?

Ex. A.7 What is the difference $\vec{F}_1 - \vec{F}_2$? What is $\vec{F}_2 - \vec{F}_1$?

A.1.5 Adding and subtracting vectors graphically

There is a geometrical procedure for adding two or more vectors as in Figure A.6.

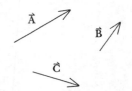

Figure A.6 Vectors to be added, using a geometrical approach.

1) Line up the vectors tip to tail (Figure A.7).
2) Draw a vector from the tail of the first vector to the tip of the last (Figure A.8).

This gives a "resultant" or "net" vector whose x and y components are the sums of the (positive or negative) x and y components of the vectors that you're adding up.

Step 1 in the procedure is crucial. If you don't place the vectors tip to tail, you are unlikely to draw the correct resultant vector. For example, look at what happens in Figure A.9 if two vectors are positioned tail to tail, which would tempt you to draw a resultant vector tip to tip.

Subtracting vectors

We can also give a geometrical interpretation of subtracting vectors if we rewrite the equation $\vec{C} = \vec{A} - \vec{B}$ in the form $\vec{C} = \vec{A} + (-\vec{B})$, where $-\vec{B}$ is a vector whose components are $-B_x$ and $-B_y$. The vector $-\vec{B}$ points in the opposite direction from $\vec{B}$, and we have $\vec{B} + (-\vec{B}) = 0$.

Figure A.10 shows a geometrical way to subtract a vector—just add the opposite of the vector.

Another way to subtract vectors is based on rearranging the equation in the form $\vec{C} + \vec{B} = \vec{A}$. Imagine what vector $\vec{C}$ you would need to add to vector $\vec{B}$ in order to produce a resultant vector $\vec{A}$ (Figure A.11).

A.1.6 The unit vector

A vector whose magnitude is 1 is called a unit vector. The symbol for a unit vector is a letter with a "hat" (carat) over it, such as $\hat{r}$; this is pronounced "r-hat". All unit vectors have the same magnitude (1), but they do not have the same direction. A unit vector pointing in the direction of any vector may be constructed simply by dividing the vector by its magnitude (a scalar):

$$\hat{r} = \frac{\vec{r}}{|\vec{r}|} = \frac{<x, y, z>}{\sqrt{(x^2 + y^2 + z^2)}}$$

$$\hat{r} = \left< \frac{x}{\sqrt{(x^2 + y^2 + z^2)}}, \frac{y}{\sqrt{(x^2 + y^2 + z^2)}}, \frac{z}{\sqrt{(x^2 + y^2 + z^2)}} \right>$$

You should convince yourself that this vector $\hat{r}$ does indeed have magnitude equal to 1. In the VPython programming language, norm(any_vector) gives a unit vector in the direction of the given vector. It is essentially equivalent to (any_vector)/mag(any_vector).

To construct a vector of known magnitude and direction, it is simply necessary to multiply the magnitude by a unit vector pointing in the appropriate direction:

$$F\hat{r} = \vec{F}$$

There are three special unit vectors which are sometimes useful. They are called i-hat, j-hat, and k-hat, and they point along the x, y, and z axes, respectively (Figure A.12):

$$\hat{i} = <1, 0, 0>$$
$$\hat{j} = <0, 1, 0>$$
$$\hat{k} = <0, 0, 1>$$

We can write a vector $\vec{F}$ in terms of these unit vectors:

$$\vec{F} = <F_x, F_y, F_z> = F_x\hat{i} + F_y\hat{j} + F_z\hat{k}$$

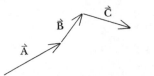

Figure A.7 Line up the vectors tip to tail.

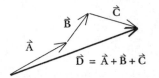

Figure A.8 Resultant or net vector goes from tail of first vector to tip of last vector.

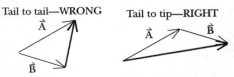

Figure A.9 Vectors to be added must be placed tail to tip, not tail to tail.

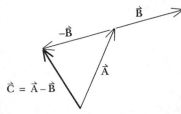

Figure A.10 Subtracting a vector by adding the opposite vector.

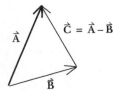

Figure A.11 What vector do we need to add to produce the resultant vector?

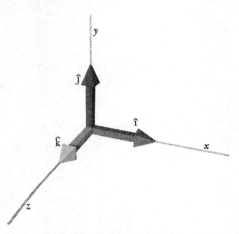

Figure A.12 The unit vectors $\hat{i}$, $\hat{j}$, $\hat{k}$.

Ex. A.8 What is the unit vector in the direction of $\hat{a}$, where $\hat{a} = \langle 4.0, -2.4, 6.2 \rangle$?

Ex. A.9 Write the vector $\hat{a}$ in terms of the unit vectors $\hat{i}$, $\hat{j}$, and $\hat{k}$.

A.1.7 Common errors in using vectors

A common error is to write an equation in which a vector is set equal to a scalar. This is not only mathematically wrong, but frequently leads to errors in reasoning. Avoid this error!

$$\cancel{\vec{F} = 3}\qquad\text{Should be } F = 3 \text{ or } \left|\vec{F}\right| = 3$$

? What is wrong with writing $\dfrac{\vec{A}}{\vec{B}}$?

You can't divide by a vector; this operation has no meaning. Perhaps what was intended was $\hat{A}/\left|\vec{B}\right|$.

? What is wrong with saying $\left|\vec{A}\right| = -12$?

The magnitude of a vector is always positive. Maybe what was intended was $A_x = -12$. Components of vectors can be negative, but the magnitude of a vector cannot be negative.

? What is wrong with talking about a "negative vector"?

Unlike a scalar quantity such as temperature, a vector doesn't have a sign; it has a (positive) magnitude and a direction. Again, components of a vector can be negative, but not a vector itself. Even the notation $-\vec{B}$ means "a vector opposite to $\vec{B}$," not a "negative vector."

A.2 Vector multiplication

Vectors can be added and subtracted, and they can be multiplied by a scalar. Two vectors can also be multiplied, but two different kinds of vector multiplication are defined: the dot product and the vector product. For completeness, both kinds of vector multiplication are reviewed here but are also discussed in the main body of the textbook at the points where they are first needed. The dot product is introduced in Chapter 4 in the context of work, and the vector product in Chapter 9 in the context of angular momentum.

A.2.1 The dot product

The dot product is an operation involving two vectors. This is encountered this in the expression for work:

$$W = \vec{F} \bullet \Delta\vec{r} = F_x\Delta x + F_y\Delta y + F_z\Delta z$$

The result of a dot product operation is a scalar (like the quantity work). The magnitude of the dot product can also be calculated as:

$$\vec{F} \bullet \Delta\vec{r} = F\Delta r\cos\theta = F_{\parallel}\Delta r = F\Delta r_{\parallel}$$

where θ is the angle between the two vectors, placed tail to tail. In the VPython programming language, dot(vector1,vector2) gives the dot product of the two vectors.

A.2.2 The cross product

The cross product is another operation involving two vectors. The cross product $\vec{A} \times \vec{B}$ is a vector in a direction perpendicular to the plane defined by $\vec{A}$ and $\vec{B}$. In Figure A.13, both $\vec{A}$ and $\vec{B}$ lie in the plane, and the vector $\vec{A} \times \vec{B}$ is perpendicular to the plane.

The cross product is encountered in the expression for angular momentum:

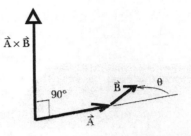

Figure A.13 The cross product produces a vector that is perpendicular to the two original vectors.

$$\vec{L} = \hat{r} \times \vec{p} = <(yp_z - zp_y),\ (zp_x - xp_z),\ (xp_y - yp_x)>$$

The result of a cross product is a vector (like angular momentum). The magnitude of the cross product can also be calculated as:

$$|\hat{r} \times \vec{p}| = rp\sin\theta$$

where θ is the angle between the vectors when their tails are placed at the same location. The direction is given qualitatively by a right-hand rule.

Right-hand rule for vector cross product

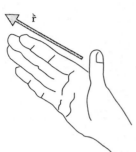

Figure A.14 Open right hand, fingers point in direction of $\hat{r}$.

- point fingers of right hand in direction of first vector $\hat{r}$ (Figure A.14)
- rotate wrist, if necessary, to make it possible to
- curl fingers of right hand through angle θ toward second vector $\vec{p}$ (Figure A.15)
- stick out thumb, which points in direction of cross product $\hat{r} \times \vec{p}$ (Figure A.15).

There are various forms of the right-hand rule, but this particular version has the advantage of incorporating the angle that is needed for determining the magnitude. The angle θ through which you curl your fingers is the angle that appears in the magnitude $|\hat{r} \times \vec{p}| = rp\sin\theta$.

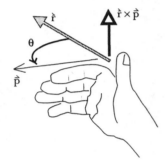

Figure A.15 Bend fingers toward lining up with $\vec{p}$. Thumb points in direction of $\hat{r} \times \vec{p}$.

Wrist rotation

The rotation of the wrist is an important part of the right-hand rule. Consider the situation in Figure A.16. With your right hand in this position, you can't bend your fingers backwards from the first vector $\hat{v}$ toward the second vector $\hat{r}$—it is a physical impossibility. You need to rotate your wrist into a position from which it is possible to bend the fingers, as shown in Figure A.17 on the following page. Pay attention to the size of the angle through which you bend your fingers. This is the angle whose sine is part of the definition of the magnitude of the cross product. This angle should never be more than 180°. If it is, you have made a mistake in orienting your hand; probably you need to rotate your wrist.

In the VPython programming language, cross(vector1,vector2) gives the cross product of the two vectors.

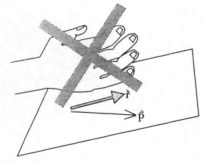

Figure A.16 You can't bend your fingers backward. You must rotate the wrist into a position that lets you bend the fingers.

A.2.3 Common errors in vector multiplication

(1) A dot product involves two vectors. Don't put a scalar in a dot product.
(2) A cross product involves two vectors. Don't put a scalar in a cross product. The result of a cross product is a vector. Don't write it as a scalar!

Figure A.17 After rotating the wrist, it is possible to bend the fingers.

A.3 Answers to exercises

A.1 (page 2) $v = 9.4$ m/s , $32°$ to the $-x$ axis, $148°$ to the $+x$ axis (vector in 2nd quadrant)

A.2 (page 2) $F_x = 18.0$ N , $F_y = 31.2$ N

A.3 (page 2) $<0.04, -3.4, 60.0>$

A.4 (page 3) 2.15×10^7 m/s

A.5 (page 3) $<4.5 \times 10^2, -3 \times 10^2, -2 \times 10^2>$ N

A.6 (page 3) 3.61×10^2 N ; 3.35×10^2 N ; 5.77×10^2 N

A.7 (page 3) $<1.5 \times 10^2, 3 \times 10^2, -2 \times 10^2>$ N

   $<-1.5 \times 10^2, -3 \times 10^2, 2 \times 10^2>$ N

A.8 (page 5) $<0.516, -0.309, 0.799>$

A.9 (page 5) $4.0\hat{\imath} - 2.4\hat{\jmath} + 6.2\hat{k}$

Appendix B

Basic Derivatives

B.1 Derivatives

In case you have not already studied calculus, or it has been a while since you last used calculus, we offer a calculation of some simple derivatives that are particularly useful in physics. You are probably familiar with the rate of change of velocity as the ratio of the velocity change $\Delta \vec{v}$ to the time interval Δt, $\Delta \vec{v} / \Delta t$. If we let the time interval Δt get very small, this ratio approaches what we call the "derivative" of the velocity with respect to the time:

$$\text{As } \Delta t \to 0, \ \frac{\Delta \vec{v}}{\Delta t} \to \frac{d\vec{v}}{dt}$$

More generally, we can approximate the rate of change of any physical quantity Q, as some other physical quantity R is varied, by the ratio $\Delta Q / \Delta R$, and if we consider a very small change ΔR, we approach the derivative of Q with respect to R:

$$\frac{\Delta Q}{\Delta R} \to \frac{dQ}{dR}$$

For concreteness, we'll take derivatives with respect to the time t, but the same principle applies if you take a derivative with respect to some other varying quantity, such as position x or y.

Derivative of a constant with respect to t

$$\frac{d(\text{constant})}{dt} = 0 \text{ because the change } \Delta(\text{constant}) = 0 \,.$$

Derivative of t with respect to t

$$\frac{\Delta(t)}{\Delta t} = 1 \,, \text{ no matter whether } \Delta t \text{ is large or small, so } \frac{d(t)}{dt} = 1$$

Derivative of t^2 with respect to t

The rate of change of t^2 with respect to t can be obtained starting from the fact that when t increases by an amount Δt, the quantity t^2 increases to $(t + \Delta t)^2$:

$$\frac{\Delta(t^2)}{\Delta t} = \frac{(t + \Delta t)^2 - t^2}{\Delta t}$$

$$= \frac{t^2 + 2t\Delta t + (\Delta t)^2 - t^2}{\Delta t}$$

$$= \frac{2t\Delta t + (\Delta t)^2}{\Delta t}$$

$$= 2t + \Delta t$$

This is nearly equal to $2t$ if Δt is very small compared to $2t$. Therefore we conclude this:

$$\frac{d(t^2)}{dt} = 2t$$

In a similar fashion it is possible to show the following:

$$\frac{d(t^n)}{dt} = nt^{n-1} \text{ for any constant value of } n \text{ (positive } or \text{ negative)}$$

Derivative with a constant multiplier

The derivative of a quantity that contains a constant multiplier C is simply C times the basic derivative:

$$\frac{\Delta(CQ)}{\Delta t} = \frac{C\Delta Q}{\Delta t} = C\frac{\Delta Q}{\Delta t}$$

$$\text{so } \frac{d(CQ)}{dt} = C\frac{dQ}{dt}$$

Derivative of sine and cosine

We can calculate the rate of change (the derivative) of the sine and cosine functions by using the trigonometric identities for the sines and cosines of sums of angles:

$$\sin(A + B) = \sin A\cos B + \cos A\sin B$$

$$\cos(A + B) = \cos A\cos B - \sin A\sin B$$

Here is the derivative of the sine:

$$\frac{\Delta(\sin t)}{\Delta t} = \frac{\sin(t + \Delta t) - \sin t}{\Delta t}$$

$$\frac{\Delta(\sin t)}{\Delta t} = \frac{\sin t\cos\Delta t + \cos t\sin\Delta t - \sin t}{\Delta t}$$

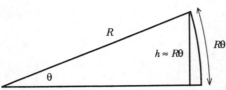

$$h \approx R\theta$$

Figure B1.1 The sine of a small angle is approximately equal to the angle, measured in radians.

The cosine of a very small angle is approximately equal to 1, and the sine of a very small angle is approximately equal to the angle, measured in radians. You can see from Figure B1.1 that if the angle θ is small, the height h is approximately equal to the arc length $R\theta$ (with θ measured in radians), in which case $\sin(\theta) = h/R \approx \theta$. As an example, $30° = \pi/6$ radians = 0.524 radians, which is very close to $\sin(30°) = 0.5$. The approximation gets even better for smaller angles.

Therefore if Δt is very small, we have approximately

$$\frac{\Delta(\sin t)}{\Delta t} \approx \frac{\sin t + (\cos t)\Delta t - \sin t}{\Delta t} = \cos t$$

Therefore the derivative of the sine is a cosine. You can go through the same kind of reasoning to find that the derivative of the cosine is a negative sine. In summary:

$$\frac{d(\sin t)}{dt} = \cos t$$

$$\frac{d(\cos t)}{dt} = -\sin t$$

Do these derivatives make sense? Note that at $t = 0$, the function $\sin(t)$ is growing, and its derivative $\cos(0) = +1$, which makes sense. At $t = \pi/2$ (90°), the sine function reaches a maximum and is momentarily neither increasing nor decreasing, and its derivative $\cos(\pi/2) = 0$, which makes sense.

The chain rule

Often we need to calculate the rate of change of a function of some variable with respect to some *other* variable, such as differentiating a function of x with respect to t:

$$\frac{d[f(x)]}{dt}$$

In this example, suppose x changes by an amount Δx when t changes by an amount Δt. Then we have the following:

$$\frac{\Delta[f(x)]}{\Delta t} = \left\{\frac{\Delta[f(x)]}{\Delta x}\right\}\left\{\frac{\Delta x}{\Delta t}\right\}$$

$$\text{But } \frac{\Delta[f(x)]}{\Delta x} \rightarrow \frac{d[f(x)]}{dx}$$

$$\text{and } \frac{\Delta x}{\Delta t} \rightarrow \frac{dx}{dt}$$

As we let Δt get very small, we obtain the chain rule:

$$\frac{d[f(x)]}{dt} = \left\{\frac{d[f(x)]}{dx}\right\}\left\{\frac{dx}{dt}\right\}$$

Here is an example of the use of the chain rule in physics (ω is a constant):

$$\frac{d[\cos(\omega t)]}{dt} = \left\{\frac{d[\cos(\omega t)]}{d(\omega t)}\right\}\left\{\frac{d(\omega t)}{dt}\right\} = \{-\sin(\omega t)\}\{\omega\}$$

B.2 Self-test

Here are some short questions on derivatives that you can use as a self-test. Try not to look back through this appendix as you work on these questions. If you find that you do need to look up something in order to answer a question, or if you find that your answer disagrees with the answers given at the end of this appendix, be sure to make a note of which question this was, to guide your further study and review.

Ex. B.1 What is the derivative of $4t^2$ with respect to t?

Ex. B.2 What is the rate of change of $(5x^{-1} + 6)$ as x varies?

Ex. B.3 Consider the behavior of the derivative of $\cos(t)$ at $t = 0$ and $t = \pi/2$. Do you get values that make sense, given how the cosine is changing near those times?

Ex. B.4 Calculate $\dfrac{d[\sin(3t)]}{dt}$.

B.3 Answers to self-test

B.1 (page 3) $8t$

B.2 (page 3) $-5x^{-2}$

B.3 (page 3)   At $t = 0$, the cosine is at a maximum, so its derivative should be zero; indeed, $-\sin(0) = 0$.

At $t = \pi/2$, the cosine is decreasing, so its rate of change should be negative; indeed, $-\sin(\pi/2) = -1$.

B.4 (page 3) $3\cos(3t)$

Greek alphabet, and its uses in this volume

Alpha	A		α	alpha particle (helium-4 nucleus)
Beta	B		β	
Gamma	Γ		γ	photon
Delta	Δ	change of; small quantity of	δ	
Epsilon	E		ε	a small quantity
Zeta	Z		ζ	
Eta	H		η	
Theta	Θ		θ	angle
Iota	I		ι	
Kappa	K		κ	
Lambda	Λ		λ	wavelength
Mu	M	mega (10^6)	μ	micro (10^{-6}); muon
Nu	N		ν	neutrino; frequency ($=f$)
Xi	Ξ		ξ	
Omicron	O		o	
Pi	Π		π	circle: circumference/diameter
Rho	P		ρ	density
Sigma	Σ	sum	σ	
Tau	T		τ	torque
Upsilon	Y		υ	
Phi	Φ		ϕ	phase angle
Chi	X		χ	
Psi	Ψ		ψ	
Omega	Ω	precession angular frequency; # of microstates of a system	ω	angular frequency